KU-167-199

The Book of

Inventions

and

Discoveries

The Book of Inventions *and* Discoveries

Associate Editor
Valérie-Anne Giscard d'Estaing

Macdonald
Queen Anne Press

A QUEEN ANNE PRESS BOOK

© Compagnie 1212 1990
Based on *Le Livre Mondial des Inventions* originally published in France.
This edition first published in Great Britain in 1990 by
Queen Anne Press, a division of
Macdonald & Co (Publishers) Ltd
Orbit House
1 New Fetter Lane
London EC4A 1AR

A member of Maxwell Macmillan Pergamon Publishing Corporation

All rights reserved. No part of this publication may be reproduced, stored in a retrieval system, or transmitted, in any form or by any means, without the prior permission in writing of the publishers, nor be otherwise circulated in any form of binding or cover other than that in which it is published and without a similar condition including this condition being imposed on the subsequent purchaser.

British Library Cataloguing in Publication Data
Giscard d'Estaing, Valérie-Anne
The book of inventions and discoveries.
1. Inventions, history
I. Title
609

ISBN 0-356-18835-3

Typeset by Rowland Phototypesetting Ltd, Bury St Edmunds, Suffolk
Printed and bound in Great Britain by BPCC Paulton Books, Ltd,

TRANSPORT

On the ground 1

The motor car 1, Two-wheelers 6, The railways 11, Urban transport 14.

By air 16

Balloons 16, Aeroplanes 17, Helicopters 22, Miscellaneous 23.

By sea 24

Shipping 24, Navigational aids 26.

WARFARE

Weapons 29

Light weapons 29, Artillery 32,
Armoured vehicles 33, Special weapons 33.

Naval warfare 36

Aviation 40

SCIENCE

Mathematics 48

Number theory 48, Geometry 50,
Algebra and analysis 52.

Physics 54

Standards 54, Hydrodynamics 56, Matter 57,
Changes of state 58, Thermodynamics 59,
Electricity and magnetism 60,
Atomic and nuclear physics 62, Particles 65.

Chemistry 66

Inorganic chemistry 66, Organic chemistry 68,
Analytical chemistry 70, Biology and genetics 71.

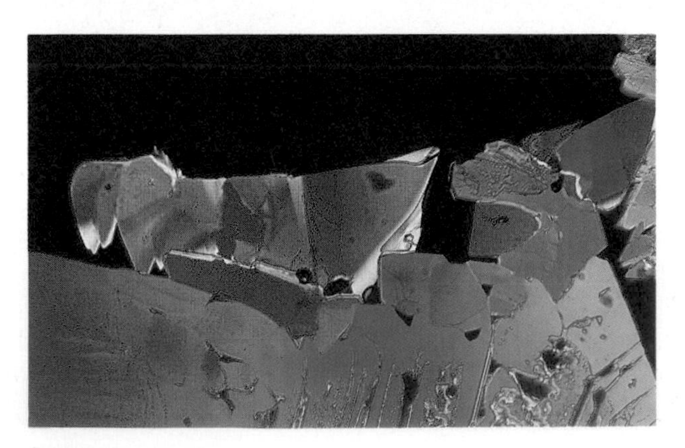

SPACE

Astronomy **74**, Rockets **80**, Satellites **82**,
Space vehicles **85**, Exploration of the cosmos **86**.

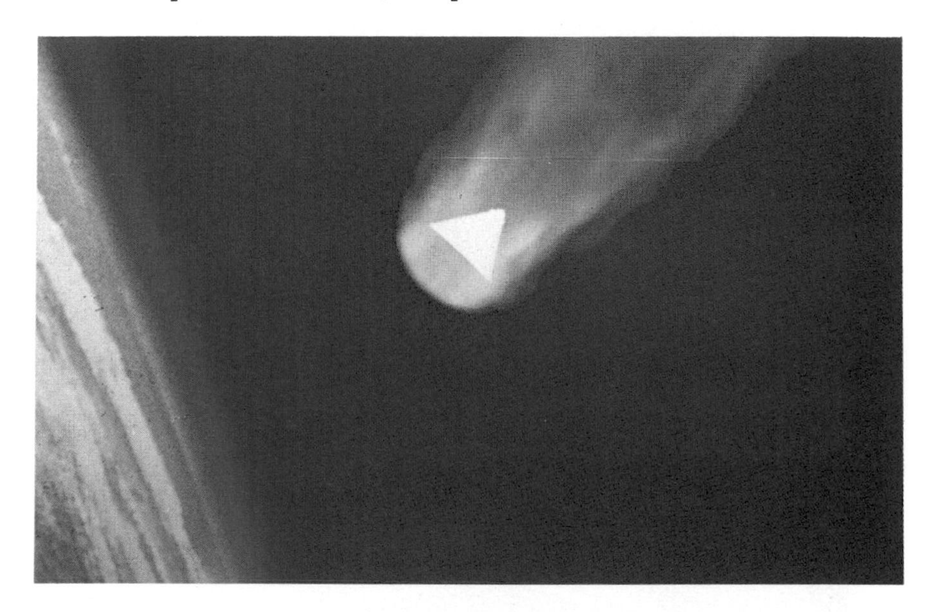

MEDICINE

Examination techniques **89**, Vaccination **92**,
The heart and lungs **94**, Surgery **96**,
Great fears, great diseases, great doctors **100**,
Gynaecology **102**, Therapeutics **103**,
Eyes and ears **109**, Dental surgery **110**.

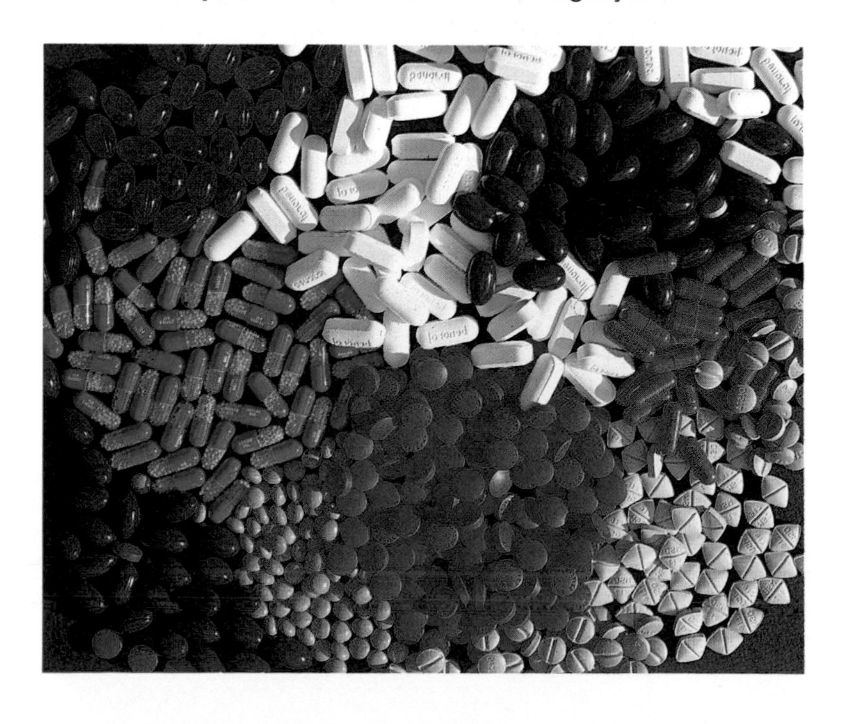

INFORMATION TECHNOLOGY

The first machines 113, Concepts 115, Software 116, Components 119, Microcomputers 122, Supercomputers 123, Computer peripherals 123, Applications 125, Robots 126.

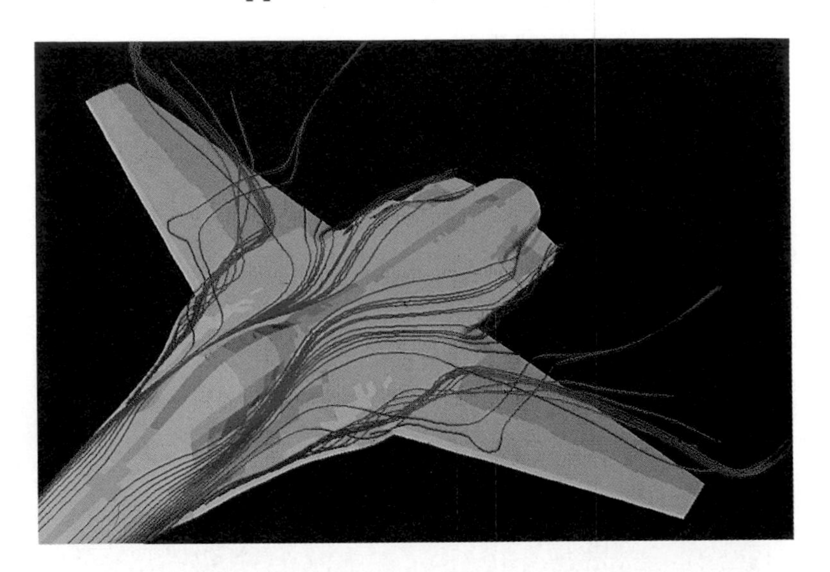

AGRICULTURE

Agriculture 128

Agricultural machines 128, Agronomy 130, Animal husbandry 132.

Gardening 135

ECOLOGY AND THE ENVIRONMENT

Pollution, oil slicks and 'green' inventions **140**,
Ozone: we cannot live without it **144**, Meteorology **146**,
Geophysics **147**, Oceanography **149**,
Volcanoes and earthquakes **150**.

POWER AND INDUSTRY

Power **153**

Hydraulics **153**, Steam engines **153** Electricity **155**,
Petroleum **156**, Engines and ignition devices **158**,
Nuclear energy **164**, New sources of energy **166**.

Industry **168**

Architecture and construction **168**, Materials **171**,
Clocks and watches **175**, Metallurgy **175**,
Textiles and new materials **176**, Laser **179**,
Holograms **179**.

EVERYDAY LIFE

In the kitchen 181, In the home 188,
In the bathroom 193, Clothes 195,
At the hairdresser's 197, Leisure 198, Commerce 200,
The inventions we cannot do without 202.

GAMES, TOYS, SPORTS

Games and Toys 204

Sport 209

Ball games 210, Swimming, sailing, watersports 212,
Shooting 215, Surfing and gliding 216,
Winter sports 218, Motor sports 220, Tennis 220,
Theme parks 221, Mountaineering 221.

THE BIZARRE

Some Bizarre Inventions 222

MEDIA AND COMMUNICATIONS

Language and Literature 228

Artificial languages 228, Writing materials 228, Printing 230, Visual arts 233, 1990: World literacy year 234.

Audiovisual 236

Radio 236, Television and video 237.

Telecommunications 240

Post 240, Telegraphy 240, Telephone 241, Communication cables and satellites 243.

THE ARTS

Music 245

Instruments 245, Sound reproduction 247,
Musical composition and performance 251.

Photography 252

Cinema 258

They Changed Our Lives 263

Index 276

TRANSPORT

On the ground

The motor car

Origins

The idea of a self-propelling vehicle for road transport dates back to antiquity. It reappeared again during the Renaissance especially in Leonardo da Vinci's work. The first 'automobiles' were built in the 18th century. Varied and ingenious devices were used to drive the wheels: coil spring (Jacques de Vaucanson, 1740), wind-engine (J. H. Genevois, 1760), compressed air (W. Medhurst, 1799).

So who is the true inventor of the motor car? Read on!

Model T Ford (1908)

Built at Detroit in 1908 by the American **Henry Ford** (1863–1947), the Model T Ford was put into mass production on the assembly line in 1908. It was a far cry from the hand-crafted automobile built by Henry Ford at Detroit in 1896, with its four-stoke engine and Kane Pennington cylinders. Assembly-line production, a Ford innovation for the car industry, was an application of the principles of Frederick W. Taylor, the creator of Scientific Management, who stressed strict specialisation, elimination of all superfluous motion, and maximum utilisation of plant and equipment. It was this system that made it possible for 18 million Model Ts to roll out of the Ford factories between 1908 and 1927.

The Mercedes (1901)

Built in **1901**, the **Mercedes** was a much improved version of the first car, invented by **Gottlieb Daimler** (1834–1900) and **Wilhelm Maybach** (1846–1929) in 1889. It was named after their sponsor's daughter.

This car caused a sensation at the time, as much because of its appearance – it was the first car that didn't look like a horse-drawn carriage – as its technical qualities. Its top speed was 86km/h *53.4mph*.

After the merger of the Daimler and Benz companies in 1926, the make became Mercedes-Benz.

The Beetle (1936)

The Volkswagen VW 1936, better known as the **Beetle**, is the longest-lasting model. The prototype was devised by **Ferdinand Porsche** between 1934 and **1936**. About 30 vehicles were produced by Daimler-Benz in 1937 and the first stone of the Wolfsburg factory, where the car was to be mass produced, was laid by Hitler in 1938.

About 23 million Beetles have been produced to date, and the model is still being manufactured in Mexico.

The 2 CV (1948)

It was **Pierre Boulanger**, Citroën's managing director, who conceived the **2 CV** in 1935. He wanted to build an economical and practical car which would be high enough for someone to get in without removing their hat.

André Lefèbvre (1894–1963), father of the front-wheel drive car, worked on the prototype. The first trials were held in 1938, but were interrupted by the war. The official launch was in **1948**. As early as 1949, the model was so popular that prospective buyers had to submit a written application and faced a two-year wait when successful!

Austin Mini (1959)

Its British inventor, **Sir Alec Issigonis**, died in 1988, 29 years after the launch of the first

Alex Issigonis' Mini could comfortably carry four passengers and all their luggage. Here 'The Incredible Austin Se7en' (sic) is launched in 1959.

model. With this engineer, born in Smyrna in 1906 of a Greek father and a Bavarian mother, died the last man of ideas who, on his own, was able to create a car and impose it on the market. Vital statistics: 998cc, 3m *3yd 10in*, long 1.4m wide *1yd 1ft 7in*; 5600000 models sold.

Its success probably stemmed from the Mini being so amazingly modern and ahead of its time. As opposed to the 2 CV and the Beetle, it changed names along the way: BMC 850 in **1959**, then Morris and Austin Mini, and now just Mini.

Gas turbine car (1950)

The first car to be powered by a gas turbine was built by **Rover** in **1950**. On 26 June 1952, one of these vehicles reached a speed of 242km/h *150mph*. Other manufacturers have since taken up the process: General Motors, Chrysler, Boeing, Fiat and Renault, whose Etoile Filante (Shooting Star), attained a speed of 308.9km/h *192mph* in 1956.

Electric car (1891)

The first electric cars were developed in the United States in **1891**. The Electobat, manufactured by the **Morris** and **Salom** firm of Philadelphia, was only produced in a small series.

In 1892 the Morrison, manufactured by William Morrison of Des Moines, Iowa, was launched.

Present models (1990)

The development of electrical vehicles has come to a stumbling block: the battery (*see* Energy). Some leading world manufacturers are at present working on electrical engines that can be adapted to their commercial vehicles: the Peugeot C15 vans, the Renault Master or the VW Golf for example.

These cars are still produced on a small scale. At present there are 35000 of them on the road in Britain, 15000 in the USA, 8000 in Japan, 2000 in France and 600 in Switzerland.

Richard Trevithick built a number of road steamers like this one as well as rail locomotives in the early years of the 19th century.

THE INVENTION OF THE MOTOR CAR: A LONG STORY

It was the steam engine that made it possible to produce the first truly usable cars. In fact, the 'truck' build by Nicolas Cugnot in 1771 is generally considered to be the first motor vehicle.

On Christmas Eve 1801, Cornishman Richard Trevithick ran his steam car on a stretch of road near Camborne where it reached a speed of 13 to 14.5km/h *8 to 9mph*.

Up to the end of the 19th century, with a few minor exceptions, all motor vehicles were powered by steam. In France, the most prominent manufacturers were Onésiphore Pecqueur, Charles Dietz and Amédée Bollée.

The first attempts at propulsion by an internal combustion engine were made by the Swiss Isaac de Rivaz as early as 1807. His example was followed at distant intervals in England (Samuel Brown, 1826) and in France (Etienne Lenoir, 1862).

The petrol engine only came into its own after 1876, when the four-stroke cycle started to be used. It was in 1883 that the Frenchmen Malandin and Delamare-Debouteville fitted a road vehicle with a four-stroke engine for the first time. But it was the Germans Karl Benz from 1885, and Gottlieb Daimler and Wilhelm Maybach in 1889, who built what can be considered to be the first true modern motorcar.

The first American petrol-driven motorcars were built by the brothers Charles and Frank Duryea in 1892.

Mechanics

Transmission

Direct transmission (1899)

In **1899** **Louis Renault** (1877–1944) equipped his first car, completed in 1898, with a transmission coupled directly to the engine, and gear-shifting by selector rod. The transmission on the 1899 Renault had three speeds and a reverse gear. The fastest speed, third gear, was reached directly, the primary and secondary propeller shafts turning at the same speed.

Automatic transmission (1910)

The automatic transmission invented by the German **Fottinger** in 1910 was nothing more than a torque converter.

Preselector gearbox (1917)

In **1917**, **Wilson**, a major in the British army, invented the preselector gearbox for use in the battle tanks that had just been developed. After the war the box was fitted to all Armstrong-Siddeleys and, later on, to some Talbots in France. However, it was mainly used in large industrial vehicles and buses.

Fully automatic transmission (1971–1987)

A belt driven system called Variomatic, which provides fully automatic variations of gear ratio, appeared in **1971** on the Dutch car Daf before being fitted on the bottom of the range Volvos and being tested at Fiat.

In **1987**, after several Japanese cars were fitted with a CVT transmission derived from the Daf system, a Selecta version of the Uno range appeared at Fiat's in which the V belts of the old Van Doorne system were replaced by a continuous metal transmission. The Ford Fiestamatic is also fitted with this system which ensures smooth variations whilst still giving engine braking. The result is a comfortable drive and lower fuel consumption.

Economical automatic gearbox (1988)

A new automatic gearbox, ALT (Automatic Layshaft Transmission), is being developed by the English firm **Ricardo Consulting Engineers**. A prototype has already been fitted on a Ford Fiesta. Derived from the Maxwell ALT gearbox, which has been used since 1980 on many commercial vehicles, the gearbox has been modified to make it more compact. Furthermore, as it can be built on the production lines used at present for manual boxes, its cost price should be reasonable. To change gears, it uses a series of small friction clutches controlled by a computer which responds to road conditions.

All-electronic gearbox (1989)

The **1989** version of the Ferrari Formula 1 cars have a revolutionary gearbox. Fitted on John Barnard's instigation, it is a completely electronic box. The gear lever is fitted on the steering wheel and the driver only has to press a tab with the index finger to change the ratio.

Electronic clutch (1988)

Officially introduced at the *Mondial de l'Automobile* in Paris the Valéo electronic clutch couples the advantages of both the manual and the automatic gearboxes. By doing away with the third pedal, it frees the driver from having to operate the clutch while giving him complete freedom of choice when it comes to ratio, gear shifting, level of acceleration, and down-shifting with or without acceleration.

Only a few months after it was launched, the new clutch was chosen by Lancia, world champion rally car manufacturers, to equip the integral Delta for the 1989 season, as well as by Ferrari, who began using it in autumn 1989.

Differential (1827)

The differential was invented by the Frenchman **Onésiphore Pecqueur** in November **1827** for a steam engine. In a motorcar, the differential is the device which transmits the power from the engine to the wheels whilst allowing them to turn at different speeds as the car goes round bends. Then, the outside wheel has to turn faster than the inside wheel.

Back axle with integral trailing arm (1974)

First seen in **1974** on the Scirocco, the back axle with integral trailing arm, invented by the **Volkswagen** engineers, is fitted on all the new generation VWs. It was largely responsible for the Golfs popularity.

The brake (1895)

Automobile braking was first ensured by means of brake shoes such as were used in carts (**1895**). In 1899 the transmission shaft band and wheel brake appeared. These were commanded by a hand lever (brake drum).
The combined hydraulic (foot brake) and mechanical (hand brake) controls formed the subject of the 1924 Perrot-Lockheed patent. The American Chrysler Corporation was the first company to utilise this patented system.

Disc brake (1902)

The disc brake, invented by an Englishman, **Dr Lanchester**, was primarily used to equip military vehicles. The victory of Jaguar at the Le Mans 24-hour race in 1953 with a vehicle equipped with Dunlop disc brakes, which had been patented in 1945 and used in aeronautics, led to the spread of the invention.

Anti-pollution piston (1987)

Thanks to a new type of piston, perfected by the British company **A.E.**, in association with **Sonex Research Inc.** from Indianapolis (USA), 'clean' engines which do not use a catalytic convertor could soon be mass produced in both countries, if the rest of the trials, begun in **1987**, confirm the first results.

Four-wheel drive

The last phase of the 1914–18 war created the need for an all-terrain vehicle. Therefore, in the early 1920s, French and German manufacturers tackled the problem. In 1926 **Georges Latil** introduced his T.L. tractor with four driving and directional wheels. The road model was soon adapted by foresters. For their part, Citroën and Unic were putting **Adolphe Kégresse**'s patents into practice to produce vehicles with caterpillar tracks and driven front wheels.

In 1937, Mercedes introduced its G5 (G for *Gelände*, land) which was capable of remarkable performances. With its 45hp fuel injection engine it could climb 55 per cent ramps with a load of 770kg *7.7cwt* and turn in a 7m *22ft 11½in* diameter circle.

The Jeep (1940)

The **Jeep** was born on 10 June **1940**, the brainchild of three American military engineers who imposed a crazy deadline on the presentation of projects: 75 days! The model selected was that designed by the **Willys-Overland** engineers, and intensive production started in 1941 in association with Ford. More than 585 000 vehicles were built during the war. The origin of its name is rather mysterious: it could have come from the name of one of Popeye's friends, Jeep; or from the letters GP meaning General Purpose that Ford put on its model; or perhaps from the initials of the American military police units, GIP, who used them for the American landing in North Africa.

On rails and road (1985)

In **1985** two English firms, **Bedford Bruff** and **United Service Garages**, perfected dual purpose breakdown vehicles for road and rail which had four driving wheels and could move at 100km/h *62.14mph* on road and 60km/h *37.28mph* on rails and across country.

Four-wheel drive saloon (1983)

The first mass-produced all-weather (as opposed to all-purpose) saloon was the **Audi 80 Quattro**. Today all manufacturers have started to produce their own versions.

Four-wheel steering (1986)

This will be the next major revolution in the car industry. The four-wheel steering system (4WS) gives greatly increased comfort and safety. The ancestor of cars with four directional wheels is Amédée Bollée's steam car of 1876. After that, the same formula was used on a few prototypes. In 1967 an American, Mickey Thompson, raced a four-wheel steering car at Indianapolis. In 1965, the Japanese firm Mazda took out its first patents. But it was in **1986** that the first commercial model was introduced: the Skyline from **Nissan**. In 1987 Honda offered the system as an option on its sports model Prelude XX. The Japanese are investing considerable sums of money to develop this system and are well ahead of the rest of the field.

Viscodrive (1986)

This system, which equips four-wheel steering German cars (VW Golf Synchro and BMW 325 4 × 4), was invented by the English engineer **Harry Ferguson**. When the front wheels are skidding, **Viscodrive** takes over by increasing the power to the back wheels. To ensure satisfactory distribution of the torque, it can be incorporated to a transfer box or a differential.

Suspension

Hydropneumatic suspension (1924)

The first hydraulic suspension system was that of the French engineer **Georges Messier** who, in **1924**, made an 'oleopneumatic suspension with position adjustment' (patented in 1920). It was fitted in 1926 on Messier cars without springs. It was in 1953 (20 years after Messier's death) that Citroën perfected its famous hydropneumatic suspension (combining a gas and an oil-based liquid) which was fitted first on 3000 of their 15hp front-wheel drive cars before being extended to their other models from 1955 onwards.

Volvo CCS suspension (1986)

Presented in Turin in **1986**, the **CCS suspension**, created by **Volvo**, is controlled by a microprocessor which enables it to react 3000

AN ELECTRONIC ROAD MAP IN YOUR CAR

In 1981 Ron Dork, an engineer at General Motors, fitted the first prototype of the electronic road map, the Loran, on his Buick. It consisted of a computer system receiving data on the position of the car via navigation satellites identical to those used by ships. The computer instantly translated the information onto a screen on which the position of the car is displayed on a memorised road map.

The European project Carminat, on which Philips, Sagem, Renault and TDF are working, aims to obtain concrete and marketable results in the near future. Carminat combined three previous projects.

Carin, introduced in 1985 by Philips, is based on the storage of an enormous library of maps on a CD-ROM compact disc. Using this data bank, Carin can indicate to the driver the best route between two points (in town or on the road) and then guide him along. Data about the vehicle's route are obtained by a tachometer and a compass. A different version proposed by Bosch uses the ABS sensors to work out the journey.

Atlas is designed to give the car all the necessary data about traffic and road conditions. On its own, it would have had to convey too much information which would have required the use of too many frequency bands. But coupled with Carin's memory, only one FM frequency is needed.

Minerva is an in-car computer that deals with data about the state of the tyres, petrol consumption, fault finding, etc.

More ambitious still, the Prometheus system, launched under Daimler-Benz's initiative, is intended to give complete driving assistance. At present, 11 manufacturers and more than a hundred companies are collaborating on this project with a view to realising it early in the next century.

times per second transmitting each impulse to the wheels thus allowing them to adapt to the road. This new suspension keeps the vehicle steady in any conditions.

Electronically variable shock-absorbers (1987)

Created by the German company **Boge**, they were first fitted on the 1988 model of the BMW M3, which was launched in February **1987**. Today they are an option on the BMW 635 and 750 as well as on the Lancia Thema 832. With the help of sensors and microcomputers, the shock-absorbers adjust to road conditions in a thousandth of a second.

Hydroactive suspension (1989)

The hydroactive suspension of the **Citroën XM**, on the market since spring **1989**, is controlled by an electronic calculator which instantaneously changes its setting to suit the road and the style of driving, using data provided by five sensors. Representing a new concept in active safety, this suspension automatically becomes firm before the driver suffers the drawbacks of too much softness, intelligently anticipating the car's reactions.

ABS (1972)

The Anti-Blocking System (**ABS**) perfected by the German firm **Bosch** has been mass-produced since 1978. The ABS allows the driver, by slamming on the brakes, to stop the car in the shortest possible time, be it on a straight road or on a bend, whether the road is wet, gravelled, icy or dry. This system, according to the specialists, represents the greatest step forward since the invention of disc brakes and safety belts.

Anti-skid system (1985)

Introduced in **1985** by the Swedish company **Volvo**, on its latest top of the line models, the ETC (Electronic Traction Control) system intervenes in case of skidding during acceleration. As soon as a drive wheel turns more quickly than a front wheel, this control device reduces the supply of petrol little by little until all four wheels are turning at the same speed. It thus considerably reduces the risk of loss of adhesion on a slippery road, even if the driver accelerates strongly. The driver is warned of the danger by means of a small light on the dashboard.

Front-wheel drive

Hooke-type universal joint (1926)

The invention of the Hooke-type universal joint Tracta by the French engineers **Pierre Fenaille** and **Jean Grégoire** in **1926** made front-wheel drive cars possible. This joint allows transmission of the engine power to the front wheels.

The Citroën 7 (1934)

The true father of front-wheel drive is the French engineer **André Lefèbvre**. Hired by André Citroën on 1 March 1933, he managed to perfect and build, in only one year, the first front-wheel drive car. The model 7 was officially introduced to Citroën agents on **24 March 1934**. After the bodywork and the accessories were further improved, it won universal acclaim at the 1934 Paris Motor Show. The system has so many advantages that almost all cars now have front-wheel drive.

Transmission (1900)

It was in **1900** that a Frenchman, **Louis Bonneville**, perfected the first 'automatic transmission by epicyclic train', an invention which was mentioned in the 18 August 1900 issue of the American magazine *The Motor Car Journal*.

In 1950 another Frenchman, **Gaston Fleischel**, invented and patented a complete range of possible controls for gearboxes with epicyclic trains, which the Speciality Equipment and Machinery Corporation (Maryland, USA) decided to use in that same year.

Fleischel's patent led to a strongly disputed lawsuit dealing with industrial property which ended in 1953 when a group of American car manufacturers bought all the shares of the Speciality Equipment and Machinery Corporation.

Bodywork and accessories

Seat belt (1903)

The seat belt is derived from a patent registered in **1903** by the Frenchman **Gustave Désiré Liebau**, which dealt with 'protective braces for use in motor cars and other vehicles', and from a slightly different model which an American military doctor, Colonel Stapp, tried out on a vehicle travelling at a speed of more than 200km/h *124mph*. First used in aeronautics, the seat belt went through various stages before being fitted on mass produced cars by Volvo after the firm conducted tests from 1959 to 1963 that led them to choose the three-point type.

Automatic-fastening belt (1988)

Toyota has fitted an electric system that automatically fastens the seat belt on some of its models for the American market as seat belt regulations are particularly strict in the USA.

Automatic-release seat belt (1989)

Kim Nag-Hyun of the Seoul Polytechnic (South Korea) presented, in New York in March **1989**, an electric system which automatically releases the seat belt catch 30 seconds after the car has been stopped by an impact.

Airbag (1981)

This anti-shock air cushion known as the **airbag** is a **Daimler-Benz** invention. This comes in the form of an air pocket incorporated into the vehicle's steering column. In the event of a frontal collision, the airbag inflates automatically in a mere several hundredths of a second, putting the equivalent of a mattress between the steering wheel and the driver. Mercedes is the first company to equip certain of its models with this system.

Eurobag (1988)

Volvo has perfected an inflatable bag particularly intended to protect the driver's face: the **Eurobag** which required several years' study. Engineers built an electronic head which has 52 sensors that collect 800 pieces of information in the tenth of a second the impact lasts. Volvo has made the head available to other car manufacturers.

Controls for the disabled (1989)

An English firm based in Hemel Hempstead has won the safety prize, awarded by the Royal Corps of Transport, for its new lever control system for disabled drivers. The system has also been approved by the West German car industry. It consists of a four-directional joy stick, fitted on cars with automatic gearboxes, which controls the steering,

The Eurobag cushion automatically inflates in a fraction of a second to protect the head.

This dump outside San Francisco clearly demonstrates the problem of recycling tyres.

brakes and accelerator and requires the use of only one hand.

Electronic detector for reversing (1989)

Reversing is still a nightmare for many drivers, especially in the dark, the rain or the snow. It will now be made easy by the electronic detector **Back Sensor** from South Korea. This consists of two detectors which are fitted under the back lights or on the bumper. As you manoeuvre, it warns you about any kind of obstacle, be it another vehicle or a pedestrian.

The windscreen (1903)

The first **windscreens** appeared in **1903** and were generally very high, as the cars of the period reached a height of over six feet. Made of ordinary glass, these early windscreens were very dangerous and were considered optional accessories.

Laminated windscreens (1920)

The first windscreens made of laminated glass (invented in 1909 by the French chemist **E. Benedictus** and marketed as of **1920** under the name Triplex) were reserved for top of the line models for some time. The first car manufacturer to use laminated glass windscreens for a series model was Volvo in 1944 for the PV44.

Windscreen wiper (1916)

The first mechanical **windscreen wipers** appeared in the United States in 1916. In 1921, the Englishman W. M. Folberth invented windscreen wipers that worked automatically, using compressed air supplied by the engine. The first electric windscreen wipers were manufactured in the United States by Berkshire.

Sensitive windscreen wiper (1983)

The first windscreen wiper to adjust its speed automatically to the intensity of the rain was developed in **1983** by the Japanese firm **Nissan** and fitted to all their cars.

Invisible windscreen wiper (1978)

Rain'x was invented by the American chemical engineer **Howard Ohlhausen** while he was a pilot, and has been used by thousands of motorists in the United States since **1978**. Rain'x works like an invisible windscreen wiper. It is a transparent hydrophobic substance which repels rain by forming minute droplets which cannot cling to glass. Rain'x improves visibility, and above a certain speed a coat of it applied to the windscreen makes the use of conventional wipers unnecessary.

Gas generator (1883)

The first gas generator that could be used to supply internal combustion gas engines was produced by the English engineer **Emerson Dowson** in July **1883**. A gas generator turns solid fuel into a gas fuel. Dowson's generator had a 60 per cent efficiency. It was used by the Crossley brothers in England and by Deutz in Germany.

Resin for windscreens (1972)

In **1972** the American engineer **Gerald Keinath** patented his technique for repairing laminated windscreens with a transparent resin injected with a piston then solidified before being polished by an ultra violet lamp. Since then this process has been greatly improved, allowing repairs even in the case of significant impact with blistering or radiating cracks.

A DASHBOARD ON THE WINDSCREEN

Here is an application of military technology to cars. The information needed to pilot a fighter plane is projected onto the windscreen so that the pilot does not have to look down at the dashboard. This system of so-called head-up displays will shortly be introduced by General Motors onto some of their models.

Double glazing to stop misting (1989)

The problems of driving in very cold weather and a study of drivers' reactions led **Volvo** to use double glazing in their cars. This overcomes the inconvenience of misting without having to resort to a fan which often makes the car feel cold or draughty. The double glazing also reduces loss of heat from inside the vehicle.

Bumpers (1905)

In **1905 F. R. Simms** patented the first bumpers to be made from rubber. The Simms Manufacturing Company of Kilburn, London, fitted the bumpers onto a Simms-Welbeck in the summer of 1905. Bumpers had previously been fitted to a Czech vehicle, the Präsident which was built in 1897. Unfortunately, the bumpers fell off after 15km *9 miles*, and were never replaced.

Pneumatic tyres (1888)

A Scottish veterinary surgeon working in Belfast, **John Boyd Dunlop** (1840–1921), invented the pneumatic tyre in **1888**. This was one of the most significant leaps forward as far as locomotion by wheels was concerned.

Dunlop had the idea of fitting air-filled tyres to his son's bicycle.

The same idea had already been put forward by a Belgian, Dietz, in 1836, and by a Scottish engineer, Robert W. Thomson, in 1845, but it was not put into practice. Abandoning his old profession, Dunlop patented his invention and founded the first tyre factory, where he utilised Goodyear's vulcanisation process. Through the mediation of a German subsidiary, Dunlop tyres were put on the first mass-produced motorcycles, the Hildebrand & Wolfmüller in 1894. The Dunlop firm immediately received complete support from most manufacturers. On the other hand, it was the French Michelin brothers, André (1853–1931) and Edouard (1859–1940), who, in 1895, first used pneumatic tyres on an automobile.

Removable tyre (1891)

Invented in **1891** by the French firm **Michelin**, the removable tyre proved to be revolutionary. A blow out, which formerly meant calling a specialised repairman, could now be fixed by the rider in less than 15 minutes. This invention was an immediate success.

Anti-puncture liquid (1988)

Perfected in the United States by **World Promotions**, it is a liquid mixture of glycol and acrylic fibres in suspension which, as soon as a puncture occurs, concentrates in the affected area and stops the air escaping. The liquid is introduced into the deflated tyres and is evenly spread after a 10km *6.2 mile* drive (the tyres must be pumped up again, of course!). It is efficient for punctures up to 5mm *⅕in* in diameter.

Rear-view mirror (1906)

In **1906** a French inventor, **Alfred Faucher**, registered the first patent concerning a 'warning mirror for motorcars'. He had also fitted his car with a 'hand to signal changes of direction', the ancestor of our direction indicators, and with the first rear warning light activated by touching the brake pedal. About ten years later, side mirrors were introduced to complement the central rear-view mirror which was fitted inside the car.

Electronic anti-glare rear-view mirror (1988)

Perfected by the company **Stewart Automotive Ltd** of Greenock in Scotland, the **Eclipse** rear-view mirror, using microchip technology, automatically tilts within one tenth of a second of being hit by the dazzle of headlights, going back to its original position only after the light conditions are normal again.

Rear-view television (1983)

In **1983** the **Mir International** company solved the problem of reversing for heavy good vehicles. It offered a camera positioned at the back of the vehicle which relays a picture onto a small screen fitted on the dashboard.

INVENTORS VS THIEVES: THE FIGHT GOES ON!

The first anti-theft device dates back to 1934 and was invented by the German industrialist Abram Neiman, a car and racing enthusiast. One day, his own car was stolen from the front of his factory. He was so infuriated that he decided to find a system to stop thieves. This resulted in the famous Neiman which has become almost synonymous with the anti-theft device. Since then, as the situation has not really improved, inventors have continued to be creative. Among the most recent inventions let us mention:
– etching the car registration number on all the windows. Perfected and patented in 1982 by Edward Bruneau from a technique previously tried in England
– The Steering Wheel by the Chinese Jaw Jia Jang (1989), which locks the steering
– the Key Buster of the Jetico Trading Co (Taiwan, 1989), which replaces door and ignition keys by a keypad on which a code has to be typed. If the wrong number is used, the locks automatically jam shut
– the Gaslock by the Mexican José Octavio Velasco which stops the petrol coming through and jams the starter.

The foolproof anti-theft device has still to be invented. . . .

Roads

Motorway (1924)

The first modern motorway was created in Italy in **1924**. Eighty kilometres *49.7 miles* of two-lane *autostrada* in the Milan area, designed by **Piero Puricelli**. Earlier attempts at building roads on lines similar to the motorway system had been made in the United States as early as 1914, then in Germany in 1921, but only over short distances. However, the first four-lane motorway, applying the most modern rules of road safety, is American. It was built in Pennsylvania between 1937 and 1940.

Black-ice detector (1985)

The British firm **Zero Products** has perfected a very reliable and accurate black-ice detector which may prevent a great number of accidents. After being tested by the Essex police, it is now fitted on Lotus cars. It is a useful accessory with which Zero Products hopes to conquer the European market.

Traffic lights (1868)

The first traffic light was set up at the junction of Bridge Street and Palace Yard, in London, on **10 December 1868**. It was a gas light mounted at the top of a 7m *22ft 11½in* steel pole. One side was red and the other green and a lever system made it rotate. Red meant *Stop* and green *Be careful*. That light was quite dangerous: the policeman whose task it was to turn it was badly injured when it exploded on 2 January 1869.

The first electric red light was set up under Alfred A. Benesch's instance at the junction of 105th Street and Euclid Avenue in Cleveland (Ohio, USA), on 5 August 1914. The manufacturer, the Traffic Signal Co of Cleveland, had also fitted it with a bell that rang when the light changed. The first three-colour traffic light was set up in New York in 1918.

Macadam (1815)

Macadam was invented by a Scotsman, **John McAdam** (1756–1836), in **1815**.

At the end of the 18th century, the state of European roads was appalling as they had not been renewed since the Middle Ages.

Born in Ayr, Scotland, John McAdam had made his fortune in the United States. On returning to Scotland he started, at his own expense, a series of studies on road surfaces. In 1815 he was appointed Surveyor General of Bristol roads, and was at last able to put his theories into practice. He created a road surface made from crushed stones and sand compacted by road-rollers. Although the name still exists, the McAdam road surfacing system has been almost completely abandoned. Nowadays, mixtures of cement, cinders or slag are used. Tar is no longer used either, being replaced by asphalt.

Parking meters (1935)

The first 150 parking meters were put into operation on **16 July 1935** in Tulsa (Oklahoma, USA). A journalist called **Carlton Magee** had thought up the parking meter system and shortly afterwards founded the first company that was to build them, the Dual Parking Meter Co.

Space Maker car parks (1976)

In New York, even more than in London or Rome, finding a parking space for your car is practically impossible. This is the reason why the American engineer **Arnold Rosen** thought of adapting the platforms used for planes by the airport maintenance staff. Called **Space Maker**, his system doubles the capacity of car parks. It has been widely available in the United States since 1976 and has been well received in England and Scandinavia.

A HIGHLY SOPHISTICATED SIMULATOR

In Berlin, Daimler-Benz has set up a unique driving simulator. It can not only recreate all the driving conditions of a real car (a Mercedes) but also makes it possible to study the concept and behaviour of new vehicles before they are built. The simulator includes two very powerful computers, six video projectors, 30 loudspeakers, a digital image generator and a six-cylinder hydraulic system.

Two-wheelers

Origins

Although no actual example is known to have existed prior to the 18th century, it is likely that the ancient civilisations had already envisaged two-wheeled locomotion. Drawings of two-wheeled vehicles have been discovered in China. The Egyptian obelisk taken from the temple of Luxor in 1836 to ornament the Place de la Concorde in Paris has among its hieroglyphs a representation of a man astride a horizontal bar which is mounted on two wheels. The obelisk dates from the reign of Rameses II (13th century BC).

The bicycle

Celeripede and Velocipede (1790)

The two-wheeled vehicle era started with **Count de Sivrac**'s celeripede in **1790**. This consisted of a two-wheeled wooden frame without a steering mechanism and propelled by no other means than the rider's feet pushing against the ground. The celeripede was renamed velocipede or dandy-horse when attempts were made to improve its appearance by making it look like a lion, a horse or even a dragon.

Draisienne (1817)

The **draisienne** was introduced in **1817** in the Luxembourg Gardens in Paris by the German baron **Karl von Drais von Sauerbronn**. It brought two-wheelers back into fashion. The draisienne had a swivelling steering mechanism controlled by a sort of rudder, the ancestor of the handlebars. It was propelled by being 'walked' along the road. After some major improvements, the draisienne became quite popular, especially in England from 1819 where it was called a hobby-horse.

Around 1839 a Scottish blacksmith, **Kirkpatrick Macmillan**, added pedals which drove the rear wheel through a system of cranks.

Velocipede (1861)

In Paris, in **1861**, the blacksmith **Pierre Michaux** and his son **Ernest** had a brilliant idea. While repairing a dandy horse, they decided to attach what was subsequently called a pedal-and-gear mechanism to the front-wheel axle. The innovation worked and by 1865 the firm of Michaux & Co. had sold more than 400 vehicles.

Rover safety bicycle (1885)

In 1870 the first ordinary bicycle was built by James Starley in Coventry; it became known as the pennyfarthing because of its huge front wheel.

Harry J. Lawson patented his 'safety' bicycle in 1876 which led to **John Kemp Starley**'s **Rover safety bicycle** being designed in **1885**. This model brought together the main features of the modern bike: wheels of equal size, geared up chain drive, direct steering with inclined forks and the diamond shaped frame.

Bicycle with propeller shaft (1989)

A firm from north-east China, **Molun**, has introduced a bicycle that uses a propeller shaft. There are several hundred million cycles on Chinese roads, but they are rather old-fashioned. Molun are aiming to produce one million bicycles with propeller shafts.

Vertical pedal bicycle (1978)

This bicycle was patented in 1978 by its two inventors, the Korean **Man Te Seol** and the American **Marione Clark**. The vertical pedal bicycle enables the rider to attain remarkable speeds without getting tired thanks to its 29 gears and the unusual movement of its pedals, which puts much less strain on the legs.

Dérailleur gears (1889)

The first two-speed gear changing system fitted on the rear hub appeared in **1889** under the brand name The Cyclist. Tested by Paul de Vivie in 1905, it was improved in 1911 (Panel's patent for the rear *dérailleur*) and in 1925 (Raymond's patent for the front *dérailleur*).

Flexible cycle (1984)

Manufactured in the United States in **1984** by **ACS** (American Cycle Systems), this bicycle has rims which automatically regain their original shape after being distorted, thanks to Zytel, the nylon resin invented in 1954 by **Du Pont de Nemours**.

All-aluminium bicycle (1986)

The **Paris-Peugeot** and **Péchiney** companies have perfected a new featherweight all-aluminium bicycle. Originally, a Cegedur-Péchiney patent made it possible to mass-produce equipment previously limited to racing models.

POCKET-SIZE BICYCLE FOR PEOPLE IN A HURRY

The Dahon can be folded or unfolded in less than a minute. Comfortable and strong, its road-holding is good and it can fit into a small case. Its inventors, the Hon Machinery Company of Taiwan, hold seven patents to protect its various qualities.

Although it only weighs 8.980kg *19lb 13oz* (1.350kg *2lb 15.6oz* for the frame) this bicycle is very strong. Not many people know that the frame of a bicycle has to undergo considerable stress. When climbing a mountain pass or sprinting, it is under the same sort of stress as the wing connections on an Airbus' fuselage.

The bicycle has been on the market since 1987.

Rimfire (1987)

'Ride into the future' is the slogan used for **Rimfire** invented in **1987** by **Satellite Wheel Safety Light**, of Tuscan, Arizona (USA). The principle is simple, and luminous: six phosphorescent lights are fitted on both wheels of the bicycle. As this goes along, the wheels look like circles of fire. The lights are powered by a small rechargeable battery. The Rimfire is very attractive but, above all, it

The Winnebiko is the bike for the year 2000. It features 54 speeds, five computers, a telephone, a security system, a speech synthesiser, a microfiche file and a flute. The bike is pictured here with its American inventor, Steven Roberts.

should reduce the great number of collisions between cars and bicycles caused by insufficient lighting on the bicycles.

Motorcycles

Origins (1818)

On Sunday **5 April 1818**, in the Luxembourg Gardens in Paris, a draisienne fitted with a steam engine at the back was introduced. The only drawing of the period which has reached us, however, does not show how the power could be transmitted to the wheels. The official name of this astonishing vehicle was the *velocipedraisiavaporiana*.

The Italian Murnigotti was the first, in 1879, to register a patent for a two-wheeled vehicle with a 0.5hp, four-stroke engine. The machine was never built.

Motorcycle (1885)

Historians still argue over the question of who the father of the first motorcycle really is. Keeping strictly to the concept of a two-wheeled vehicle, this honour falls to two Germans, **Wilhelm Maybach** and **Gottlieb Daimler**. In **1885** they built a motorcycle with a wooden frame and wooden wheels, powered by a four-stroke internal combustion engine. The engine produced 0.5hp and went at 18km/h *11mph*.

The English, on the other hand, claim that Edward Butler invented the motorcycle a year earlier. But this claim derives only from the patent for a tricycle with a gasoline engine, which was not built until three years later.

The thing that these two inventions had in common was that neither one was developed any further in its country of origin; for economic reasons in Germany and because of over strict regulations in England.

First mass-produced motorcycle (1894)

In **1894**, two Germans, **Heinrich Hildebrand** and **Alois Wolfmüller**, built a motorcycle that was mass-produced with over a thousand units. It was a two-cylinder 1488cc motorcycle.

De Dion three-wheeler (1895)

The **Marquis de Dion** was the first person to see the potential that the Daimler engine could have on a light economical vehicle. The first tricycle with a four-stroke engine came out in **1895**. It was a three-quarter horsepower engine with electric ignition. 15000 De Dion tricycles were produced up to 1902.

WHEELS WITHOUT HUBS OR SPOKES

Introduced at the beginning of 1989, the Swiss Franco Sbarro's invention created a sensation. It is a revolutionary wheel which does away with the conventional hub as well as the solid central disc. The rim alone is retained, and that is attached to the outside of the machine and carries the tyre. This gives the impression that the tyre is held on by magic. The effect is striking, and the saving in weight considerable as each wheel only weighs 600g *1lb 5.12oz*. The wheel's inventor, who had it tested on motorcycles first, is convinced that it will also be suitable for mass-produced cars and racing vehicles, the weight reduction guaranteeing considerable fuel economy. Further experiments are needed, however, to establish whether, from the safety point of view, it might not have unpredictable reactions under certain riding and braking conditions.

Motorbike (1897)

The motorbike, named the *Motocyclette* by its French inventors **Eugène** and **Michel Werner**, was exhibited for the first time at the Paris Salon in **1897**. These two Russian-born journalists had already produced several machines – a phonograph, a cine-projector and a typewriter – when in 1896 they undertook to mount a small engine, designed by H. Labitte, on a bicycle. First, they placed the engine horizontally above the back wheel, then in front of the handlebars with a leather belt linking it to the front wheel to drive the bike. It was an immediate success and thousands of these models were built from 1898 onwards in France and, under licence, in England and Germany.

Military motorcycle (1899)

The military motorcycle was born at the turn of the century, almost at the same time as the civilian one. The first model known dates back to **1899**, the year in which the Englishman **F. R. Simms** introduced a de Dion tricycle adapted for use in the Boer war. The tricycle had the front of a quadricycle fitted with a Maxim machine-gun.

The motorcycle's true military vocation was discovered in 1914 when the British and the American forces fitted them with sidecars to be used as ambulances or with a machine-gun for despatch riders.

From the road motorcycle modified by the army was born the cross-country motorcycle. The first military motorcycles not derived from civilian ones appeared in 1939: they had a driving wheel on the sidecar and were produced by BMW and Zündapp in Germany; FN and Gillet in Belgium; Gnome and Rhône in France, and Norton in England. The most original one was definitely the French Simca-Sevitane which was completely waterproof and could be turned into an engine to power a landing craft.

Franco Sbarro's design dramatically alters the traditional form of the tyre.

Sidecar (c.1910)

The idea of attaching a sidecar to a bike was discussed as early as 29 April 1894, in the magazine *Le Cycle*. But this addition had to wait until motorcycles became more solid and powerful; sidecars became popular around **1910**. An articulated assembly was perfected in America in 1916. It allowed the motorcycle to lean when going into a curve or around a corner. A motorised sidecar wheel was developed in 1939 and adopted by the best-known makes in all industrial countries.

Snow scooters (1959)

In 1935, after several years' research, a Canadian mechanic, **J. Armand Bombardier** (1907–1964), invented a control wheel and a caterpillar track that were both revolutionary. This marked the start of the production of snowmobiles. One of the first models had bodywork similar to a contemporary car and could carry seven passengers.

At the beginning of **1959**, Bombardier invented the **Ski-Doo** snowmobile. That model gave birth to a new sport and marked the start of a fantastic expansion of the industry. Today there are about three million Ski-Doos on the snow.

Two-stroke engine motorcycle (1900)

The first two-stroke engine fitted on a motorcycle was perfected by the Frenchman **Cormery** who had the invention patented in Paris on **20 August 1900**.

In 1901 another French manufacturer, Léon Cordonnier, registered a patent for his Ixion engine which also marked the beginning of the rotor arm. During the same period, the Englishman Alfred A. Scott was working on the first two-cylinder two-stroke engine which obtained its British patent on 11 February 1904 but which was not built until 1908, when six were made.

It was not until 1911, when water-cooling appeared, that this vehicle reached its most competitive form. The two-stroke engine was only perfected in the 1930s due to research carried out by the German company DKW.

Finally, separate lubrication appeared on some makes such as Scott in England in 1914 and DFR in France in 1924.

Four-stroke engine motorcycle

All the improvements carried out since 1895 have been geared to reducing the size and bulk of the motorcycle engine whilst increasing its power and, as a result, its rate of revolution.

Cooling

The de Dion engine was the first one to be fitted with fins to help improve cooling by air.

As early as 1912, James Booth adopted water cooling on an eight-cylinder 5500cc engine; that motorcycle weighed 370kg *3cwt 155lb 14oz*.

Valves

The de Dion engine with automatic valves of 1895 came first, then an engine with side valves was introduced by Peugeot in 1903, followed by mechanically operated inlet valves. The push-rod operated overhead valve was introduced as early as 1899 on the French Buchet engine. The distribution system with overhead camshaft really started to develop from 1930, followed by positive controls.

Five-valve cylinder head (1988)

As early as 1977, **Yamaha** started research into a means of making four-stroke engines more efficient than two-stroke ones. This led them to look for a multi-valve cylinder head. After many experiments they arrived at lentiform combustion chambers with five valves (three inlet valves and two exhaust valves). As opposed to the 'roof-shaped' combustion chamber, Yamaha's five-valve chamber does not require a hemispherical piston head to achieve a high rate of compression, and the distance between the spark-plug's electrode and the piston head remains generous. Maximum power is achieved and the torque curve is excellent. The five-valve cylinder head invented by Yamaha is fitted on the 750FZ and the 1000FZR Genesis.

Single-cylinder engine (1897)

The first single-cylinder engine dates back to the invention of the first motorcycle by the **Werner** brothers in **1897** (see above). Its popularity has wavered. By using it in 1976 on the XT500, Yamaha created a new type of motorcycle, the Trail.

Six-cylinder engine

The most famous of these engines are quite recent, such as the one which powered the Grand Prix Honda 250. Its rate could go up to about 18000rpm and it became world champion in 1966 and 1967. Then came the renowned Honda CBX 1000 and Kawasaki 1300. The only one still in existence is the Benelli 900 Sei (1983). Starting from a Kawasaki 1300, Godier-Genoud has built a turbo-charged model which approaches 170hp. It is currently the most powerful engine around.

Four-cylinder motorcycle (1901)

Holden, an English colonel, built a motorcycle with four opposed cylinders as early as **1901**. Its connecting rods drove the back wheels without the use of a drive belt, like the connecting rods on a locomotive.

Turbocharger (1981)

The first mass-produced motorcycle with a turbocharger was the **Honda CX500 Turbo** in **1981**.

Rotary engine (1908)

In **1908** an Englishman named **Umpleby** built an engine with three combustion chambers and only one piston which was quite close in principle to the Wankel engine (*see* Energy) which only appeared 50 years later. Umpleby's engine was never mass-produced.

In 1974 the Japanese manufacturer Suzuki produced a motorcycle with a Wankel engine, the RE5. At present, the British Army is conducting experiments on a rotary engine built by Norton.

Rocket-propelled motorcycle (1983)

The Dutch dragster champion **Henk Vink** built a motorbike propelled by a rocket. Estimated power: 2500hp. It covers 400m *437yd 1ft 9.6in* in just over five seconds from a standing start, crosses the line at more than 400km/h *248.6mph* and is stopped by parachutes.

Motorcycle with reverse gear (1988)

The Goldwing 1500 is the largest and most luxurious machine ever built by Honda. This Rolls-Royce of the two-seaters comes with a new six-cylinder 1520cc 100hp engine. It is the only modern bike with a reverse gear (to make manoeuvring easier).

The Honda Goldwing 1500cc: six-cylinder 1520cc engine, reverse gear, adjustable hydraulic suspension, three disc-brakes, a ventilation system, a computer and a stereo radio.

10 TRANSPORT

Two German companies, Schuberth and BMW, created these helmets after studying road accidents.

Transmission (c.1900)

The drive belt, chain and camshaft, which are the three main means of transmission of the engine's power to the back driving wheel, appeared with the first motorcycles, around **1900**.

Drive belts

The leather belt, flat or with links, and the reinforced rubber belt, were the most used until 1914. Nowadays only Harley Davidson and Kawasaki still use this technique on some of their models.

Chain (1897)

Adopted almost from the first days of the motorcycle (**1897**), it changed very little until 1972 when the Duplex chain (a double chain) invented by the Englishman **Reynolds** appeared. In 1982, Yamaha's Ténéré was given an O-ring chain, which was waterproof and self-lubricating.

Variable transmission (1910)

As early as **1910**, before gearboxes appeared, variable transmission by pulleys with variable cheek spacing appeared at **Terrot**'s in France and one year later, at **Rudge Witworth**'s in England.

This system was not suitable for increasingly powerful machines and was abandoned for motorcycles. An automatic version was fitted on mopeds, the first one being by Motobécane in 1966.

Gearbox (c.1914)

True gearboxes became common on motor bikes just before the First World War. At first the gearbox was controlled manually via a lever and a serrated quadrant fixed on the tank. The pedal control appeared in England in 1923 on the *Vélocette* but only took precedence after the Second World War.

Electric starter motor (1913)

In **1913**, an American firm, **Indian**, introduced the Hendee Special, with a V-shaped, two-cylinder, 998cc engine. This was the first motorcycle equipped with an electric starter motor, the Dynastart. Too advanced for its time, the Hendee Special was a complete failure, and all Indian models with an electric starter motor were recalled to the factory for removal of this accessory, as unperfected as the fragile batteries supplying its current. The electric starter motor did not make its appearance commercially until the 1960s, on mass-produced Japanese motorcycles.

Front suspension (1903)

The first motorcycle front suspension to be marketed was the very complex **Truffault** fork which was seen in December **1903** at the Paris Salon on a Peugeot motorcycle.

Telescopic fork (1904)

In **1904** a well-known French make, **Terrot**, patented what can be considered to be the first telescopic fork. It was used until 1908, the year in which, in England, Scott also introduced a telescopic fork which did not have hydraulic dampers. The pendular fork used by Terrot in France then by Alcyon and Triumph in England appeared in 1909.

Parallelogrammatic fork (1907)

The parallelogrammatic fork was the best-known type of front suspension. Invented in **1907** it was used until after the First World War, in various guises.

Telescopic fork with hydraulic damper (1935)

In **1935**, the German firm **BMW** conceived the first telescopic fork with hydraulic damper. It was universally adopted when it came out and it is still being used. Only BMW, the firm that invented it, uses a different system which they introduced in 1955; the Earles type suspension. That system was very popular until the 1960s, and was fitted on almost all scooters as well as the first Hondas.

Rear suspension (1904)

The suspension principle was invented for velocipedes as far back as 1898. For motorcycles, the first development was the French **Stimula** of **1904**, which had rear suspension with a cantilevered rocking arm and a spring under the seat. In 1911, the German firm NSU inaugurated rear suspension with a rocking arm and two near vertical shock absorbers with helicoidal springs, as used today. The English ASL of 1912 inaugurated pneumatic suspension, with inflatable front and rear shock absorbers. In addition, the seat contained an inflatable cushion.

All the basic principles of rear suspension had been invented by the beginning of the century, but the technology was not far enough advanced to apply them. It was only after the Second World War that sliding suspensions appeared, then rocker suspensions with greater and greater wheel clearance. Finally, in 1979 and 1980, the first variable geometric suspensions were marketed: Honda's Pro Link, Kawasaki's Uni Track, Suzuki's Full Floater, etc. These suspensions contain a combination spring and shock absorber, activated by an intricate arrangement of articulated levers.

Rear single-sided swing arm Pro Arm (1988)

The fruit of **Elf** technology, this rear single-sided swing arm has been used by the engineers at **Honda** to equip the new NVT 650. To reduce bulk, the universal joint that drives the back wheel goes through the swing arm. Furthermore, the oscillating single-sided swing arm Pro Arm, combined with a back wheel with a central nut, allows the wheel to be taken off quickly.

Cast wheels (1972)

Cast wheels appeared as early as **1972** under the impulse of the Frenchman **Eric Offenstadt**: they were cast in a magnesium alloy and gradually replaced those with spokes. They made possible, among other things, the use of tubeless tyres, which are safer in the event of a puncture.

ABS for motorcycles (1987)

The anti-blocking brake system, perfected by the German firm **Bosch** for cars was fitted for the first time by **BMW** in **1987** onto their top of the range motorbikes. Several years were required to master a technique originally aimed at four-wheeled vehicles. But in both areas, the advent of the ABS system is certainly as important as that of the disc brake was some 20 years ago.

Anti-blocking MC ALB (1988)

This anti-blocking system, which comes into action when the driver slams on the brakes, was developed by **Honda** and is not electronic. Fitted in the hubs, the very compact MC ALB system automatically controls the degree of braking as well as the tyres' gripping threshold, preventing the rider from executing any sudden manoeuvres which would result in a fall. It comprises high and low pressure pipes linked to the master cylinder and the brake fluid tank. The system is activated automatically by a speed sensor.

Deltabox frame (1988)

A direct derivation of the one used on the 250cc and 500cc winners of the 1986 world championships, this frame consists of rectangular sections made from very high quality alloys. This technique gives optimum resistance to buckling and better road-holding. Strong and light (12.2kg *26lb 14.4oz*) the **Deltabox frame** is used on the Yamaha 1000 FZR Genesis.

The Railways

Origins

The first British parliamentary decree relating to the creation of a railroad concerned a mining line running between Middleton and Leeds in 1753. The passenger railway was not created until 1830.

Rails (1602)

If we leave out the grooves used by Roman chariots, the first wheel guiding rails seem to have appeared in **1602**, in mines around Newcastle in England. These rails were made of wood. In 1763, Richard Reynolds made the first cast-iron rails, manufactured as a consequence of a slump in the iron industry. The first metal raised rails date back to 1789 and were invented by the Englishman William Jessop. Steel rails were invented by Bessemer in 1858. Nowadays, all rails are made of steel. On main lines, new rails are delivered already welded in units of up to 800m *875yds.*

Points (1789)

In **1789**, **William Jessop** also perfected a points system. After the advent of metal tramroads, around 1765, forerunners of the points had been invented, but they did not have any moving parts. Jessop's ingenious innovation was to incorporate a moving tongue-rail to this primitive device. His system marked the start of the invention of the points switching system, which remains a collective achievement.

Trevithick's steam engine (1802)

In **1802–3** the Cornish engineer **Richard Trevithick** (1771–1833) built the first steam locomotive at the Coalbrookdale ironworks. Soon afterwards, urged on by Samuel Monfrey, an ironmaster near Cardiff, he built a second one.

Tests began on 2 February 1804 which proved conclusive. Trevithick's locomotive, with a six tonne convoy, ran on the Pen-y-Darren line (15km *9.32 miles*). A few passenger carriages were added. Empty, its speed was 20km/h *12mph*; loaded, 8km/h *5mph.*

Trevithick invented high-pressure steam machines and also designed several prototype steam cars.

Blenkinsop locomotive (1812)

The first steam locomotive to be mass-produced was built by an Englishman named **John Blenkinsop**, beginning in **1812**. An engine with an ordinary, nontubular steam boiler, the Blenkinsop was designed to carry goods and travelled at a very low speed. Its special characteristic was that it ran on a toothed rail.

From 1812, the Blenkinsop ran between Leeds and Middleton.

Stephenson's Rocket (1829)

In **1829**, the Englishman **George Stephenson** (1781–1848) developed the first high-speed steam locomotive. As early as 1813, he had built a steam locomotive equipped with driving wheels joined by connecting rods. These wheels were smooth and rolled on rails. His *Rocket* won the Rainhill trials in 1829, a contest organised by the contractors for the Liverpool–Manchester railway line. Under the competition's conditions (to pull a 40-tonne train), the *Rocket* attained a speed of 26km/h *16mph*, but with no train to pull it could reach up to 56km/h *35mph.* This exploit marked the birth of the railway.

Crampton locomotive (1846)

In **1846** an Englishman, **Thomas Russel Crampton**, built the first two locomotives embodying a principle he had conceived three years earlier. Crampton's idea was to build a high-speed locomotive modelled on Stephenson's Long Boiler but without the latter's major disadvantage; namely, a lack of stability resulting from the overhanging position of the boiler furnace in relation to the wheel axles. Crampton shifted one driving axle to the rear of the firebox, leaving only two axles under the cylindrical body of the boiler. His two locomotives, put into service in Belgium on the line running between Liege and Namur, easily reached a speed of 100km/h *62.14mph.*

In 1848 Crampton built the *Liverpool*, a locomotive of huge size intended for use on the London–Wolverton line of the London and Northwestern Railway. The engine reached a speed of nearly 127km/h *78.92mph* but was nevertheless taken out of service because it put too much strain on the rails.

Passenger carriages (1830)

Although the Stockton and Darlington Railway had, on occasion, carried passengers from its inception in 1825, they had to sit in the coal trucks.

The first passenger carriages as such, which looked like stage coaches, appeared in September **1830** when the Liverpool to Manchester line was opened. This was the first inter-city service in the world.

Long Boiler Locomotive (1841)

The Long Boiler model was perfected by the Englishman **Robert Stephenson** in **1841**. Its main characteristic was longer chimneys. Some of the original models were still in circulation in the 1920s. These machines were the forerunners of the modern steam engine.

Wide tracks (1835)

The 1.435m *4ft 8½in* track width had barely been fixed, when an English engineer, **Isambard Kingdom Brunel**, thought it to be too narrow and developed a 2.13m *7ft* wide track.

In **1835**, I. K. Brunel founded the Great Western Railway which was the first to link London to Bristol – on a 2.13m *7ft* wide track, of course.

Hitler also thought that the existing tracks were too narrow. He therefore planned to build his own wide-track network. In 1942, he gave instructions for the creation of a European super railway, going from Paris to Rostov and Istanbul, with offshoots to Brest,

The opening of the Stockton to Darlington railway in September 1825 with George Stephenson's Locomotion.

Leningrad, Kazan and Baku. The tracks were meant to be 3m *9ft 8in* wide. Electric locomotives weighing 1000 tonnes and developing 30 000hp would have pulled passenger trains at 240km/h *149.14mph*. . . . Unlike Brunel's, Hitler's train never came into being.

Railway signals (1849)

The running frequency of trains was originally based on the time interval that separated them. After numerous accidents caused by one train catching up with another, a block-signalling system was adopted by the **New York & Erie Company** in **1849**. In 1856 an English engineer, Edward Tyer invented an electric signalling device that was adopted for use in the lower Blaisy tunnel on the Paris–Dijon line.

The automatic block signal was invented by Thomas Hall in the United States in 1867. The signals were then operated by the train. In 1871, Franklin Pope installed on the Boston & Lowell Railroad the first signalling system to be centrally controlled.

In 1878, Edward Tyer invented a system which is still in use in some countries, including Great Britain, to prevent trains on single track lines colliding: the electric staff.

Rack-railway (1862)

In **1862**, the Swiss **Niklaus Riggenbach** invented the rack railway which was to be used on slopes over six per cent. He was inspired by John Blenkinsop's system, whereby a cog on the locomotive engaged with a toothed rail.

In 1868, Sylvester Marsch built the Mount Washington line in New Hampshire (USA) ascending slopes that reached gradients of up to 30 degrees. In 1870, the first rack railway in Europe was built at Righi, Switzerland. In 1885, a Swiss engineer from Lucerne, Abt, invented a system of triple gears with staggered cogs for the Harz Mountain railway in Germany.

Restaurant car (1863)

The first restaurant car was put into service in the United States between Philadelphia and Baltimore in **1863**. In Europe, the first dining car in which food was prepared ran between Leeds and King's Cross, London in 1879; from 1880, a restaurant car service was also available between Berlin and Frankfurt. Such a service had already been tried out in 1867 in Russia. The first regular restaurant car services in France date back to 1 June 1883 on the Paris–Caen and the Paris–Trouville routes.

The first buffet car was created in England in 1899 on the Great Central Railway. It was a forerunner of the snack-bars which have generally replaced the traditional restaurant car.

Sleeping cars (1865)

The first (American) patents for the sleeping car system were registered by **George M. Pullman** in 1864–5 and the maiden voyage of the first de luxe 'Pullman' sleeping car, the Pioneer, dates back to **1865**.

In 1867, Pullman founded the Pullman Palace Car Company for the construction and exploitation of luxury sleeping- and restaurant-cars.

It was George Pullman who first offered comfortable surroundings to ordinary passengers.

Today there are only three Pullman trains left in the world: the Manchester Pullman (London–Manchester), the Merseyside Pullman (London–Liverpool) and the Yorkshire Pullman (London–Leeds), the last two trains having been put back into operation on 13 May 1985. As for couchettes, they appeared in the United States in 1836.

Car-sleepers (1955)

This type of service was created by British Rail which in **1955** put into service the first **car-sleeper** between London and Perth. In 1956, a car-sleeper link was established between Ostend and Munich as well as between Hamburg and Chiasso.

Air brakes (1869)

Originally, the brakes used on railway carriages consisted of a simple brake shoe that pushed against the wheel. These brakes were manoeuvred by hand. In **1869** an American, **George Westinghouse** (1846–1914), patented automatic air brakes. Compressed air was distributed to each waggon by a central air tank that fed auxiliary air tanks through a triple valve. When the pressure was lowered, a supply of air was released to the brake cylinder. This system was tried out on a passenger train for the first time in 1872. It contributed to the improvement of rail transport and the same principle has remained in use up to the present day.

Electric locomotive (1879)

After various trials by the American Davenport and the Englishman Davidson in 1839, a small train driven by an electric locomotive – developed by **Werner von Siemens** and **Johann Georg Malske** – circulated within the walls of the Berlin Fair in the summer of **1879**. Despite its small dimensions, this locomotive is considered to be the starting-off point for electrically driven vehicles. Preserved by the Siemens Company, this machine still exists.

In 1902, a 200km/h *124mph* speed record was set by an electric locomotive between Zossen and Marienfelde, in Germany.

The first electric railway in the world was that of Giants' Causeway in Ireland, inaugurated in **1884**.

Pacific locomotive (1892)

The famous *Pacific* was put into service for the first time in the United States on the Missouri Pacific Railroad in **1892**. The *Pacific* developed 2200hp and a test pressure of 16kg *227.57lb*. It weighed 93 tonnes.

Atlantic locomotive (1900)

The *Atlantic* locomotive appeared about **1900** on the Philadelphia–Atlantic City line (USA), hence its name. The use of the *Atlantic* spread to Europe in 1901. It then had 1500hp, weighed 64 tonnes, and developed steam pressure of 16kg *227.57lb*.

Diesel locomotive (1912)

The first diesel locomotive was built in **1912**, in Winterthur, Switzerland, by the firm, **Sulzer**. It weighed 85 tonnes and developed 1200hp; that is weak when compared to the power of steam locomotives at the time.

In Switzerland there are still 70 private companies linking villages by rack-railway.

Magnetic trains

For the past 20 years, a great number of projects dealing with magnetic levitation trains have been studied in large industrial countries. The trains are suspended a few centimetres above a (very special) track and are propelled by linear induction motors.

The challenge is to go faster than the French TGV within an acceptable cost range. It remains to be seen which line will be the first to dare choose a Maglev rather than a 'wheels on rails' system.

United States (1967)

The first magnetic levitation vehicle was tried out in the United States in **1967**, at the Pueblo (Colorado) trials centre. In 1973, a linear induction motor engine, developed by Garrett, reached the speed of 402.5km/h *250.11mph*.

Transrapid: 412.6km/h *256.39mph*

As early as 1971, West Germany produced an 'electroglider' using the magnetic levitation principle. But the first true Maglev vehicle, the **Transrapid**, came out a year later. It has been developed by seven large German companies in association with Lufthansa and the German State which has pledged subsidies equivalent to £300 million over 15 years.

The 54m *59yd* long train is supported by 64 electromagnets; another 56 help with lateral guidance. Its official maximum speed of 412.6km/h *256.39mph* was recorded on 21 January 1988. The Japanese MLU has only managed to reach 405.3km/h *251.85mph*.

25km/h *15.54mph* in Britain

Since 1983, a short line (2km *1¼ miles*) using magnetic levitation has linked Birmingham International station to the airport. But the maximum speed on that line is only 25km/h *15.54mph*.

HSST versus MLU in Japan

Magnetic levitation has been used in Japan since **1971** when Japan Air Lines (JAL) produced the HSST, and Japanese National Railways the ML500.

In 1978 the HSST reached a speed of

The German Transrapid has reached a speed of 412.6km/h 256.39mph.

307km/h *190.77mph* and the ML500 301km/h *187mph*. The latter is said to have reached 407km/h *252.91mph* in 1979 and 517km/h *321.26mph* in 1980. The HSST works on a principle similar to that of the German Transrapid but, according to the specialists, it is more difficult to perfect.

Another type, the MLU 001, literally floats on a field of nobium-titanium superconducting magnets positioned under the floor of the train. In 1987 it reached a speed of 405.3km/h *251.85mph*.

High-speed trains

Japanese bullet-trains (1964)

The **Japanese National Railways** inaugurated their first high-speed train line, Tokyo–Osaka (515km *320 miles*) on **1 October 1964**. The maximum speed of these bullet-trains is 210km/h *130.49mph*, although during one trial, held on 17 October 1985, a speed of 272km/h *169mph* was reached.

At present, this high-speed network (*shinkansen*) numbers three lines: Tokyo–Hakat (1069km *664 miles*), Tokyo–Niigata (297km *184 miles*), Tokyo–Morioka (492km *305 miles*).

TGV (1978)

Research geared to creating a high speed train, the **TGV** (*Train à Grande Vitesse*), began in France in 1967, the year in which the first gas-turbine train, the TGS, was produced. This TGS gave birth to two great families of trains: the ETG (*Eléments à Turbines à Gaz*) in 1970 and the RTG (*Rames à Turbines à Gaz*) in 1973, a year after the TGV 001, also using a gas turbine, came out. This reached a speed of 318km/h *197.61mph* in 1972. The first electric TGV was delivered in **1978**. One of the trains, No 16, reached 380km/h *236.13mph* in February 1981, a few months before the new Paris–Lyon line was put into commercial service. In 1985, the TGV's top speed was 270km/h *167.78mph*.

British HST (1973)

To increase the speed of trains on non-electrified main lines, **British Rail**, in **1973**, created an experimental diesel-electric train comprising two 1680kW locomotives with seven or eight slip coaches in between. It could do 200km/h *124.28mph* in commercial service and was named the High Speed Train or HST. In 1979, a mass-produced version was put into operation between London and Bristol. Today there are about one hundred of those trains, known as the 125, in the whole country.

Another high speed train had been studied in 1975–6 by British Rail: the much talked about APT (Advanced Passenger Train). But too many unsolved technical problems meant the idea had to be abandoned in July 1986.

The German ICE (1985)

The German **ICE** (Intercity Experimental) is the Deutsche Bundesbahn's (DB) high speed train. Like the TGV, it is designed to travel at high speed on new lines. Two of these are being built at present between Hanover and Würzburg and between Mannheim and Stuttgart.

DB took delivery of the prototype ICE train in April **1985**. Comprising two locomotives with three slip coaches in between, the train reached speeds of 317km/h *196.98mph* in November 1985 and 345km/h *214.38mph* in 1986.

The ICE is fitted with asynchronous motors, powered by triphase current. The train has continuous power of 2800kW, with a maximum of 4200kW. Under the name of Intercity Express, it will run from 1991 at a cruising speed of 250km/h *155.35mph*.

Urban transport

Origins (1661)

Blaise Pascal (1623–62), the French philosopher, was the inventor of the public transport system. In **1661**, he proposed a system of coaches that would 'circulate along predetermined routes in Paris at regular intervals regardless of the number of people, and for the modest price of five sols'.

On 19 January 1662, the King's Counsel authorised the project's financiers, the Marquis de Sourches and the Marquis de Crenan, to begin running coaches in the city of Paris and its suburbs. The first coach went into service on 16 March 1662.

This public transport system attracted a good deal of curiosity, but its success was short-lived, for the coaches, ill-adapted to the tortuous, crowded medieval streets, were far too slow. Lacking patrons, the price rose to 6 sols, and the company went bankrupt 15 years later.

Tram (1775)

The tram was invented by the Englishman **John Outram** in **1775**. This public transport vehicle ran on cast-iron rails and was drawn by two horses. It was not used in the city.

In 1832, John Stephenson built the first urban streetcar between Upper Manhattan and Harlem in New York (USA). It ran for only

Although many tramways were closed after the Second World War, they survive in Hong Kong.

At the turn of the century, the motor omnibus was beginning to drive the horse-drawn version from the road. By 1914 George Shillibeer's bus had been completely replaced in London.

three years. In 1852 a Frenchman, Emile Loubat, thought of embedding the rails in the road surface. That same year, he used his idea to build the Sixth Avenue line in New York. The cars were horse-drawn and open at both ends.

Electric tram (1888)

The first operational electric tramway line was built in **1888** by the American **Frank J. Sprague**. He obtained a concession for a 27km *17 mile* line to Richmond, Virginia. Ten years later 40 000 trams were in use in the United States. Sprague's 'first' had been preceded by a few prototypes: that of Siemens and Halske in Berlin in 1879, and that of Edison in Menlo Park in 1880.

Bus (1819)

The first omnibus line was inaugurated on **4 July 1819**, in Great Britain, by **George Shillibeer**, a coach builder. The buses had 22 seats and were drawn by three horses, and the conductors wore midshipmen's uniforms. By 1856 the London General Omnibus Company was the largest in the world.

In 1825, 150 years after the first public transport system was abandoned, a former soldier in the French Imperial Army, Colonel Stanislas Baudry, thought of using vehicles derived from the stagecoach (which could hold some 15 passengers, including a conductor) to provide public transportation. Baudry made his vehicles available to Parisian customers of his hot-water bath house, which was in the suburbs. But, after noticing that many people who lived in the suburbs also used his coaches, he expanded the service. His terminus in the city was located at the Place du Commerce, in front of the shop owned by a certain Monsieur Omnes, whose sign included the words Omnes omnibus. Baudry found the word omnibus (which means 'for everybody') appealing, and decided to use it for his transport line.

Motorised bus (1831)

Walter Hancock, an Englishman, provided his country with the first motorised bus in **1831**. This bus, powered by a steam engine, could carry ten passengers. The *Infant* was put into service experimentally between Stratford and the City of London that year.

Steam-powered buses were replaced by the petrol driven bus, first built in 1895 by the German firm Benz. The Benz bus was put into service on 18 March 1895, on a 15km *9 mile* route in the northern Rhineland. There was room for six to eight passengers and two drivers, who remained outside.

Taxi (1640)

Nicolas Sauvage, a French coachman, opened the first taxi business in **1640** on the rue Saint-Martin in Paris. He started with a fleet of 20 coaches. In 1703, the police laid down laws for their circulation and gave each an easily readable number, thus introducing the first form of vehicle registration.

The real use of the automobile as an individual means of public transport with a meter registering both speed and distance, is attributed to Louis Renault who, in 1904, launched small, specially designed two-cylinder cars.

The underground (1863)

The first underground railway was inaugurated in London on **10 January 1863**. It was 6.4km *3 miles* long and used steam traction.

Otis mini urban train (1985)

The American firm **Otis**, which specialises in lifts, has created a new mini urban train system. It derives both from the lift and the hovercraft: it is pulled by a cable and glides on air cushions. The 'smallest urban train in the world' has been running at Serfaus in Austria since the end of **1985**. Its single track is only 1300m *1422yd 7in* long. The train (there is only one) comprises two 'cabins' which can hold 270 passengers each. There are two other similar systems: one in Tampa, Florida (USA), the other in Sun City (South Africa).

Funicular railway (1879)

In 1835, the Swiss inventor Egben suggested a convenient way of getting to the top of Mont Blanc. First, the ice-cap had to be cleared, then a huge trench dug to install a type of funicular and then one had to wait for the ice to cover the whole installation. It was not until

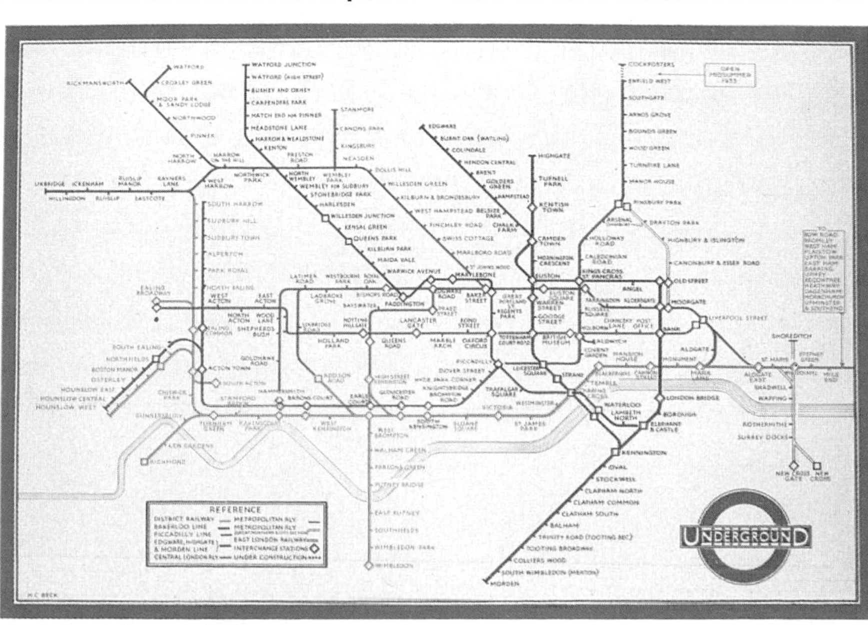

The first urban underground line in the world – the Metropolitan – was opened in 1863 and ran from Paddington to Farringdon. The ingenious tube map was designed by Harry Beck in 1933.

1879 that the first funicular railway started operating on the slopes of Mount Vesuvius where the gradient is over 60 per cent. It was pulled by two steam engines of 45hp each.

First sub-glacier funicular (1990)

It is being built above Les Deux-Alpes in France. The train will go under the Jandri glacier along a 1650m *1 mile 45yd* long gallery and will come out at an altitude of 3410m *11 185ft* opposite les Ecrins mountain.

Cable car (1908)

In the 15th century, so the story goes, a squad of soldiers from the Japanese Imperial Army who were surrounded at the top of a mountain, managed to escape using a primitive cable car. The first picture of a cable car was drawn by Faustus Wranczi in the **17th century**. In the last century, the forerunners of the modern cable car took over from railways in a large number of mines and quarries. The first cable car specifically dedicated to passenger transport was inaugurated on 27 July **1908** in Switzerland; it was nicknamed the 'Wetterhorn lift'.

By air

Balloons

Hot-air or Montgolfier balloon (1783)

The first flight by a hot-air balloon took place at Annonay, near Lyon, France, on 4 June 1783. This hot-air balloon was constructed by two brothers, the paper manufacturers **Joseph** (1740–1810) and **Etienne** (1745–1799) **de Montgolfier**. The *Montgolfière* was made of pack-cloth covered with paper. The balloon carried a portable stove in which wool and straw were burnt to produce hot air, that is, a gas lighter than air. This first balloon attained an altitude of 1000m *1094yd* and landed after ten minutes. When the news of the flight reached Paris, it caused a sensation.

On **19 September 1783**, the Montgolfier brothers repeated their first experiment in front of the King and his court at Versailles. This time the balloon carried a suspended cage containing the first air passengers: a sheep, a duck, and a cock. This flight was also witnessed by Benjamin Franklin, then United States minister to France.

First human flight (1783)

Jean Pilâtre de Rozier (1756–85) and the **Marquis d'Arlandes** (1746–1809) made the first human flight on **21 November 1783**. Ascending in a basket supported by a hot-air balloon, the first two aeronauts left the Bois de Boulogne, Paris and landed 25 minutes later close to the centre of the city.

Pilâtre de Rozier, the first pilot, was also to become aviation's first victim. He was killed on 15 June 1785, while attempting to cross the English Channel. Soon after departure, the balloon caught fire and crashed three miles from Boulogne.

The modern Montgolfier (1963)

The revival of interest in Montgolfiers is due to research done by the US navy in the early 1950s, and it was in 1963 that touring balloons first appeared. Balloon coverings are now made of very light, synthetic materials; the basket is usually wicker, and the pilot heats the air with the help of a flame from a propane gas burner.

Hydrogen balloon (1783)

It was French physicist **Jacques Alexandre Charles** (1746–1823) who at one go invented all the rules governing modern ballooning. The balloon he designed, inflated with hydrogen, went up in the Jardin de Tuileries in Paris on **1 December 1783**, with Charles and the Robert brothers on board. The American balloonist Wise perfected Charles' balloon by introducing the use of a rip cord. Modern sporting balloons differ from that of Charles only in the use of a non-flammable gas, helium, and in the quality of the covering used.

It was in a mixed hot air/helium balloon, the Dutch Viking built by the British company Cameron Balloons that, in August 1986, the Dutchman Hen Brink (accompanied by Evelien and Willem Hageman), beat the speed record for crossing the Atlantic by balloon: 51h 14min to cover more than 4000km *2486 miles*.

A collection of hot-air balloons pictured in Paris on 26 September 1909.

Airship (1852)

The airship was invented in **1852** by the Frenchman **Henri Giffard**, who perfected a balloon furnished with a means of propulsion. This hydrogen-filled airship was driven by a 3hp steam engine, and it flew for the first time on 24 September 1852, covering a distance of 28km *17.3 miles* at a speed of 6.9km/h *4.3mph*.

In 1883, the French Tissendier brothers attached an electric motor to a conventional airship.

The first closed-circuit (with return to the takeoff point) dirigible flight was accomplished on 9 August 1884 by the French army captains Charles Renard and Arthur Krebs. Their dirigible, *la France*, was powered by an electric engine.

The Zeppelin (1890)

Count Ferdinand von Zeppelin undertook experiments as early as **1890** and took out a series of patents for a streamlined dirigible that would not lose its shape but remain rigid. A moveable hangar enabled it to be taken out without risk, no matter what the wind direction was. The first ascension of the LZ1 took place on 2 July 1900.

This airship comprised a 128m *420ft* cylinder with an aluminium frame covered with specially impregnated cotton.

Between 1900 and 1939, 52 000 people had travelled 2 million kilometres *1 242 800 miles*

On Friday 7 January 1785, Blanchard and Gefferies made the first successful airborne crossing of the English Channel. They left Dover at 3pm and landed near Calais 45 minutes later.

by zeppelin. One accident brought the fashion to an end: the *Hindenburg* burst into flames on arriving at New York on 6 May 1937.

Skyship (1984)

The Skyship 600 fitted with a 12m *13yd 4in* long gondola can carry 12 to 15 passengers. Its manufacturer, **Airship Industrie**, claims the airship can fly at a speed of 102km/h *63.4mph* and has a 55 hour endurance and hence a range of 5500km *3418 miles*. Its polyester envelope is made in France by Zodiac. The craft is powered by two Porsche turbo engines. The US Navy is studying the possibility of a Skyship 5000. At present there are about ten skyships in operation worldwide.

Helicopter-balloon (1992)

The Canadian company **Hystar Aerospace** has conceived its future helicopter-balloon in answer to the needs of foresters. At present, practically insoluble transport problems are compelling foresters to cut down whole sections of forest at a time.

The Canadian balloon will be similar in shape to a flying saucer, and topped by a rotor. According to its manufacturer, the helicopter-balloon will be much more reliable than the helistat which was devised by the American engineer Frank Piasecki in 1956, but only flew for the first time in 1984. The helistat has four rotors which are impossible to synchronise.

Aeroplanes

Origins (1809)

People have always dreamed of emulating birds. In the 15th century, Leonardo da Vinci observed how birds flew and designed artificial wings and even helicopters. This study was taken up by Borelli in 1680.

Englishman, **Sir George Cayley** (1773–1857), is the true father of the aeroplane, pioneering the theory of heavier-than-air flight. In **1809** he perfected a fixed-wing glider with a stabilising tail and in 1853 he built a fixed-wing glider which flew 500m *1640ft*. He was also the first person to think of using propellers to obtain the necessary force to drive the plane.

In 1871, the French engineer Alphonse Pénaud managed to make a model aeroplane propelled by a twisted rubber-band system fly more than 50m *54.7yd*. This enabled him to set out flight equations for the first time.

In Britain in 1843 William Henson had patented a steam-powered flying machine, but it was his partner John Stringfellow who, in 1848, managed to make a model aeroplane fly a few dozen metres. In 1856, the Frenchman Jean-Marie Le Bris had made the first gliding flight.

First flight (1890)

The glory of getting off the ground for the first time aboard a motorised machine, and making a 'flying leap' goes to the Frenchman **Clément Ader** (1841–1925): on **9 October 1890**, at Armainvilliers, Seine-et-Marne in France, he 'flew' a distance of about 50m *54.7yd*, at 15cm *5.9in* from the ground, on board the *Eole*, a bat-shaped aircraft fitted with a 20hp steam engine. Ader's patent, registered on 19 April 1890, uses the term *avion* (aeroplane) for the first time. His *Avion III*, which still exists, was equipped with two 20hp steam engines. It 'leapt' over a distance of 300m *328yd 7ft 2in* at Satory on 14 October 1897 but was damaged. Unable to obtain any financial support, Ader had to give up his research into military uses of flying machines. In the 1880s, the German Otto Lilienthal (1848–1896) was practising piloting a glider so as to compile information that would enable him to tackle motorised flying.

First sustained flight

It is undeniable that, if the terms used are agreed, the first sustained powered flight in the world was accomplished by two Americans: the **Wright** brothers, **Wilbur** (1867–1912) and **Orville** (1871–1948). They flew their first glider on the Kitty Hawk dunes in North Carolina in 1900. They then built a biplane, the *Flyer I*, and on **17 December 1903**, after drawing lots, Orville became the first man to fly aboard a powered machine. During that first trial the aeroplane covered 36.60m *40yd* in 12 seconds. Wilbur and Orville alternately accomplished two flights of 13 and 15 seconds

The Wright brothers' aeroplane was patented in 1907.

each. During the fourth flight, Wilbur covered 284m *310yd* in 59 sec. The *Flyer*'s take off was assisted by a rail along which rolled a carriage carrying the aeroplane.

Seaplane (1910)

The Frenchman **Henri Fabre** (1882–1984) is the father of marine aviation. He built a flat-bottomed seaplane with floats. On **28 March 1910** he took off in it from the Etang de Berre in the South of France. The machine was of the 'canard' type with wings and engine at the back.

In 1905, Gabriel Voisin had already conceived the idea of a glider mounted on parallel floats and towed by a motorboat. Later on, the seaplane was developed in France and in the United States where Derhaut and Curtiss came up with the flying-boat at the same time.

Boeing 247 (1933)

The **Boeing 247** – the first airliner – was put into service by United Airlines in the United States in March **1933**. It carried ten passengers. For the first time, travellers could cross the United States in less than 20 hours. This low-wing, twin-engine aircraft had retractable landing gear, wing de-icers, constant-speed, full-feathering propellers (which permitted the automatic pilot mechanism to be used and also guaranteed maximum engine efficiency under all conditions), and the ability to maintain flight on a single engine.

Douglas DC-3 (1935)

Equipped with the same advanced technology as the Boeing 247 but able to carry more passengers, the **Douglas DC-3** was the first transport airplane to fly over the Himalayas, between India and China. American Airways was the first company to put it into service, flying between New York and Chicago, in **1935**. During the Second World War, when it was known as the Dakota, it was frequently used as a military transport aircraft. Over 10000 of these commercially profitable airliners were built, and they are still in service, notably in Africa and the Third World.

Viscount V-630 (1948)

The Viscount V-630, constructed by the British firm, Vickers-Armstrong, was a low-wing monoplane powered by four Rolls-Royce turboprop engines. Its cruising speed was 440km/h *273mph*. This airliner made its first flight on 29 July **1948** and was put into service for a period of only two weeks (29 July–14 August 1950) on the London–Paris route. It became the first turboprop to be used for commercial service. However, it was almost immediately dropped in favour of the larger and more powerful Viscount V-700.

Havilland Comet (1949)

The Comet 1, built by the British firm **De Havilland**, made its first flight on 27 July 1949. Powered by four turbojet engines, each with 2020kg *4453lb* of thrust, this aircraft could cruise at 788km/h *490 mph*, at an altitude of 12000m *39360ft* carrying 36 passengers. Entering service with BOAC on 2 May 1952, this airliner could fly from London to Johannesburg (10750km *6680 miles*) in 17h 6min. The Comet was withdrawn from airline service following a series of accidents.

Boeing 707 (1954)

The first trials of the B-367-80 (prototype of the Boeing 707) were made in Seattle, Washington (USA) on **15 July 1954**. The Boeing 707-120 aircraft made its first commercial flight with Pan American Airways on 20 December 1957. This aircraft, whose wings were placed at a 35 degree sweep, was powered by four turbojet engines. It had a wing span of 39.9m *131ft* and could carry 179 passengers at a cruising speed of 912km/h *566.7mph*. More than 3000 Boeing 707s were built. Larger and more powerful than any airliner of the period, the Boeing 707 became the standard long-range airliner.

Caravelle (1955)

The revolutionary idea of mounting jet engines at the rear of the fuselage, rather than on the wings originated with the French Caravelle, which made its first flight on **27 May 1955**. Rear mounting of the jet engines allowed undisturbed airflow over the wings, as in gliders, and brought increased aerodynamic efficiency, better stability, and decreased cabin noise. The Caravelle was also one of the first airplanes to have a rear integrated stairway. This efficient, rear-mounted engine system was soon adopted by almost all aircraft manufacturers.

The plane was put into service in 1958 but by 1970 it had been overtaken on the technical front by other airliners and was no longer being manufactured.

The seaplane built for the multi-millionaire Howard Hughes is the largest wooden aeroplane. Despite its eight 3000hp engines, it flew just once; the seaplane's owner took it a few hundred metres over the port at Los Angeles on 2 November 1947.

Boeing 747 (1969)

The first flight of a Boeing 747 took place on **9 February 1969**. The aircraft was put into service by Pan American Airways on 21 January 1970. With this aircraft, Boeing launched a new generation of large-capacity planes which became known as jumbo jets. This aircraft has a wing span of 60m *196.8ft* and it can cruise (with a full payload) over 7402km *4600 miles* at Mach 0.89.

Tupolev 144 (1968)

The Soviets' supersonic Tupolev 144 has a wing span of 28.8m *94ft 5in*, a length of 64.45m *211ft 4in* and can carry up to 126 passengers. It made its maiden flight on **31 December 1968**.

On 1 November 1977, the Moscow–Alma-Ata line was opened to the public. This run was never satisfactory and an accident in June 1978 brought commercial application of the aircraft to a halt.

Tupolev 204 (1990)

The new Soviet plane, the TU 204, flew for the first time on 2 January 1988. It is a very modern short to medium range aircraft, which can rival the Western medium range planes. It will be put into operation in **1990**.

Airbus (1973)

The first commercial flight of the Airbus, the first medium range jumbo jet, took place on **28 October 1973**, in Toulouse, France. The Airbus is a European aircraft as it has been conceived and manufactured by companies from France (**Aérospatiale**), Germany (**MMB**), Spain (**CASA**), Britain (**British Aerospace**) and Belgium (**Belairbus**).

A320 (1988)

The Airbus A320 entered service with Air France and British Airways in April **1988**. Designed from the outset to be the most economical airliner in its class while providing considerable improvements in safety and comfort, the A320 is by far the most modern aircraft. It makes the most of new materials (improved metal alloys and composites) as well as the most advanced technology, even surpassing that of Concorde: the A320 is the first civil airliner to use 'fly-by-wire' technology. Fly-by-wire controls driven by computers are used to ease the pilot's workload in routine operations, and safety is enhanced by the built-in protection against dangerous situations.

Airbus Industrie, in its attempt to increase the aircraft's reliability at the beginning and the end of flights, the most critical phases, has invented an automatic guidance system, the alpha-floor, which can detect, before the pilot does, the 'microbursts' of wind that are particularly dangerous for aircraft.

The A320 is a 150-seat single-aisle twin-jet aircraft with a range of 3500 to 5950km *2175 to 3697 miles* depending on the model. It flew for the first time on 22 February 1987 with Pierre Baud, Airbus Industrie's head of flight testing, at the controls.

A330 and A340 (1991)

After the A320, the next step in the development of the Airbus family is the **A330**, a high capacity medium to long range widebody twin aircraft (328 seats) with a range of 9300km *5779 miles*, and the **A340**, a four-engined long range aircraft (13200 to 14300km *8202 to 8886 miles*) with 262 to 294 seats, depending on the model. The A330's maiden flight is planned for June 1992, and the A340's in July **1991**.

CONCORDE AND SOON . . . SUPER CONCORDE

On 2 March 1969, the supersonic plane Concorde had its maiden flight. The first commercial flight, for Air France, linking Paris to Rio de Janeiro took place seven years later, on 21 January 1976. On the same day, a British Airways Concorde inaugurated the London to Bahrain link. The Franco–British plane's birth certificate had been signed on 29 November 1962, by Aérospatiale and British Aircraft Corporation.

This commercial plane – the first supersonic civil aircraft – is the fastest passenger aircraft in the world. It has a 25.56m *83ft 10in* wingspan and is 62.17m *203ft 11in* long. It can carry between a hundred and 139 passengers and take off with a total payload of 185000kg *407925lb*. Its maximum range is 6200km *3853 miles*. It can 'do' Paris to New York in three-and-a-half hours, flying at an altitude of 18000m *59040ft*.

Concorde was conceived to fly 45000 hours. Before the plane received its certificate in 1975, it had to complete 5300 hours of test flying, of which 2000 were at supersonic speed. Air France owns seven of these aircraft of which five are in use. The most used plane since the supersonic network opened has totalled 11400 hours, so they still have a lot of life left in them.

Concorde suffered considerably from being born at a time when the oil crisis was just beginning. However, the management of British Aerospace and Aérospatiale have been planning to give Concorde a successor which, based on an estimate of requirements in about 20 years time, would mean 300 to 500 new aircraft. This Future Supersonic Transport Aircraft, according to the studies already carried out, would be fitted with 200 seats and could cover 12000km *7457 miles* in one stretch at a speed of Mach 2.2 to Mach 2.5 (about 2500km/h *1553.5mph*).

Pedal aircraft (1977)

It was the American industrialist **Paul McCready** who renewed popular interest in human propulsion for aircraft. His *Gossamer Condor* was, in **1977**, the first aircraft in the world to give a convincing demonstration of human-powered flight.

Crossing the Channel

In **1979** McCready's *Gossamer Albatros*, piloted and 'pedalled' by Bryan Allen, managed to cross the Channel. It seemed as if no one could go any further, but the famous MIT (Massachusetts Institute of Technology) decided to take up the challenge. With the help

The Airbus A320 is the first civil airliner to use 'fly-by-wire' technology.

of sponsors such as the Smithsonian Institute, Anheuser-Busch Inc. and United Technology, a team from MIT led by Mark Drela got to work.

Chrysalis and co

A first prototype, the *Chrysalis* biplane, took to the air as early as 1979. On 11 May 1984, the *Monarch*, also a biplane, won the Kremer prize by covering 1500m *1641yd* at an average speed of 32km/h *19.88mph*. The project's promoters than decided to concentrate their efforts on achieving the mythical feat: crossing the Aegean Sea. The *Light Eagle*, built in 1986 after a great deal of research involving high technology, beat McCready's record on 22 January 1987. Seven years and seven months later, it covered 58.7km *36 miles 838yd* as opposed to 35.9km *22 miles 542yd*. But that was on a closed circuit.

Daedalus (1988)

Drawing from their experience of the *Light Eagle* the team were able to build ***Daedalus***. Its technical specifications are: wingspan 34.14m *112ft*; wing area 30.84m² *332sq ft*; length 8.84m *29ft*; unladen weight 31.75kg *70lb*; speed 24 to 28km/h *15 to 17.4mph*.

On 23 April **1988**, piloted by a Greek racing cyclist, Kanellos Kanellopoulos, ***Daedalus*** took off from Heraklion in northern Crete, reaching 3h55 and 118km *73 miles 572yd* later the Island of Thira (or Santorini).

Boeing 767 (1978)

This twin-jet medium-range aircraft was a completely new concept. The programme started in July **1978**. It was the first plane to be launched by **Boeing** since the inception of the 747 in 1966. The first deliveries took place in August 1982.

In April 1988 a Boeing 767 beat the world distance record for a non-stop flight by a twin-jet plane, covering 14 468km *8990 miles* from Halifax (Canada) to Mauritius, a trip lasting 16h 35min.

Glasair (1979)

The **Glasair** was the first plane in kit form and the one that performs best. In the United States more and more people wish to own a plane but do not have abundant funds. It is for those people that **Thomas S. Hamilton** and **Robert M. Gavinsky** invented, in 1979, the Glasair Taildragger. The Glasair III, which came out in 1986, holds two world records and has a 2250km *1398 mile 264yd* range.

Voyager (1986)

In December **1986** this amazing twin-engine plane, invented by the American **Burt Rutan** and piloted by his brother Dick assisted by Jeana Yeager, Dick's wife, flew non-stop around the world in a little over nine days. That's 42 000km *26 099 miles* aboard a propeller plane which, for the best part, consists of fuel tanks as it weighs 900kg *1984lb 8oz* unladen and 4210kg *9283lb* when its 17 tanks, which are situated in the wings and the fuselage, are full. *Voyager* is of course built from ultra-light materials which this experiment (or adventure, rather!) served to test once again. It also led on to military applications with the AT3, the prototype of which was tested in 1988.

A Fokker triplane from a 1925 KLM poster.

Fokker 100 (1987)

This 42 tonne twin-jet Dutch aircraft can carry 100 passengers over 2500km *1553.5 miles* and has been used by Swissair since autumn **1987**. It flew for the first time in November 1986. The Fokker company was founded in Germany by the Dutchman **Anthony Fokker** on 22 February 1912. After the First World War, Fokker went back to Holland where he created the Dutch company on 21 July 1919. One of the first ever airliners was the three-engine Fokker F. VII, which in 1924 flew between Amsterdam and Batavia (now Djakarta).

Joined-wings aircraft (1990)

This new joined-wings design invented by the American engineer **Julian Wolkovitch**, gives a weight saving of 30 to 40 per cent, whilst retaining the same aerodynamic properties. This weight saving would allow an increase in the number of passengers carried. Its special shape, furthermore, gives the plane better air penetration as well as added protection in the event of an accident. A JW-1 prototype is being built by ACA Industries for NASA. The first flight is scheduled for **1990**. Wolkovitch is also working on an unmanned joined-wings aircraft.

GULFSTREAM

The Gulfstream is an American supersonic aircraft which is proposed as an alternative to Concorde: a business plane with ten to 12 seats that could fly at Mach 1.5.

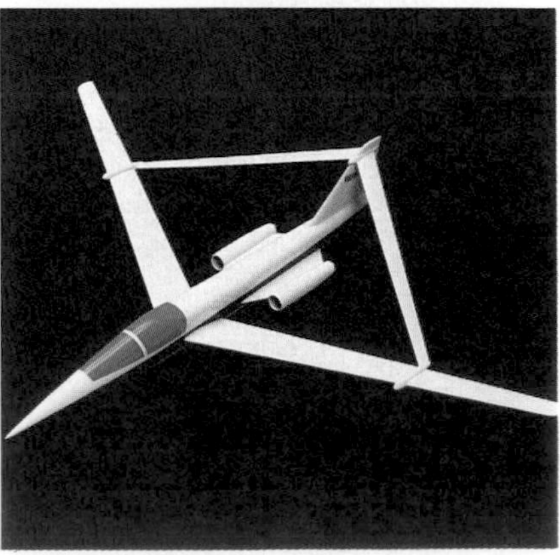

A prototype of the joined-wings aircraft.

SOON NEW YORK–TOKYO IN THREE HOURS WITH THE HYPERSONICS

These orbital planes or rocket planes – Hotol, Orient Express and Sänger – might come into being before the end of the millenium.

Unveiled at the Le Bourget Salon 1987, Aérospatiale's AGV (*Avion à Grande Vitesse* – high speed plane) is powered by four engines with 30 tonnes thrust each (or six engines with 20 tonnes thrust). At an altitude of 30 000m *98 400ft* it will carry 150 passengers at the speed of Mach 5.5 which is more than 5000km/h *3107mph*. And all this while still using conventional runways for take off and landing.

The British Hotol, created by Rolls-Royce and British Aerospace, will also be able to take off from ordinary airports. It would have a hybrid propulsion with reactors that can 'breathe' the surrounding air during the phase of the flight that goes through the atmosphere. The estimated manufacturing cost is £2.5 billion.

The Germans, for their part, are also planning an autonomous space aeroplane: Sänger. At the end of 1987, Aérospatiale presented a new project, the STS 2000.

The race is definitely on in Europe, but it seems that the various governments already know that they will have to work together. Today, their aim is to ensure they each have a prominent place when it comes to planning and manufacturing the future European hypersonic craft.

As for the Orient Express from the American company McDonnell Douglas, it would link Washington to Tokyo in two and a half hours; would carry 305 passengers and fly at Mach 5.

All these space aircraft projects raise numerous technical problems, notably the speed/overheating ratio as the sound barrier is also the heat barrier. But the most serious problem at present is propulsion. Perhaps the ramjet engine, invented by the engineer René Leduc in 1913, will become, because of its extreme simplicity, the ideal candidate to power a hypersonic craft.

Optimists forecast that prototypes will appear in the late 1990s, others say it will only be in the next millennium.

Canard wings aircraft

The configuration of these planes is the opposite of the usual one as they have large back wings and small fore wings, hence their name: *canard* is the French word for duck. Some early aircraft, such as Fabre's seaplane already had this arrangement. It had been abandoned then taken up again in the 1980s for amateur aircraft. As they were very successful, **Beechcraft**, a company based in Wichita, Kansas (USA) invented a twin-engine canard plane which could carry 20 passengers. The first prototypes of the Starship were stolen in 1987. Still in the United States, Burt Rutan, *Voyager*'s inventor, also created several canard-style planes.

TUPOLEV 155 (1989)

On 18 January 1989, the 20 minute long test flight of the liquified natural gas propelled Tupolev 155 proved that it is possible to fly with this type of fuel which is cheaper and less polluting than kerosene.

For its part, the West German manufacturer Messerschmitt-Bölkow-Blohm, a partner of Airbus Industrie, is planning to perfect a hydrogen propelled system for the Airbus planes. Hydrogen has the advantage of being much lighter than kerosene, and having three-times greater combustion power.

Hotol, the British rocket plane, will be able to take off from conventional airports. But which of the European projects will get into the air first?

Helicopters

Origins (1480)

The helicopter, in model form, flew for the first time in 1784. The French Launoy and Bienvenüe presented their model to the Academy of Science. This very simple machine consisted of two two-bladed propellers arranged at the opposite ends of a spindle so as to contra-rotate, powered by a taut whalebone bow.

In fact as early as **1480** Leonardo da Vinci had designed a machine, a sort of airscrew, which had wings rotating around a vertical axis. Borelli in 1680 and Paucton in 1768 studied his theory. In 1862 the Frenchman Ponton d'Amécourt (to whom we owe the word 'helicopter'), and in 1877 the Italian Enrico Forlanini built crafts that were powered by steam engines. These experiments solved many problems and finally paved the way to making piloted machines.

First take-off (1907)

The first take-off by a manned helicopter was accomplished by the Frenchman **Paul Cornu** on **13 November 1907** at Lisieux in France. The machine weighed 260kg *573lb 4.8oz* and was powered by a 24hp Antoinette engine.

Rotor (1908)

The history of the rotor is intimately linked to that of the helicopter. In 1907 a Frenchman, Paul Cornu, had hovered a few feet off the ground for a few seconds. In America in **1908** the Russian-born aeronautical engineer **Igor Sikorsky** (1889–1972) tackled the problem of the blade and rotor mechanism: the rotor ensuring both the lift and propulsion of the aircraft. This problem found many solutions during the evolution of the helicopter. A major invention was the variable cyclic pitch which makes it possible to change the position of the blades as they rotate so as to correct the uneven lift created by the forward motion of the craft.

In 1939, Sikorsky's work led him to fit the VS-300 with a single main rotor. This helicopter, in 1941, beat the world record for range with a one-and-a-half hour flight. (In 1937, a two-rotor system had been fitted on the German helicopter, the Focke-Achelis FA61.)

New solutions

Since then there have been many different rotors: the first streamlined antitorque rotor Fenestron perfected by **Aérospatiale** in **1967**; the Starflex, also from Aérospatiale (1973), with a rotor head made of a type of fibre glass and blades made of composites; the tilt-rotors which allow the prototype Bell XV to perform as well as an aeroplane (*see* Helicopter-aeroplane), Sikorsky's contra-rotating rotors, which are still in the experimental stage, etc.

Autogiro (1922)

As early as **1922** the Spaniard **Juan de la Cierva** (1895–1936) had started to work on an autogiro. In 1924, he fitted a four-blade rotor above the cockpit of a Deperdussin monoplane. On 18 September 1928, la Cierva and a passenger, Henri Bouché, crossed the Channel from England to France, aboard the C8II. The autogiro does not allow vertical flight but it was a first step in the tackling of slow flight. Many of these machines were built in England, France, the United States, the USSR and Japan.

Midway between the aeroplane and the helicopter, the autogiro was invented in 1922.

First helicopter flight (1924)

Sixteen years after the aeroplane, the first helicopter to fly a distance of one kilometre over a closed circuit was the one flown by the Frenchman **Etienne Oehmichen** on **4 May 1924**. In France, from 1920 to 1925, a Spaniard the Marquis Paul Pescara built three helicopters which managed to take off. It was the first time that the complete flight programme of the helicopter, including autorotation, was studied and its principle established. The theory was only put into practice in 1936 when Louis Bréguet and René Dorand invented the Gyroplane Laboratoire which, piloted by Maurice Claisse, went through the complete routine of hovering, flying sideways, long cruising flight and above all the first precision landing using autorotation with the engine off.

First operational helicopters (1940)

The first mass-produced helicopter was the Focke-Achelis FA 223. The prototype's maiden flight had taken place in **1940**. The Bell 47 was the first helicopter in the world to receive a civil aviation certificate of airworthiness (8 March 1946). The first operational Soviet helicopter was the MIL Mi-1, built by the Russian company Mikhail Mil and mass produced from 1951.

Jet powered helicopter (1953)

The only craft of this type to be mass produced was the French SO 1221 Djinn, a two-seater with an unladen weight of 369kg *813lb 10.3oz* which flew for the first time in **1953**.

Turbojet helicopter (1955)

The first helicopter to be powered by a gas turbine was the SO 1120 Ariel III. The flight took place at Villacoublay on 18 April 1951. But the turbojet's great popularity dates back to March **1955** when Jean Boulet took off for the first time aboard the Alouette II in France. Invented by **Charles Marchetti** and **René Mouillé** for Aérospatiale, the Alouette II is the first helicopter to be created round a turboshaft engine. It is a five-seater machine whose 'offspring' are still in production.

Recent helicopters

The most prominent recent machines are the middle-size twin-engine ones (3.8 to 4.5 tonnes) created for the business market and oil-rig shuttle services. They are fast, quiet, comfortable and can carry eight to 12 passengers over 500 to 800km *311 to 497 miles* at a speed of between 250 and 280km/h *155.4 to 174mph*. The Bell 222, and Sikorsky S-76C and the Dauphin 2-N vie for this side of the market. In 1984, a Germano–Japanese competitor appeared, the BK117. Turboshaft engines are now perfectly reliable, quiet and economical. Another step forward was the certification (the first in the world) of a full defrosting system for the rotors of Aérospatiale's Puma. The greatest area of progress at the moment is in the growing use of more economical composites to make the blade and rotor heads.

Helicopter-planes (1988)

The first tilt rotor aircraft, the V22 Osprey Tilt Rotor was introduced by the **Bell** company in Dallas, Texas (USA) in January 1988. The great attraction of this type of aircraft is that they take off like helicopters. Once airborne, the rotors go back to a vertical position and then act as propellers. The first applications were military but a civil version, able to carry about 30 passengers, is being studied.

The Eurofar (1988)

Some European countries – France, Germany, Italy, Great Britain and Spain – joined forces to build a competing machine, the Eurofar. The first full-size prototype is due to appear in 1994. Mass production should start at the beginning of 1995, the first machine to be completed by the end of 1998.

Miscellaneous

Rocket belt (1961)

The rocket belt was perfected by America's **Bell Aerosystem Company** and exhibited in 1961. The device was invented by Wendell F. Moore. To date, experimental flights with the belt have not exceeded 200m *219yd* horizontally and 20m *65.6ft* vertically. The rocket belt consists of two vertical tubes and a fuel tank strapped onto the user's back. Two motorcycle-type handles are used to control it. This belt allows vertical take-offs and stationary flight, and it can rotate 360 degrees.

'Driven' by a 'flying man', the rocket belt made a remarkable sight at the opening ceremonies of the Los Angeles Olympic Games in August 1984.

Automatic pilot (1914)

The first night flights were made possible thanks to progress made in flight instrumentation and in the willingness to replace a human being at the controls of the plane. The first efficient system was produced by the American **Elmer Sperry** and further perfected by his son, Lawrence, on a Curtiss seaplane in 1912.

In **1914 Lawrence Sperry** presented his device at a competition on aeroplane safety which took place in Paris. In order to demonstrate the stability of the plane, he made the flight with both arms in the air, while his passenger held on to the wing of the plane.

The latest automatic-pilot systems have benefited from developments in computer technology.

Pressurisation (1920)

With the advent of jet transport and the consequent increase in flight altitudes, machines had to be pressurised so that passengers could breathe normally. The first airliner to be pressurised was the Boeing 307 Stratoliner, dating back to before the Second World War, but the first trials took place in **1920**.

Gyroscope (1852)

In **1852**, the French physicist **Léon Foucault** (1819–68) set out to prove the earth's rotation. He hung a 60m *65yd* long pendulum from the centre of the Pantheon's dome thus demonstrating that a pendulum swings on a fixed plane. He then invented a mechanical device including a rotor and called it a *gyroscope*. At the beginning of the 20th century gyroscopes were used as stabilising devices on ships. These heavy and cumbersome systems have nowadays been replaced by automatic pilots.

Gyrocompass (1904)

A true innovation in the domain of navigation was the gyrocompass, invented at the beginning of the 20th century. After Foucault's observations, another Frenchman, G. Trouvé, made a spinning top powered by an electric motor in 1865. After that, the aim was to make gyroscopes for ships. Three names are linked with the development of this aspect: two Germans, Anschütz and Schüler, and an American, E. Sperry. **Anschütz**'s first gyrocompass was patented in **1904**; in 1911 Sperry made a gyrocompass suspended by a wire. This type of equipment remained unchanged until the Second World War, from which time radio-navigation began to be developed.

Air hostess (1930)

The first air hostesses were in fact stewards. Before the First World War, the large Zeppelin airships already had staff on board.

In 1919, at the beginning of commercial aviation, it was often the radio operators who

Imperial Airways was formed on 1 April 1924 and played a pioneering role in opening up routes through little-known territory. The services relied on a mixture of landplanes and flying boats and passengers had to make frequent changes along the way.

served drinks or picnic-style meals. It was not until 1927–8 that the first stewards started to appear on the British Imperial Airways aircraft.

But it was in **1930** that Boeing Air Transport, which was to become United Airlines, hired stewardesses for the first time. Ellen Church and seven other young women made up the first team on the San Francisco to Chicago line. In Europe, the first air hostess was Nelly Diener, a Swiss woman hired in 1934 by Swissair.

Joystick (1906)

It was the French engineer and airman **Robert Esnault-Pelterie** (1881–1957) who invented the control lever or joystick in **1906**. Unfortunately he did not register a patent and an American firm used his invention in spite of all his efforts to have his rights acknowledged.

Parachute (1802)

The first parachute was patented on **11 October 1802**. It was invented by the Frenchman **André-Jacques Garnerin** (1769–1823) who made a descent in Grosvenor Square, London in September 1802.

A parasol in the hand

Already in ancient times, Chinese acrobats used bamboo and paper parachutes to amuse their audience and, a few centuries later, Leonardo da Vinci (1452–1519) drew a sketch of a parachute.

From a height of 800 metres

In 1783 a French physicist Sebastien Lenormand dropped from a first floor window holding a parasol in each hand. He was the one to give the parachute its name *para* from parasol, and *chute* which means fall in French. But it was **André-Jacques Garnerin** who made the first true parachute jump on **22 October 1797**. He went up in a balloon above the parc Monceau in Paris. When he reached an altitude of about 800m *2624ft*, at 5.35pm, he cut the rope which held the balloon and the basket together. The basket went down, hanging from a parachute.

The first drop from an aeroplane was made by the American Captain Albert Berry, who jumped from a biplane above Saint Louis, Missouri on 1 March 1912. The USSR became the first country, in 1935, to use the parachute for military purposes.

In-flight refuelling (1923)

An American, Wesley May, should be credited for having performed the very first in-flight refuelling. On 12 November 1921, May jumped from the wing of a Lincoln Standard to the wing of a JN-4 with a fuel tank strapped to his back. He then climbed on top of the engine and poured the fuel into the upper wing tank.

The first in-flight refuelling using a pipe took place over San Diego, California, on **26 June 1923**. An airplane piloted by US Army **Lieutenant Seifert** refuelled an aeroplane piloted by **Captain Smith** and **Lieutenant Richter**.

28 February–2 March 1949. After flying exactly 94 hours 1 minute, US Air Force Captain James Gallagher completed the first nonstop, around-the-world flight. His B-50 *Lucky Lady II* was refuelled in the air several times during the flight.

Ejector seat (1946)

This system was invented with the high speed plane, as jumping from one of these using a conventional parachute would be far too dangerous. Considered possible as early as 1918 by Colonel Holt, preliminary designs were studied more closely in 1939 by the Germans and the Swedes.

The first experiment in which a man replaced the dummy used so far was carried out on **26 June 1946** by the Englishman **Bernard Lynch** who jumped out of a Meteor in a seat built by the British company Martin-Baker. He was at an altitude of 2500m *8200ft* and was flying at a speed exceeding 500km/h *310.7mph*. The first pilot to be saved by his ejector seat was a Swede who had crashed into another plane on 30 June 1946.

The first time a pilot was ejected at supersonic speed was on 26 February 1955 from a F-100 Sabre.

Retractable undercarriage (1911)

As is often the case, the idea was conceived in the first years of aeronautics: as early as 1876, the French pioneers Alphonse Pénaud and Paul Gauchot had thought of reducing the aerodynamic resistance produced by the wheels.

The first retractable undercarriage appeared in **1911**, on the German monoplane **Wiencziers**, though the undercarriage was a folding one rather than a retractable one. A few years later, the American pilot and engineer Glenn Martin fitted his Martin K-3 Kitten fighter with an undercarriage which could be retracted towards the rear. In 1920, on the American monoplane Dayton-Wright RB racer, the wheels disappeared up into the fuselage. The year 1922 saw the first plane with an undercarriage retracting into the wings: it was the American racer Verville-Sperry R3.

But it was in 1929 that the Frenchman Georges Messier perfected the first retractable undercarriage operated by hydraulic controls.

By sea

Shipping

Steamboats

A boat with a steam boiler (1730)

When use of Newcomen's steam pump had spread into the coal-mining regions of England during the 1730s, a mechanic named **Jonathan Hulls** used it to equip a tugboat. Placing a crank at the end of the beam of Newcomen's machine, he transformed the back-and-forth movement of the piston into a rotating movement which was transmitted to the paddle wheel of the boat. But the mechanical irregularity of this atmospheric engine and the large quantity of coal that it consumed made Hulls' project impractical and it was forgotten.

Steamboat with a paddle wheel (1793)

Watt's invention of the rotative steam engine, which permitted an increase in the driving power of the engine and used less combustible fuel, led to a breakthrough in the progress of steam-powered navigation. It was in America that steam navigation was to undergo its greatest improvements.

Steamboat with oars (1787)

A curious demonstration, witnessed by George Washington and Benjamin Franklin, took place on the Delaware River during the summer of **1787**. Two American builders, **John Fitch** and **James Rumsey**, introduced a boat with oars fixed to a horizontal wooden rod, operating in the same way as ordinary oars but powered not by men but by a steam engine.

The *Clermont*, the first river steamboat (1807)

The *Clermont* was built by the American **Robert Fulton** in New York in **1807**. It was the first successful commercial steamboat. It measured 50m *164ft* long by 5m *16ft* wide, had a capacity of 150 tons, and its paddle wheels were 5m *16ft* wide. Few people, however, believed that this powerful riverboat would succeed, and even though its trial runs had taken place with no major problems, no passengers showed up for its maiden voyage. Only Fulton and his crew made that first run up the Hudson River from New York to Albany. During its night-time voyage, the *Clermont* spread terror: Fulton fired the boiler with pine boughs, which produced a lot of smoke and sparks. This column of flame, along with the noise of the engine and the

This trireme, built in Athens, is identical to those constructed 2500 years ago by the ancient Greeks. The warship has three banks of oars on each side.

paddle wheels crashing on the water, terrified the people living along the banks of the river.

The *Savannah*, the first transatlantic steamer (1818)

In **1818**, Captain **Moses Rogers** of Savannah, Georgia, planned to build a steamboat intended for regular service between America and Europe. A corporation launched the operation, acquiring a handsome sailboat and installing a steam engine and paddle wheels. These paddle wheels could be dismounted and folded on deck.

The *Savannah* left port on 26 May 1819, and arrived in Liverpool 25 days later. Its engine had functioned only 18 days of that time, as the captain wanted to take advantage of a favourable wind to economise on coal.

After this successful run, the *Savannah* made its way into the Baltic Sea to Kronstadt and St. Petersburg, where it was visited by Tsar Alexander I. But the *Savannah* ended up in obscurity. After returning to the United States, it was reconverted into a passenger sailboat and ended its adventurous career on the Long Island coast. It would up sinking in the harbour during its last voyage.

An English steamer, *Enterprise*, travelled to India in 1825.

Wind-powered cargo ships (1980)

The last large commercial sailing ship disappeared in 1936. The high price of fuel has now reopened the possibility of using the wind as an auxiliary source of power.

In **1980** the Japanese shipyard **Nippon Kokan** launched the *Shin-Aitoku-Maru*. This is a cargo vessel fitted with two upright wings – each one opening like a book – with a total area of 194.4m² *2098sq ft*. These wing-sails can save up to 50 per cent in fuel costs. Five other cargo vessels to be used in the coastal trade have been launched since. Janda (the Association for the development of the Japanese navy) then instigated the building of a sea-going cargo vessel. The *Usuki Pioneer* was finished in November 1984. Its two computer-controlled metal sails should give this 16.25m *533ft* long cargo ship a saving of about 30 per cent in fuel costs. In the United States the 3100 tonne coasting vessel *Mini-Lace*, was fitted in **1980** with a canvas sail which can be made smaller by winding it round the rotating mast. It gives a 20 per cent fuel economy.

Wind-driven boat (1870)

In 1870, an old 23-foot lifeboat, renamed the *City of Ragusa*, was equipped by its skipper, **John Buckley**, with a sail and a six-bladed wind engine intended to turn a double-bladed screw propeller in the water. This device's lack of efficiency did not prevent the boat and its crew of two men from crossing the Atlantic.

Hydroplane (1897)

Count Lambert de Versailles presented plans for the first hydroplane as early as **1897**. In 1907, the Brazilian airman Alberto Santos-Dumont (1873–1932) reached a speed of 100km/h *62.14mph* during trials on the Seine. His hydroplane consisted of three aluminium and wood 'cigars' covered with silk.

The hydroplane is a flat-bottomed boat which rises out of the water at high speed. At this point, only the bottom part of its 'steps' remain in contact with the water. Driven by an aerial propeller, the craft can travel along shallow rivers but over calm water only.

Hovercraft (1955)

This type of vessel, also known as a Surface Effect Ship (SES), or Air Cushion Vehicle (ACV) was patented in **1955** by the Englishman **Sir Christopher Cockerell**.

His **hovercraft** is a boat that rides on a cushion of air. It does not touch the water but 'hovers' over it.

The seven tonne prototype, the SR.N1 made its first public appearance in 1959 and caused a sensation. It was highly manoeuvrable and capable of a speed of over 100km/h *62.15mph*.

On 25 July 1959, the 50th anniversary of Bleriot's cross-channel flight, the first hovercraft crossing was achieved.

Ice-breaker (1898)

The *Ermak* was the first English icebreaker. It was built in **1898** by the Armstrong shipyards,

Boating enthusiast Jim Wilkinson has built a boat driven by a windmill.

Vagabond II is a remarkable yacht designed by Janusz Kurbiel: it is capable of breaking up an ice floe 50cm 1½ft thick.

following the design of Russia's Admiral Makarov.

However, the first ice-breaker was Russian; the *Païlot* left the shipyard in **1864**.

Lenin (1960)

The *Lenin* is the first atomic-powered surface ship. It is an experimental ice-breaker developed by the Russians around **1960**. The heart of the *Lenin*, its nuclear reactor and the associated containment system, weighs 3000 tonnes. In free water the ship can run at 18 knots and can break an ice-cap 2.5m *8.2ft* thick at the constant speed of 2 knots.

Atomic-powered ships (1958)

Laid on the stocks on 22 May 1958, in the United States, the *Savannah* was the first commercial atomic-powered ship. Launched in 1959, it was given the name of its famous ancestor, the first steamer ever to cross the Atlantic. It could transport 9500 tonnes of cargo and 60 passengers. Attended by a crew of 100, it could develop 20000hp and reach speeds of 20.5 knots.

Although the *Savannah* was undoubtedly a technical success, she was a commercial failure. The operating company had to take her out of service in 1967.

Magnetic boat (1985)

It was at the Tsukuba exhibition in Japan that, in **1985**, a model of the magnetic boat was first shown to the public. It has no engine (in the strict sense of the term), no sail and no rudder. It is propelled by a system of superconducting magnets arranged along the hull. These create a powerful magnetic field in sea water, which is a good electricity conductor. Its Japanese inventor **Yoshiro Saji**, a 60-year-old physicist, has been working on it for about 15 years.

Sailing liner (1986)

The *Wind Star* (and its twin sister, the *Wind Song*) is the first sailing liner ever built. Produced by the **Ateliers et Chantiers du Havre** (ACH) at the end of **1986**, the *Wind Star*, commissioned by the American company Windstar Sail Cruises, has a sail area of 2000m² *21528sq ft*. The masts rise to 58m *190ft 3in* above sea level and the sails are completely computer-controlled. It can carry 180 passengers with a crew of 75. The ship's speed is between 10 and 15 knots. Four such sailing ships are envisaged, two for the West Indies and two for Polynesia.

The largest sailing liner in the world, the *La Fayette*, was launched on 22 December 1988, also by ACH.

Navigational Aids

Anchor (3000 BC)

The first anchors, used by Chinese and Egyptian sailors around 3000 BC and later by the Greeks and the Romans, were stones or bags containing sand or pebbles that could simply by thrown overboard. Pliny, Strabo, and other Roman authors attribute the invention of the metal anchor to several different seafaring peoples. The first approach tried was the single-palm grapnel used around 600 BC.

An important improvement was made in the 18th century when better quality, less brittle iron was used, and when the arms were given a new camber. Around 1770, iron-stock anchors completely supplanted wood-stock anchors. (The stock is a bar perpendicular to the shank of an anchor; its purpose is to make the anchor swing so that one of the flukes grips the bottom.)

In 1821, the Englishman Hawkins engineered the mooring hawse-hole anchor with palms. With this system the stock is no longer necessary, the arms being mounted in such a way that they automatically lean to the same side to grip the bottom. This type of anchor was modified between 1872 and 1887 by the Englishmen C. and A. Martin, S. Baxter, and W. Q. Byers.

The CQR anchor, or plough anchor, which the Englishman G. I. Taylor patented in 1933, is of an extremely original design. At the extremity of the shank, a double ploughshare is mounted, and the gripping power this provides is twice as great as that of a standard anchor.

Compass card (1876)

It seems that the first compass or mariner's card was made by Flavio Giova, an Amalfi craftsman. After centuries of trial and error, the **Thomson** compass was perfected in **1876**: this dry compass had thin cylindrical bars mounted on silk thread and tied to very thin paper – it weighed no more than 20g *.7oz*. The liquid compass, a result of the work of two Englishmen Dent (1833) and Ritchie (1855) came into general use around 1880. After the First World War, it was strongly challenged by the gyrocompass perfected from the work of the French physicist Léon Foucault and of G. Trouvé.

Canal lock (14th century)

The double, or chamber, canal lock is an invention attributed by some to 14th century Dutch engineers (later to be improved upon by the Italian genius Leonardo da Vinci) and by others to da Vinci himself around 1480. By regulating the conditions of passage, the invention of the chamber lock solved the problem of getting vessels from one reach of water to another and simplified the often dangerous procedure of lowering and lifting boats.

The 209m 688ft long Great Eastern of 1858 foreshadowed today's liners.

THE LARGEST OF THE LARGEST OCEAN LINERS IN THE WORLD

The *Sovereign of the Seas*, ordered by a Norwegian company from the Chantiers de l'Atlantique, had hardly been launched when a new, even bigger project was announced: the Tikkeo Cruise Line has ordered, from the Belfast shipyard, Harland and Wolff, a 377m *1236ft 7in* long liner (111m *364ft* longer than the *Sovereign*!) which can accommodate 3026 passengers. It will have 12 swimming pools and tennis courts and a 1500-seat cinema. Designed by naval architects Knude and Hanson from Copenhagen, it will have a cruising speed of 29km/h *18mph*. Its name: simply *Ultimate Dream*.

But this dream might in fact not be the ultimate one as a new, even bigger project comes to challenge it: the *Phoenix*. Will this floating island, measuring 380m *1246ft 5in* in length and carrying 5600 passengers, truly be the largest? This ship is Norwegian, Tage Wandberg is the architect who designed it and it was commissioned by Knut Utstein Kloster who, with his company, World City Corporation, had bought the *France* in 1979. From 1993, the *Phoenix* will be based in Florida and cruise the Caribbean Sea.

The ancestor of the ocean liner was the packet-boat – reflected in its French name *paquebot* – which carried the mail between Dover and Calais. The first transatlantic liner was Isambard Kingdom Brunel's *Great Western*. It was propelled by a paddle-wheel and travelled from Bristol to New York in 15 days and 5 hours.

Propellers

Screw-propeller (1837)

In **1837** a Swedish engineer **John Ericsson** patented a screw propeller in America. His propelling system, consisting of two screws, was used on a tug boat, the *Francis Ogden*. In the same period, Francis Pettit Smith, a Kent farmer, perfected a similar system and founded a company, the Steam Propulsion Company, which in 1838/9 built a full-sized, sea-going screw ship called the *Achimedes*, which attained 10 knots.

In 1843, Isambard Kingdom Brunel built the first transatlantic steamship powered by a screw-propeller: the *Great Britain*, measuring 98m *321ft 5in* in length. Two years later, the Royal Navy tied two ships back to back: the *Alecto* which was a paddle-steamer and the *Rattler* which had a screw-propeller. With its paddle wheel at full power the *Alecto* was still pulled by the *Rattler*; the screw-propeller had proved its superiority.

First propellers (1785)

In **1785**, an Englishman **Joseph Bramah** patented a 16-bladed propeller to drive boats. But the first experiment was made by an American, John Stevens, and an Englishman, Sir Marc Isambard Brunel, who coupled two four-bladed propellers to their one-cylinder steam boiler to sail up the Passaic River in New Jersey (USA).

Mechanical log (1801)

The mechanical log was created by the Englishman **Edward Massey** in **1801**. It standardised an already ancient, but imprecise, system. Originally quite rudimentary, the log was simply a wooden block that was thrown into the water towards the bow of the ship and then recovered when it had reached the stern. The apparent speed of the vessel was calculated by taking into account the time the log took to pass from one end of the ship to the other.

Outboard engines (1905)

Two models appeared in **1905**: an American engine designed by a Norwegian **Ole Evinrude**, who called it outboard because it had the characteristic of being screwed vertically to the boat's outer hull; and a German engine built by **Fritz Ziegenspeck**, who named it the Elf-Zett.

Octant (1730)

The first known octant was made by an English mathematician and astronomer, **John Hadley**, in **1730**.

Lighthouses

Origins (285 BC)

In 285 BC, one of the Seven Wonders of the World was built on Pharos (an island, in Alexandria harbour, Egypt). This was the lighthouse built according to the instructions of Egypt's King Ptolemy II. This lighthouse is said to have measured over 130m *426ft 5in* in height. A wood fire was kept burning all night at its top. It was destroyed by an earthquake in 1302.

The decisive advance in lighthouse construction was the invention, by the Englishman John Smeaton in 1759, of a cement that could set in water.

As to the light, it was provided by wood fires until the 18th century. In 1780, a Swiss, Argand, designed the flat-wick oil lamp called the Argand burner. In 1901, a new development appeared; this was the petroleum burner, invented by Arthur Kitsen.

The progress made in optics had an impact on lighthouse design. As of 1752, the parabolic reflector designed by the Englishman William Hutchinson increased the power of the light signal.

The most decisive step in the use of lenses was conceived in 1821 by the French engineer Augustin Fresnel. An optical lense was used to focus the light into a beam aimed at the horizon. Escaping light was collected by concentric rings of prisms.

The use of electric lighting (beginning in 1859 in Dungeness, Great Britain) greatly improved the effectiveness of lighthouses.

Weighted keels (17th century)

In order to prevent sailing ships from capsising under the lateral force of their sails, it was always necessary to concentrate the weight of the ship as low as possible. When fighting ships had no cargo, they had to be filled with stone ballast that acted as counterweight to the sails.

Towards the end of the 17th century, King Charles II had the bottom of one of his yachts covered with lead plates.

In 1796, the Royal Navy had two frigates built, the *Redbridge* and the *Eling*, which had lead ingots fixed to the outside of the hull. In 1844 the American sloop, *Maria*, designed by Robert L. Stevens, had lead-lined wooden hull rails. The first sailing ship to be completely lead-ballasted on the outside of the hull was the *Peg Woffington*, owned by the Scot George L. Watson in 1871.

Keel of the *Australia* (1983)

Thanks to an original keel, the America's Cup was won by the yacht *Australia*, in **1983**, after 132 years of American domination of the contest.

The designer of the *Australia*, **Ben Lexcen**, envisaged fins under the keel to prevent a loss of lateral pressure. Thanks to this, it was possible to design an even more efficient keel, broader above than below. This keel is beginning to be fitted on cruise ships, allowing them to reduce their draught by 20 per cent to 30 per cent with no loss in efficiency.

Tide tables (13th century)

The first known tide tables were drawn up in the **13th century** by the **monks of Saint Albans** to give the height of the Thames at London Bridge.

The first printed tables were published in the Nautical Almanac by the Breton Brouscon in 1546. One of the most famous tables is that of Richard and George Holden, published in Liverpool in 1773. For a long time, it was used to determine the tides of that habour, although these are very irregular. In 1858, the Nautical Almanac, published by the British Admiralty, gave a tide table for all of the harbours in the world. As of 1910, it gave tables for only 26 of them.

A MYSTERIOUS INVENTION: THE COMPASS

No one knows when, where or by whom the compass was invented. It was in China, around the year 1000, that the magnetic needle appeared as an aid to navigation. In China, these first instruments pointed to the South. It was only two centuries later that the compass was mentioned for the first time in Europe in the works of an English monk, Alexander Neckam.

Even the origin of the words magnetite, magnetic and magnetism is obscure. They might be connected to an ancient part of Thessaly (Greece) called Magnesia. A legend tells of a shepherd by the name of Magnes whose metal tipped crook and hobnailed shoes stuck to the ground and enabled him to discover the magic mineral.

As early as the 12th century, the magnetic needle or steering compass, which in Europe pointed to the North, became essential for sailing in bad weather. In the 14th century, as the steering compass came into general use in the Mediterranean, trade was made easier whatever the weather all the year round.

Mariners travelling through the shallow waters of Northern Europe could use a sounding-lead to pinpoint their position. By measuring the depth of water under the hull, and using maps with very precise measurements, they could pinpoint where they were. The compass, therefore, only appeared in Northern Europe in the 15th century.

The compass enabled Christopher Columbus to undertake his westward journey to India which led him to rediscover America, a continent the Vikings had discovered several centuries before and which was probably shown on Soliman the Magnificent's maps that Colombus would have consulted.

Weapons

Light weapons

Origins

It would appear that weapons were some of the earliest inventions. Without them our ancestors would not have been able to survive within a hostile environment, to protect and find food for themselves. The first weapon was undoubtedly a stone, followed by the club, the hunting spear in 500000 BC, the axe in 250000 BC, the spear and the knife and then the bow and arrow which were depicted in rock paintings in the Sahara around 30000 BC. The first decisive step was taken with the discovery of metal, the second with the invention of gunpowder.

Firing tube (14th century)

With the invention of gunpowder appeared a device which was both an early version of the cannon and the forerunner of the rifle. It consists of a metal tube closed at one end. The gunpowder and shot were loaded into the mouth and the powder was lit by a small opening or priming hole in the side of the tube.

Bayonet (16th century)

According to legend, the bayonet was invented, or at least manufactured for the first time, in 1590, in the French town of Bayonne, after which it was named. It was used commonly as a weapon towards the end of the 17th century.

Match-lock (15th century)

Until the middle of the 15th century, light weapons were fired by holding a lighted fuse to the priming hole or pan of the weapon with the right hand. About this time, the fuse coil was invented which freed the right hand and enabled the user to maintain a firmer hold on the weapon.

Wheel-lock (16th century)

The wheel-lock produced the spark necessary to set off the gunpowder, so that the user no longer had to worry about keeping the fuse alight in order to fire the weapon. **Johann Kiefuss**, a watchmaker from Nuremberg, is generally thought to have invented the wheel-lock in 1517, but it has also been attributed to Leonardo da Vinci. It was a complicated and costly device which was particularly popular with horsemen as it enabled them to fire one-handed.

Flint-lock (17th century)

The flint-lock worked on a similar principle to the wheel-lock: the charge was ignited by a spark produced by a flint moving against the strike-plate. The flint-lock was adopted by the French Army around 1660 and it was later perfected and incorporated into the British Army's Brown Bess.

Cartridge (17th century)

The paper cartridge was an early invention. It was in general use by the Swedish army c.1630, but it did not become widely used until the 18th century. Originally it contained gunpowder only and did not combine bullets and gunpowder until 1738.

Percussion lock (1807)

In **1807 Alexander John Forsyth**, minister of Belhevie near Aberdeen in Scotland and a keen hunter, invented the method of firing a percussion cap to ignite the priming. Among other things, it eliminated the misfires caused by wet weather. The well-known toy 'cap pistols' use a small amount of fulminate placed between two paper discs.

Integrated priming (19th century)

When rifles could be breech-loaded, inventors had the idea of incorporating the primer into the cartridge. There were many different suggestions, including the pin-fire cartridge invented by the French gunsmith Casimir Lefaucheux in 1836, and rim priming invented by the French gunsmiths Béringer and Nicolas Flobert in 1845. The invention was used intensively during the American Civil War (1861–5) and then spread via Europe to the rest of the world.

Minié bullet (1836)

The Minié bullet, invented by the French army officers **Henri-Gustave Delvigne** and **Claude-Etienne Minié** in 1836, was the first to enable the rifled gun to be used efficiently. Gunsmiths realised that the range of the rifle could be substantially increased but the problem of loading the bullets still had to be solved. The problem was how to obviate tightness, (ie. retarding friction) caused by rifling. The bullet, with a diameter slightly less than that of the barrel, was automatically propelled by the pressure of the gasses. It increased the range of an infantry rifle from 200m *218.8yd* to more than 1000m *1095yd* and was the most widely used ammunition during the American Civil War.

Metal cartridge (1861)

There were many drawbacks to cartridges made of paper and other low-resistance materials. Powder often leaked or was affected by poor weather conditions etc. During the second half of the 19th century, many inventors tackled the problem but the first ammunition with a completely metal case was not tried out in the field until the American Civil War.

In 1865, Gosselin of France presented cases in drawn brass to the Commission of Vincennes, but the level of knowledge of the working of copper was so limited that they did not become widely used. In the same year, another Frenchman, Schneider, and two Englishmen, Boxer and Daw, developed ammunition which was made entirely of metal with centre priming. However the case did not consist of a single piece of metal but of a cartridge made of a thin sheet of brass in the form of a coil. No further progress was made until after the Franco–Prussian War (1870–71).

Centre priming (1865)

Centre priming was invented in **1865** by both the English army officer, **Boxer**, and the Frenchman **Schneider**. It was further improved by the Americans who, from 1867 onwards, manufactured cartridge cases made from a single piece of brass. The development of this type of cartridge led to the invention of repeating rifles.

The invention of nitro powder in 1884, made it possible to reduce the calibre from 11mm to 8mm and subsequently to 7.5mm, 6.5mm and 6mm. These cartridges are still in use today and are effective over a distance of up to 1000m *1094yd*. Several years ago, some very small-calibre cartridges were brought out i.e. 5.56mm which have a high muzzle velocity.

Modern research is directed towards developing caseless ammunition; the first trials were carried out during the Second World War by Germany in 1944. It has already been developed by the West German companies. Heckler & Koch and Mauser, and the former have even presented the prototype of an assault rifle using this type of ammunition at a NATO exhibition. A few problems still have to be resolved before this type of cartridge becomes fully operational.

Breech-loading (1841)

The idea had already been in existence for some time when Samuel Pauly, a Swiss gunsmith living in Paris, presented the first true breech-loading rifle to Napoleon I in 1812. The weapon also used closing cartridges made of brass. But the Emperor was obviously preoccupied at the time. . . .

The breech bolt rifle was developed in 1835 by the German **Johann Nikolaus von Dreyse** (1787–1867), a former student of Pauly. The weapon appeared in its final form in **1841** and was adopted by the Prussian Army. In 1866, it enabled them to win an easy victory over Austria at Sadová, and over the Danes in Schleswig-Holstein.

Many old muzzle-loaded weapons, were subsequently converted to models with a moveable breech.

Jacketed or armour-plated bullet (1878)

The invention of nitro powder made it possible to obtain considerable muzzle velocities which were unsuitable for lead bullets. The Swiss army officer, **Rubin**, invented the composite bullet around **1878**. It consisted of a lead nucleus encased in a jacket of a more resistant metal such as cupro-nickel, brass or steel, which was sometimes electroplated with copper or other metals. This type of bullet is used exclusively for military purposes and on rifle ranges. A lead-tipped bullet, known as a semi-plated bullet, is used for hunting.

Dum-Dum bullet (1897)

These hollow-tipped bullets, fitted by the English to several models of the .303 military cartridges to improve their stopping power, were developed in 1879. Most of the tests on this type of bullet were carried out at the arsenal at Dum-Dum on the outskirts of Calcutta during British colonial rule. The Dum-Dum bullet, which causes terrible wounds, was banned by The Hague Convention in 1908, but a certain number were, however, used during the Boer War (1899–1902) between the English and the Dutch settlers of South Africa, and in the First World War.

.22 Long Rifle (1847)

In **1847** the Parisian gunsmith **Nicolas Flobert** had the idea of fitting a shot pellet on to a case. He had just invented the modern version of the cartridge and in particular the .22 or 5.5mm which was exported to the United States where it was greatly improved and extremely successful. This led to the development of the .22 Long Rifle.

Repeaters (1860)

In 1854, Smith and Wesson, the directors of the American company Volcanic, bought the rights for the Hunt & Jennings rifle which, with its tubular magazine, was the first repeating rifle to be produced but which was unusable as it was. They therefore gave B. Tyler Henry the task of making the invention a viable proposition. Unfortunately his research did not produce any results worthy of note, and it was not until **Oliver Winchester** bought the business from the two Americans that the Henry rifle came into being in **1860**.

Winchester (1866)

In **1866** the company was named after Winchester, and the Henry rifle was developed even further and given the same name. The frame of the early models was made of a yellow metal, brass, and they were often called Yellow Boy. These are the rifles which are always associated with the history of the Far West.

Although the 1866 Model was relatively fragile, its successor, the 1873 Model achieved considerable fame, mainly because the latest models used the same 44-40 calibre ammunition as the equally famous Peacemakers, the Colt Single Action revolvers. Winchesters are still manufactured today in a range of calibres and in an improved form based on the 1894 Model.

Mauser rifle (1871)

After the Franco–Prussian War of 1870, the Dreyse rifle was replaced by the rifle invented by German gunsmith, **Paul Mauser**.

In 1898 taking advantage of the research carried out on repeating rifles and the invention of nitro powder. Mauser developed his famous Gewehr 98 with a magazine which held five 7.92mm cartridges. It is still in use today.

Revolver (1835)

The idea of repetition by rotation has very early origins. Evidence of it can be seen on a bronze horse pistol dating from around 1680 and attributed to the London gunsmith J. Dafte, which is preserved in the Tower of London. A carbine of a very similar design has

Paul Newman brandishes a Colt Peacemaker.

also survived and can be seen in the Milwaukee Public Museum in the United States.

In 1814 J. Thomson, a shopkeeper from Islington, north London, patented a flint-lock pistol with a revolving magazine feed mechanism with nine chambers and a single barrel. In 1818 an American, E. Collier, with the help of Captain Artemus Wheeler and Cornelius Coolidge, followed up with a hunting rifle and a five shot revolver with a rotating breech. Other inventors followed before a patent was registered on **22 October 1835** by a young man of 21 named **Samuel Colt** (1814–62), whose name has gone down in history.

The most famous of the many revolvers currently in existence are undoubtedly those manufactured in America, for example the Colt, Ruger and Smith & Wesson.

Although the best-known calibres are the .22, S & W Special 38 and S & W Magnum 35, there are also a few chambered monsters that take high powered cartridges such as the S&W Magnum 44 and the Casull 44.

Colt (1836)

The American, **Samuel Colt** was not so much the inventor as the populariser of the revolver to which he made many improvements. He obtained the American patent on **25 February 1836**, by which time he already held the patent in England. In spite of its qualities, this first revolver did not sell well and the business went bankrupt. It was only after 1850 that the Colt achieved worldwide popularity.

Automatic revolver (1895)

The automatic revolver was invented and patented on **16 August 1895** by **G. V. Fosberry**, a colonel in the British Army. It made use of the recoil action to reset the cocking piece and turn the cylinder one notch. It was brought onto the market in 1901 under the name Webley-Fosberry in the English regulation calibre of .455 and in Colt Automatic 38.

Semi-automatic rifle (c.1890)

The first semi-automatic rifles, which used either the recoil or the action of the combustion gasses to operate the repeater mechanism, appeared towards the end of the last century as a result of an invention by a Mexican general, **Mondragon**. His invention was further developed by the French gunsmiths, the **Clair brothers** from St-Etienne, who acquired the patent.

In spite of the fact that a few of these weapons were used on a limited scale during the First World War e.g. the Meunier rifle or A6, Models 17 and 18 of the RSC rifle and the Mauser automatic aircraft carbine, as well as for civilian purposes, e.g. the Winchester 351 SL and the 401 SL aircraft gun, it was not until much later with the Pedersen device and then especially the invention of John Garand of the United States that the weapon was brought into service in 1932.

Assault rifle (1944)

The first assault rifle to be produced in any quantity was the **1944** German **Sturmgewehr 44**, a name invented by Hitler. It was in fact a lightweight machine gun. In 1947, Mikhail Kalashnikov of the USSR used it as a basis for his famous AK 47.

During the 1950s the Belgians developed the first true assault rifle the F.A.L. which fired 7.62mm cartridges. The weapon was sold in hundreds of thousands throughout the world.

In 1956 Eugène Stoner of the United States invented the AR-15 which was very similar to the Kalashnikov in terms of its use but very different in terms of design. The Americans appreciated its true value during the Vietnam war, and the AR-15 (the design of which had been bought by Colt) was adopted at the end of the 1960s by the American Army under the name of the M 16. A new version of the rifle, the M 16 A2 was brought into service for the Marines in 1985.

Bullpup (c.1980)

The bullpup rifle was designed as a modern weapon adapted to new forms of combat, but the system dates back to the beginning of the century when the Burton automatic rifle was invented in the United States in 1917. The overall length of this compact rifle is much reduced, although the barrel is of the same length as on a traditional rifle so that its accuracy is in no way affected. The first versions of the rifle were designed in Great Britain in about 1955, but the design was not widely adopted until the end of the 1970s with the AUG manufactured by Steyr-Daimler Puch for the Austrian Army in 1977, the FAMAS designed by GIAT for the French Army in 1979 and the Royal Ordnance's SA 80 for the British Army in 1985.

Submachine gun (1915)

The first submachine gun was manufactured and brought into service by the Italians in **1915**. It was in fact two combined weapons which used the regulation ammunition for the pistol i.e. the Glisenti 9mm. The whole thing was named the Villa-Perosa after the factory where it was produced, *Officine di Villa-Perosa*. Its inventor, **Abiel Revelli**, wanted to produce a lightweight machine gun which could be fired on a tripod.

By separating the two weapons and mounting each one on a wooden support in the same way as the carbine, the engineer, Tullio Marengoni, turned it into a *Moschetto automatico* which only fired in bursts. The weapon appeared in 1918 and was immediately followed by an improved version with a double firing level which enabled it to fire shot-by-shot. Both models were produced by the Beretta company. At the same time, in Germany, Theodor Bergmann was developing the MP 18.1 which was the basis for most of the later versions.

Submachine guns are currently being replaced by automatic rifles in many armies. However, they still have considerable advantages for use in close combat, maintaining law and order and special operations.

The most famous examples of the submachine gun are the American Thompson, the German MP 38 and MP 40, the British Sten, the French Mat 49 and the Israeli Uzi.

Mini submachine gun (1932)

The mini submachine gun first appeared in **1932** in the form of the Mauser 96 (1932 Model). It is a very small weapon for special use in certain police and commando units and for the close protection of VIPs. Among the most famous are the American Ingram, the Israeli Mini-Uzi, the Czechoslovakian Skorpion and the Polish WZ 64. Unfortunately they are known to have been used in a number of terrorist attacks.

Pistol with a firing selector

These pistols make it possible to choose to fire shot-by-shot or in bursts, but given its dimensions, the latter are often difficult to control. The first versions were the Mauser C 96, known as the Schnellfeuer Pistole, used by certain SS units during the Second World War, and its Spanish copies, the Astra 900 and its derivatives.

In the 1960s the Russians brought out a similar weapon, the Stetchkin, with the same calibre as the 9mm Makarov, and in 1977 Beretta brought out its R 93, the usefulness of which is not immediately obvious.

Automatic pistol (1872)

The first weapon of this type was patented by **Plessner** in **1872**, followed shortly afterwards by Lutze of the United States in 1874. Then came the invention of the French gunsmiths, the Clair brothers: a rather strange gas-operated weapon which they demonstrated to the army but that was not adopted.

The first true automatic was invented by Hugo Borchardt in 1893, and followed by the invention of Theodor Bermann a year later. These German weapons were paralleled in Austria by the first versions of the Mannlicher.

Lüger (1898)

In **1898 Georg Lüger** of Germany developed the weapon named after him, based on the invention by Borchardt. The Lüger pistol was chosen in 1902 by the German Navy, and six years later by the Imperial Army as the P.08. It was adopted as a regulation weapon by many countries, and continued to be manufactured until 1943. The Lüger began to be produced commercially again in the 1970s.

Browning (1896)

About the same time in the United States **John Moses Browning** (1855–1926) designed his Model 1900 for the Belgian National Force in **1896**. He went on to develop a whole series of weapons of the same type, in particular the Colt Government Model 1911 (followed by the 1911 A1) which became the regulation weapon of the American Army. The models derived from the famous Browning GP.35, developed by the successors of John Moses Browning are still being manufactured today.

Silencer (1908)

The silencer was invented in **1908–9** by the American **Hiram Stevens Maxim** (1840–1916). It only worked on shot-by-shot rifles and pistols. H. S. Maxim developed it and subsequently adapted it for car exhausts.

Stun Gun (1984)

The stun gun is a new American invention which has the advantage of immobilising rather than killing or wounding by means of a powerful electric charge of between 40 000 and 50 000V which momentarily dazes and paralyses the victim. It has been tested by the

Harris County Constable Department which is responsible for the transportation of dangerous mentally ill patients. The Stun Gun, like its predecessor the Taser, which operated on the same principle but was more unwieldy, is also used by the American police to control prisoners, ward off attacks and so on.

Glock 17 (1986)

Invented by the Austrian, **Gaston Glock**, most of the components of this automatic pistol are made of plastic except, of course, the barrel. However, this does not, as was originally thought, enable it to pass undetected through the metal detector control systems used in airports. It is a 9 × 19 calibre (= 9mm Parabellum) lightweight weapon (650g *1lb 6.9oz*) which is extremely accurate and reliable.

Machine guns

Mechanical machine gun (1850)

The mechanical machine gun appeared around **1850**. It consisted of either a single barrel with a handle which, when turned, propelled and fired the cartridge, or of several barrels which were loaded and fired in rapid succession. The most popular model, invented by the American Richard J. Gatling (1818–1903), appeared in 1862. The same principle was used for the Vulcan cannon where an electric motor enables 6000, 20mm shells to be fired per minute.

Automatic machine gun (1884)

The first continuous firing automatic machine gun, operated by the effect of the recoil action, was invented in **1884** by the American **Hiram S. Maxim**. It demonstrated its true value in 1894 when it was adapted for the use of nitro powder. Maxim's basic principle is used today in most types of machine gun.

Researchers introduced different methods of operation, based on his invention. In 1892 the American John Moses Browning, invented the first gas-operated machine gun. In 1893, a former officer in the Austro-Hungarian army, Captain von Odkolek, patented a very similar invention and sold the rights to the French based company, Hotchkiss. Founded by Benjamin B. Hotchkiss (1828–85) (an American living in France), the company had already gained a substantial reputation with its rapid firing cannons. From the beginning of the century, the company manufactured a number of models culminating in the famous 1914 Model used by the French Army as well as the armies of many other countries.

Automatic rifle or light machine gun (1902)

In **1902 Madsen** of Denmark developed a light machine gun which could be operated and carried by one man. Other models were subsequently produced by other companies. In fact, a distinction was fairly rapidly established between the light machine gun which was easily transported but operated by several men, such as the famous MG 34 and MG 42 which also existed in the form of a heavy machine gun, and the development of the mass weapon operated by one or at the most two men i.e. the automatic rifle.

The first automatic rifle worthy of the name to be brought into service was without doubt the French army's FM 15, also known as the the CSRG after its inventors Chauchat, Suterre and Riberyrolles, and the company, Gladiator, which had produced the prototypes. Although it was a fairly crude and unreliable weapon, it was produced on a large scale and was even adopted, in the absence of anything better, by the American Army in 1917.

Some of the most famous examples of the automatic rifle are the BAR of John Moses Browning; the ZB 26 developed in Czechoslovakia by the Holek brothers and its English counterpart, the famous Bren Gun; the British Lewis Gun with its circular magazine, and the French FM, 1924 M29, developed at the Châtellerault factory by a team under the supervision of Colonel Reibel.

Heavy machine gun (1938)

This is the name given to heavy weapons which have to be transported by motor vehicle and operated by an entire team. The most famous version and the first to receive its baptism of fire, in **1938**, was the German four barrel anti-aircraft gun, known as the 2cm Flakvierling, manufactured by the German company, **Mauser**. Nowadays, the firing of these weapons is usually controlled by radar.

Motorised machine gun

Towards the end of the 1940s, the American company **General Electric** revived the Gatling principle, but replaced the need for 'elbow grease' with an electric motor. The McDonnell Douglas Helicopter company has developed another model of motorised machine gun known as the Chain Gun. Both weapons have also been adapted as light machine guns. General Electric's 62mm Minigun was used by troops airlifted by helicopter during the Vietnam war, and the 7.62mm Chain gun is used in Kenyan Army helicopters and has been adopted by the British Army as a secondary weapon for their tanks.

Artillery

Cannon (14th century)

The first pieces of ordnance, the forerunners of the cannon, appeared with the of gunpowder in the **14th century**. They went under various names, bombards, blunderbusses etc. However, the first document recording the use of the cannon is of Arab origin and dates back to 1304.

Mortar (1917)

The mortar, a very short wide-barrelled piece, appeared about the same time as the pieces of ordnance. The modern version of the mortar was invented by the Englishman, Stokes, who developed an 81mm calibre weapon which was brought into service by the British Army in 1917.

In the 1930s a Frenchman, Edgard Brandt developed an entire series of mortars ranging from 45mm to 155mm. The Stokes-Brandt mortars have been continually improved and form the basis of all modern mortars.

Shrapnel shells (1784)

In **1784** an Englishman, **Henry Shrapnel** of the Royal Artillery, invented the exploding shell.

With his invention, Shrapnel wanted to maximise lethality by having hollow iron balls filled by bullets explode on target. Shrapnel was adopted by the British Army in 1803 and used during the wars fought by the Empire, particularly at Waterloo. Shrapnel subsequently became, and remained until very recently, the classic missile of the artillery. During the First World War, it was used against aircraft.

Recoil brake (1897)

Invented in **1897** for the 75mm French field gun, the hydraulic recoil brake marks the

Artillery played an important part in the capture of Rouen by the English in 1419.

A British cannon in 1918.

birth of modern artillery. Until then, pieces had recoiled by several metres every time they were fired. The 75mm field gun was developed by a team of French artillerymen, General Sainte-Claire Deville, Captain Rimailho and Colonel Deport.

Recoilless gun (c.1910)

This weapon was invented by an American, **Davis**, at the beginning of the 20th century and brought into service for a short time on several English aircraft during the First World War. It had two opposing barrels and a central chamber for the propellant charge. Towards the end of the 1930s, the German companies Krupp and Rheinmetall manufactured several models of the 75mm recoilless gun, one of which was tested by German parachutists in Crete.

Armoured vehicles

Armoured vehicles

Horse drawn chariots were already being used in ancient times to carry soldiers armed with bows and javelins. They were used by Cyrus, King of Persia, against Croesus, King of Lydia, at the Battle of Thymbreae in 540BC. The drawings of Leonardo da Vinci also show various projects for armoured chariots, covered with a conical protection of wood.

Armoured car (1902)

In **1902** the French company **Charron, Girardot & Voight**, brought out the first armoured car combining all the modern technical possibilities: a motorised vehicle, armour plating and turret mounted weapons.

Tank (1908)

In **1908** an Englishman, **Roberts**, presented, near London, an armour plated, tracked vehicle. In 1912, the Austrian, Gunter Burstyn, developed a similar vehicle but armed with a gun. However, the military had little faith in his invention.

The trench warfare that took place from 1914, revived the idea of an armoured land vehicle which was studied by the British Royal Navy and the French Artillery. In France, the idea was promoted by Colonel (later General) Jean Baptiste Estienne. The project for this new vehicle was top secret.

The first battle in which tanks participated took place in France on 15 September 1916 at Flers-Courcelette.

Underwater tank (1944)

In **1944** a team of German engineers took only a few weeks to develop a tracked pocket submarine which was equally capable of travelling on the seabed and on land. This tracked submarine was difficult to detect and carried two torpedoes powerful enough to sink a warship. It was an invention that the Germans did not have time to use.

AN ANTITANK ANT

Known as the Fire Ant, this tiny vehicle, no bigger than a child's pedal car, was developed by engineers of the Sandia Laboratories (USA). Completely automated, with four driving wheels, it is able to locate and destroy a heavy armoured vehicle within a range of 500m *547yd.*

Special weapons

Bazooka (1940)

Several inventors claim the credit for the bazooka, or rocket launcher, including the Swiss Mohaupt brothers who, at the beginning of the Second World War, sent plans to America for a projectile which subsequently became the T 10 anti-tank grenade and then the active charge of the bazooka. The plans were delivered to New York on 31 December **1940** by Colonel Delalande.

The bazooka consists of a single tube of thin metal which acts as the launcher for a rocket equipped with an explosive head with a hollow charge. It can seriously damage armoured vehicles.

Anti-tank gun (1918)

The first anti-tank gun to be used effectively at the front was invented by **Fischer**, a lieutenant-colonel from Bavaria, who constructed it while installing a modification of the 37mm revolving gun onto a captured French St-Etienne machine gun carriage. With improved ammunition, it could produce a muzzle velocity of 506mm *1660.69ft* per second and penetrate 16mm *.63in* of steel at a distance of 450m *500yd.* Anti-tank guns were gradually replaced by hollow charge projectiles.

Hollow charge (1883)

In **1883** the German, **Max von Forster**, published the results of five years' research carried out on the effects of an explosive charge in the form of a hollow cone.

The principle was developed by Lepidi of France in 1890, Lodati of Italy in 1933, and finally by the Swiss born Mohaupt brothers. In Germany, the method was used from 1940 onwards for the charges used to destroy the Belgian forts in Liège. It was then applied to rifle grenades and later to rockets.

Stalin's Organ (1942)

These batteries of rockets with explosive heads, propelled by solid fuel were used extensively by the Russians for bombardments during the Second Wórld War. Their German counterpart was the multi-tube rocket launcher known as the *Nebelwerfer.* The main disadvantages of this type of weapon was their lack of precision and the fact that they were easily detectable as they gave off large amounts of smoke when fired.

Powder and explosives

Gunpowder (7th century)

During the early years of the Tang Dynasty, the alchemist and pharmacologist, **Sun Simao**, gave the first description of how to make gunpowder from a mixture of saltpetre, sulphur and charcoal. It is however likely that it had been invented several decades earlier, but that alchemists had closely guarded the secret of their discovery.

Gunpowder has been used in warfare from the 10th century onwards. Manufacturing techniques were subsequently improved many times to keep pace with the development of firearms. In the 13th century, the wars fought against the Arab countries of central Asia led to the discovery of this explosive by Europeans, and its use then spread into the Mediterranean countries. However, it was not until the 15th century that the European countries produced their first powder guns.

The safety fuse or slow match wick (1831)

The safety fuse or slow match wick was invented in **1831** by the Englishman, **William Bickford** (1774–1834). The company, Davey-Bickford, still exists today.

Incendiary compounds

In AD 647, during the Siege of Constantinople by the Arabs, the Syrian architect, **Callinicus**, showed the Emperor Constantine the secret of Greek fire, according to an old Chinese tradition. His compound consisted of sulphur, tar, naphtha, oil, fat, resin and charcoal. Firing stone which consisted of gunpowder mixed with saltpetre and sulphur appeared c.1450.

Experiments with phosphorous began in the 19th century. In 1890 thermite was invented. This is a substance with a base of powdered aluminium and metallic oxides which only ignites at high temperatures and is therefore easier to handle. It used to be used for welding tramlines.

In 1910 a French engineer, Sazerac de Forges, invented an incendiary bomb designed to destroy German airships which he tested by throwing it from the first level of the Eiffel Tower. The flame-thrower appeared during the First World War, and the napalm bomb was invented in 1943.

Molotov cocktail (1939)

The Molotov cocktail is used in urban guerilla warfare and as a close-range anti-tank device. It was invented in **1939** by the Finns during their Winter Campaign against the Russians. It would appear, however, that the Molotov cocktail was also used by the Republicans during the Spanish Civil War. This elementary incendiary weapon is quite simply a bottle filled with a petrol based mixture, a device improvised by those fighting barehanded against armoured vehicles.

Guncotton (1847)

In 1847, a German chemist, **Christian Friedrick Schönbein** (1799–1868), who in 1839 isolated ozone by water electrolysis, developed the manufacture of guncotton or nitrocotton. Guncotton consists of nitrocellulose and is an extremely powerful explosive.

Nitroglycerine (1847)

In **1847**, the Italian chemist, **Ascanio Solaro**, produced **nitroglycerine** by pouring a half volume of glycerine drop by drop into a mixture of one volume of nitric acid and two volumes of sulphur acid. Nitroglycerine, which was extremely powerful and exploded at the slightest jolt, was responsible for the most terrible accidents. Two Swedes, the Nobel brothers, carried out a study of nitroglycerine. One of them was killed by this dangerous explosive but the other, Albert, courageously continued his experiments.

Dynamite (1866)

During one of his **1866** experiments, **Alfred Nobel** made a discovery. Nitroglycerine from a broken flask was absorbed by *kieselguhr*, a form of clay used as an insulating substance. Nobel found that the absorbed nitroglycerine retained its explosive qualities but was considerably more stable and much easier to handle. Dynamite was born.

AN UPDATE ON CHEMICAL WEAPONS

Chemical warfare was already being practised in the Middle Ages with the use of asphyxiating missiles made from powdered Euphorbia (a poisonous plant) added to vine charcoal. Poisoned bullets and quicklime were also used.

But modern chemical warfare began in 1915 when the Germans released 180 tonnes of chlorine gas upwind of the French positions, with the result that 15000 men were put out of the war. Chlorine was gradually replaced by phosgene and then, on 12 July 1917 at Ypres, the Germans used shells loaded with mustard gas. From June 1918 a quarter of the shells fired by the French also contained mustard gas. Among the German victims was a 29 year old corporal, Adolf Hitler, but the experience did not deter him, in 1933, from supporting research in the field of nerve gasses.

The first of these gasses, tabun, was discovered in 1936 by the German company I. G. Farben during research on insecticides. Sarin was discovered in 1939, and soman in 1944 by the German Richard Kuhn, who won the Nobel rize for Chemistry in 1938, not for this discovery but for his work on vitamins.

These nerve gases are among the most dangerous of their type. They cause the organic muscles to contract which leads to convulsions and respiratory standstill. In 1987 some 5000 Kurdish men, women and children in Halabja in the North of Iraq died in this way.

Bacteriological weapons, which are being researched as much as chemical weapons, have been used since Roman times and made sporadic appearances up to the 16th century. Today these weapons are included in the arsenals of many world powers. Since 1987 various conferences, committees and commissions have undertaken the difficult task of developing projects which aim to impose a total ban on the use of both chemical and bacteriological weapons. However, it will require at least ten to 12 years of negotiations to obtain the necessary agreements. These terrible weapons will therefore continue to remain a very real threat until the year 2000.

THE NEUTRON ANTI-EXPLOSIVES DEVICE

A new technique has been developed for the detection of explosives by the Science Applications International Corporation of San Diego, California (USA). This offers increased protection to airports. Objects inspected, such as luggage, are subjected to a bombardment of neutron rays. If there is nitrogen present, as is the case with explosives, the nitrogen particles absorb the neutrons and give off gamma rays in return. The technique was introduced in certain airports in July 1989.

Nuclear weapons

Atomic bomb (1945)

The atomic bomb was the work of a team of American scientists. **Arthur H. Compton**, **Robert Oppenheimer**, and the Italian- and Hungarian-born physicists, **Enrico Fermi** and **Leo Szilard**, who were both living in the United States. The bomb was exploded, experimentally, for the first time on **16 July 1945** at Alamagardo, 350km *217 miles* to the South of Los Alamos in New Mexico.

A letter written by Albert Einstein (1879–1955) on 2 August 1939 influenced the United States in its decision to built an atomic bomb. The American president, Franklin D. Roosevelt, did not find out about the letter until 11 October of that same year, over a month after war had broken out on the 1st September.

On 2 December 1942, in their secret laboratory below the terraces of the football ground at the University of Chicago, the research team led by Enrico Fermi succeeded in releasing the first chain reaction in a uranium-graphite atomic pile (*see* Energy).

On 6 August 1945, a uranium 235 bomb was dropped on the Japanese city of Hiroshima killing 80000 and wounding 50000. On 9 August, a plutonium bomb completely destroyed Nagasaki. Japan surrendered and the Second World War was over, but was replaced by the fear of nuclear war.

H Bomb (1952)

On **31 October 1952**, the United States exploded the first H bomb on the Pacific atoll of Eniwetok. The H bomb was invented by **Edward Teller** as a result of research carried out on the A bomb between 1949 and 1951.

In August of the following year, Russia exploded a similar bomb, followed by Great Britain in 1957, China in 1967 and France in 1968.

Neutron Bomb (1958)

Research on the neutron bomb began in the United States on the initiative of **Samuel Cohen**, around **1958**. The experts prefer to refer to it as the enhanced radiation warhead or the neutron shell.

'STAR WARS'

The deployment of the first American anti-missile space shield has been somewhat delayed for financial reasons. Although the programme may be entirely satisfactory from a technical point of view, there are a few problems in terms of cost. The American Defence Department has abandoned the idea of creating a shield to protect the whole of the United States in favour of a simplified version which protects military installations only. The defence programme could become operational from 1996. Several successes are responsible for the revival of this very controversial programme, in particular the performance of the MIRACL high-power laser which destroyed in mid-flight a target missile travelling at more than twice the speed of sound. The launch of the prototype of the detection satellite Delta-Star is planned.

The Strategic Defense Initiative (SDI), otherwise known as 'Star Wars', was launched by President Ronald Reagan in 1983. The initial forecast was for a budget of 29 billion dollars. Some of the biggest names in American industry are involved in the research, including General Motors, Lockheed, McDonnell Douglas, Boeing and Rockwell International.

SKYNET-4B

This was the first exclusively military satellite and was launched by Ariane-4 in December 1988. Skynet-4B, a British communications satellite, is part of the Skynet-4 programme which will become fully operational when it consists of four satellites. Skynet-4B, built by the Space and Communication division of British Aerospace, has a total mass weight of 1433kg *3159lb 12¼oz* during launching and 790kg *1741lb 15oz* in its geostationary orbit. It should remain operational for seven years.

Neutron rays are particularly deadly, but the device is a 'clean bomb' in the sense that being less powerful than the H bomb, it destroys all forms of life without affecting the material environment and without causing pollution. Therefore, once the neutrons have dispersed, it is possible to take over enemy installations.

Detection systems

Sonar (1915)

In **1915**, following the sinking of the *Titanic* three years previously, a French professor called **Paul Langevin** developed a system for detecting icebergs and, by extension, submarines, which formed the basis of the **Sonar** system (the acronym for Sound Navigation and Ranging) developed in England in the 1920s. The military derivative, the Asdic (Allied Submarine Detection and Investigation Committee) was installed in 200 Royal Navy destroyers and escort vessels in 1939, and was extremely useful to English and American ships during the final stages of the war. More sophisticated versions are in use today.

In 1986–7 Sonar was used in a non-military exercise: another attempt to discover the Loch Ness Monster. This involved a team of scientists on board ten boats equipped with highly developed Sonar equipment.

Radar (1940)

Radar (the acronym for Radio Detection and Ranging) was developed in England in **1940**, although a great deal of preliminary research had been carried out by, for example, Heinrich Rudolph Hertz of Germany in 1886, the Serbian–American inventor Nikola Tesla in 1900, and the German engineer Christian Hülsmeyer who patented a 'detector for objects with a continuous radio wave' in 1904.

In about 1934, Henri Gutton, a research worker in the French Wireless Company, CSF, developed the magnetron which later became the main component of future radar systems. The director of the company, Maurice Ponte, who was married to an Englishwoman, was able to have Gutton's invention sent to England during the war thereby enabling the English to develop their own projects.

At the same time the project of another Frenchman, Pierre David, an electromagnetic system for the detection of aircraft, was tested successfully at Le Bourget. The system, based on an idea conceived in 1928, made it possible to detect an aircraft at a distance of 5000m *3107 miles*.

Under the pressure of events, research continued during the early stages of the Second World War and resulted in the invention of radar by a British technical research team under the supervision of **Sir Robert Watson-Watt**. Radar proved a determining factor in the anti-aircraft defence system during the Battle of Britain.

From 1991 Great Britain will have a new form of radar capable of detecting missiles within a field of 360 degrees. It will be the first radar in western Europe able to detect, simultaneously, a threat from both East and West.

Rita (1985)

In **1985** the French company, **Thomson-CSF**, in association with the American company **GTE**, won a contract for more than four billion dollars to equip 25 divisions of the American Army with a tactical transmissions system. RITA, an integrated automatic transmissions network, was chosen in preference to the British system Ptarmigan.

Missiles

Ballistic missiles (1942)

The first major steps in the field of missiles took place in Peenemünde, now in West Germany, in 1937. Most of the theoretical de-

The Rita system equips 25 US army divisions.

velopments took place between 1939 and 1945, and only technological limitations prevented some of these from being put into practice.

The first trial of the **V-2**, the first operational ballistic missile in history, took place on **13 June 1942**. Its form has been adopted for the modern ground-to-ground missile. It was developed by a research team directed by **General Walter Dornberger** assisted by **Wernher von Braun** and **Hermann Oberth**.

The main categories of modern missile are the lightweight ground-to-air portable missiles, fired from the shoulder, such as the British Blowpipe, or which can be transported on light vehicles: the British Rapier and the American Stinger for example; and the intermediate range ground-to-air missiles, for example the Russian SAM, and the long range land-to-sea, sea-to-sea and air-to-sea missiles, the most famous of which is the French Exocet, employed during the Falklands War in 1983.

Anti-tank missiles belong to a separate category. They are usually 'guided by wire'. The best known are the Franco–German systems, Milan with a range of 2km *1 mile 427yd* and HOT with a range of 4km *2 miles 855yd*, and the American TOW with a range of 3km *1 mile 1520yd*.

Intercontinental missiles

The long range intercontinental ballistic missile, **ICBM**, is the weapon which will maintain the balance of world power. These missiles with their nuclear warheads can be launched from underground silos or submarines and are preprogrammed to destroy cities. They have an incredible range, in the order of 9000km *5593 miles*. The most famous are the American Minuteman, which is soon to be replaced by the four-stage MX with a range of 10000km *6214 miles*, and the Russian SS-19.

In April 1987, the static trials took place of the engine of the first stage of the American strategic intercontinental missile of the future, Midgetman.

Intermediate range missiles

Intermediate range ballistic missiles (IRBM) generally have three nuclear warheads and a range of 5000km *3107 miles*. The best known are the American Pershing, the Russian SS-20 and the French S-3.

The submarine-launched ballistic missiles (SLBM) – the Anglo–American Polaris, the Russian SS-N-20 and the French M-20 – have the same characteristics as the IRBM.

Cruise missiles

Cruise missiles are high speed, low altitude pilotless aircraft. They were invented by the Germans in 1944 in the form of the V-1, the *Vergeltungswaffe*, or Vengeance Weapon which was powered by a ram-jet engine.

In 1982, Boeing delivered the first Air Launch Cruise Missiles (ALCM) to the American Air Force. Twelve of these tiny planes with arrowhead collapsible wings are attached to the underside of the wings of a B52.

The 'invisible' AGM-129/A (1989)

On **2 March 1989**, the American Air Force tested the new 'invisible' cruise missile, the AGM-129/A, manufactured by **General Dynamics** and **McDonnell Douglas**. It has a greater speed than that of the previous generation of cruise missiles and is virtually undetectable by radar.

AS-X-19

The Russians are in the process of developing a new supersonic cruise missile which will be transported by the new variable geometry Tupolev bomber, Blackjack. The **AS-X-19** missile will replace the subsonic cruise missile, the AS-15, on board the Blackjack.

THE DSP BLOCK 14

This is a new American satellite for advance anti-missile warning. It is equipped with an infra-red telescope which is able to detect the trails of burning gasses emitted by the rocket during take-off. In this way, the user receives an almost immediate warning that a strategic missile has been launched, and is able to take the necessary action.

Over a dozen devices of this type, though an older model, have been in orbit since 1971. They have made it possible to observe the launching of more than 166 missiles during the course of the Iran–Iraq War.

Anti-ship missiles

The sinking of the Israeli frigate *Eilat* in 1967 by a small Egyptian patrol boat armed with Styx anti-ship missiles, triggered off and accelerated the programme for the development of such missiles in the western world. The Kormoran was developed jointly by France and West Germany; the MM38, the first version of the Exocet, by France; Otomat by France and Italy; Harpoon by the United States; Penguin by Norway and Gabriel by Israel. They all travel at high subsonic speeds. Any future attempts to counter the anti-missile systems will have to develop missiles with a greater range and with a high penetrating power i.e. missiles that will reach supersonic speeds (Mach 2). Given the range of speeds and altitudes to be covered the statoreactor is the most satisfactory solution (*see* Energy).

Since the autumn of 1988, the American company Northrop has been producing a new missile, Tacit Rainbow, capable of destroying enemy radar systems.

Naval warfare

Battleship (1850)

A French naval engineer, **Stanislas Henri Laurent Dupuy de Lôme**, developed the first high speed fighting strip, the *Napoléon*, in **1850**. This was followed by *La Gloire*, an armoured frigate, presented to Napoleon III in 1857. Both were steam-powered and screw-propelled.

In 1859 Dupuy de Lôme was responsible for the construction of two high tonnage ships, the *Magenta* and the *Solferino* which, although similar to earlier warships, were armoured.

In 1854, the American John Ericsson developed the revolving turret which extended the field of fire of a battleship's guns to 360 degrees. His idea was put into practice with the construction of the famous *Monitor* which played a decisive role in the American Civil War, and in particular at the Battle of Hampton Roads.

Dreadnought (1906)

The disappearance of rigging from sailing ships made it possible to install turrets on the midship centre line, so that as early as **1906**, the modern form of the battleship was established by the English ship *Dreadnought* which revolutionised naval history. Five armoured rotating turrets housed ten 304mm *12in* breech-loading guns.

Modern battleships

After the Vietnam War, during which the famous American battleship *New Jersey* fired 5688 406mm *16in* shells, i.e. seven times the number it had fired during the Second World War, ships of this type seemed doomed for the scrap-heap. However, the American authorities decided otherwise. These battleships have been thoroughly modernised and are now equipped with missiles, rapid-firing guns, helicopter platforms etc.

Submarine (1624)

Origins

The Dutch physicist, **Cornelius Drebbel** (1572–1633), inventor of the thermometer and tutor to the children of James I, was responsible for the creation of the first submarine. In fact he applied the theories of the English mathematician William Bourne, who had defined the principle of ballast tanks in 1578. Bourne had also had the idea of a hollow mast to provide ventilation. This was the principle of the *schnorkel* which was used much later in the German Type XXI submarines.

In **1624**, Drebbel had an ovoid, wooden submarine built which was propelled by 12 oarsmen in addition to the crew. The trials took place on the Thames between Westminster and Greenwich, much to the amazement of the general public. It would appear that Drebbel has the idea of renewing the air on board chemically, using an alkaline solution, which intrigued the Anglo-Irish Physicist, Robert Boyle (1627–91). Apparently nobody ever obtained details of this strange mixture.

Bushnell's *Turtle* (1776)

The hull of this single-seater vessel looked like two turtle shells stuck together. This hand-cranked craft was equipped with screw propellors and a sort of brace used to fix a box, containing a large charge of gunpowder and a detonator, to the hull of the ship to be blown up. It was piloted by the American army sergeant, Ezra Lee, who had received instruction from the inventor, **David Bushnell**. In **1776**, the *Turtle* attacked the English ship, *Eagle*, outside New York harbour, but was unsuccessful as the wooden screw of the *Turtle* could not pierce the ship's copper-lined hull. However, a year later it blew up an English schooner and the inventor was rewarded personally by George Washington.

***Nautilus* (1797)**

In **1797** an American mechanic, **Robert Fulton**, designed a propeller-driven submarine which was intended to place explosive charges under the hulls of enemy ships. The *Nautilus*, built in 1798, was tested on the Seine but failed to impress Napoleon. Fulton went to England where his reception was no better. The British Admiralty also turned it down as they believed the submarine would enable weaker nations to sink British warships.

WARSHIPS OF HISTORY

Type	Date	Characteristics
Galley	**3rd–2nd century BC**	The Cretans were already familiar with the galley which, propelled by oars, dominated the Mediterranean for centuries. It had its hour of glory at Salamis on 28 September 480 BC when the 470 Greek vessels of Themistocles defeated the Persian fleet of Xerxes which consisted of more than 1100 ships.
Longship	**8th century**	Invented by the Vikings, the Longship had a characteristic high prow and stern. The prow of the vessel was often decorated with the head of a dragon.
Cog	**13th century**	Always associated with the period of the crusades and great discoveries. Not ideally suited to war at sea, it was the predecessor of the vessels which ruled the oceans from the 16th to the 19th century.
Galleon	**16th century**	The *Henri Grâce à Dieu* or *Great Harry*, built by the English in 1514, and the *Grande Françoise*, prefigured modern ocean liners. The perfection of the design of this type of ship was achieved at the end of the 18th century by the skill of the French naval engineer, Jacques-Noël Sané, nicknamed the Vauban of the Navy.
Fire ship	**16th century**	It made its appearance at the beginning of the 16th century. It was a ship, no longer fit for service, which was filled with inflammable material and launched against enemy ships when the wind was favourable.

HMS Dreadnought *was the first to be fitted with steam turbines, which made her the fastest battleship afloat.*

David (1864)

The first feat of arms performed by a submarine carrying a torpedo – actually a simple barrel of gunpowder towed by a long rope – dates back to the American Civil War. During the night of **17 February 1864** a submarine, the *David*, designed by the American naval captain, **Horace L. Hunley**, and piloted by a crew of nine men, sank the Federal frigate, the *Housatonic*, which was taking part in the blockade of Charleston. However, the experience was not repeated due to the fact that 34 Southerners were killed in the incident and because of the measures taken by the Northerners to protect their ships.

Submersible (1899)

Earlier inventions had been 'submarines' in the true sense of the world. The *Narval* built in **1899** by **Maxime Laubeuf** (1864–1939), a marine engineer in the French Navy, was a submersible boat which had ballast tanks on the outside of its thick hull in order to withstand pressure more effectively. It was also far superior in certain respects, including being able to sail on the surface. It was operated underwater by an electric motor and on the surface by an oil-fired steam motor.

In 1875 an Irish-born American engineer, John P. Holland, designed an early submarine which combined most of the basic control systems of the modern submarine, and had obtained his first command from the US Navy Department in 1895. The *Holland* was delivered in 1900, followed by five more. These were the first fully operational and efficient submarines in the world, and the Royal Navy commissioned five.

The first truly 'seaworthy' submarine on the open sea was the *Argonaut* designed by another American, Simon Lake, in 1894 and tested in 1898.

The 'classic' submarine

These diesel-electric propelled submarines which are still in use today were developed in 1901. The infamous German U-Boats of the First and Second World Wars and the French *Daphné* are of this type.

The first submersibles were used mainly for shelling on the surface after releasing torpedoes underwater. After the appearance of the German Type XXI, which were able to dive rapidly and were equipped with schorkels which avoided them having to resurface to recharge their batteries and replenish their air supply, everything was done below the surface. The Type XXI was the prototype for a whole generation of classic post-war submarines.

Schnorkel (1938)

Originally designed for an American submarine in 1897, the *schnorkel* ('snort') was installed on several Dutch submarines in 1939. The German engineer Helmut Walter improved the system and adapted it for German submarines.

The first sea trials were carried out in 1942 by the U-107, under the command of Lieutenant-Commander Hessler, son-in-law of Admiral Dönitz.

The schnorkel consists of a retractable tube of about 8m *26ft 3in* long and 30cm *11.8in* in diameter which can be raised at right angles to the submarine, thus enabling it to cruise below the surface using its diesel engine. The schnorkel has two pipes, one to take in fresh air, and the other to extract expel gases.

On 3 August 1958 USS Nautilus, *the first nuclear submarine, surfaced at the North Pole.*

Nuclear submarine (1955)

The first nuclear submarine was constructed by the **US Navy** at the instigation of **Admiral Hyman G. Rickover**. The USS *Nautilus* was included in the naval budget in 1951, was put on the slipway in 1952 and launched on 21 January 1954. Its first sea mission started on **17 January 1955**, and in May of the same year it beat all records by covering the distance between New London and Puerto Rico, i.e. 1397 nautical miles or 2587km in 84 hours. On 3 August 1958, it was the first submarine to surface at the North Pole in a lead or channel between the ice floes. After ten years it had covered 330 000 nautical miles and used up three 'hearts' with a 6kg *13lb 3.7oz* uranium capacity. To cover the same distance, a standard diesel-electric propelled submarine would have needed 38 million litres of gas oil. With a submerged tonnage of 4091 tones, USS *Nautilus* carried a crew of 103 at a speed of 20 knots.

The modern nuclear hunter-killer submarines have a submerged speed of the equivalent of about 35 knots. This is due to the carefully designed droplet shape of their hull with its single propeller.

Typhoon (1980)

The Russian submarines of the ***Typhoon*** class are the most impressive strategic submarines ever to be brought into service. They have two nuclear reactors which ensure their virtually unlimited autonomy, and appear to be constructed from two thick hulls joined together. Two vessels of this type have been in service since September 1980. Some sources have suggested that the Russian Navy could produce eight of these submarines between now and the beginning of the 1990s. The superiority of the Soviet submarines, which are generally speaking better armed, silent and capable of sailing at greater depths, is currently a subject which preoccupies certain experts.

Pocket submarine (1987)

Crack naval commando units are currently equipped with 15 or so pocket submarines which enable them to approach enemy ships and coastlines with minimum risk of detection. There is a two-seater and a six-seater version, with a maximum speed of 6 knots.

These submarines, known as SDVs (Swimmer Delivery Vehicles), enable their crews to attach an explosive charge discreetly to the hull of a ship or to take commandos ashore with a minimum of noise. Once its mission has been completed, the SDV heads for the open sea where it is taken aboard a downgraded nuclear missile launching submarine, specially adapted for the purpose.

Ohio (1990)

The ***Ohio*** is the latest nuclear submarine for launching the American SLBM missiles. Ten vessels are planned and construction is already underway. Like the *Typhoon*, it is 170m *558ft* long and will initially carry 24 Trident 1 missiles with a range of 8000km *5000 miles* and by the end of the 1990s, 24 Trident 2 missiles with a range of 11 000km *6835 miles*.

Aircraft carriers (1911)

The first plane to take off from a warship was piloted by an American, Eugene Ely, on 14 November 1910. He took off in a Curtiss biplane, equipped with a 50hp engine, from the American cruiser, USS *Birmingham*.

On **18 January 1911**, **Eugene Ely** landed the same plane on a specially equipped platform on the quarter-deck of the battleship *Pennsylvania*. He took off again a few minutes later: the aircraft carrier had come into being.

Landing remained a problem for a long time, and the development of the oblique landing runway was a major step forward. During the Second World War, the Americans built 120 aircraft carriers which moved naval combat into the air.

Russian aircraft carriers (1975)

In 1970 the USSR undertook construction of the first aircraft carriers in her history. Since its entry into service in May **1975**, the *Kiev* has been followed by three other ships of its type. They are called 'aircraft carrier cruisers' by the Russians and are powerfully armed against ships, missiles, aircraft and submarines. The *Kiev* holds a total of 28 aircraft, 16 of which are Forger-type vertical take-off and landing aircraft.

BIK-COM-2 (1983)

Since **1983**, the USSR has undertaken the construction of a traditional aircraft carrier with an oblique flight deck and steam catapults for the use of unadapted planes. Known in the West under the name BIK-COM-2, the new Soviet carrier could have a displacement of approximately 75 000 tonnes at maximal load, a floating length of 300m *984ft 7in* and a deck width of 73m *239ft 10in*.

Propulsion will be ensured via two nuclear reactors which provide a total power of 200 000hp and allow a speed of 30–32 knots.

Nimitz (1975)

The three largest aircraft carriers in the world are in the Nimitz class, used in the US Navy, and with an overall length of 330m *1083ft*. Their eight 280 000 cont. hp nuclear reactors propel 85 000 tonnes at a speed of more than 30 knots. They have a crew of 6300 men and carry 90 aircraft and helicopters. The fourth American nuclear aircraft carrier in the Nimitz class, the *Theodore Roosevelt*, which is the fifth nuclear aircraft carrier and the fifteenth aircraft carrier in the US Navy, was launched on 25 October 1986 at Newport in Virginia (USA). It is 398m *1306ft* long and can carry 90 aircraft.

Cruiser (c.1920)

The cruiser is not an armoured vessel, but is quick and heavily armed. It took over during the **1920s** from the scouts of the First World War. The cruiser was intended to act as a scout for squadrons and convoys, but its function became gradually extended and generalised to the point where it played an important part, particularly from the beginning of the Second World War.

The biggest and most powerful cruisers currently in service throughout the world are the Kirov class of the Russian Navy.

Self-propelled torpedo (1864)

The self-propelled torpedo was invented in **1864** by the captain of a frigate in the Austrian Navy, **Luppis**, and later improved and constructed at Fiume (then an Austrian port) in Yugoslavia in 1867 by the British engineer, Robert Whitehead (1823–1905). The torpedo is a tiny independent 'submarine' carrying a heavy explosive charge in its bows and equipped with a self-steering mechanism which enables it to move itself towards the enemy vessel.

At present, the torpedo which is up to 7.3m *23ft 11in* long with a diameter of 53cm *20½in*, weighing more than a tonne and loaded with 270kg *595lb 5.6oz*, can cover a distance of 13km *8 miles* at a speed of 50km/h *31mph*.

Piloted torpedoes (1936)

These 'slow speed torpedoes' were developed just before the end of the First World War by two Italian naval officers, **Tesei** and **Torschi**. The torpedoes carried a two-man crew at a shallow depth with a charge of TNT which they had to attach to the hulls of ships anchored in enemy harbours. These first frogmen, equipped with aqualungs, attacked the English fleet in Gibraltar in September 1941, and again in the port of Alexandria in December of the same year, causing serious damage to the battleships *Valiant* and *Queen Elizabeth*.

In 1944 the Germans brought the *Marder* into service. This was a double torpedo of which the upper one contained a single-seater cockpit.

The Japanese developed the *Kaiten*, a sort of miniature submarine used in suicide attacks. It carried an explosive charge and was piloted by volunteers who crashed them into Allied ships.

Torpedo boat (1860)

The first torpedo boats appeared between **1860** and 1865. They were basic small craft equipped with a long moveable pole in the bows. The pole carried an explosive charge which was set off below the surface near the objective after a silent approach under cover of darkness.

The first successful use of the torpedo boat dates back to the Russo–Turkish War of 1877–8.

The counter torpedo boat or destroyer

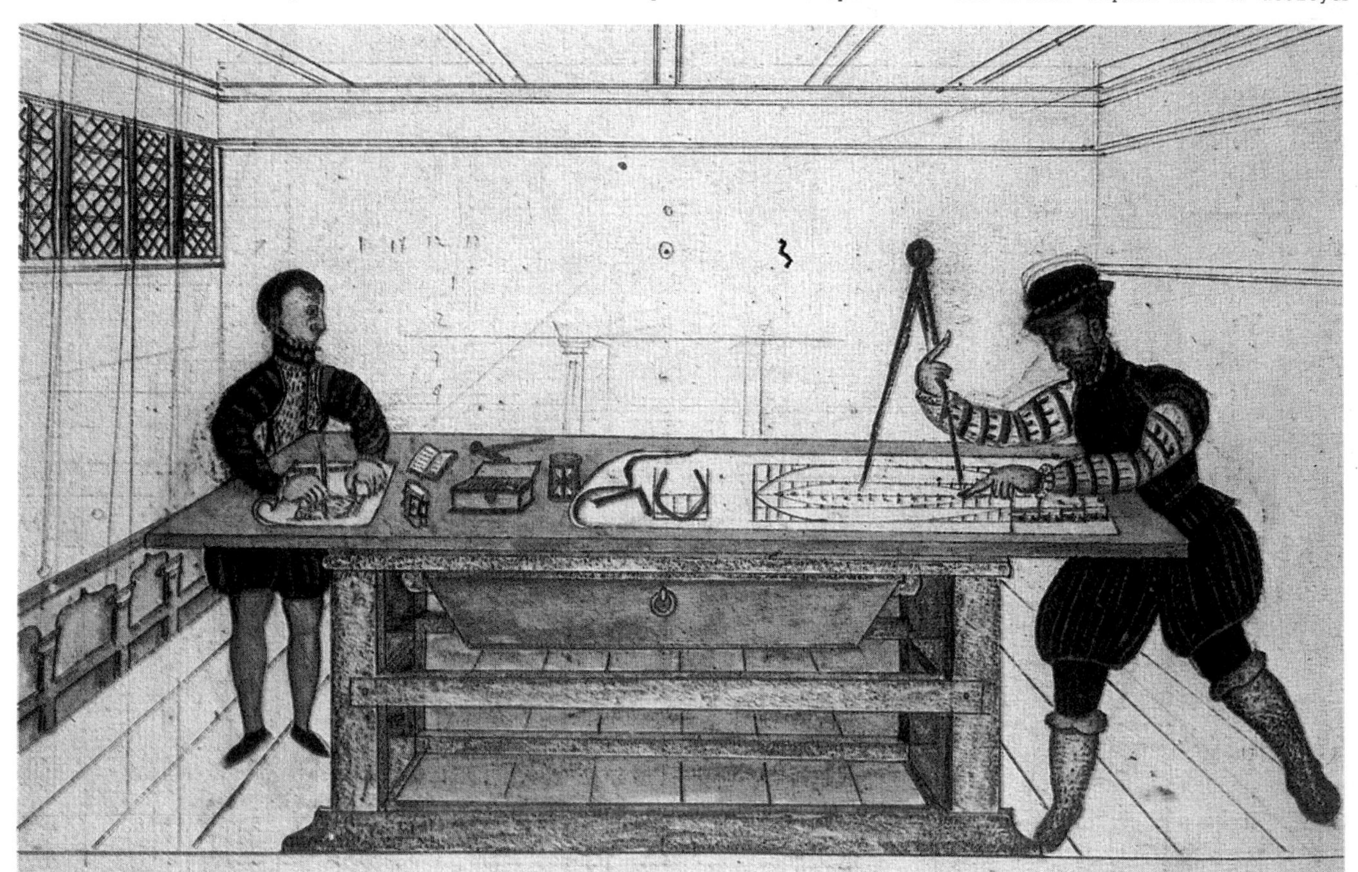

Tudor shipwrights design a new galleon for the fleet: fast sailing and seaworthiness were the chief requirements. Over a thousand tonnes of timber might be used for the hull alone.

appeared in 1893, the logical answer to the torpedo boat which it eventually replaced.

Sea mine (1861)

Although it had been described in various works at the beginning of the 17th century, the sea mine did not really make an appearance until the American Civil War of **1861–5**. It was widely used during the First World War.

The Second World War saw the development of the magnetic mine which was exploded under the magnetic effect of iron hulls; the acoustic mine, mainly activated by the noise of the propellers, and the low pressure mine which operated on the basis of the suction effect and the low pressure caused by the movement of the ship.

It was as a result of mine warfare that the minesweeper was brought into service during the First World War, and the mine detector developed towards the end of the 1960s.

Mobile mine (1980)

Since **1980** the great naval powers, and the United States in particular, have begun to develop a new form of mine: the mobile mine. Nuclear powered submarines, and especially missile launchers, constitute the main objective of this type of mine.

Torpedo mine (1995)

The Swedish Navy is currently researching a new 'intelligent' anti-submarine weapon which could be brought into service during the 1990s. It is a combination of the mine and the torpedo and will be able to distinguish between allied and enemy submarines. The torpedo mine rests on the seabed from where it uses an acoustic device to record the sounds of passing ships and compare them with programmed sounds in a data bank.

Aviation

Kite (2nd century BC)

Although it is difficult to say exactly where and how the kite came into being, it is known that it existed in China several centuries BC. Some sources quote its inventor as being **General Han Si** who was the first to use the kite for military purposes in the **2nd century BC**.

In 1903 kite trains were used by the Russian and British navies and armies. In 1906 British Army kites carries observers to a height of 400m *1312ft 9½in*. In 1909 the French army was equipped with kites which, with captured enemy balloons, were used during the First World War to observe the front lines.

Military airship (1902)

A French engineer, **Henri Julliot**, invented the first military airship in **1902**. It was a semi-rigid airship nicknamed *le Jaune* because of its colour (yellow), 57m *187ft* long and equipped with a 40hp engine. At the beginning of the First World War, airships were intended to be used for reconnaissance missions, and eventually on bombing missions. At the end of the War, they came into their own when they were used to protect convoys and combat submarines.

During the Second World War, the US Navy successfully used some 170 airships to escort convoys in the Atlantic.

Today, the airship is back, and has been used in various ways for several years: as a tactical platform, for anti-aircraft surveillance etc.

At the end of 1986 the US Navy launched an appeal for offers to develop a prototype for a large airship to be used for surveillance and advance warning. The new airship, the Sentinel 5000 constructed by Airship Industries in association with Westinghouse, is due to begin trials in 1990.

Reconnaissance (1911)

Aircraft were used for military operations for the first time in **1911**. In October of that year, the Italians used Blériot aircraft to carry out reconnaissance missions in Libya. Planes were also used for this purpose during the Balkan Wars of 1912 and 1913. Reconnaissance planes were subsequently transformed into bombers and fighters.

AWACS (1977)

The Boeing Airborne Warning and Control System (**AWACS**) brought out in **1977** makes it possible to carry out airborne surveillance, control and command simultaneously. It is a system which meets the requirements of tactical forces and aerial defence. A new vertical scanning radar, specially developed for the E-3 system, makes surveillance possible at all altitudes.

Mainstay

The Russians also have an AWACS system, the **Mainstay**, as it is referred to by NATO. It is derived from a four-engine heavy transport aircraft, the Ilyushin-76, and will be produced at the rate of five or six aircraft per year. With its four Soloviev turbojets, the Ilyushin takes off over a distance of only 850m *2790ft*, cruises at a speed of between 750 and 800km/h *466 and 497mph* and can cover an overall distance of 6700km *4163.4 miles*.

Fighter gun (1914)

The first fighter gun, a fixed weapon on the centre line of the plane, was an idea put forward in 1911 and tested in 1912 on a Blériot aircraft equipped with a 37mm revolving gun placed in front of the propeller. In **1914** two Frenchmen, engineer **Raymond Saulnier** and pilot **Roland Garros**, developed a system for firing through the propeller, the blades of which were protected from any impact by steel corners.

In Germany the Dutch manufacturer and pilot Anthony Fokker improved the system of the two Frenchmen and invented a mechanically synchronised firing system whereby

In the early years of aerial warfare, bombs were launched by hand.

the operation of the machine was interrupted by a series of cogs and rods when one of the propeller blades was in front of the gun barrel. The system was used at the front in July 1915.

Aircraft machine gun (1914)

The first machine gun was used on an aircraft by an American, **Captain De Forest Chandler**. The aircraft was a Wright biplane piloted by **T. de Witt-Milling**, and the gun was a Lewis light machine gun which had to be held at an angle between the knees of the operator. The first planes to use this type of equipment were two Breguet reconnaissance aircraft which took off from the base at Dugny, north of Paris, in August 1914 equipped with a Hotchkiss 1908 machine gun.

First aircraft guns (1917)

The first guns to be mounted in a plane were the German 20mm Beckers, introduced in 1917 on the Gotha bombers, at the request of the pilots and against the advice of the senior officers. The weapons, which were of modern design, reappeared in Switzerland after the war and gave rise to the group of guns known as the the *Oerlikon*.

The Spitfire is perhaps the most famous British military aircraft of all time, and certainly one of the most agile.

Spitfire (1936)

The forerunner of the Spitfire was the Supermarine S.6B, which won the Schneider Trophy in 1931. The familiar Second World War Spitfire was the result of a private venture, created by a team led by **Reginald Mitchell**. Without this plane, the RAF would have had nothing to match the German fighters. The prototype Spitfire flew in March **1936**.

Bombers (1911)

The first bomb attack from an aircraft was carried out by an Italian lieutenant, **Guilio Gavotti**, on **1 November 1911**, from an Etrich-Taube monoplane on an enemy column during the occupation of Tripolitania (Cyrenaica). The bomb was a spherical Cipelli-type device from which the pin was simply removed and the bomb thrown out of the aircraft.

On 3 August 1914 the effectiveness of aerial bombing was demonstrated when, barely six hours after war had been declared, a German Taube dropped three small bombs on the French town of Lunéville.

Thirty one years later, virtually to the day, on the 6 August 1945 the huge American bomber, the Boeing B-27 Superfortress *Enola Gay*, dropped the first atomic bomb on the Japanese city of Hiroshima.

MiG aircraft (1940)

The famous Russian MiG aircraft was created in 1938 by the research unit of **A. Mikoyan** and **G. Gurevich**. The first fighter, the MiG-1, was produced in **1940/41**. The MiG-9, the first jet aircraft to be mass produced in Russia, took to the air on 24 April 1946. The MiG-15 with its arrowhead wings appeared in 1947. The delta wing MiG-21, made in India from 1959, was the first Russian aircraft to be manufactured in a non-communist country.

MiG-29 Fulcrum (1983)

The single-seater supersonic Russian fighter, the MiG-29 Fulcrum, was the star of the 1988 Farnborough Air Show. The plane has been in existence since **1983** and its prototype was flying in 1977, but it had always been shrouded in secrecy. A MiG-29 was photo-

The Red Arrows is the acrobatic squadron of the RAF's training school which was founded in 1912 and is the oldest flying school in the world.

graphed in July 1986 in Finland. It can fly at speeds greater than Mach 2.2 and has a very short take-off distance of about 400m *437yd 21½ft.* There are about 500 MiG-29s currently in service, and the aircraft would appear to be closely based on the American F-18.

Jet fighter (1941)

The first jet aircraft were developed simultaneously by an English pilot and engineer, **Frank Whittle**, and a German engineer, **Ernst Heinkel**. In 1930, at the age of 23, Whittle had registered a patent for a jet engine which he tried to adapt for fighter planes from 1937. The Gloster E28/39 was flown at Cranwell on **15 May 1941**. But Whittle was beaten by the Heinkel company who carried out the first secret flight on 27 August 1939 with a Heinkel He 178 piloted by E. Warsitz. The Heinkel He 280, which was the first aircraft to be designed as a jet fighter, made its maiden flight on **5 April 1941**. Meanwhile, Whittle had vastly improved his engine, which made it possible for Rolls-Royce to manufacture its first turbojet engine.

During the Second World War, the only Allied jet aircraft to become operational was the twin-engined Gloster Meteor. A squadron was used to intercept the V-1 flying bombs. The first German jet fighter, the Messerschmitt Me 262, which became operational on 3 October 1944, could fly at a speed of 869km/h *540mph* at a height of 9000m *30 000ft.*

Supersonic plane (1947)

The American **Bell XS-I** (later restyled X-1), piloted by Charles Yeager, was the first aircraft to break the sound barrier on **14 October 1947**, propelled by a rocket engine at a speed of 1105km/h *714.6mph.* These planes do not take off directly from the ground, but are transported by other aircraft carriers from which they subsequently take off. The rocket engine operates, by definition, without an intake of external air. Its fuel consumption at low altitudes is extremely high. Modern supersonic aircraft achieve much greater speeds, reaching six times the speed of sound.

VTOL (1954)

This type of aircraft shares features with the

BOMBERS

Type	Date	Built By	Country	Characteristics
Handley Page 0/100	1916	Handley Page Ltd	GB	Strategic bomber used for night flying.
Breguet XIV	1917	Breguet	France	Archetype of all bombers until 1945. Louis Breguet used Duraluminium in its construction.
DH.4	1917	De Havilland	GB	Single-motor bomber able to carry a bomb load of 203kg *448lb.* Also built in USA.
Tu TB 1	1927	Tupolev	USSR	First monoplane multi-engine bomber. Able to carry a bomb load of 1000kg *2205lb* at 208km/h *129.25mph.*
Tu TB 3	1932	Tupolev	USSR	First four-engine bomber.
JU 87 Stuka	1938	Junkers Flugzeug-und Motoren-werke AG	Germany	Dive-bomber. Stuka was an abbreviation of *Sturzkampfflugzeug.* It was able to carry a bomb load of 1800kg *3969lb.*
B-29 Superfortress	1942	Boeing	USA	Operational in 1944. A B-29 was used to drop the first atomic bomb on Japan on 6 August 1945.
Arado AR 234 Blitz	1943	Arado Flugzeuwerke GmbH	Germany	The only jet bomber to serve during the Second World War.
A-4 Skyhawk	1956	McDonnell Douglas	USA	Suitable for use on aircraft carriers due to its small dimensions.
B-58 Hustler	1956	Convair	USA	First supersonic bomber.
A-5 Vigilante	1958	North American	USA	Introduced aerodynamic and electric innovations. Fighter and strategic reconnaissance plane. The fastest aircraft in the US Navy.
F-111	1964	General Dynamics	USA	First swing-wing bomber.
Tupolev T-26	1969	State Industry	USSR	Brought into service in 1974. Carries standard, nuclear and thermonuclear bombs. Can operate with supply planes and has a capacity for ten hours' autonomous flying.
Tupolev Blackjack	1988	State Industry	USSR	Quadrijet with a maximum speed of Mach 2. Missile launcher.
B-1B	1986	Rockwell International	USA	Variable geometry supersonic quadrijet. Brought into service in 1986. There have been a lot of problems with this aircraft.

helicopter and the aeroplane. Its propeller, or propellers, can be used as either a rotor or a standard propeller. In the more recent types of VTOL, such as the Osprey V-22 (see below), the pusher engines which drive the aircraft are able to pivot in such a way that the turbofans function as rotorblades.

The first VTOL craft (Vertical Take-Off and Landing) was the 'Flying Bedstead' constructed in **1954** by **Rolls-Royce**. Two vertically mounted reaction control jets lifted it off the ground.

The Osprey V-22 takes off like a helicopter thanks to the two sets of rotor blades on its wings.

Convair XFY-1 (1954)

On **2 June 1954**, the first vertical take-off and landing of the Convair XFY-1, piloted by J. F. Coleman, took place at Mofett Naval Air Station in California (USA).

This fixed-winged aircraft, equipped with a 5500hp Allison turboprop engine, had the peculiarity of landing on its tail with its nose pointing upwards. It was known as a 'Pogo Stick'.

Hawker Siddeley Harrier (1960)

The Hawker Siddeley Harrier made its maiden flight in October **1960**. It was designed by the engineer, **Sir Sydney Camm** for the British company, Hawker Siddeley. It was the first fixed-wing VTOL fighter plane to become fully operational, and has been in service with the RAF since 1969.

Osprey V-22 (1988)

The first aircraft with pivoting rotors was presented by the **Bell Aircraft Corporation** of the USA in January **1988**. In September 1989 the Osprey V-22 took off vertically and the pilot then tipped the rotors to 45 degrees and completed half of the test in that position. The flight lasted 55 minutes, the V-22 climbed to 700m *2296ft* and achieved a speed of 240km/h *149mph*.

The V-22, which is currently reserved for military purposes, could equally well be developed for civilian use. Its manoeuvrability, the fact that it can land in the middle of a city, the number of passengers it can carry etc, make the VTOL craft and the Osprey in particular the plane of the future.

Fastest plane in the world (1964)

On the **29 February 1964**, Lyndon B. Johnson, President of the United States, revealed the existence of an extraordinary aircraft: the **Lockheed A-11**. Under the name of SR-71, the craft became a strategic reconnaissance plane designed to replace the Lockheed U-2. It could reach a speed of Mach 3 at an altitude of 21000 *68922ft*. So that it could sustain a continuous speed of Mach 3, the SR-71 was constructed in titanium which resists temperatures of kinetic heat of the order of 300°C *572°F*.

On 27 July 1976, an SR-1 A beat the world closed loop record over a distance of 1000km *621 miles 714yd* at a speed greater than Mach 3. On 28 July it beat the world speed record for straight line flying at a speed of 3529.56km/h *2193.27mph*. This record remains unbeaten.

Inverted-wing aircraft (1984)

The X-29A was officially presented by **Grumman** at Calverton, USA on **27 August 1984**, in the form of the prototype no. 2. The following day, tests began on the prototype no. 1. The wing unit of the aircraft is inverted which gives it an advantage in aerodynamic terms and a better lift at low speeds. It can be lighter, smaller and less costly than a conventional aircraft.

The idea of an inverted-wing unit is not a new one since the first operational aircraft of this type was the German experimental quadrijet bomber, the Junkers Ju 287, tested in 1944 and recuperated by the Russians at Dessau.

The B-1B undergoes trials in Florida to test its resistance to extremes of temperature.

IN PROGRESS. . .

Lockheed has a project in progress for the development of a short ground run, vertical take-off aircraft which is capable of supersonic speeds. Fitted with a hybrid engine with an adjustable thrust based on research by Rolls-Royce, it will have a take-off distance of less than 200m *218yd 28.8in* fully loaded (19000kg *41895lb*) and be able to take off vertically carrying a reduced load. It will land vertically and reach speeds of up to Mach 1.5. This single-seater aircraft with tapered wings, double drift and nose-planes, will have a tricycle undercarriage which will distinguish it from the Harrier with its quadricycle undercarriage.

THE B-2: THE INVISIBLE STEALTH BOMBER

The maiden flight of the B-2, the famous invisible bomber developed by the research division of the US Air Force in conjunction with the American company Northrop, should have taken place in the autumn of 1988. But dates given for the launch of such revolutionary aircraft are rarely observed.

What is so special about this much-talked-about 'invisible' aircraft? It is a long-range strategic aircraft which looks like a flying wing; there is no fuselage in the conventional sense of the word, and the cockpit and two engines are incorporated into this 'wing.'

The term invisible obviously does not mean that the plane cannot be seen, but that its overall shape, the materials from which it is built, its paintwork etc enable it to escape detection by radar as it throws back virtually no electromagnetic waves. The bomber was designed to be completely undetectable: for example, so that there is no release of heat which would enable the B-2 to be picked up by infra-red detectors, the aircraft is not supersonic.

The stealth programme is obviously a costly one. A single B-2 bomber costs $516 million. Including the time spent on research (John Patierno, the 'father' of the B-2, will have worked on the project for 17 years), and the costs of research and development, it will be the most expensive aircraft in the world. The Pentagon has ordered 132 of them.

Another stealth programme, for the development of fighters, exists in the United States, but it is much more discreet. The F-117 A Stealth Fighter is a twin-engine, single-seater fighter built by Lockheed. It made its maiden flight in June 1981 and was brought into service in 1983.

Each B-2 bomber costs more than $500 million.

Variable wing camber (1985)

The F-111-AFTI, Advanced Fighter Technology Integration, developed by **General Dyamics** and produced by the Boeing Military Airplane Corporation, made its maiden flight on **18 October 1985**, at the Edwards Air Force base in the USA.

Its originality lies in its variable wing camber. The wing profile can be adjusted by the pilot who only has to register the altitude, the Mach number, and the manoeuvre he wishes to perform. The automatic pilot system selects the optimum camber.

C-5B Galaxy (1986)

The C-5A developed by Lockheed made its maiden flight on 30 June 1968 and until the presentation of the Antonov-225 *Mriya* it was the biggest aircraft in the world. The Military Airlift Command subsequently launched the **C-5B** to increase the strength of its strategic forces. The first of the 50 C-5Bs was delivered in **1986**.

A 5-K 'Kong Yun' (1988)

This is the first prototype of the Chinese fighter plane which made its maiden flight on **17 September 1988** at the Hangzhou airbase near Shanghai. The navigation and attack system was developed by the French company, Thomson-CSF. The assessment of the system was completed in 1989. Chinese pilots have been trained in the use of the equipment in France.

MRIYA

The biggest plane in the world is currently the Antonov-225 *Mriya*, presented on 30 November 1988 at Kiev. It weighs 600 tonnes on take-off, is 77m *253ft 8.6in* long, and can transport virtually anything. In particular it will be able to carry the Russian space shuttle to Baykonyr. It has a crew of six and a cruising speed of 800 km/h *497.12mph*.

FIGHTER PLANES

Type	Date	Built by	Country	Characteristics
Vickers FB.5	**1914**	Vickers	GB	Nicknamed Gunbus, this two-seater biplane was used by the best British fighter squadrons.
Fokker E.1	**1915**	Fokker Flugzeugwerke GmbH	Germany	Monoplane, based on the Morane N, installed with the Fokker firing system i.e. propeller synchronised.
Polikarpov 1-16	**1936**	State Industry	USSR	Small, robust monoplane fighter with retractable landing gear and closed cockpit.
Messerschmitt Bf 109 (Me 109)	**1936**	Messerschmitt AG	Germany	Single-seater monoplane. The first of the outstanding fighter planes of WWII. Speed 570km/h *354.2mph*. Archetypal Second World War fighter. Like the Spitfire, it was modified many times during the war.

Supermarine Spitfire	1936	Supermarine Division, Vickers Armstrong Ltd	GB	Great rival of the Messerschmitt. 20334 aircraft built up to 1947. Speed 570km/h *354.2mph.*
Gloster Meteor	1943	Gloster Aircraft	GB	Used againsed the V-1s. First British jet plane.
Me 262	1944	Messerschmitt AG	Germany	First operational German jet fighter. Speed 869km/h *540mph.*
Me 163 Komet	1944	Messerschmitt AG	Germany	First rocket-powered fighter. Maximum speed 900km/h *559.26mph.*
Lockheed F-80 Shooting Star	1944	Lockheed	USA	Used during the Korean War, it was the first fighter to win a fight between two jet planes by shooting down a MiG-15.
F-102 Delta Dagger	1948	Convair	USA	The first delta-winged plane. Very advanced in terms of electronics. A modified version, the Delta Dart, exceeded Mach 2 in 1959.
F-86 Sabre	1950	North American	USA	Jet plane with arrowhead wings. 9500 aircraft were built. Extremely fast and manoeuvrable.
MiG-15	1950	Mikoyan and Gurevitch	USSR	Jet plane with arrowhead wings. Rival of the American fighter planes.
Super Sabre F-100	1953	North American	USA	The first fighter to exceed Mach 1 in horizontal flight.
Mirage III	1956	Dassault	France	Remarkable delta-wing fighter, with a French engine, the Altar 101. It exceeded Mach 2.
MiG-25	1965	Mikoyan	USSR	Code name: Foxbat. Still the fastest fighter in the world. It has been detected at a speed of 3395km/h *2109.6mph* i.e. Mach 3.2.
Grumman F-14 Tomcat	1970	Grumman Aerospace	USA	Its variable sweep wings enables it to fly at speeds of between 200 and 2500km/h *124.3 and 1553.5mph.*
F-15 Eagle	1972	McDonnell Douglas	USA	Multi-purpose fighter for use in aerial combat, interception and ground attack. 620 planes built at the beginning of 1981.
F-16	1975	General Dynamics	USA	Remarkable close-range fighter. Light, easy to manoeuvre. Its post-combustion turbofan engine produces a thrust of 11000kg *24255lb.*
F-20 Tigershark	1982	Northrop	USA	Advantages: short take-off, faster climbing speed, ability to turn instantaneously, high load capacity.
Mirage 2000	1984	Dassault-Breguet	France	To replace the Mirage III, V and F1. Used for tactical intervention, back-up and reconnaissance.
Rafale	1996	Dassault-Breguet	France	The French land and ground forces hope to replace their Crusaders and Etendards with the Rafale in 1996.
A-12	1995	McDonnell Douglas and General Dynamics	USA	This fighter plane of the 21st century will share some of the features of the stealth bomber.
ATF	1994	McDonnell Douglas	USA	The Advanced Tactical Fighter is planned to replace the F-15 Eagle.
EFA	1995–6	British Aerospace	GB, W. Germany, Italy, Spain	European competitor of the Rafale. It made its first experimental flight in August 1986.

Aircraft of the future

Swedish Griffon (1992)

The maiden flight of the prototype of the Swedish fighter plane, the JAS 39 SAAB Griffon, took place on 9 December 1988. One of the distinguishing features of the Griffon is its small size. It is about half the weight (i.e. 8 tonnes) of the aircraft it is replacing – the JA 37 Viggen – while still being able to carry the same load of weapons.

The Swedes have used carbon-based composites in the plane's construction which account for about 30 per cent of the structure. The aircraft will be brought into service in **1992**.

Japanese FS-X (1993)

The new Japanese fighter plane the FS-X, which was launched in 1989, should be operational by **1993**. The plane, inspired by the F-16 developed by General Dynamics, was constructed by **Mitsubishi Heavy Industries** using mainly composite materials (carbon fibre reinforced with polyacrylnitrile plastic), which made it possible to reduce the weight by 40 per cent. In addition, by using a process which consists of layering 150 sheets of carbon fibre, the Japanese are able to construct the fuselage and the wings in a single piece of composite material.

ATF (1994)

The Advanced Tactical Fighter or **ATF** will be one of the American planes of the future to be used in aerial combat. It will replace the F-15 Eagle built by McDonnell Douglas but, like its predecessor it will be a single-seater, twin-engine fighter. It may also replace the F-14 Tomcat based on board the US Navy aircraft carriers. In October 1986, two projects were chosen, one by Lockheed and the other by Northrop.

The final choice between two prototypes will be made in 1990 and production should begin in **1994**.

The ATF should have a cruising speed of Mach 1.5 and be able to reach a maximum speed of Mach 2.5. It is an interceptor with a long flying range which will use the invisible and undetectable stealth technology. It will also be equipped with a navigation and attack system which will enable it to detect and destroy its adversaries.

A-12 (1995)

This 21st century fighter plane will replace the A-6 Intruder and could also be used by the US Air Force to replace its land based bombers. The new aircraft, which is to be designed by **McDonnell Douglas** and **General Dynamics** within the framework of the ATA or Advanced Tactical Aircraft, will have some of the characteristics of the stealth bomber. It must be able to take off from an aircraft carrier in order to attack targets on land. The first of these aircraft should be in service by **1995**.

Rafale (1996)

The Rafale is an experimental prototype intended to develop the technology to be used for the next French fighter plane. Built by **Avions Marcel Dassault-Breguet Aviation**, the aircraft made its maiden flight on 4 July 1986. It is a twin-jet plane weighing 9.5 tonnes when empty and 20 tonnes when fully loaded, and has delta wings with nose-plane ailerons at the front. Many new materials such as carbon-fibre, Kevlar, titanium, aluminium-lithium were used in its construction. In 1988, the Rafale had already completed 305 flying hours. It should be brought into service in **1996**.

The European Fighter Aircraft should be in service by the middle of the 1990s.

EFA (1996)

The European competitor of the Rafale, the prototype of the **EFA**, the European Fighter Aircraft, developed jointly by Great Britain, Italy, West Germany and Spain, will be flown in Germany in 1991. The eighth and final prototype should be flown in 1993, and the plane should be brought into service in **1995–6**.

Hornet 2000

This is the American counter proposal for the European 21st century fighter plane.

Fighter helicopters

The first military helicopter (1939)

The Russian-born American manufacturer **Igor Sikorsky** resumed his research on helicopters just before the Second World War. In the spring of **1939** Sikorsky developed his first prototype, the VS-300. This was improved in January 1941 and became the VS-316 A which, in May 1942, went to Wright Field, Ohio, to undergo military standardisation tests. It was named the XR-4. Orders came from the United States and Great Britain where the helicopters were given the name *Hoverfly* and were brought into service in 1945.

The S 75 ACAP helicopter (1984)

On **16 August 1984** the American helicopter, the Sikorsky S 75 ACAP, made its first public flight which lasted 20 minutes. Constructed from composite materials, the 75 S's main interest lies in the fact that it has a lighter structure than conventional helicopters, but nevertheless has a better resistance to bullets.

Apache (1984)

The AH 64 A Apache, built by **McDonnell Douglas**, is the most powerful, the heaviest and the most expensive anti-tank helicopter in the western world. It is equipped with 16 laser guided missiles, 76 rockets and an automatic 30mm gun, and flies at a speed of 360km/h *223.7mph*. The Apache has been used by the US Army since **1984** and has replaced the AH I Cobra. An improved version of the Apache, particularly in the field of detection, is to be tested at the end of 1993.

X-Wing: the helicopter-aircraft (1987)

In August 1986 the American company **Sikorsky** presented one of the two helicopters that it is developing within the context of the RSRA X-Wing project. It has a quadriblade rotor which enables it to take off vertically, but which stops rotating at a certain speed, each blade becoming a temporarily fixed wing unit. Propulsion is maintained by jet engines. The flight trials took place at the end of 1986 and the trials for conversion in flight, i.e. the changeover from a helicopter to a fixed wing aircraft, took place in **1987**.

NH 90 (1991)

A programme is underway involving **Aérospatiale** of France, **Messerschmitt-Bölkow-Blohm** of West Germany, **Agusta** of Italy and **Fokker** of the Netherlands (Great Britain has withdrawn), to produce the **NH 90**, a medium tonnage aircraft which is intended to replace the Super Puma. There will be an NH 90 military transport helicopter and a naval version for use as an anti-submarine craft. The first flight is scheduled for October **1991** and deliveries of the helicopter should take place in 1995 and 1966.

The NH 90 will weigh between eight and nine tonnes. It will be able to fly in all weathers, reach a height of 6000m *19692ft* and cover a distance of over 700km *435 miles*.

In October 1988, the Germans started test flights for the BO-108 helicopter built by MBB. It is intended to replace the 22 year old BO-105 and should reach a maximum cruising speed of 270km/h *167.78mph*.

EH 101 (1992)

The British company **Westland**, and the Italian company **Agusta**, are collaborating in the development of the **EH 101**, a heavy 30-seater, three-engined helicopter weighing 14 tonnes. The certification should be obtained in 1990 and the Royal Navy should receive its first EH 101 in **1992**.

Sea Dragon (1987)

In August **1987** the Americans sent the aircraft carrier *Guadalcanal* into the Gulf loaded with Sikorsky MH 53 E mine sweeper Sea Dragon helicopters, the mine detecting version of the Super Stallion. The Sea Dragon is the largest helicopter in service. It is 27m *88ft 7.34in* long and its blades have a diameter of 22m *72ft 2.4in*. The helicopter's efficiency is largely due to its speed, which reaches up to 300km/h *186.42mph*, and its ability to fly for periods of up to four hours. The mine sweeping can be either mechanical, where the mine has risen to the surface and is destroyed with guns, or is carried out using magnetic and acoustic systems.

TRANSPORT PLANE

A project for a tactical transport plane intended for use by air forces by the end of the century is being studied by the FIMA group i.e. Aéropatiale, British Aerospace, MBB, Lockheed, Aeritalia and Cada.

Mathematics

Number theory

Numeration (3000 BC)

Numbers were first written 3000 years before Christ, as is attested to by the clay tablets discovered in Susa and Uruk (currently Warka in Iraq) and those from Nippur (Babylonia, 2200 to 1350 BC). The Babylonian system of numeration is based on 60. Our time divisions are a vestige of this. There was no zero; missing units were simply indicated by a space.

The ancient Mayan system was a system based on 20 which combined the number of fingers and toes. Their system was already a position system and included a final zero which was not an operator.

In the 5th century BC the Greeks used the letters of the alphabet. For units of one thousand, the nine first letters accompanied by an inferior accent to their left were used (α equals 1 and ,α equals 1000). This system, which had no zero, was used for a thousand years. The Hebrews and Arabs adapted this to their own alphabets. Calculations were then made using abacuses. Numbers were represented by pebbles (the word calculation derives from the word *calculus*, meaning pebble).

Modern numeration (5th century)

Toward the 5th century AD decimal position arithmetic appeared in India: it used ten figures from 0 to 9 such as we know today. In 829, the scientist, Mohammad Ibn Musa al-Khwarizmî (780–850), published a treatise on algebra in Baghdad in which he adopted this decimal system. A French monk called Gerbert became interested in the Arabic figures during his voyage (980) to Cordoba, in Spain, and was able to spread the use of these symbols when he became Pope Sylvester II in April 999. However, it was not until Leonardo Fibonacci, known as Leonard of Pisa, through his *Liber Abaci*, written in 1202, that the Arabic science began to spread throughout Europe. In 1440, thanks to the invention of printing, the shape of these ten figures was definitively fixed.

Zero (4th century BC)

Babylonian numeration was perfected in the 4th century BC by the appearance of the zero in mathematical texts. The zero was placed either at the beginning of a number or within a number, but never at the end.

The word zero comes from *sunya* which means 'nothing' in Sanskrit; it became *sifr* in Arabic and was Latinised into *zephirum* by Leonardo Fibonacci. It was fixed at zero in 1491 by a Florentine treatise.

Prime numbers (3rd century BC)

After Euclid, who demonstrated in the 3rd century BC that all prime numbers are infinite, the riddle of Erastosthenes (c.284–c.192 BC) was the first method used to investigate these numbers systematically. Nonetheless in 1876 a Frenchman, Edouard Lucas, developed a method to study the primary nature of some large numbers.

The greatest known prime number – 2 216 091 . . . 1 (65 050 digits) – was discovered by chance in 1985, by a team of technicians in the oil firm Chevron, in Houston, Texas (USA). While they were trying out a super-computer, this new prime number came to the fore: it would fill at least ten pages of this book.

Irrational numbers (4th century BC)

Aristotle (4th century BC), by demonstrating the impossibility of writing the number $\sqrt{2}$ as a fraction, revealed the existence of **irrational numbers** (foreseen by Pythagoras).

Complex numbers (17th century)

We owe to the Italian mathematician **Raphael Bombelli (1526–73)** the first definition of complex numbers, then called 'impossible' or 'imaginary' numbers, in the work published shortly before his death, *Algebra* (Bologna, 1572). He defined these numbers from his study of cubic numbers, and introduced the square root of -1. A complex number is the sum of a real number and an imaginary one e.g. $(a + ib)$.

Until 1746, people used these imaginary entities without really knowing how they were structured. In that year, however, the French mathematician Jean Le Rond d'Alembert established their general form $a + b\sqrt{-1}$, assuming the principle of the existence of n roots in an equation of n degrees. The Swiss mathematician Leonhard Euler (1707–83) introduced the notation 'i' to designate the square root of -1, a notation taken up again by Gauss in 1801.

Transcendental numbers (18th century)

It was in the 18th century that mathematicians came up with the notion of the transcendental number. We call a number transcendental if it cannot be expressed as an algebraic equation in rational coefficients.

In 1844 the French mathematician Joseph Liouville gave the first example of a transcendental number. Then in 1873, the Frenchman Charles Hermite showed that 'e' was transcendental, and in 1882, the German Ferdinand von Lindemann demonstrated the same for π.

The number π (3rd century BC)

By using polygons of 96 sides, inscribed and excribed to the circle, the Greek scientist, **Archimedes** (287–212 BC) demonstrated that the number π is located between

$$3 + \frac{10}{71} \text{ and } 3 + \frac{10}{70}.$$

Thus, when Ptolemy (Greek mathematician of the 2nd century AD) adopted the value of 3.1416 for π he noted, to justify it, that it was nearly the mean of the two Archimedean boundaries.

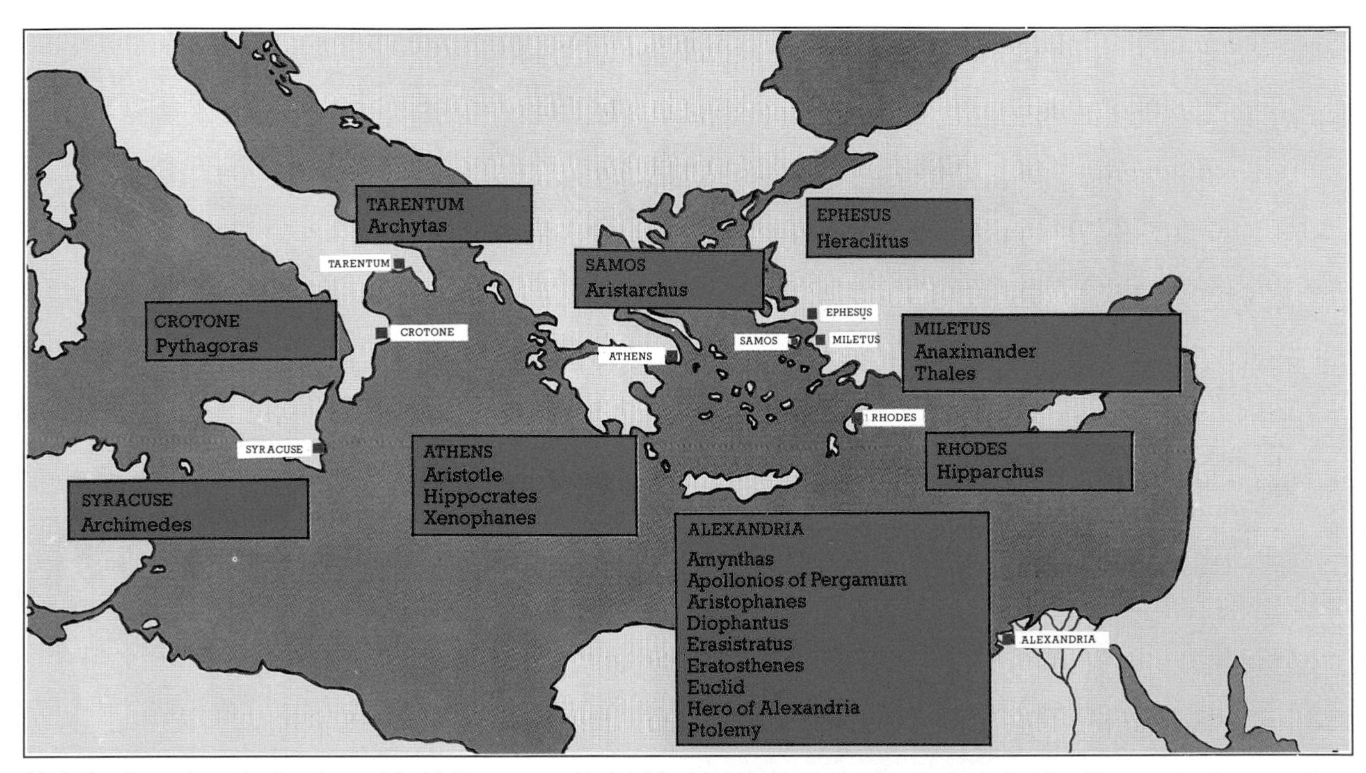

Much of modern science developed around the Mediterranean and in Asia Minor, but who knows what the scientists looked like. Above are imaginary portraits of Thales, Pythagoras, Euclid, Archimedes and Ptolemy.

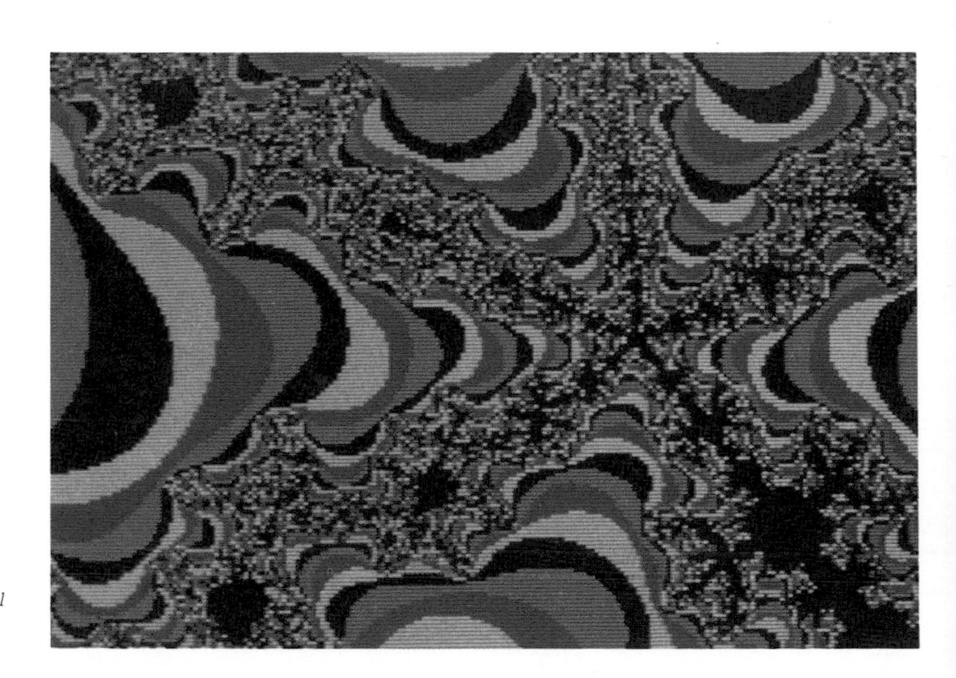

Fractal images were invented in 1962 by the Frenchman Benoît Mandelbrot. They are created by a computer that repeats the same mathematical formula over and over again. Each part of the pattern has the same structure as the whole thing.

Phidias used the golden number to calculate the proportions of the Parthenon in the 5th century BC.

In 1874, an Englishman, William Schanks, calculated the 707 decimal places of π engraved in the Palace of Discovery in Paris. The first 527 are exact, but the remainder are wrong. Since then, through the use of the computer, π has been calculated to millions of decimal places.

Golden number (3rd century BC)

The golden number, the solution of the equation

$$\frac{1}{x} = \frac{x}{1 + x}$$

is equal to

$$\frac{1 + \sqrt{5}}{2} \ (\approx 1.618)$$

and exists in an asymmetrical sharing when the ratio between the largest of the two parts and the smallest is equal to the ratio between the whole and the largest. This number was known before **Euclid**, but it was he who, in the **3rd century BC**, made of it a famous problem by seeking to divide a straight line in mean and extreme reason or 'golden section'.

The harmony based on this golden number has been studied in numerous arts: in architecture (Phidias, who worked on sculptures in the Parthenon in the 5th century BC, Alberti in the 15th century; Le Corbusier in the 20th century), in music (Pythagorean research on sound intervals), and in painting (Leonardo da Vinci, Raphael).

THE IMPOSSIBILITY OF SQUARING THE CIRCLE

From the 5th century BC, Greek mathematicians had posed the following problem: how can you construct from a circle with a given radius, using only a ruler and compass, a square whose area is equal to that of the circle? This problem, known under the name of the problem of the quadrature of the circle involves constructing from a section of length 1 (the radius of the circle) another section the length of the root of π (the side of the square).

At the beginning of the 19th century, mathematicians proved that numbers that could be constructed with ruler and compass were precisely those numbers that were not transcendental. So when Carl Louis Ferdinand Lindemann (1852–1939) proved in 1882 the transcendence of π, he could proudly deduce the answer to the problem of squaring the circle, a problem which had been the object of 24 centuries of research. If π is transcendental, it cannot be the root of an equation, therefore it cannot be constructed, therefore you cannot square a circle.

Geometry

Thales theorem (7th–6th century BC)

Before Thales, each surveyor or geometer found gadgets with which to measure distances, surfaces, etc. The Greek philosopher and mathematician of the Ionian school, **Thales of Miletus** (c.624–c.546 BC), had the very clever idea of measuring heights by using shadow at the time that the 'shadow is equal to the object'. That is when the sun's rays are projected at 45°. In order to measure the height of the Great Pyramid, he refined his method by using the rays at any given hour. He might have left it at that, but he wanted to formulate a theory based on his experiments. The use of the sun's rays caused him to study parallel lines and the ratios between the lengths projected and the initial lengths. He then drew up his theorem, called Thales theorem: 'parallels are projected from a straight line onto another line of proportional length.' Thales of Miletus thereby introduced the deductive and demonstrative aspect of mathematics to **geometry**.

Pythagoras' theorem (6th century BC)

Using the work of Thales on parallel lines and projections, and in the same spirit of demonstration, **Pythagoras**, the Greek philosopher and mathematician of the **6th century BC**, became interested in orthogonal projections and demonstrated the theorem which bears his name. This theorem establishes a relationship between the lengths of the sides of a right-angled triangle: that the square on the hypotenuse is equal to the sum of the squares on the other two sides. This relationship had been known since surveyors had begun to practise, but Pythagoras was the first to demonstrate it.

Euclidian assumption (3rd century BC)

The Greek mathematician of Alexandria, **Euclid (3rd century BC)**, worked mainly on a synthesis of his predecessors' work. In his *Elements*, he structured the knowledge of his era by redemonstrating everything from five assumptions which were considered to be true although they could not be demonstrated. The foremost and best known is: given a point exterior to a straight line, only a line which is parallel to this straight line can be made. The contrary of this assumption had been imagined by Aristotle, but from a strictly didactic point of view. Until the 19th century, mathematicians thought that the demonstration of this assumption was possible. Thus, in the 18th century, numerous mathematicians tried in vain to demonstrate it by absurdity. Two possible negations appeared: at least one point exists by which there is no straight line parallel to a given straight line. The other was: at least one point exists by which at least two distinct parallel lines pass. The fact of having clearly expressed these two contrary thoughts allowed the following century to create two new kinds of geometry.

Trigonometry (3rd–2nd century BC)

During antiquity, trigonometry developed as a technique annexed to astronomy. Thus, the Greek astronomers, **Aristarchus** of Samos (c.300 BC) and **Hipparchus** of Nicea (c.200 BC), were the precursors of trigonometry. The Alexandrian, Ptolemy (AD c.80–c.160), summarised all the knowledge of the era in his treatise, the *Almageste*.

It was the Arabs who, during the 9th and 10th centuries, developed trigonometry as a separate science. Al Khwârizmî (780–850) established the first sine tables, and Habasch al Hasib those of the tangents. The *Perfection of the Almageste* (877–929) by al Bâttâni was a veritable treatise on modern trigonometry and was much more complete than Ptolemy's *Almageste*. The studies were taken up and elaborated upon by the German mathematicians, Johan Müller (1436–76) and Georg Rhaeticus (1514–76). Abraham de Moivre (1667–1754) and Leonhard Euler (1707–83) made a radius and an angle correspond to

The Italian painter Piero della Francesca was inspired by the works of Archimedes and Euclid. He wanted to lend to shapes the mathematical perfection of regular bodies and to arrange the relationships between space, colour and light through the use of perspective.

each complex number, thus allowing trigonometry to be dealt with by means of the exponential complex. Trigonometry was thereby integrated into an algebraic theory.

Conic sections (3rd century BC)

Conic sections have been studied in very different ways over the centuries, and are a good example of how geometry has evolved from antiquity to the present.

Apollonius of Perga (c.262–c.180 BC) in his treatise on conic sections studied various ways of cutting a cone. He then showed how parabolas, hyperbolas and ellipses would be obtained: he also originated these terms.

In the 17th century, Descartes translated conic sections into equations and showed that they could be obtained from second degree equations.

Blaise Pascal (1623–62) developed the modern aspect by approaching conic sections from an analytical point of view. In the 20th century, they are part of the general theory of quadratic forms.

Co-ordinates (17th century)

The use of numbers for the univocal location of a point on a surface had been known since the time of Archimedes (3rd century BC). But it was not until the **17th century** that co-ordinates were used in a systematic way in problems of geometry. Legend has it that the French philosopher and mathematician **René Descartes** (1596–1650) had the idea while watching an insect flying by his window. This discovery allowed geometry problems to be treated by algebraic method; thus, with the French mathematician **Pierre de Fermat** (1601–65), analytic geometry began, where equations and curves were linked.

Vectors (1798)

The Danish geometer **Caspar Wessel**, in **1798**, and the Swiss **Jean-Robert Argand** in **1806**, wrote two papers on complex numbers. Both of them had the idea of not only representing complex numbers by points A on a plane, but also of identifying them by the vector from the origin 0 to the extremity A on a Cartesian plane. Thus the notion of the $\overrightarrow{0A}$ vector was born: the sum of two complex numbers then allowed the sum of two vectors to be worked out. Vectors, quantities specified by both magnitude and direction, are thus geometrical objects, which can undergo the same sort of operations as number sets.

Structure of vectorial space (1844)

The German mathematician **Hermann Grassmann** (1809–77), in his 'theory of linear extension' of **1844**, defined vectorial spaces in more than three dimensions. At the same time the Irish mathematician Sir William Rowan Hamilton (1805–65), with his study of quaternions, elaborated the first vector system. These definitions were very useful to physics at the time when the theory of relativity was

Renaissance mathematicians like Luca Paccioli de Borgo, a Franciscan and a friend of Leonardo da Vinci, developed the work carried out by the Arabs in the early centuries AD.

developed, in which space-time is considered as a vectorial space in the fourth dimension.

Non-Euclidean geometries (18th century)

In the **18th century Giovanni Girolamo Saccheri**, **Johann Heinrich Lambert**, **Taurinus**, **Reid** and numerous other mathematicians tried to work out the logical consequences of the negatives of Euclid's postulate; but they came up with no complete theories. At the beginning of the 19th century, these theories took shape and developed into two different geometries, both possible and practicable.

Elliptic geometry (19th century)

The German physician and mathematician **Johann Karl Freidrich Gauss** (1777–1855) created a geometry in which a plane was defined as the surface of a sphere with an infinite radius.

The German **Bernhard Riemann** (1826–66), a student of Gauss at Göttingen, continued his work and proposed a revision of classical geometry, allowing elliptical geometry to be considered as a particular case of a more general theory.

Hyperbolic geometry (19th century)

The Hungarian mathematician **Jéanus Bolyai** (1802–60) and the Russian mathematician **Nicolai Ivanovitch Lobatchevski** (1792–1856), created a geometry in which the plane is a hyperbolic surface. Each of its points is shaped like a saddle.

Definition of geometry (1872)

Various works on non-Euclidean geometries at the beginning of the 19th century stirred up passionate debate and violent polemics; they revolutionised the philosophy of knowledge, in fact, more than geometry itself.

Thus it was necessary to bring this work together and create a theory large enough to accommodate these co-existing but different worlds of geometry. The German mathematician **Christian Felix Klein** (1849–1925) in his opening address to the Erlangen congress (the 'Erlangen programme', **1872**), defined geometry as the study of 'groups' or 'collections' of transformations, leaving certain geometrical objects invariable, such as medians and perpendiculars. In studying the structure of these groups, C. F. Klein integrated geometries into one algebraic theory. Thus in the 20th century algebra and geometry form part of mathematics.

Algebra and analysis

Origins (3rd century)

The word *algebra* comes from the Arabic word *al-jabr*: to reduce. It is generally considered that the work of Diophantus of Alexandria (3rd century) was the first step in the history of algebra. It was in his *Thirteen Books of Arithmetic* that mathematicians such as the Frenchmen Pierre de Fermat (1601–65) and François Viète (1540–1603), found the starting point for their own research. François Viète is often considered to be the inventor of modern algebra. He founded the algebraic language in 1591 in his work, *Ars analytica*. The language uses letters not only to indicate unknowns and indeterminates but also to form words or algebraic expressions.

Algebraic symbols (15th–17th century)

The Egyptians used symbols: addition was indicated by two legs walking in the same direction, and subtraction by two legs walking in opposite directions. Conversely, no symbolism is found with the Greeks, where each reasoning process was written out fully.

It was the Germans and the French, in the 15th, 16th and 17th centuries, who introduced symbolic calculation. Our + and − signs, which appear in 1498 in an arithmetic book by the German, Jean Widmann D'Eger, were promulgated by the German Michael Stifel in 1544 in his treatise on algebra, *Arithmetica integra*. The root sign $\sqrt{\ }$ was invented by Cristoff Rudoff in 1526. The sign × is more recent, and William Oughtred (1637) is credited with using it for the first time. The greater than > and smaller than < signs are due to Thomas Harriot (1631). Finally, the French philsopher René Descartes created in 1637 the use of figures placed as exponents to designate powers, and the Englishman John Wallis (1656) was responsible for the idea of negative exponents.

Fermat's last theorem (1637)

The great French mathematician **Pierre de Fermat** (1601–65) asserted in **1637** that he had demonstrated the following theorem: 'There do not exist positive integers x, y, z, n such that $x^n + y^n = z^n$, where n is greater than 2.'

In the 17th century, mathematicians were not very interested in demonstrating their theorems. So Fermat never published his proofs, simply stating this last thorem which he believed he had demonstrated. A number of such statements by Fermat were later found to be correct and important, so scientists were accustomed to treat the great mathematician's assertions with considerable respect.

The thorem stated by Fermat in 1637, however, which is called his 'Last Theorem', has never been completely proved after three hundred years of research and is the mathematical problem for which the greatest number of incorrect proofs have been published. We only know that the theorem is valid for all the values of n between 1 and 30 000, but a complete proof is remarkably elusive. All mathematicians think that the general theorem is true and hope to end up by proving it.

In 1988 a Japanese mathematician, Yoichi Miyaoka, said that he had arrived at a proof.

Algebraic equations

1st and 2nd degree equations (1700 BC)

We find in the Rhind papyrus (Egypt), which dates back to 1700 BC, some examples of the solution of 1st and 2nd degree equations attached to specific problems. Chinese literature offers examples of solutions of these two systems of equation by two unknown writers, in the work *Nine Chapters on the Art of Calculus* (c.200 BC).

The history of 2nd degree equations goes back to the Babylonian civilisation of the second millennium (1800 BC). The Babylonian arithmeticians knew how to solve all equations of the 2nd degree but did not express them in real number sets. The Greeks in the 3rd century BC made the solution of 2nd degree equations the basis of all their geometry, and in order to make them work for real sets, they replaced the Babylonian calculations with constructions by ruler and compass. The Greek algebraists, however, calculated within the set of positive rational numbers, which left many equations unsolved. It was not until the 16th century and the identification of complex numbers that all 2nd degree equations could be solved.

3rd and 4th degree equations (16th century)

The Italian school of the 16th century brought solutions to the resolution of 3rd and 4th degree equations. The three pioneers were successively Scipione del Ferro, Niccolo Fontana, called Tartaglia (around 1500–57), and Girolamo Cardano (1501–76).

THE FISCHER–GRIESS MONSTER

In 1980 mathematicians finished the complete classification of all finite and simple groups. A group is said to be *finite* if it possesses a finite number of elements, and *simple* if it is, in a certain manner, irreducible.

This classification has involved generations of mathematicians since the 19th century, and takes over 5000 pages of tightly-packed proofs in mathematics journals. It concludes that there are 17 infinite families of regular groups, and 26 groups that are completely isolated and are called sporadic.

The existence of sporadic groups is a remarkable phenomenon. The smallest one has 7920 elements. The biggest one is called the Fischer–Griess Monster, named after the mathematicians B. Fischer (a professor in Bielefeld, West Germany) and R. Griess (of the University of Michigan) who had predicted its existence. This monster has an extraordinarily large number of elements: $2^{46} \times 3^{20} \times 5^{9} \times 7^{6} \times 11^{2} \times 13^{3} \times 17 \times 19 \times 23 \times 29 \times 31 \times 41 \times 47 \times 59 \times 71$, that is to say, around 8×1053 elements.

On the 14 January 1980, R. Griess, one of the discoverers, sent a rather original New Year greeting to all his colleagues expert in simple group theory. In this message, he announced that he had succeeded in constructing the monster exactly, and by hand (without the help of computers). In April 1980, Griess explained that the monster appeared as a rotation group in a space of 196 883 dimensions!

Abel and the 5th degree equation (1826)

While he was still a student, the Norwegian mathematician **Niels Henrik Abel** (1802–29) attacked the formidable problem of solving an algebraic equation of the 5th degree. One day he thought he had cracked it, but soon saw that his proof had an error in it. He then decided to prove that it was in fact impossible to solve the general quintic equation. In **1826**, he proved his revolutionary result: that it is indeed impossible to solve equations of degree greater than four by means of radicals.

Abel decided to visit the two greatest mathematicians of the time, the German Johann Karl Freidrich Gauss and the Frenchman Augustin Cauchy (1789–1857), and show them his results. But they did not want to acknowledge the genius of the 24 year old, and could not be bothered by understand his extraordinary paper. Unrecognised and despairing, Abel returned to Norway, where he died of tuberculosis at the age of 27.

Theory of probabilities (1656)

The theory of probabilities was born from the study of games of chance. The word hazard, transmitted through Spain, comes from the Arabic *az-zahr* meaning 'the die'. Blaise Pascal and Pierre de Fermat were the first, in their correspondence, to want to 'mathematise' games of chance.

The Dutch scientist **Christiaan Huygens** (1629–95), aware of this correspondence, published in **1656** the first complete account of the calculation of probabilities, *De Ratiociniis in ludo aleae*. Subsequently, Jacques Bernoulli (1654–1705) wrote the work *Ars conjectandi*, a deeper study of the science than that of Huygens. Finally, it was the French mathematician Pierre-Simon de Laplace (1749–1827) who produced work on the application of mathematical analysis to the theory of probabilities which had a philsophical element.

Statistics (1746)

In **1746**, a professor from Gottingen, Germany, **Gottfried Achenwall**, created the term statistics. In fact, the gathering of data goes back to antiquity. For example, the Chinese Emperor Yao organised a census of agricultural production in 2238 BC

The Belgian Adolphe Quetelet, in 1853, was the first to think that statistics might be based on the calculation of probabilities.

The appearance of powerful calculators gave birth to methods of analysis of multidimensional data which are currently enjoying wide popularity.

Matrices (1858)

The study of systems of equations and linear transformations led the English mathematician **Arthur Cayley** (1821–95) to establish tables called matrices, and to define operations on these tables.

Matrical calculation was thus developed, and he published it in a thesis in **1858**. His work permitted the further study of the structures of linear transformation groups and prepared the ideas which were put forth by Christian Felix Klein in his Erlangen programme (1872).

Group structure (19th century)

During the course of the **19th century**, many groups were studied at the same time that work was being carried out on various equations. The French mathematician Augustin Cauchy studied groups of root permutations of algebraic equations. Evariste Galois (1811–32) continued this work and developed a theory of groups which unfortunately remains unfinished; he was killed in a duel at the age of 21. His theory, despite some oversights, remained a guide for further research.

During the period 1870 to 1880, the Norwegian mathematician Sophus Lie (1842–99), formulated the theory of continuous groups of transformation or 'Lie groups', notably to study differential equations. These groups were used by Klein in his Erlangen programme. Since then, this group structure has been used in all modern mathematics and physics, in research on atomic structures.

Mathematical logic (1854)

The English autodidact **George Boole** (1815–64) was the creator of symbolic logic. In 1847 he published a short treatise (*Mathematical Analysis of Logic*) in which he maintained that logic must be connected to mathematics and not to philosophy.

In **1854**, in his treatise *An Investigation of the Laws of Thought*, Boole set out the results of his thinking. Thus began what is today called Boolean algebra, a system of symbolic logic which codifies non-mathematical logical operations, using the numerical values 0 and 1 only. Logic took its current form with Gottlob Frege, in 1879, whose work was made known to the public by the philosopher Bertrand Russell in 1903.

The set of real numbers (19th century)

In the 4th century BC the Greek mathematician and astronomer Eudoxus tried to put into the form of a set numbers not limited to the rational, which he sensed were insufficient. He was not successful, neither were any number of later mathematicians of antiquity, who were reluctant to deal with irrational numbers.

It was not until the **19th century** that the Danish-born German mathematician **Georg Cantor** (1845–1918) studied irrational quantities and the notion of 'continuity', taking account of the 'continuum' aspect of the law of real numbers, formed by an infinity of distinct points, each representing a number. This gave rise to many paradoxes which challenged intuitive ideas. Cantor, aware of the break with traditional good sense, had to struggle for several years to convince his contemporaries of this arithmetic of the infinite. When he died, on 6 January 1918, his work had become universally accepted.

Theory of groups (19th and 20th centuries)

Since the work of the German philosopher and mathematician Gottfried Wilhelm von Leibnitz (1646–1716) it appeared indispensable to create notations and symbols to systematise logic. In his algebra, George Boole

Like the ancient Greeks, mathematicians in the Middle Ages relied on the ruler and compass to solve equations.

made the union and the intersection of groups correspond to addition and multiplication, on the one hand, and the 'or' and 'and' on the other. Thus, parallel to logical symbols, symbols and a theory of groups appeared.

The Italian mathematician and logician Giuseppe Peano (1858–1932) introduced in 1895 the symbol ϵ which signifies 'belonging to', $\cup$ for the combination of two sets and $\cap$ for their intersection. The German Schröder, in 1877, introduced **C** for inclusion.

With the work of Cantor on real groups, paradoxes appeared which led Peano to define the cardinal number of a group. The French mathematician Emile Borel (1871–1956) introduced the notion of innumerable groupings which opened up 20th century research to topology and the theory of measurement.

Analysis

Logarithm (1614)

Archimedes (3rd century BC), in his *Study of the grains of sand*, calculated the number of grains of sand necessary to fill the universe, and was close to becoming the inventor of logarithms. The Frenchman Nicolas Chuquet (1445–1500) invented arithmetical and geometrical progressions, as well as negative exponents, but it was the Scotsman **John Napier** (1550–1617) who invented logarithms in **1614** while doing research on a new method of numerical calculation. His system allows the replacement of multiplications by additions, and divisions by subtractions, using the smallest numbers. However, he did not find his results satisfactory and so developed, together with his friend, the Englishman Henry Briggs (1561–1631), the decimal or common logarithm.

Functions (17th century)

Gottfried Wilhelm von Leibnitz had the idea of treating problems by analogy; in effect, he was interested in the similarities between various problems. In particular, he noted in his correspondence with the Swiss mathematician Jean Bernoulli (1667–1748) that some variables, such as time and distance, might be linked and expressed one as a function of the other. He thus used functions and expressed them using the form ψx. This discovery was made while researching new methods of calculation which developed during the next century under the name of infinitesimal calculation.

Infinitesimal calculus (18th century)

Jean Bernoulli, professor of mathematics in Basel, Switzerland, explained and made known the calculation methods of Leibnitz and introduced them to France around 1691. He was, in particular, the professor of Leonhard Euler (1707–83), who ordered and developed the work of his predecessors. The Swiss Euler provided the first general theory of variation calculus, clarified the notion of function, and reassembled all these results in his *Institutiones calculi differentialis* (1755) and *Institutiones calculi integralis* (1768–70).

Independently of the work of Leibnitz and of Euler, the English physicist and mathematician Sir Isaac Newton (1642–1727) formulated a theory of calculus which deals with exactly the same problems. Therefore, at practically the same time, infinitesimal calculus made its appearance in various scientific communities. In the 19th century, integral calculus made considerable progress thanks to the work of the German Georg Friedrich Bernhard Riemann (1826–66).

Topology (19th–20th centuries)

Topology is that part of mathematics which studies the ideas, *a priori* intuitive, of continuity and limit. Until the beginning of the 19th century, mathematicians had used these ideas without defining them correctly. The German mathematician David Hilbert (1862–1943) sought to make them axiomatic and introduced 'neighbourhoods'. The Frenchman Maurice Fréchet and the Hungarian Frederick Riesz, at the beginning of the 20th century, defined respectively the notions of 'metric theory' and of 'topology'. Finally, around 1940, the definition of 'filters' by the French mathematician Henri Cartan rounded off the history of the idea of limit. Jules Henri Poincaré (1854–1912) is considered to be the inventor of algebraic and differential topology.

Physics

Standards

Weights and measures

The first units of measure were most often derived from the human body: the thumb (inch), the forearm (cubit), the distance between the tip of the king's nose and the tip of his middle finger in medieval England (yard). Or they had some relation to physical activity, such as the league, which equalled an hour's walking. The Celts used the capacity of two cupped hands united as a measure, while the ancient Egyptians derived their measure for liquids from the mouthful or draught.

For centuries units of measure spread throughout Europe in haphazard fashion. However, developments in the areas of science and technology created a need for a single, coherent system.

In France, weights and measures became standardised with the institution of the decimal metric system. In the English-speaking countries, the *Système International d'Unités* was not adopted until the second half of the 20th century: Britain's conversion to the metric system was only completed in 1980.

Metric system (1795)

The principle of mandatory units for weights and measures was established in France, during the Revolution. A decree issued on **7 April 1795** instituted the metric system, established the names of the units and, for the first time, legally defined the metre as a fraction of the distance between the North Pole and the Equator, measured on the meridian from Barcelona to Dunkirk by the Frenchmen Jean Delambre and Pierre Mechain in 1791. The unit of weight became the kilogram.

The *Système International d'Unités* (SI) (1960)

This system of units defined, in **1960**, the seven base units from which the other units derive. They are:

- length (the metre)
- mass (the kilogram)
- time (the second)
- electric current (the ampere)
- thermodynamic temperature (the Kelvin, which equals the degree Celsius, but the Kelvin scale begins at absolute zero and not 0°C; 0°C = 273.16 K)
- amount of substance (the mole)
- luminous intensity (the candela).

The SI metre is no longer defined as the distance between two marks on a platinum-iridium bar, kept at the International Bureau of Weights and Measures at Sevrès in France, but as the light-metre (as from 1986), which is to say the 299 792 458th part of the distance travelled by light in a vacuum in one second.

Planck's constant (1898)

The German physicist **Max Planck** (1858–1947), after discovering the quantum of interaction *h* (*see* Quantum theory), noted in **1898** that *h* could be used to establish an absolute scale of units.

These units have no proportional relation to the ordinary physical world. If you wanted to test this scale physically, with the aid of a particle accelerator constructed by present day techniques, you would need an accelerator the size of our galaxy, that is about 100 000 light-years in diameter.

It is thought that the Planck constant is the limit on this side of which quantum effects appear in the gravitational field. It is not understood at present, however, how these effects could intervene.

Foundations of physics

Cosmological model (antiquity)

In the **5th century BC**, the Pythagoreans (i.e. the disciples of the renowned Greek scholar and philosopher Pythagoras) together with the Greek Eudoxus of Cnidus, imagined a system of concentric spheres, whose rotational axes, variously inclined, passed through a common centre: the earth. This cosmological system was systematised by the last astronomer of antiquity, Ptolemy (AD c.80–c.160) of Alexandria.

The Copernican system (16th century)

The Pole **Nicolaus Copernicus** (1473–1543), doctor of law, canon and passionate astronomer, seems to have developed in the early **16th century** the cosmogony for which he would become famous: that the earth itself rotates and, like the other planets, it rotates around the sun.

Copernicus knew the Church, and he would no doubt have imagined the general outcry from the theologians when his theory ruined their certainty that the earth, and thus man, the 'image of God', was the centre of the universe. So Copernicus did not hasten to publish his book, *De revolutionibus orbium cœlestium libri VI*, which he put into the hands of his friend Georg Rhaethicus. The work appeared a few days before Copernicus died, on 24 May 1543.

A little-known but very interesting fact to remember: 18 centuries before Copernicus, the Greek Aristarchus of Samos (c.310–c.230 BC) had been the first to conceive of a heliocentric universe. Unlike his contemporaries, he thought that the earth and the other planets went round the sun, and not vice versa. Moreover, he had noted the rotation of the earth itself.

Copernicus demonstrated that the earth turned on herself and around the sun. The Polish astronomer believed that the earth was at the centre of the universe.

Laws of dynamics (17th century)

In his *Discourses and Demonstrations Concerning Two New Sciences*, published in 1638, **Galileo** stated the principle of inertia according to which a body not subject to the action of external forces has a rectilinear and uniform movement. And in the *Principia (Philosophiae naturalis principia mathematica)*, published in 1687, **Sir Isaac Newton** (1642–1727) stated the fundamental principle of dynamics, according to which a body subject to an external force gathers acceleration in proportion to that force.

Quantum theory (1900)

At the end of the last century, no law had been discovered to account for the phenomenon of heat and light radiation by a solid, white-hot body. In **1900** the German physicist **Max Planck** guessed that radiation did not occur in a continuous fashion but in small discrete units, separate quantities or quanta. This discovery, which enabled scientists to explain heat radiation, turned physics upside down, especially in the sphere of classical mechanics which became inoperable in the area of infinitely small quantities.

Thanks to this theory, Albert Einstein explained in 1905 the photo-electric effect by

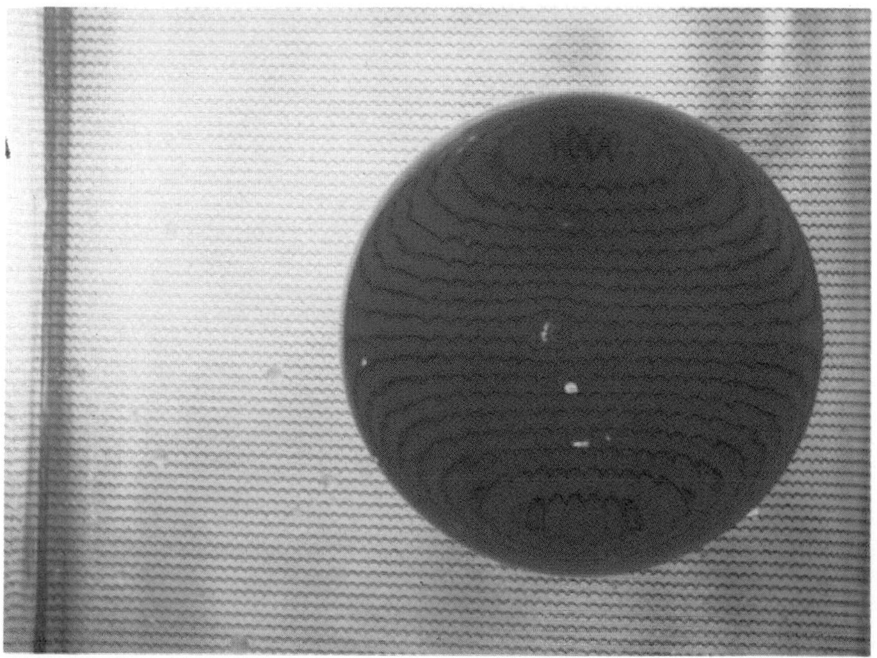

A drop of oil floats in a mixture of alcohol and water of the same density: under the action of surface tension, the drop takes on a perfectly spherical form.

showing that light, which is both wave and particle, moves by quanta, tiny packets of light, which were later called photons.

The Danish physicist Niels Bohr built on this quantum theory a model of an atom, describing in 1911 the movement of electrons inside the atom. This model enabled him to achieve remarkable results in the fields of the spectroscopy of gaseous matter and of X-ray physics.

Wave mechanics (1924)

The Frenchman **Louis de Broglie** produced the wave theory of matter in **1924**, which also derived from quantum theory.

Statistical determinism (1927)

In **1927** the German **Werner Heisenberg** stated that the absolute principle of determinism in classical mechanics (the same causes produce the same effects) was no longer true of wave mechanics. He introduced the idea of statistical determinism, which allowed the calculation of probabilities only.

Today quantum mechanics has become the basic tool of modern physics, but it has not yet yielded up all its secrets.

Fundamental Forces

The elementary particles of matter may act among themselves in various ways, because of forces or what we call 'interactions'. Four fundamental forces are known at present, of which two were known to antiquity – gravitational and electromagnetic forces – and two of which are the fruit of 20th century research: weak and strong interactions.

The process of interaction, according to modern theories, is like the superimposition of elementary interactions occurring when a particle is exchanged, characterised by the force called an 'intermediary boson'. The particles interact a bit like rugby players, passing the ball from one to another. In the case of electromagnetic force, the particle that is exchanged (the boson) is the famous light-particle invented by Einstein in 1905, the photon.

In 1986, the physicist Ephraïm Fischbach and his colleagues at the University of Washington suggested the possibility of a fifth force, but its existence has not been confirmed since.

Electromagnetic force (1820–64)

It is this force which links atoms and molecules to form ordinary solid bodies. Thus if your elbow doesn't sink into the wood of your desk while you are writing, it is because the electrons in the atoms of your desk and of your elbow push against each other by means of electromagnetic interaction.

The relationship between electricity and magnetism was discovered in 1820 by the Danish scientist Christian Oersted in an experiment during which he noticed that a magnetic needle was deflected by an electric current. The French physicist André Marie Ampère later generalised these observations, but it was the Scotsman James Clerk Maxwell who, in 1864, formulated the general laws of electromagnetism and showed that light was nothing but an electromagnetic wave.

Since the 1930s, a number of physicists such as the Englishman P. A. M. Dirac and the Americans Richard Feynman and Julian Schwinger have developed the modern theory of electromagnetic interaction between electrons (with the exchange of photons).

WHAT IS RELATIVITY?

The principle of relativity goes back to Galileo (17th century) and is stated as follows: no experiment can measure speed in relation to a fixed space, from the laboratory in which the experiment takes place.

Albert Einstein, guided by the experiments of the Americans Michelson and Morley (1887), issued a hypothesis in 1905 concerning the constancy of the speed of light: the velocity of light is independent of the motion of the observer who measures it.'

Allied to the principle of relativity, this hypothesis in Einstein's hands had revolutionary consequences. For example, Einstein showed that two observers moving relative to one another would assign different times to the occurrence of an event. This surprising effect together with a number of others formed Einstein's theory of relativity in 1905. All other theories in physics are today based on this theory and, especially, on the general theory of relativity (1915) which concerns gravitational field.

Force of gravity (1687/1915)

Responsible for the movements of great masses on a large scale (the rotation of the earth around the sun, the movement of the galaxies, the expansion of the universe etc.), the force of gravity, whose laws were first stated by **Sir Isaac Newton** in **1687**, is the weakest of known forces. Today it has a very special status because it has been interpreted since **Einstein** (**1915**) as a manifestation of the curvature of space-time. With regard to gravitational force, the boson which plays a role analogous to the photon is called the *graviton*.

Weak force (1934/1974)

It has been known since **Enrico Fermi**'s work in 1934 that this force is manifested during certain radioactive processes, such as the spontaneous disintegration of the neutron. In 1974, the physicists **Sheldon Glashow** (US), **Abdus Salam** (Pakistan) and **Steven Weinberg** (US) decided that the existence of three exchange particles, the intermediary bosons W+, W– and Z0, played a role analogous to that of the photon and the graviton for weak force. (For this work they won the Nobel Prize for Physics in 1979.) These three particles were observed at the European Centre for Nuclear Research (CERN) in Geneva by the physicists Carlo Rubbia (Italy) and Simon van der Meer (Holland), who won the Nobel Prize for Physics in 1984. Their discovery was certainly one of the most important in particle physics in the second half of the 20th century.

Strong force (1935/1965)

Strong interaction, discovered by the Japanese **Hideki Yukawa** in **1935**, is principally responsible for the cohesion of atomic nuclei: it is this which maintains the links at the centre of the atom between the protons and neutrons which constitute the nucleus. More precisely, the protons and neutrons are made up of quarks and the strong force is what binds the quarks to each other. A difficult theory known as quantum chromodynamics has been elaborated by various physicists since **Nambu** (1965), and suggests that quarks interact among themselves by the exchange of eight bosons which are called gluons, massless particles which transmit the forces that bind quarks together.

Theory of super-strings (1984)

'Super-strings' in their present form were invented by the Englishman **M. Green** and the American **J. Schwartz**. Previous work was done by the Frenchmen **A. Neveu** and **J. Scherk**. Ordinarily in physics, one thinks of elementary particles as points, without dimensions. The new idea is to replace the concept of the particle, an object with zero dimension, with the concept of a string, which has one dimension. Then one could think of interpreting particles and their associated waves as excited states of a vibrating string, and thus arrive at a classification of particles and a unification of the four fundamental forces (see above).

For example, it has been shown that the lowest state of vibration of a string may be identified with the graviton, the hypothetical particle of the gravitational field. A notable characteristic of super-strings is that they necessarily evolve in a space-time having more dimensions than the four usually allowed to space-time.

Since the summer of 1984, when Green and Schwartz proved a very important result for the theory of super-strings, dozens of physicists have been working on the subject, and thousands of articles have been published. The theory is being developed at this very moment.

Hydrodynamics

The Principles of Archimedes

Archimedes (c.287–212 BC) was the first to formulate the principle of floating bodies which bears his name: all bodies weighed when immersed in fluid (liquid or gas) show a loss of weight equal to the weight of fluid they displace. After having discovered this principle, as we know, Archimedes cried Eureka! ('I've found it!'). Less well-known are the circumstances of this discovery: the ruler of Syracuse, Hiero II, a naturally suspicious man, had given some pure gold to a jeweller so that he could melt it down and make a royal crown. Archimedes was put in control of this work. He then had the idea of immersing in a receptacle filled to the brim with water first the crown, then gold and silver equal in weight to that of the crown. After each experiment, so the story goes, he weighed the water that had overflowed. He finally showed that the figure of the first weighing had fallen between the figures of the next two: it weighed less than pure gold and more than silver. Thus it was proved that the crown had been made of a mixture of gold and silver.

Hydrostatic paradox (1586)

The Flemish physicist and mathematician **Simon Stevin** (1548–1620) known as Simon of Bruges, inspector of dykes for the States of Holland and in that capacity directly interested in the internal forces of liquids, was the first to make a truly scientific study of these forces.

In **1586**, Stevin's three books on mechanics were published. These contained his famous hydrostatic paradox: the pressure of a liquid at the bottom of a receptacle depends solely on the depth of the liquid and not on the shape of the receptacle. Conversely, the weight of the liquid depends on the shape of the receptacle.

Fundamental principle

Taking the fundamental relationship as a starting point, **Blaise Pascal** (1623–62) also deduced his fundamental principle: pressure applied to any one point of an incompressible fluid at rest is transmitted without loss to all other parts of the fluid.

Characteristic numbers

Reynolds number

The English physicist and engineer **Osborne Reynolds** (1842–1912) carried out research in the field of hydrodynamics, studying the flow of viscous fluids. Reynolds number is a dimensionless coefficient expressing the relationship between the inert forces and the viscous forces.

Froude number

The English physicist and engineer **William Froude** (1810–79) was the first to study by experimental means the resistance of a fluid to motion. To carry out his experiments, he devised the first model tank.

Mach number

The Austrian physicist and philosopher **Ernst Mach** (1838–1916) was the first person to recognise the role of velocity in aerodynamic flows. The Mach number (M) is the ratio of the inertial forces to the square root of the forces of pressure. If M is greater than 1, the plane is supersonic (or greater than the speed of sound in air which is 240 metres per second *760.6mph*). If M is less than 1, the plane is subsonic.

The centre of gravity (3rd century BC)

The greatest scientist of antiquity, the Greek **Archimedes** (born in Syracuse in 287 BC), first determined the centres of gravity of homogeneous solids defined geometrically, such as the cylinder, the sphere and the cone. Archimedes developed this idea in his *Book of Balances*. In this book he also displayed his rigorous law of levers.

Falling bodies

The Italian physicist and astronomer **Galileo** (1564–1642) demonstrated by experiment that falling bodies accelerate at a rate which is independent of their nature and composition. According to legend, he threw a wooden ball and a lead ball from the top of the tower of Pisa, and noted that they reached the ground at the same time (disregarding a slight resistance in the air). In 1686, **Isaac Newton** came to the same conclusions as the result of experiments with oscillating pendulums. And in 1922, the Hungarian baron and physicist **Lorand Eötvös** showed, with the aid of torsion pendulums that the 'universality' of falling bodies was correct to within a thousand millionth or more.

None of these experiments were satisfactorily explained until 1915, when **Einstein** discovered the general theory of relativity.

The laws of universal attraction (1685)

One evening in 1665, **Isaac Newton** was meditating in his garden. The moon was full. An apple fell at his feet, and Newton wondered why the moon, too, did not fall on the earth. This question led the man of science to state in his *Philosophiae naturalis principia mathematica* (**1685**) the laws of universal attraction. However, the discovery of the laws which govern the universe cannot be reduced to this one anecdote, related by Voltaire, and anyway not verified.

Before Newton, the German astronomer Johann Kepler, starting out from the notebook of the Danish astronomer Tycho Brahe, had established between 1601 and 1618 the three laws which bear his name and which enable us to calculate a planet's orbit around the sun and its rotation period. At the same time, Galileo issued many observations on the trajectory of projectiles and for the first time, in 1609, had turned a telescope on the sky. Before these men, Copernicus and then Giordano Bruno, condemned to death and burned at the stake by the Inquisition in 1600, had fallen foul of official theory which made the earth the centre of the universe.

It thus fell to Newton to unify the astronomical knowledge of his time in the famous law of universal attraction.

THE MYSTERY OF THE EARTH'S DIAMETER

The great Greek astronomer and historian Eratosthenes (3rd century BC) was intrigued by a manuscript in the library of Alexandria, of which he was curator. The manuscript maintained that in the town of Syene (now Aswan), situated 800km *500 miles* south of Alexandria, the sun did not cast any shadow from a obelisk at midday on the summer solstice, 21 June.

Eratosthenes decided to find out whether the same phenomenon also occurred at Alexandria, on the same day at the same hour. To his great surprise, he found that it was not the case; on the contrary, the obelisk in Alexandria did cast a shadow, whose length he measured. How was this possible?

The astronomer reasoned as follows: the parallel rays of the sun must necessarily cast identical shadows from the two obelisks, equally identical, on condition that the latter were parallel. If the shadows were not identical, it was because the obelisks were not parallel, and thus the verticals themselves, at Syene and Alexandria, were not parallel either. The only possible conclusion was that the earth's surface was curved. Having measured the length of the shadow cast by the obelisk in Alexandria at midday on 21 June, and knowing the distance between Alexandria and Syene, Eratosthenes could calculate that the earth's diameter must be about 13000km *8000 miles*. This was the first correct estimate of the earth's diameter.

The general theory of relativity (1915)

In **1915** the German-born physicist **Albert Einstein** (1879–1955) put the finishing touch to an extraordinary theory that was very difficult for its time. This theory showed how the gravitational law of Newton could be made compatible with the laws of relativity.

The theory of relativity interpreted the field of gravity not as the usual field of forces, but as a manifestation of the space-time curve. For example, the sun curves space-time around itself, and thus the earth is obliged to circle the sun along a curved line, which is roughly elliptic.

Today the general theory of relativity is largely accepted. It has been so thoroughly verified by experiments and observation that physicists think that even the most mysterious objects evolving from this theory, such as black holes, must really exist in nature.

Gravitational waves (2000?)

In the general theory of relativity, the gravitational interaction between two bodies such as the earth and the sun moves in waves, at a speed equal to the speed of light. These are thus called gravitational waves. There are a number of experiments being made throughout the world in order to observe these waves. They are emitted by objects far off in the universe, such as exploding or colliding stars. No further observation can be made yet, because gravitational waves are of an extremely weak intensity. Astrophysicists, however, in observing the object called the binary pulsar PSR 1913+16 have proved that gravitational waves really exist. The proof of their existence constitutes the most important contemporary test of the theory of relativity.

Matter

Gases (17th century)

The Flemish doctor and chemist **Jean Baptiste van Helmont** (1577–1644) was the first to recognise the existence of different gases, such as, carbon dioxide and oxygen, which were identified as such much later. Until the 17th century, the knowledge of these states of matter were purely empirical.

The Greeks designated the immense dark space which existed before things began as 'chaos'. Van Helmont invented the word 'gas' from the sound of their word.

Oxygen and nitrogen (18th century)

It was in **1777**, in a paper that was not published until 1782, that **Antoine-Laurent de**

In 1988 oxygen was served as a tonic in the bar of the Takashimaya department store in Tokyo.

Lavoisier (1743–94), the founder of modern chemistry – in the wake of the English chemist Joseph Priestley and the Swede Carl Wilhelm Scheele – named life-giving air oxygen (*oxygène*, literally: that which produces an acid), and non-vital air nitrogen (*azote*, literally: that which does not maintain life). In 1772, the Scottish physician and botanist Daniel Rutherford (1749–1819) had distinguished between 'noxious air' (nitrogen) and carbon dioxide in his doctoral thesis, *De aere mephitico*.

Radon (1900)

In **1900** the German **Ernst Dorn** discovered the last inert gas, radon, in the radioactive disintegration products of radium. Radon is a dangerous gas. In the United States it was discovered in 1986 that it could contaminate houses in certain regions. At present, 12 per cent of American houses contain enough radon to make their inhabitants run the same risk of lung cancer as if they had smoked half a pack of cigarettes every day of their lives.

Water (1781)

The Swiss doctor Paracelsus (1493–1541) was the first person to draw attention to the existence of hydrogen.

In **1781**, the English chemist **Henry Cavendish** (1731–1810) had the idea of burning oxygen and hydrogen together. He measured the quantities of the two gases and observed that they were converted into a quantity of water whose weight was the same as the sum of the weights of the two gases.

Lavoisier repeated and completed Cavendish's experiments. He had the idea of vaporising the water and separating the vapour into its two constituents, which he then combined to form water again. This series of experiments led him to state his famous law of the conservation of matter in a chemical reaction which states that the sum total of matter in the universe cannot be changed.

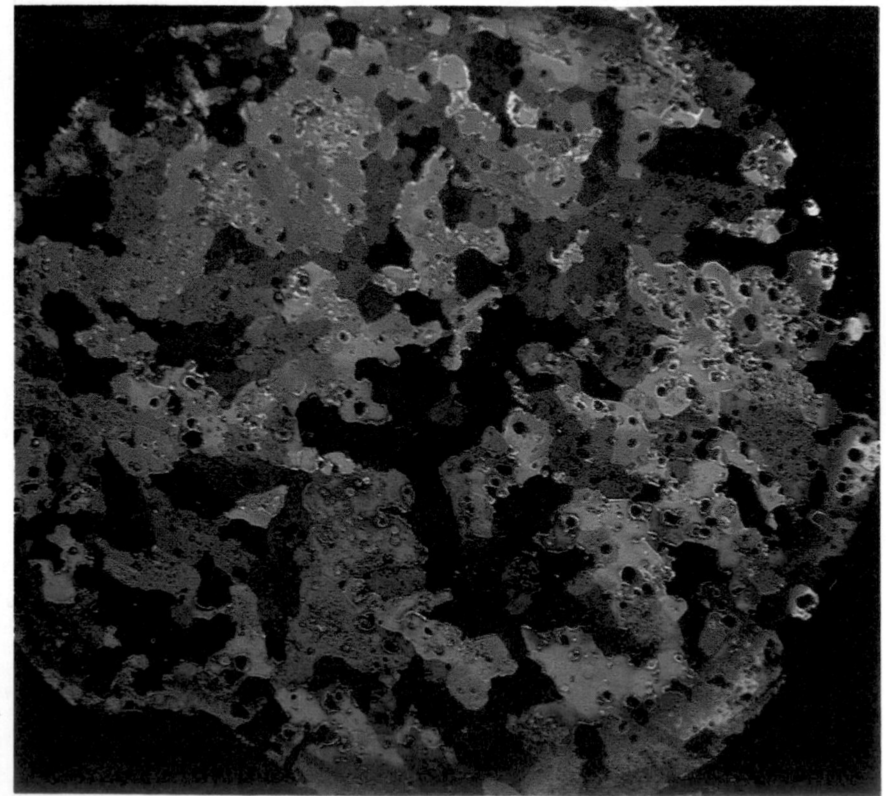
At 0°C water freezes and changes its state to become solid. Above is an ice crystal in close up.

Changes of state

Atmospheric pressure (1643)

The Italian physicist and mathematician, pupil of Galileo, **Evangelista Torricelli** (1608–47) was the first to measure air pressure, in **1643**. To prove the existence of atmospheric pressure, he used mercury, which is 13 times denser than water, so that he could work on more easily measurable heights. He filled a crucible and a glass tube with mercury, and inverted the open end of the tube over the crucible. The mercury level went down in the tube and stabilised at 75cm *30in* from the open surface of the crucible. Torricelli deduced from this that air exercised pressure on this surface, balanced by the hydrostatic pressure exercised by the 76cm *30in* of mercury in the tube. Torrecelli's tube was the first barometer. The term barometer was invented by the Frenchman Edme Marlotte.

Vacuum (1654)

The German engineer **Otto von Guericke** (1602–86), a native of Magdeburg, developed the first vacuum pump. In 1654 von Guericke presented it to the nobility of the Holy Empire.

He joined together two hollow bronze hemispheres, each 50cm *19.7in* in diameter. Creating a vacuum in the sphere, he harnessed two horses to each hemisphere. The horses pulled: nothing happened. He had two more horses added: still nothing. It took eight horses harnessed to each hemisphere to pull them apart.

A man of science, von Guericke was also a showman. During his experiments, he demonstrated that when burning candles were snuffed out in a vacuum, animals quickly expired, and bells no longer rang. These 'miracles' made his fellow countrymen take him for a magician.

Expansion of gases (1661)

In **1661**, the Irish self-educated scientist **Robert Boyle** (1627–91) demonstrated that the variation in the volume of a gas is inversely proportional to the variation in its pressure. He was interested in the experiments of van Helmont and Pascal and oriented his research to the compressibility of gases. Using a simple graduated J tube and mercury, Boyle proved that the volume of the air imprisoned in the tube can be reduced by half by doubling the pressure exerted by the mercury.

Wright of Derby's painting shows a demonstration of the effects of vacuum on a bird: without oxygen, it dies.

Charles' Law (1798)

In **1798** **Jacques Charles** (1746–1823), a French physicist and ballooning enthusiast (he was the first person to think of filling balloons with hydrogen) pronounced the following law: if a gas is held at a constant pressure, its volume is directly proportional to its absolute temperature.

Gay-Lussac's Law (1804)

In **1804**, the French physicist and chemist **Joseph Louis Gay-Lussac** (1778–1850) discovered that the volume of gas at a given temperature *t* is related to the volume at 0°C/32°F (if the pressure is constant) in the same way as the relationship discovered by Jacques Charles.

Ideal gas (19th century)

The French physicist **Emile Clapeyron** (1799–1864) was the first to use the notion of ideal or perfect gas. Ideal gas is a purely theoretical perfect fluid, which represents the limit of a real gas state in which the temperature tends towards absolute zero, and where the pressure becomes very high.

Liquefaction of gases

Origins (1818)

In **1818**, the Englishman **Michael Faraday** (1791–1867) discovered a means of liquefying gas (i.e. transforming it from the gaseous state to the liquid state) by increasing the pressure when cooling the gas. The son of a blacksmith, Faraday started out as an errand boy in a bookshop and later became a bookbinder. An avid reader, he taught himself physics. To perfect his scientific knowledge, he enrolled in an evening course taught by Sir Humphrey Davy and became his assistant at the Royal Institution.

In 1818, Faraday oriented his research in a direction that was entirely new for the day: studying the effects of pressure and cold on gases. Faraday successfully liquefied hydrogen sulphide and sulphuric anhydride but did not succeed with oxygen, hydrogen or nitrogen.

Liquid oxygen (1877)

In **1877**, the Frenchman **Louis-Paul Cailletet** (1832–1913), ironmaster in Burgundy, invented a pump which enabled him to obtain and maintain atmospheric pressures in the hundreds.

He managed to liquefy oxygen by causing the sudden expansion of gas in a capilliary tube in which he decreased the pressure from 300 to 1 atmosphere which dropped the temperature to −118.9°C *−182°F*.

A few days after the success of this experiment, **Raoul-Pierre Pictet** (1846–1929), professor of physics at the university of Geneva, published the results of similar research.

The Irish chemist Thomas Andrews (1813–85) also liquefied gases independently, and discovered the 'critical temperature' which Cailletat and Pictet used in their further work on liquefaction.

Liquid air (1899)

In **1895**, the inventor and industrialist **Karl von Linde** (1842–1934) succeeded in liquefying air by compression and expansion with intermediary cooling. He was thus able to prepare almost pure liquid oxygen.

In 1902 the French scientist Georges Claude (1870–1960) invented another process for liquefying air, by expanding the gas and applying an outside force. From the liquid air, he isolated oxygen, nitrogen and argon in liquid form by fractional distillation. He thus put in train the first industrial process for liquefying gases.

Liquid hydrogen (1899)

In **1899**, the Englishman **Sir James Dewar** (1842–1923) used Lind's air liquefaction process to obtain boiling liquid hydrogen.

Liquid helium (1908)

In **1908**, the Dutch physicist **Heike Kamerlingh Onnes** (1853–1926) liquefied helium in his cryogenic laboratory in Leyden. Helium was the last gas to be liquefied. Research did not end there, however. In 1971, in liquefying helium 3 (an isotope of helium 4, the gas's normal form) at less than 2.7mK, it was discovered that this also had the property of superfluidity (already discovered in helium 4); it has almost no viscosity.

Note that Onnes also discovered supraconductivity, now called superconductivity, in 1911: that is, the property of certain metals or alloys, at temperatures close to absolute zero, to lose their resistance to electrical currents.

Thermodynamics

Origins

In **1849** the Glasgow engineer **Sir William Thomson** (the future Lord Kelvin, 1824–1907) invented the term thermodynamics to describe the study of relations between thermal and mechanical phenomena, but the discipline as such may be said to have been founded by the Frenchman Sadi Carnot (1796–1832). Thermodynamics is the mathematical formulation of the parameters defining a system in the exchange of energy. These parameters are temperature, pressure and volume.

Carnot principle (1824)

After studying the steam engine invented in 1703 by the Englishman Thomas Newcomen (1663–1729), **Sadi Carnot** (1796–1832) stated in his only published work, *Reflections on the Motive Power of Fire* (**1824**), that mechanical energy could be produced by simple transfer of heat.

Work–heat equivalence (1842)

The German physicist and doctor **Julius Robert van Mayer** (1814–78), following Carnot, was interested in gases and the motor power of heat. In **1842** he published the results of his research in *Annalen der Chemie und Pharmaciae*. He stated intuitively the theory that was later to be called the principle of work–heat equivalence.

This principle was confirmed by the numerous and exact measurements made by the Englishman **James Prescott Joule** (1818–89), working at the same time (between 1840 and 1845), turning his famous current meter in an isolated enclosure, filled with a liquid whose temperature he measured as it rose. Thus he was able to evaluate the quantity of heat set free by clearly defined mechanical work.

Principles of thermodynamics (1852/1906)

In **1852 Sir William Thomson** stated the first two laws of thermodynamics based on the work of Carnot, von Mayer and Joule. The third principle was formulated in **1906** by the German physicist **Walter Hermann Nernst** (1864–1941).

Temperature

In **1848**, **Sir William Thomson** (Lord Kelvin) stated the zero principle of dynamics. This principle enabled him to define thermodynamic temperature and to establish an objective method of measuring it.

When two systems are each in thermal equilibrium with a third, they are in thermal equilibrium with each other. This equilibrium is expressed by their equal temperatures. If you give a conventional value to the temperature of a system in a given physical state, the other temperatures can be determined by what are called thermodynamic measures.

In 1961, the General Conference on Weights and Measures chose as the standard unit of thermodynamic temperature the kelvin (K), defined as the degree on the thermodynamic scale of absolute temperatures at which the triple point of water is 273.16K (the equivalent of 0°C). At this temperature, ice, water and steam can co-exist in equilibrium.

According to this convention the freezing and boiling points of water under atmospheric pressure are respectively 273.15K and 373.15K. The temperature interval measured by one kelvin is equal to that which measures 1°C.

Electricity and magnetism

Electrostatics

Origins

The Greeks had already observed the phenomenon of electricity, as Thales of Miletus (c.624–c.546 BC) relates in his writing around 600 BC, describing the **electrostatic** power of a fossil resin found on the Baltic seashore: amber, which the Greeks called *elektron.* Other phenomena had been recorded, such as the flash or electrotherapical response of a torpedo fish under a drop of water, but no links had been established between them.

First electrical machine (17th century)

The first electrical generator was invented by the German physicist **Otto von Guericke** in the second half of the 17th century. It consisted of a revolving ball of sulphur to which friction was applied by a person's hands. Thanks to this machine, Guericke discovered in 1672 that the static electricity could cause the surface of the sulphur ball to glow, that is, he understood electroluminescence.

In 1708 the English doctor William Wall made the connection between electroluminescence and lightning.

Conductors (18th century)

Between 1727 and 1729, the English physicist **Stephen Gray** (1670–1736) discovered conductivity and carried out the first experiment of transporting electricity over a distance, using silk threads attached to a glass tube, which was rubbed, and an ivory ball.

In **1740** the French inventor **Jean Desaguliers** proposed that solids through which electricity freely circulates, such as iron and copper, should be called conductors, and those through which it does not circulate, such as glass or amber, should be called insulators.

WHAT IS SUPERCONDUCTIVITY?

First discovered in mercury in 1911 by the Dutch scientist Heike Kamerlinghe Onnes, who succeeded in liquefying helium at 4.2° kelvin, that is −268.96°C *−425.128°F,* superconductivity is the property of certain metals or alloys, at very low temperatures, to lose their resistance to electricity. For a long time it was thought that this temperature would be in the range 0°–20°K.

In a metal at normal temperature, the free electrons, responsible for carrying the current, encounter a certain resistance to their displacement. In a state of superconductivity, the electrons associate in pairs and behave uniformly throughout the material. Therefore the group of electrons move about without friction, without resistance and thus without loss of energy. A current established in a superconductor in principle could be maintained indefinitely, even if the external source of power is cut off, as long as the temperature is maintained below the transition temperature, i.e. the temperature below which a metal becomes a superconductor.

Leyden jar (1745)

At this period electrical friction machines were perfected and became more powerful. The possibility was even discovered of accumulating strong electrical charges on a conductor isolated by glass or air. In **1745** the German **Ewald Jürgen von Kleist** made the first condenser, a few months before the invention of the famous Leyden jar, built by the Dutchman **Petrus van Musschenbroek**. The jar became a curiosity and attraction in the courts of Europe, the great and the wealthy across the continent coming to see how electric charges were produced. It is a glass jar coated inside and outside with tinfoil or other conducting material.

WHO COINED THE WORD ELECTRICITY?

It was towards the end of the 16th century that William Gilbert (1544–1603), a doctor in the reign of Elizabeth I, noted that glass, and a great number of other bodies, when rubbed had the same properties as amber. He called this force of attraction 'electricity'.

Lightning-conductor (1752)

The ground was prepared for **Benjamin Franklin**'s (1706–90) famous experiment of **1752**, when he flew a kite with a metal tip in a thunderstorm. The tip was joined to a piece of insulating silk holding the wet string, and thereby to an iron key which, dangling free in the air near the technician, filled the role of the condenser. Thanks to this contrivance, a dangerous one, Franklin could charge up a Leyden jar and prove that thunderclouds contained electricity.

Franklin made a fortune by inventing and selling lightning-conductors, and also became the most important politician in the United States. He is considered to be the first scientist of the New World.

Coulomb Laws (1785)

Between 1784 and 1789 the French engineer and physicist **Charles de Coulomb** invented his famous torsion balance, which enabled him to establish the fundamental laws of electrostatics (**1785**).

The volt battery (1799)

In 1791, the Italian physiologist Luigi Galvani (1737–98) noted by chance that the muscles of a dissected frog contracted when they were touched at the same time by two different metals. He thought that the muscles contained a fluid which he called animal electricity.

Another Italian, **Alessandro Volta** (1745–1827) repeated Galvani's experiment and his work led him to invent in **1799** the first chemical battery, made up of alternating layers of silver, of blotting paper impregnated with sulphuric acid, and of zinc. The chemical generation of electricity was born. This invention led to Volta's being made a count by Napoleon Bonaparte in 1801.

In 1831 the English physicist Michael Faraday managed to produce an electric current by induction, rotating a metal disc across magnetic lines of force. The first generators worked by steam engines soon made their appearance. These large and not very powerful machines provided an alternating current.

Electrodynamics

Ohm's Law (1827)

In **1827**, the German physicist **Georg Simon Ohm** (1789–1854) used a hydraulic analogy to formulate a precise definition of the quantity of electricity, the electromotive force, and the intensity of a current, thereby formulating the law that bears his name. Ohm likened electric current to a liquid flow and the electric potential created by an electromotive force to a difference in level.

Joule Effect (1841)

When a current passes through a homogeneous conductor, the conductor heats up. After 1882 this effect carried the name of the English physicist **James Prescott Joule** (1818–89), who formulated the law in **1841**.

The heat engendered by the passage of the current turns the filaments of incandescent lamps red. Elements are still used to produce heat in cookers, electric radiators, etc. The principle of the fuse is also based on this effect: a metal wire or low fusion temperature is inserted into an electric circuit; if the intensity increases to an abnormal degree, the wire melts and the circuit is broken.

Electric discharge in gases

An electric discharge corresponds to the passage of charges, that is to the current in an environment which is a low conductor or insulator.

Paschen's Law

Two physicists, one German, **Friedrich Paschen** (1865–1947), and the other Irish, **Sir John Townsend** (1868–1957), showed that the breakdown potential of a gas between parallel plate electrodes is a function of the product of the gas pressure and the electrode separation. This potential is minimal for a determined value of the product in question. Paschen discovered this law empirically, and Townsend expressed it scientifically.

Neon tube (1909)

In **1909** the French scientist **Georges Claude** (1870–1960), invented the neon tube. This is a tube containing neon at low pressure, in which the gas becomes luminous when a certain voltage is applied to the electrodes. The neon light is one of the best-known applications of glow discharge.

Electromagnetism

Origins (1819)

In **1819** the Danish scientist **Christian Oersted** (1777–1851) proved that the passing of an electric current creates a magnetic field in the space surrounding the conductor. This phenomenon is induction: the current induces a magnetic field.

In 1820 the French scientist François Arago (1786–1853) made the first electromagnet. In 1820 also, the French mathematician André-Marie Ampère (1775–1836) developed his theory of the magnetic reactions of live electricity. He studied the reciprocal reactions of currents and magnets.

Induction (1831)

In **1831**, while studying the results obtained by Oersted and by Ampère, the Englishman **Michael Faraday** (1791–1867) discovered the principle of electromagnetic induction. Faraday's discovery had enormous technological ramifications. Among other things, it served as the basis for electricity-producing machines known as generators: the magneto, which produces a magnetic field by a permanent magnet; the dynamo, for producing direct current, and the alternator, for alternating current.

Foucault current

The self-educated French physicist **Jean Bernard Léon Foucault** (1819–68) was the first to demonstrate the existence of electric currents inducted by an alternating magnetic field in a massive conductor. Named in honour of their discoverer, these currents create a magnetic induction whose flow opposes that of the alternating magnetic field that produces them. The mechanical force resulting from the passage of these currents is therefore always a resisting force.

Foucault currents, also known as eddy currents, have many applications, including induction-heating and electromagnetic braking systems.

Electromagnetic waves (1864)

James Clerk Maxwell (1831–79), a pupil of Michael Faraday, was one of the most illustrious physicists of the 19th century, and was the first to suppose the existence of electromagnetic waves, in **1864**. Although Maxwell was not able to prove his theory experimentally, he hypothesised that light has an electromagnetic nature.

Hertzian waves (1887)

In **1887** the German physicist **Heinrich Rudolph Hertz** (1857–94) demonstrated the existence and properties of electromagnetic waves for the first time. These waves, named in his honour, are a remarkable verification of the theory formulated by James Clerk Maxwell.

Electronics

Semi-conductors (1929)

A semi-conductor is a solid with an electrical conductivity intermediate between those of insulators and those of metals. An insulator when at low temperatures, a semi-conductor becomes a conductor when subject to increased heat or light. The odd phenomenon of electrical conductivity had already been noted by Michael Faraday (1839, in his experiments with silver sulphide) and by Karl Ferdinand Braun (1874, on galenite), without having been explained.

The first coherent theory of conductivity in solids (covering the properties of insulators, conductors and semi-conductors) was produced by **Felix Bloch** (born in 1905), a Swiss physicist who became an American citizen in 1939. In **1929** he suggested a theory of bands. Between 1925 and 1935, Bloch and scientists from various countries perfected the theory of semi-conductors.

The understanding of the semi-conductor mechanism was the basis for the invention of the transistor, which was to revolutionise electronics.

DO MAGNETIC MONOPOLES EXIST?

What is a magnetic monopole? Since antiquity, having observed that the ends of magnets attract or repel each other, men have wanted to isolate the active substance responsible for the phenomenon of magnetism. The magnetic monopole is the hypothetical particle of which this active substance is composed. No one, however, has been able to divide a magnetic bar made up of North and South poles in order to separate the magnetic poles. Do magnetic monopoles really exist?

The English scientist P. A. M. Dirac, in 1931, produced a powerful theoretical argument in favour of their existence, and even calculated their magnetic charge. And in 1982, the American Blas Cabrera believed that he had observed a magnetic monopole coming from space which would then pass through the earth. It is not known to what extent his observation was accurate.

'Electronic gate' (1986)

The Bell laboratories at AT&T and Cornell University in the USA established in 1986 the speed record for semi-conductors: they created an electronic gate prototype, capable of interrupting an electric signal in 5.8 millionths of a millionth of a second (or *pico-second*). During this time a light ray could only travel 1.6mm *.063in*. Some parts of the circuit measure just a third of a micron.

Cathode rays (19th century)

Around 1850 the German **Heinrich Geissler** (1815–79), who had been a glass blower before becoming a manufacturer of laboratory equipment, constructed glass containers equipped with electrodes, in which he created a vacuum. He thus obtained very attractive lighting effects that varied with the shape of the container and the type of gas used. This phenomenon was further observed by two more Germans, Julius Plücker (1801–68) in 1854, and Johann Hittorf (1824–1914) in 1869.

In 1879 the Englishman William Crookes (1832–1912) carried out experiments proving once and for all that rays were emitted by the cathode of such glass tubes.

In 1895, the Frenchman Jean Perrin (1870–1942) demonstrated that cathode rays are charged with negative electricity and that they are deflected by electric or magnetic fields.

In 1897, the Englishman Sir Joseph Thomson (1846–1940) calculated the ratio between the charge and the mass of the particles emitted. It then became apparent that these particles were electrons. The existence of electrons was first hypothesised in 1874 by the Irishman George Stoney (1826–1911), who named them in 1891.

Cathode-ray oscilloscope (1897)

In **1897** the German physicist **Karl Ferdinand Braun** (1850–1918) perfected the cathode-ray oscilloscope. The trajectory of cathode rays is rectilinear, but in an electric field, they are deflected in proportion to the voltage applied. The cathode-ray oscilloscope can display the transient or repeated wave forms on a fluorescent surface.

The cathode-ray tube is the device that creates an image on television screens. In this case, a signal is induced in the television's aerial by a signal given by the transmitter of the broadcasting network.

NEWS ABOUT SUPERCONDUCTIVITY

After Kamerlingh Onnes's discovery, research stayed at the level of laboratory curiosity for a long time, until suddenly progress was made in a totally unexpected way, because of materials known for their insulating power and not as conductors: ceramics. The first work with these was carried out in the laboratory for crystallography and material sciences in the University of Caen in France.

But it was in 1986 that superconductivity really took off with the results obtained in IBM's Zurich laboratories on ceramic superconductors by two scientists, K. A. Müller (Swiss) and J. G. Bednorz (West German) which earned them the Nobel Prize for Physics in 1987. Their superconductivity record at −238°C *−396.4°F* launched an era of other records (notably that of the American researcher at the University of Houston, Paul Chu, who achieved −180°C *−292°F*), which in their turn open up astonishing perspectives on tomorrow's technology: trains running by magnetic levitation, the stockpiling of enormous amounts of energy, new kinds of computer, etc.

Three major American companies have set off on the race to patent application: Bell, with the researcher Robert Dynes; Du Pont, to whom Paul Chu is attached at present; and IBM. Some people think that discoveries about superconductors could be as important in the future as was the invention of the transistor or the laser in the past.

This experiment with a floating magnet is achieved by using a superconductive lozenge placed in liquid nitrogen that keeps it at the required temperature.

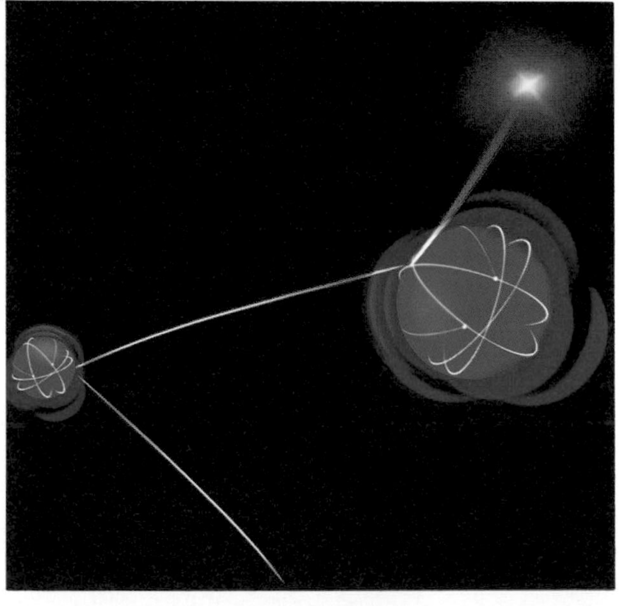

The collision of an electron (the orange light) with two atoms seen taking place in a metal.

Atomic and nuclear physics

Atom (5th century BC)

The Greek philosophers of the 5th century BC were the first to suggest that all matter was made up of invisible particles: atoms.

Atomic theories

The first modern atomic theory was created by the English physicist **John Dalton** (1766–1844), who in 1801 adopted the Classical hypothesis of the indivisibility of matter, and gave it a scientific basis. In 1810 the Frenchman **Joseph Louis Gay-Lussac**, experimenting with chemical actions in gases, established that gases combine in simple proportions by volume, and that the volumes of the products are related to the original volumes, thus refuting Dalton's theory.

The phenomenon was explained by the Italian **Amedeo di Quaregna e Ceretto**, Count Avogadro (1776–1856), who distinguished clearly between the atom and the molecule.

In 1897, the indivisible nature of the atom was taken up again by the Englishman Sir Joseph John Thomson and the Frenchman Jean Baptiste Perrin, who won the Nobel Prize for Physics in 1906 and 1926 respectively.

Rutherford's atom (1911)

The explosion of a radioactive atom discharges alpha particles with great energy. When a beam of alpha rays passes through a thin metal plate, some particles are widely deflected. To explain this phenomenon, evidenced in the experiments of the German Hans Geiger (1882–1945) and his team, the New Zealander **Ernest Rutherford** (1871–1937; later Lord Rutherford, winner of the Nobel Prize for Chemistry in 1908) in **1911** went back to Jean Perrin's hypothesis of the nuclear structure of atoms. That is that all the mass and all the positive charges are concentrated in a small central nucleus, which creates an intense field of attraction in which electrons revolve like the earth revolves round the sun.

Rutherford calculated that these positive particles are about 1836 times the mass of the electrons: he called them protons. The neutral atom helium has a nucleus made up of two protons, around which revolve two electrons.

Bohr's atom (1913)

The Danish physicist **Niels Bohr** (1885–1962, Nobel Prize for Physics in 1922), inspired by the quantum theory proposed in 1900 by Max Planck (1858–1947), suggested in **1913** a theory explaining the radiation emitted by atoms, when under electric discharge, for example.

Arnold Sommerfeld (1868–1951), a German mathematician and physicist, began in 1915 to apply relativist mechanics and quantum theory to the atom to explain the fine structure of spectral lines from hydrogen. He replaced the circular orbits suggested by Bohr with elliptical orbits.

The atom in wave mechanics (1925–6)

The Bohr–Sommerfeld model, in which the electrons were precisely located on an orbit, precluded the development of atomic mechanics that could take account of all the phenomena involving atoms.

Two physicists developed a satisfactory theory from **1925–6**. The Austrian **Erwin Schrödinger** (Nobel Prize for Physics 1933) applied to the atom the Frenchman Louis de Broglie's idea: an electron or any other particle has a wave associated with it.

The German **Werner Heisenberg** (Nobel Prize for Physics 1932) formulated his uncertainty principle which said that it was impossible simultaneously to determine the position and the momentum of a particle with absolute certainty, enabling the combination of Schrödinger's formalist mathematics with a physical interpretation which satisfied the wave–particle duality.

The periodic table of elements (1869)

A fundamental stage in the development of chemistry and of modern science in general, the table was the work of the Russian **Dimitri Mendeleev** (1834–1907) in **1869**. It enabled scientists to establish relationships between various chemical elements that had been considered as independent entities, and to understand why certain elements had the same properties.

The interesting thing about this classification is that it shows the periodic variations of chemical and physical properties in the chemical elements when they are classed in ascending order of their atomic mass.

The 110th element (1987)

The 110th element according to Mendeleev's periodic table was synthesised for the first time in **August 1987**, by researchers in the **Doubna Nuclear Research Institute** near Moscow. This new element was obtained from a U-400 cyclotron, at the end of two years' work by **Professor Yuri Aganessian**. Experiments are under way to obtain a 111th element.

Avogadro's number (1811)

In **1811** the Italian **Amedeo di Quaregna e Ceretto (Count Avogadro)** (1776–1856), professor of physics at the University of Turin, established a law that was named after him. Avogadro assumed that in analogous conditions of temperature and pressure, equal volumes of gas contain the same number of molecules.

Avogadro's hypothesis did not gain immediate recognition. It was not until some 50 years later that an Italian chemist, Stanislao Cannizzaro (1828–1910), demonstrated the necessity for adopting Avogadro's concept as the basis for a coherent atomic theory. Cannizzaro honoured Avogadro by giving his name to an atomic constant. He defined the Avogadro number as the number of gaseous molecules contained in a gram-molecule of any substance, that is, the quantity of that substance occupying a volume of 22.4 litres *9.48cu.ft* in typical conditions of temperature and pressure.

Natural radioactivity (1896)

In nature, some heavy nuclei emit natural radioactivity. The Frenchman **Henri Becquerel** (1852–1908) discovered this phenomenon in Paris in **1896** while performing experiments on uranium. In fact, it was the discovery of X-rays that led to the discovery of radioactivity.

One cloudy day in Paris, Becquerel set up an experiment designed to verify whether a sample of pitchblende (a black mineral composed of uranium and potassium) exposed to sunlight emitted X-rays. Unable to complete his work because of the weather conditions. Bequerel put his equipment away. He resumed the experiment another day when the weather had improved, placing the samples of pitchblende on a photographic plate that had not been removed from its wrapping. When later developing the plate, he was surprised to see an image appear whose contours followed the outline of the ore sample perfectly. What could possibly be the origin of the 'energy' in the mineral that was capable of leaving an impression on a photographic plate?

Radium and polonium (1898)

Becquerel took the matter up with his friends **Pierre Curie** (1859–1906) and his wife, **Marie** (1867–1934). Examining the pitchblende more closely, Pierre and Marie Curie discovered that the radiation had been caused by at least one substance that was much more radioactive than uranium. Finally, after two years of unrelenting and meticulous work, the Curies revealed the existence of not one but two elements that emitted this strange radiation: radium and polonium. The second one was named in honour of Marie Curie, née Sklodowska, who was Polish by birth. For these discoveries Henri Becquerel shared the 1903 Nobel Prize for Physics with Pierre and Marie Curie.

Radioactive disintegration (1902)

Sir Frederick Soddy, English physicist and chemist (1877–1956, winner of the Nobel Prize for Chemistry in 1921), explained the phenomenon of radioactive decay of atomic nuclei, thus paving the way for research in nuclear energy.

Geiger counter (1913)

The Geiger counter was invented in **1913** by the New Zealand physicist **Ernest Rutherford** and his German assistant, **Hans Geiger**. This

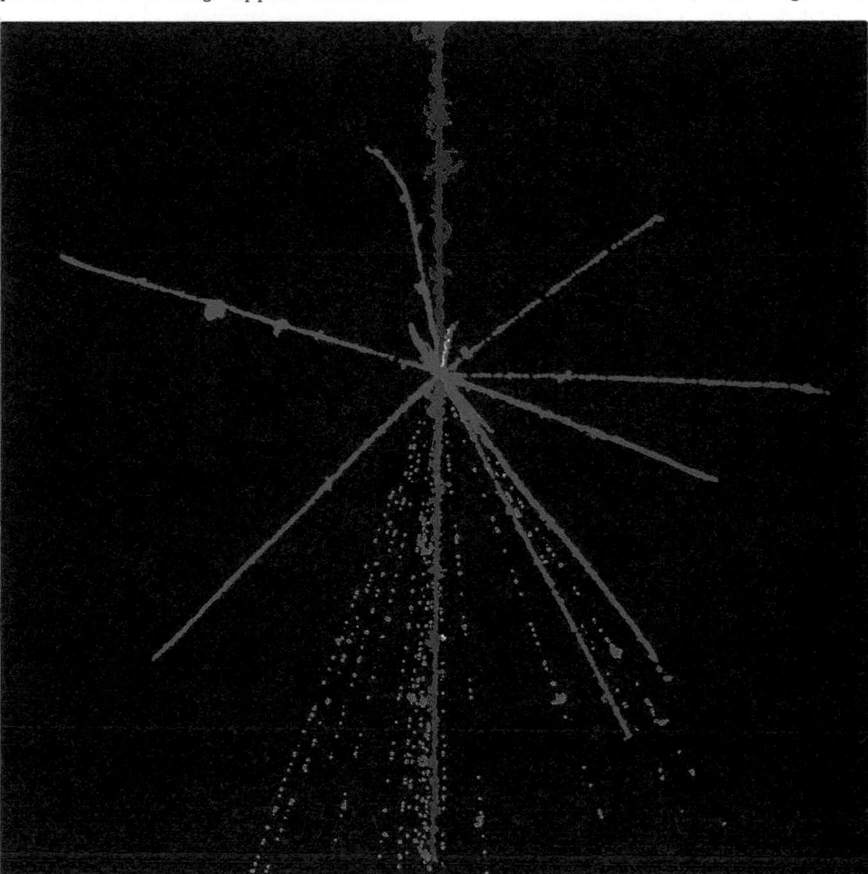

This photo was taken in 1959 by the British physicist Cecil Powell, a pioneer in the use of photographic emulsions to record traces of electrically charged particles. It shows a collision between two nuclei that leads to a rain of other particles.

enabled them to locate and count alpha particles, a constituent of the rays emitted by radioactive decay.

Artificial radioactivity (1934)

In **1934**, in Paris, **Irène** (1897–1956) and **Frédéric** (1900–1958) **Joliot-Curie** (the daughter and son-in-law of Pierre and Marie Curie), obtained radioactive phosphorus by bombarding aluminium with alpha particles (the nuclei of helium).

In their natural state, certain elements such as radium, neptunium and actinium are radioactive. Nuclear reactions, on the other hand, bring into play the disintegration of atomic nuclei, thus obtaining radioactive nuclei unknown in nature: these elements are said to have an artificial radioactivity.

This discovery earned the Joliot-Curies the Nobel Prize for Chemistry in 1935, and has enabled the fabrication of isotopes used in medicine, biology, metallurgy, etc.

Transuranian elements (1940)

The discovery of artificial radioactivity in 1934 led physicists of the time to think that there might be bodies with atomic numbers greater than Z92, the number of uranium in the periodic table. These are called transuranian bodies. The first one was discovered in **June 1940** at the University of California at Berkeley, by **E. M. McMillan** and **P. H. Abelson**. They called it neptunium, and it has an atomic mass of Z93. At the end of 1940 **G. T. Seaborg, J. W. Kennedy and A. C. Wahl** discovered plutonium, of mass Z94.

The latest element was discovered at Darmstadt (GDR) in 1982, thanks to Unilac's linear accelerator of heavy ions. Its atomic mass is Z109. It is thought that this will be the last of the elements in the periodic table, although theoreticians such as Sven Gosta Nilsson and the Swedish school think that Z114 could exist.

X-rays (1895)

In **September 1895** in Würzburg the German physicist **Wilhelm Conrad Röntgen** (1845–1923) discovered X-rays. Röntgen called the rays 'X' because their nature was then unknown. It was not defined until 1912, by another German physicist, Max von Laue (1879–1960), who managed to diffract them through a lattice of crystal.

X-rays are electromagnetic waves, which pass through material that is normally opaque to light. These rays have a very short wavelength.

The discovery of X-rays immediately created a considerable stir. Röntgen, a national hero before the century was out, was awarded the Nobel Prize for Physics in 1901.

Nuclear magnetic resonance (1946)

Now a means of medical investigation, NMR is a physical phenomenon which was discovered in **1946** by the Swiss-born American physicists **Felix Bloch** and Edward Mills Purcell, who jointly obtained the Nobel Prize for Physics in 1952.

NMR or nuclear induction is a method of measuring the magnetic field of atomic nuclei, using a particular property of the proton whose behaviour is closely linked to its environment.

It is especially useful in analysis, as it allows the precise detection of specific atoms in a number of areas: in botany, in geology, in food technology, etc.

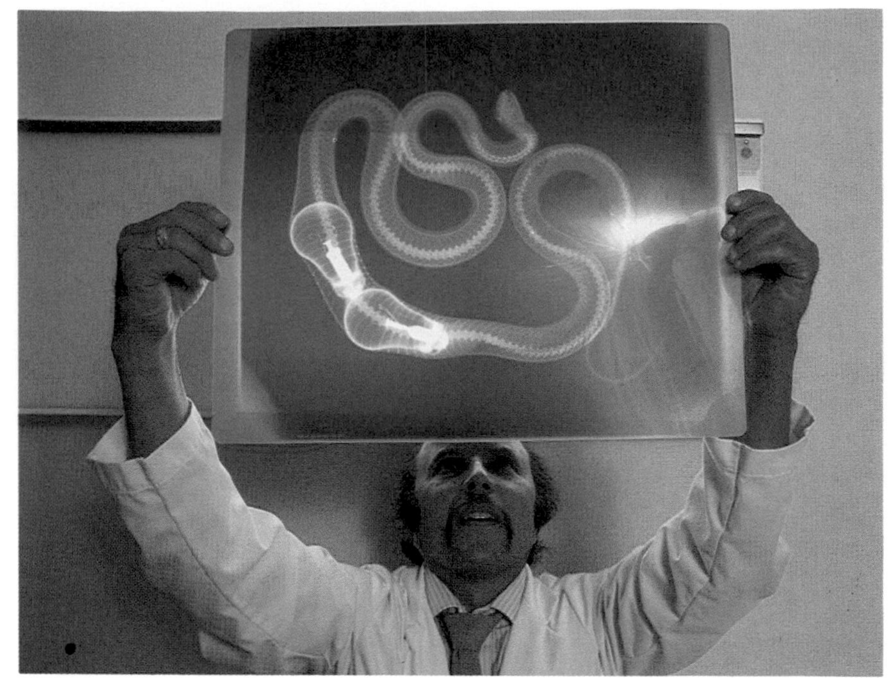

Thanks to X-rays we can admire the electric bulbs swallowed accidentally by this snake.

Carbon 14 (1947)

In **1947 William Frank Libby**, an American chemist specialising in the radioactivity of living organisms and recipient of the 1960 Nobel Prize for Chemistry, explained the formation of carbon 14 in the atmosphere. Carbon 14 is an isotope of common carbon, carbon 12. It has two more neutrons than carbon 12.

Carbon dating

The dating of ancient objects by means of carbon 14 is based on the extent of the residual activity of isotope 14 in carbon.

A number of other radioactive isotopes, contained in samples from this or that event, allow the occurrence to be dated along the same principles (measuring the residual activity of a radioactive isotope whose period is known). Argon–potassium dating is much used. Such a process, used on the charcoal of the Lascaux caves in France, enabled scientists to date the habitation of the caves to 13 000 years BC.

One of the most recent examples of carbon dating has been its use on the Holy Shroud of Turin, considered by many Christians to be

The Turin Shroud has been shown, by carbon dating, to belong to the Middle Ages.

Christ's shroud. Three institutions – the British Museum, the University of Arizona and the Federal Institute of Technology in Zurich – took part in this work, the results of which were made known at the end of 1988: in fact, the shroud dates back to the Middle Ages.

Particles

Elementary particles

An elementary particle is one which is 'indivisible', which has neither dimension nor internal structure. Of course some particles may appear to be elementary at one time in scientific history, and then later be revealed as complex. This was the case, for example, with the proton and the neutron.

At present elementary particles are organised in two categories: the leptons and the quarks, with which antiparticles are also associated. The existence of the latter was suggested by the English physicist **P. A. M. Dirac** in **1929**. Even by observing them closely by the most powerful particle accelerators, no one has been able to detect the slightest internal structure in leptons, quarks and their antiparticles.

Leptons

The electron (1875)

This was the first lepton to be discovered. Its existence was deduced from electromagnetic experiments made by **Sir Joseph John Thomson** from **1875**. Its negative electric charge was precisely determined by the American R. A. Millikan in 1916. The positron, which is the antiparticle associated with the electron, was discovered in cosmic rays in 1932 by the American C. D. Anderson.

The muon (1937)

Discovered in 1937 by **C. D. Anderson** in collaboration with **Neddemeyer**, the mass of the muon is 206.77 times that of the electron. It was created in an accelerator in 1939.

The tau (1976)

This came to the fore during an experiment on collisions between electrons and positrons, run by **Professor Martin L. Perl** and his team at Stanford (USA) in **1976**. Its mass is 3600 times that of the electron.

The neutrino (1933)

The existence of this lepton was guessed at by the Austrian-born Swiss physicist **Wolfgang Pauli** in **1933**, to explain certain phenomena in β radioactivity and it was named neutrino (tiny neutron) by the Italian Enrico Fermi. It is an extremely light particle, which can travel through dense matter like the earth without difficulty because it hardly interacts with matter. It was directly identified towards the end of the 1950s by the Americans Reines and George Arthur Cowan

It is now known that there are in fact three types of neutrino, associated respectively with the electron, the muon and the tau.

Quarks (1964)

In **1964** the physicists **Murray Gell-Mann** and **George Zweig** independently postulated the existence of quarks (although Zweig called his 'aces') as the fundamental constituents of protons and neutrons. There are now thought to be six quarks, known by their English names: Up, Down, Strange, Charm, Bottom, and Top. A team at the European Centre for Nuclear Research (CERN) found experimental evidence of Top's existence in 1984, and estimated its mass at 30–50 billion electron volts.

The physicist Greenberg proved that quarks must have a new kind of charge, which he called colour, which take the form of three different shades, conventionally as blue, red and green.

Composite particles

These are not elementary particles but are made up of elementary particles linked together. There are lots of composite particles in nature; the proton, the neutron and the pion are the most important examples.

The proton (1886)

The discovery of the proton goes back to an experiment conducted by the German physicist **E. Goldstein** in **1886**. It is the main constituent of the atomic nucleus, with a positive charge, and is made up of three quarks.

The neutron (1932)

This is the second constituent of the atomic nucleus, which was discovered by the Englishman **James Chadwick** in **1932**, following on from the work done by W. Bothe in Germany and the Joliot-Curies in France. Its name comes from the fact that it is electrically neutral. It is made up of two Down quarks and one Up.

The pion (1935)

In **1935** the Japanese physicist **Hideki Yukawa** postulated the existence of a new particle to explain the transmission of nuclear force: the pi-meson or pion. He predicted that it would have about 200 times the mass of an electron. The pion was actually discovered in 1947. It is made up of a quark and an antiquark (the antiparticle associated with the quark).

WHAT DO PARTICLE ACCELERATORS DO?

Knowledge of the innermost structure of matter is in fact the main concern of scientists, whatever their specific field. On this depends our understanding, and perhaps our control, of the universe.

In order to penetrate the organisation of matter, we must break it down to its most elementary particles. This is what researchers are doing when they use particle accelerators. By bombarding their targets with electrons, neutrons, positrons or antiprotons at a great speed, or by causing head-on collisions between these particles, scientists are able to penetrate even further into this field of knowledge.

Particle accelerators break down matter into elementary particles which can then be studied. When a proton and an antiproton collide they destroy each other and become pure energy and many new particles shoot off in all directions.

AN ACCELERATOR CALLED AGLAE

A particle accelerator that is the only one in the world to be used for studying and authenticating works of art, called AGLAE (Accélérateur Grand Louvre d'Analyse Elémentaire), was installed on 23 March 1989 12m *39 ft 4.6in* under the gardens of the Louvre in Paris. It will henceforth enable the Louvre to identify the composition of an ancient glass, or to locate the origin of a jewel, or even to authenticate a painting without touching it. No other museum in the world has such a tool. Its first task in 1989 was to study 2000 works of art.

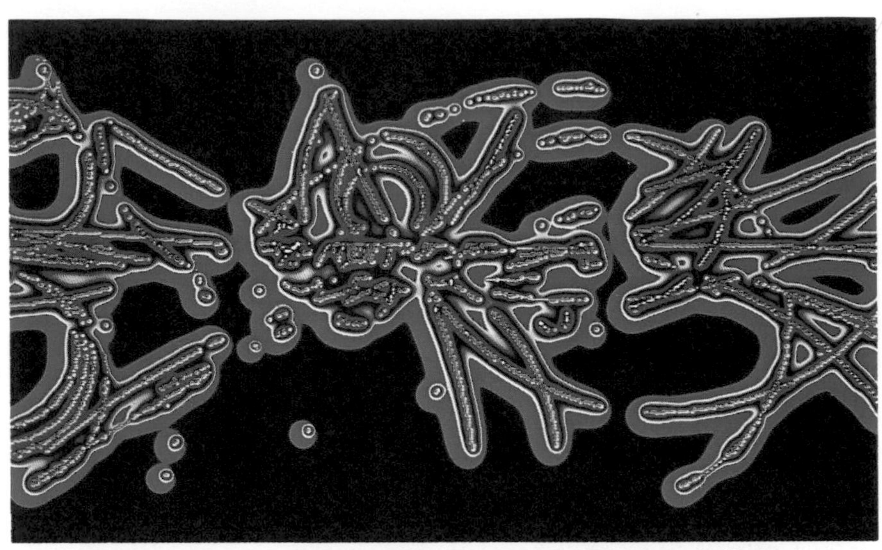

This photograph was taken when boson Z was discovered by the European Centre for Nuclear Research in Geneva in 1983. The particle's life span is so short that it is invisible, but specialists have deduced its existence by the trajectory (in yellow) of an electron-proton pair.

Bosons

These are particles associated with classic waves. They appear also as the particles exchanged during the process of fundamental interactions. Their name derives from the fact that they obey Bose–Einstein statistics.

The photon (1923)

'Invented' by Albert Einstein in 1905 to explain the photoelectric effect, the photon is a light particle. It 'carries' the electromagnetic force between charged particles. The American physicist **A. H. Compton** found evidence of its existence in **1923**.

The graviton (21st century)

This is a particle in the gravitational field. At present it only exists as an abstract concept. It is thought that gravitational waves may be formed from a large number of gravitons.

Weak interaction bosons (1983)

There are three of these, called W−, W+ and Z0, which were found to exist by a team of 200 research workers led by the Italian **Carlo Rubbia** and the Dutchman **Simon van der Meer** at CERN in Geneva, in **1983**.

Gluons (1982)

There are eight gluons responsible for strong nuclear interactions. They have never been seen, but an experiment conducted in **1982** by a group of American research workers from Brookhaven and the City College in New York seems to prove their existence.

Chemistry

Inorganic chemistry

Nitric acid (1838)

The preparation of nitric acid by the catalytic oxidation of ammonia gas was discovered in 1838 by the French chemist **Frédéric Kuhlmann**; the catalyst used in the reaction was platinum.

Using this method, the German chemist Wilhelm (1853–1932) made nitric acid on a small scale. But when another German chemist, Carl Bosch (1874–1940) replaced platinum with a catalyst based on iron, manganese and bismuth, the production of nitric acid on an industrial scale could be considered. The resulting production of nitrates replaced the natural nitrates which had been imported from Chile.

Sulphuric acid (18th century)

The first sulphuric acid plants were created in England in the 18th century. They used the leaden condensing chamber developed in 1746 by the Scottish chemical engineer John Roebuck (1718–94). For the first time a link was formed between the laboratory and industry.

In 1774 steam replaced water in the process, which changed from discontinuous to continuous.

Aluminium (1886)

In 1822 the French mineralogist Pierre Berthier (1782–1861) discovered near the village of Les Baux, in Provence, the first deposits of an ore which he named bauxite. It was in fact hydrous alumina (aluminium oxide).

In 1825 the Danish scientist Hans Christian Oersted obtained aluminium in a powdered form, and in 1827 it was created in the form of an ingot by the German scientist Friedrich Wöhler (1800–82).

In 1886 the French metallurgist **Paul-Louis Héroult** (1863–1914) and the American **Charles Martin Hall** (1863–1914) discovered independently, but almost simultaneously, the process of electrolysis, still in use, which was to give rise to the aluminium industry.

Since then, because of technological progress, aluminium has never ceased to grow in importance. From 16 tonnes in 1886, it has risen to over 16 million tonnes in 1986.

Ammonia (1908)

By applying the laws of chemical equilibrium, the German chemist **Fritz Haber** succeeded in **1908** in synthesising ammonia using the elements which make up this substance: nitrogen and hydrogen.

The industrial development of the procedure in 1909 by BASF was the work of another German chemist, Carl Bosch.

Cadmium (1817)

This silver-white metal, with its slight tinge of blue, was produced for the first time in the laboratory in **1817** by the German chemist **Frederic Strohmeyer** using its oxide present in a sample of zinc carbonate. In 1818, Strohmeyer proposed the name cadmium for this newly discovered metal because it was mainly recovered from the zinc *cadmia fornacum*, flowers formed on the walls of ovens used to distill zinc.

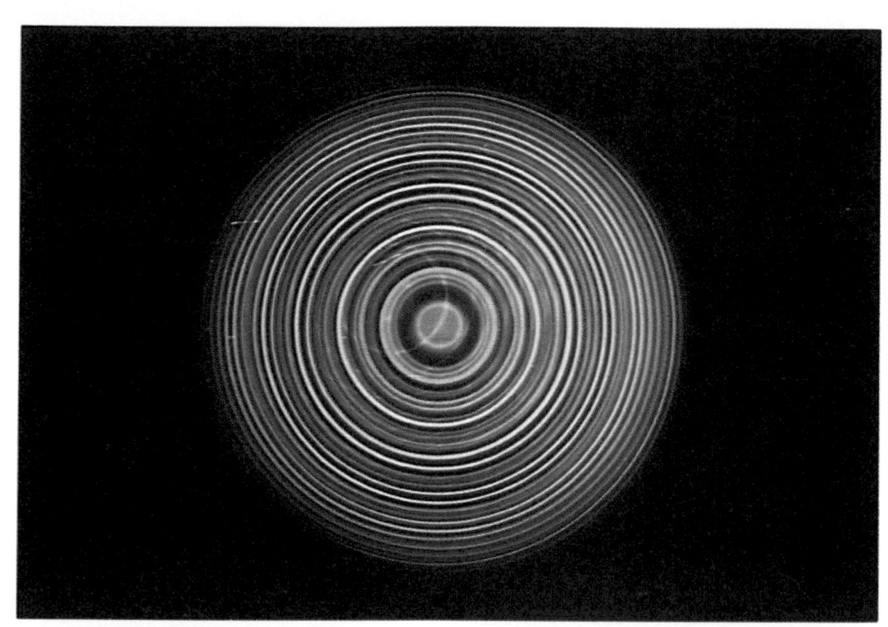

Cadmium was produced for the first time in 1817.

Industrial production of cadmium began in 1827 in Upper Silesia. The main uses of cadmium and its compounds are electroplating alloys and car radiators. In addition, cadmium rods are used to control the flux of neutrons in nuclear reactors.

Ceramics (7000 BC)

Terracotta pottery, invented simultaneously in Turkey, Syria and Kurdistan around 7000 BC, was the first technique to transform matter of mineral origin by the use of fire. With pottery, man had invented the first artificial material.

In general, antique pottery had one major drawback: it was not watertight. Thus through the centuries men attempted to perfect a watertight clay as well as the exterior varnishing.

Alexandre Brongniart, in the mid-19th century, created the precursor of modern pottery, before the appearance of industrial ceramics around the 1950s, with neo-ceramics based on the principle of calcination, and oxide ceramics. Because they hardly expand, the latter are very resistant to heat and are thus used in the aerospace industry.

Transparent ceramics (1988)

A transparent ceramic used in the new scanner-detectors has been commercially produced since **1988** by the American firm **General Electrics**. These new detectors are made up of 900 sensitive elements (from photodiodes to silicon) placed side by side and covered with a block of transparent ceramic. This material (HiLight Ceramic) acts as a scintillator when hit by X-rays.

Superconductive ceramics (1988)

The American company **Du Pont** has applied for a patent for a superconductive ceramic, which is more stable and easier to manage than the non-resistant materials currently in use for carrying electricity.

Cobalt (1735)

This metal was isolated for the first time in **1735** by the Swedish chemist **Georg Brandt** from copper minerals found in the Harz region. The word cobalt is derived from the Middle High German word *Kobalt* which means goblin because the vapours given off during the fusion of copper minerals, which contain cobalt, are toxic.

In 1910 it was discovered that this metal improved high-speed steel, and metallurgists became interested in it. Cobalt is used in alloys that are resistant to high temperatures, in magnetic materials and in hardwearing alloys that resist corrosion.

Heavy water (1932)

In **1932** the American chemist **Harold Clayton Urey** discovered deuterium and heavy water. He was awarded the Nobel Prize for Chemistry two years later for this discovery.

In 1933, using Urey's method of preparation, the American physicist and chemist Gilbert Newton Lewis successfully prepared a few millimetres of almost pure heavy water by fractional distillation of ordinary water.

The term heavy water, a chemical compound similar to water, is used to describe deuterium (the heavy isotope of hydrogen) oxide. This oxide is contained in all water in virtually constant proportions – that is, one molecule of heavy water to approximately 1000 molecules of water.

Germanium (1885)

While analysing argyrodite ore, the German chemist **Clemens Winkler** discovered in **1885** a new metal, which he named germanium, after its country of origin. Germanium is easy to purify and is used as a solid semi-conductor in transistors and rectifiers.

Germanium dioxide is used in the composition of glass with a high refractive index. It is also a good catalyst in the polymerisation of polyester.

Semi-crystals (1984)

A geometric rule requires that the elementary figure of a crystal possess symmetrical axes of the order of two, three, four or six only. Symmetry of the order of five is impossible in practice, for the simple reason that a pentagon does not have its edges touching in space.

In **1984**, however, four research workers announced the discovery of a structure in matter symmetrical to the order of five. The discovery was made jointly by **Dan Schechtmann** and **Ilon Blech** of the institute of technology, Technion, in Haifa; by **Denis Gratias** of the centre for metallurgical

The new ceramics have remarkable properties: this knife will never become blunt.

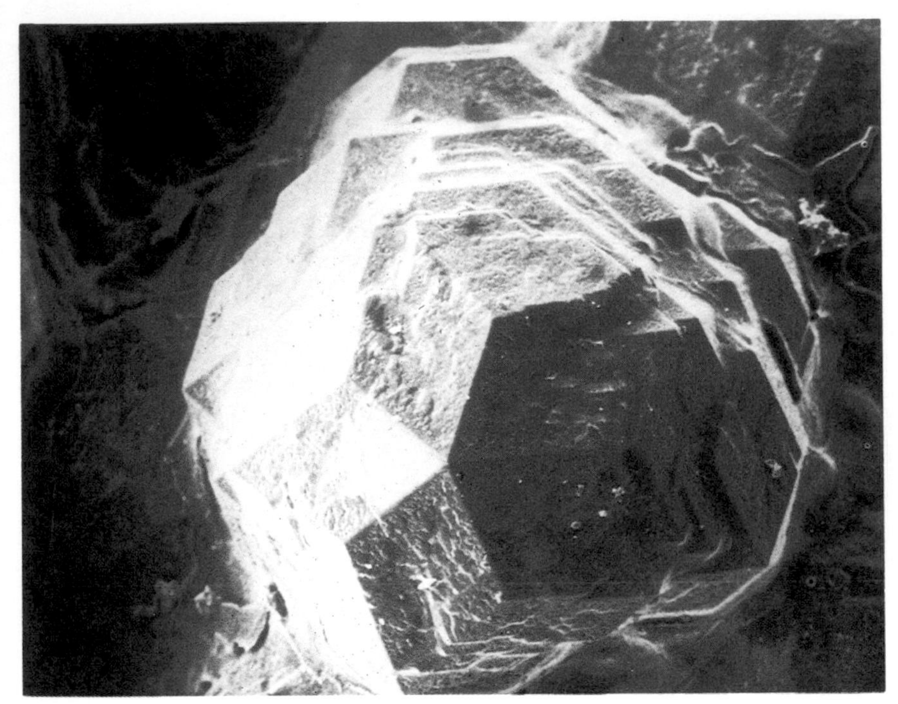

New light, resistant alloys may be produced thanks to the progress being made in the field of semi-crystals.

chemistry at CRNS, Vitry-sur-Seine, and by **John Cahn** of the National Bureau of Standards in Washington. They chose to call this new stable state of matter semi-crystal, not entirely a crystal and not entirely amorphous.

In 1985 mathematical models created by computer showed that this impossible structure was feasible in a six-dimensional space, and in July 1986 a team at the Péchiney research centre, in France, managed to create a cupro-manganese alloy made up of single crystals observable by the naked eye.

This discovery enables scientists to envisage the fabrication of new alloys whose resistance and lightness could completely alter the industrial scene. Makers of aeroplanes are watching these new developments with particular attention.

Silicon (1823)

This is the most common element on the surface of the globe (28 per cent) after oxygen, with which it is associated in the form of silica (flint) or silicates (clay). It was isolated in its pure state for the first time in **1823** by the Swedish chemist **Jöns Jakob Berzelius** (1779–1848), one of the creators of modern chemistry. It has important uses in metallurgy, where it increases the resistance of steel to corrosion. It is also used in the making of light alloys, such as alpax (aluminium–silicon), which have many uses: pistols, casings, bicycle wheels, electric appliances, etc.

It is increasingly used to make semiconductors in the electronics industry, and in the plastic industry.

Tungsten (1781)

Discovered in **1781** by the Swedish chemist **Carl Wilhelm Scheele**, tungsten is a hard, malleable element. Its name means heavy stone in Swedish. Tungsten is the most fire resistant element known to date, with a melting point of 3410°C *6170°F*.

Tungsten is used in the manufacture of lamp filaments, and in the composition of silver, copper, zinc, pewter, argyrodite and germanite based minerals. The main sources are found in China, Korea, the United States and the USSR.

Organic chemistry

Liquid crystals (1929)

The French physicist **Friedel** in **1929** discovered products which could be presented in a stable state and were intermediary between solids and liquids. One family of these products is now well-known: that of liquid crystals. Unlike solid crystals, their molecules can change direction under certain circumstances, in particular, when submitted to a very weak electric current, their transparence is altered. It is this property which is used in watch display panels; any figure can be formed from just seven segments.

In 1971, the Swiss company Hoffmann-La Roche perfected the first panel of liquid crystals.

From 1973, liquid crystals made their appearance in many objects of everyday use: watch-dials, calculator screens, electronic games, portable computers, etc.

Synthetic diamonds (1955)

Ever since it has been known that the diamond is nothing more than ordinary carbon in a practically pure state and in a form of very regular crystals, attempts have been made to produce the gem artificially.

In 1894 the French chemist Henri Moissan thought that he had manufactured diamonds by heating carbon to a very high temperature under great pressure. However, it wasn't until **1955** that the American company **General Electric** produced true synthetic diamonds by heating carbon to 2600°C at a pressure exceeding 100000 atmospheres.

These tiny diamonds (hardly longer than a millimetre) are often black and are used in industry. But it is possible by increasing the temperature at which they are formed to create transparent diamonds for use in jewellery. Due to the cost of production and the time it takes, these artificial diamonds are more expensive than natural ones, and not as beautiful.

Synthetic jade (1984)

On **30 November 1984** the American company **General Electric** announced an important first: the creation in a laboratory of a sample of synthetic jade. This was the result of the work of two chemists: **Robert C. DeVries** and **James F. Fleischer**.

Formaldehyde (1868)

Discovered in **1868** by **August Wilhelm von Hofmann** (1818–92), it is used in the textile industry, in papermaking, tannery, dyeing, photography and joinery (wood glues). It is also a base of synthetic resins and insulating foam which is injected into partitions.

Moreover formaldehyde is used, because of its antiseptic properties, as a disinfectant in solution, known as formol. Formol is easily polymerised into metaldehyde, in which form it is used as blocks for lighting barbecues.

Nicotine (1828)

In 1562 the Frenchman **Jean Nicot** extracted a juice from tobacco (brought from America by Christopher Columbus), that accelerated the healing of wounds.

However, it was not until **1828** that **R. Posselt** and **R. Reimann** isolated the substance nicotine in its pure state. Ingested, it is a violent poison which has been used for criminal ends, and its toxicity is utilised in plant medicine to destroy parasites.

Lead tetraethyl (1923)

Lead tetraethyl, put onto the market in America in **1923** by the **Ethyl Corporation**, is an anti-knock agent which, when added to petrol, increases the octane level and thus its efficiency. It also prevents deterioration of the engine. However in 1965 the American scientist Clark C. Patterson denounced its harmful effects. Inhaled or accidentally ingested (from polluted plants for example) lead tetraethyl enters the blood stream and affects the central nervous system.

Urea (1828)

This organic substance exists naturally in the blood (0.20 to 0.50g to the litre) and in urine (2.5g to the litre), where it was discovered. Its synthesis in **1828** by the German chemist **Friedrich Wöhler** marked a turning point in demonstrating for the first time that an organic substance can be obtained independently of any living organism.

RUBBER AND PLASTIC

Invention	Date	Characteristics
Rubber	1736	Natural rubber was discovered by the French naturalist Charles Marie de La Condamine, on an assignment in Peru.
Nitrocellulose	1833	Obtained by two French chemists T. J. Pelouze and H. Braconnot. First industrial use as cotton-powder by the German C. F. Schoenbein, 1847.
Vulcanisation	1839	This technique, which stabilises the properties of rubber and makes it usable, was invented by the American Charles Goodyear.
Artificial rubber	1860	The isolation of isoprene by the Englishman Charles G. Williams. First successful use: G. Bourchardat (France), 1880, and Tilden, 1884.
Cellulose acetate	1869	Obtained by J. Schutzenberger. Patents for industrial production: Charles Cross and Edward J. Bevan, 1884.
Celluloid	1870	Invented by the American Hyatt brothers for making billiard balls, a mouldable mixture of camphor and nitrated cellulose.
Rayon	1891	In 1891 Charles Cross obtained the patent for fabricating cellulose xanthate, the basis of cellulose. In 1921 Edward Bevan obtained a patent for obtaining rayon from acetate.
Cellophane	1908	Invented in 1908 (trademark registered in 1912) by J. E. Branderburger, who regenerated cellulose xanthate in order to make not fibres but a transparent film: cellophane.
Bakelite	1909	The first synthetic plastic material, patented by a Belgian named Hendrik Baekeland, living in the USA. This phenoplastic remains interesting because it will not melt after its first heating and setting, and is a good insulator.
PVC	1913	In 1913 the German Professor Klatte patented the polymerisation of a gas, vinyl chloride. It was put into industrial production in 1931 by I. G. Farben. PVC is used in many things: soles of shoes, fibres, bottles. Its near relation is vinyl polyacetate (used in microchips).
Polymers	1922	The structure of polymers, made up of very long chains of thousands, even hundreds of thousands, of atoms, these chains spreading out or forming links with each other, was discovered in 1922 by the German chemist Hermann Staudinger (Nobel Prize for Chemistry 1953). In 1980 the Norwegian John Ugelstad, of the University of Trondheim, invented a process for making polymers from particles of the same size, which is particularly interesting for medical science.
Plexiglass	1924	In 1924 Barker and Skinner made an organic glass, the polymethacrylate of methyl, commercialised in 1934 by Rohm under the name Plexiglass (from 'plastic flexible glass'). The organic glass in spectacles comes from the same family.
Polystyrene	1933	Perfected by Wuff (Germany). Good-looking, it is used for pens and toys. A family of styrene polymers ensued, among them the stretch polystyrene invented by BASF in 1951.
Polyamides (nylon, kevlar)	1935	In 1935 W. H. Carothers perfected at Du Pont (USA) the first polymide fibres. He patented polymide 6-6 or nylon in 1937. In 1965 Stephanie Kwolek, at Du Pont, created kevlar, a descendant of bakelite and nylon. One of the latest Du Pont polymides is kapton. Although polymides are best known as fibres, they also make excellent technical plastics. The latest to be created, Dinyl and Pebax, are French.
Polyurethanes	1937	Invented by Otto Bayer. Many different forms and uses.
Teflon	1938	Known as Teflon, the tetrafluorethylene polymer was discovered by Roy J. Plunkett, an engineer at Du Pont (USA) in 1938 (patented in 1939). The idea of spreading it over a metal was invented by Marc Gregoire in 1954 (used for saucepans, etc.).
Polyesters	1938	First synthesised in 1901 (Smith). First used in 1927. First heat-resistant polyesters (Ellis, 1938). Reinforced by glass fibres (US Rubber, 1942). There is an enormous variety of polyesters. PET belongs to the family of linear thermoplastic polyesters, and was obtained by the Englishmen Dickson and Whinfield in 1940, and commercialised in 1946 under the name of Dacron.

Polyethylenes	**1939**	The first grams of low density polyethylene (PED) were obtained by Fawcett and Gibson for ICI (UK) in 1935. Industrial production began in 1939. Used notably in radar construction which enabled the British victory over the Luftwaffe in 1940. Excellent insulator. High density polyethylene (PEHD) obtained by Karl Ziegler of Germany (Nobel Prize for Chemistry 1963) in 1953. In 1985, DSM (Holland) and Allied (USA) produced a polyethylene fibre 30 times more resistant than the best steel of the same weight.
Polypropylene	**1954**	The polymerisation of propylene perfected by Giulio Natta (Italy) in 1954.
Peba	**1981**	A new family of synthetic materials created by Gérard Delens (ATOCHEM), intermediary between rubber and plastics.

PVC has been used to form these food copies.

Analytical chemistry

Mass spectrography (19th century)

The discovery of the electron in 1897 by Sir Joseph John Thomson proved that the atom was not single and indivisible but made up of several particles. Researchers very soon delved into the innermost structure of matter and came up with evidence for the existence, at the heart of the same natural substance, of two or more atoms chemically similar but of different atomic mass. They called these isotopes, and the process used was called mass spectography. The differences between such atoms came from the number of neutrons contained in their atomic nuclei, which varied from one isotopes to another. Thus uranium ore has three isotopes: uranium 238 (92 protons, 146 neutrons), uranium 235 (92 protons, 143 neutrons) and uranium 234 (92 protons, 142 neutrons). It is the same for most natural bodies, except for a few such as aluminium which only exists under the form 27.

For his work on the nature of isotopes the Englishman **Frederick Soddy** (1877–1956) gained the Nobel Prize for Chemistry in 1921.

Mass spectrograph (c.1919)

By separating isotopes from a body, it became possible to make a very precise analysis of them. The mass spectrograph was created by the Englishman **Francis Aston**, an assistant of J. J. Thomson. It enabled Thomson to analyse in **1919** the neon atom, and to show that it is made up of two isotopes: neon 20 and neon 22. Aston then continued with the analysis of many bodies, and received the Nobel Prize for Chemistry in 1922.

The mass spectrograph, which uses the difference in deflection of isotopes in a magnetic field, remains a powerful analytic tool. In effect, it allows the composition of all bodies to be precisely determined. For example, in the search for traces of explosive in the case of a murder attempt, it is possible to use this technique to find the exact nature of the explosive used, as certain isotopes are explosive and others are not.

MOLECULE PERFUME: A NEW CHEMICAL WEAPON

A little spray – invisible, inoffensive and odourless (except to specially trained dogs): this is the new chemical weapon for dealing with the traffic in works of art.

Invented by two French researchers, Claudine Masson and Marie-Florence Thal, the new process (its chemical make-up is kept a secret) is based on the principle of pheromones, chemical substances secreted by an animal which influence the behaviour of others of its species. Marked with this molecular signature, discreet and indelible, paintings and other works of art are less easy to smuggle across frontiers.

A SPECTROMETER IN THE TOWER OF LONDON

The laboratory of the royal armoury at the Tower of London has had a spectrometer since 1988 that helps archaeologists in their work of dating and their research into the authenticity of objects. This machine, made in the USA, can analyse within minutes the properties not only of metals, but also of paintings and flash up their age on a screen. An example is the Victoria Crosses that were thought to be made from bronze from Russian cannons captured during the Crimean War (1854–6), and which were revealed to date from the First World War. . . .

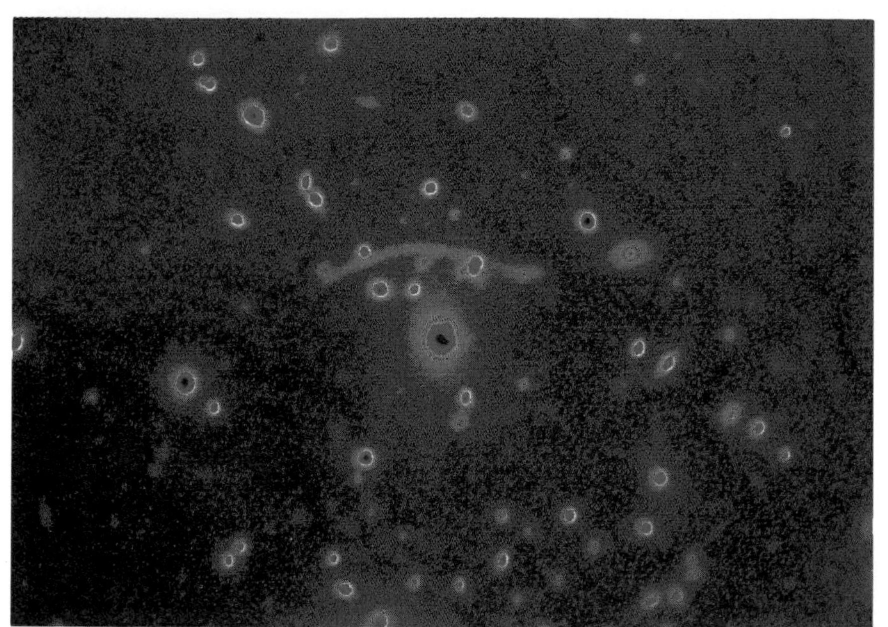

Researchers in France succeeded in 'photographing' (using a mass spectograph) the circle produced by the 'hidden mass' in the universe, whose existence was predicted by Albert Einstein.

Biology and genetics

It was the naturalist **Jean-Baptiste de Lamarck** who, in **1802**, invented the word 'biology'.

Origins of genetics (1910)

The science of heredity, genetics is a branch of biology whose laws were discovered by **Gregor Mendel** in 1865. Although it had long been in practical use in breeding cattle, and developing edible plants as well as creating varieties of flowers, genetics did not really come into its own until **1910**, with the work of **Thomas Hunt Morgan** on the fruit-fly. Morgan was moreover the coiner of the word *gene*.

Nucleic acids (1953)

Deoxyribo-nucleic acid (DNA)

In 1944, the American bacteriologist **O. T. Avery** of the Rockefeller Institute in New York, demonstrated that the transmission of hereditary characteristics from one bacterium to another takes place due to the intermediary DNA molecules. Due to this discovery, it was recognised that the genes, whose primordial role in the phenomenon of hereditary transmission was known, are made up of DNA.

In **1953**, using diffraction via X-rays, the electron microscope and chromatography, the American **James Watson** and the Englishman **Francis Crick** described the exact structure of the DNA molecule in the British magazine *Nature*. Their work earned them the Nobel Prize for Physiology and Medicine in 1962. The DNA molecule exists in the form of a double-strand helix made up of a series of units called the nucleotides.

Ribo-nucleic acid (RNA)

This is also composed of a set of nucleotides, each comprising a sugar (ribose), and four bases, adenine, cytosine, guanine and uracil. **RNA** plays an important part in the synthesis of proteins.

The cell (1665)

The English scientist **Robert Hooke** (1635–1703) was the first to use the word cell (from *cellula*, little room) to describe the miniature empty structures which he observed in **1665** with the aid of a rudimentary microscope, when he cut into a piece of cork. Cork being a dead tissue, Hooke in fact was looking at only the outer walls of cells. At the same period, **Antony van Leeuwenhoek** in Holland was using a somewhat better microscope to observe some isolated cells, such as those in drops of blood, sperm and bacteria.

It was not until 1824 that the Frenchman René Dutrochet established that living tissue was made up of juxtaposed cells. Then in 1833, the Scotsman Robert Brown (1773–1858) described the cell nucleus. Today we describe a cell as made up of a cytoplasm and a nucleus enclosed by a membrane.

Chromosomes (1888)

Chromosomes are short rods, usually curved (measuring 0.005mm *.000194in* in humans), that are found in the cell nucleus. The German anatomist **G. Waldyer** named them in **1888**. The essential constituent of chromosomes is DNA.

Clones (1981)

Clones are genetically identical organisms derived originally from a single individual. The best-known clones are cuttings: a section of a plant which is artificially or naturally removed and which is able to take root and provide a completely new plant.

The Swiss **Professor Illmensée** and his team tried the same kind of experiment but this time on animals. By stimulation of the cells from a mouse ovary, he obtained in 1981 three perfectly functional baby mice, who had been born without a father. These young mice are a sort of carbon copy of their mother, as they have exactly the same genetic make-up.

Genetic code (1943)

The first experimental arguments in favour of the transmission of hereditary characteristics stemming from DNA information go back to

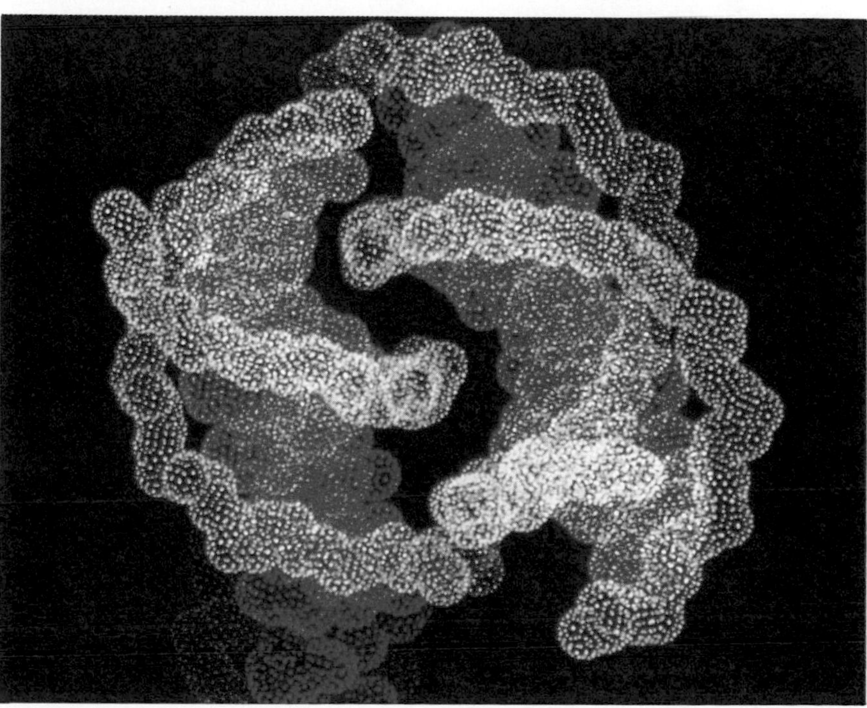

Two views taken at 90° from each other of the double helix of a DNA molecule.

CALVES MADE TO MEASURE

In January 1989, thanks to a process perfected by Granada Inc., an American biotechnical company, seven genetically identical calves were born, products of cloning from a single embryo.

For this first experiment, a super-bull provided the sperm, inseminating an equally 'top class' cow. So in the future, perfect cows and bulls could be created . . . prefiguring, for example (perhaps), sports teams cloned to measure.

1943 with the work of the Italian-born American microbiologist, **Salvador Luria**. We have since learnt that each hereditary characteristic is coded in the form of a message in a tiny fragment of DNA or the gene, and that the translation of this message produces a specific protein which materialises the hereditary characteristic. The synthesis of the protein requires the intervention of RNA on which the gene's message is recopied. This message must then be translated in order to produce the protein.

In vitro culture (1907)

In vitro cultivation is a method by which a living organism, animal or vegetable, or a part of this organism (cells, tissues, organs), is sustained outside its natural environment.

The culture in vitro of animal cells and tissues was invented in **1907** by the American **R. G. Harrison**, and perfected in 1910 by the Frenchman Alexis Carrel. This method with regard to vegetable cells and tissues was invented between 1931 and 1938 by the Frenchmen Roger Gautheret and Pierre Nobécourt, also by the American Philip White.

In vitro cultures are used in research, to obtain medical products (serums, vaccines, antibiotics), to produce plants that are protected against viruses, and to propagate plants from buds or cuttings.

Darwinism (1859)

In his book entitled *The Origin of the Species by Natural Selection*, published in **1859**, the English naturalist and biologist **Charles Darwin** (1809–82) studied the problem of evolution of the species. According to him, the species are not unchangeable, the result of distinct creations, but are progressively transformed by selection of the individuals who are best adapted to their environment (survival of the fittest).

Heredity (1865)

In **1865** the botanist **Gregor Johann Mendel** (1822–84) born in Moravia in 1822, demonstrated that hereditary characteristics are transmitted via distinct elements, which are today called genes. Two sets of experiments allowed him to reach this conclusion. The first set consisted of crossing peas of stable lines that differed among themselves in a couple of characteristics; for example peas with smooth or wrinkled seeds with green or yellow cotyledons. The crossing of such plants produced first-generation plants in which dominant characteristics were revealed; for example, the hybrids obtained by the crossing of smooth yellow peas with wrinkled green peas were smooth and yellow. But when the hybrids were crossed, the parental green and wrinkled characteristics appeared in a quarter of the second-generation hybrids.

From these observations, Mendel deduced that hereditary factors determining traits went in pairs, and that the recessive characteristic was only expressed in the plant when both parents had that characteristic. Such experiments allowed Mendel to determine that in peas the smooth and yellow characteristics are dominant, while the green and wrinkled characteristics are recessive. He also drew up laws that bear his name and which, by revealing the segregation of characteristics, prove that hereditary factors behave independently: they join together and separate across generations and hybridisations, according to the statistical norms of chance.

Lamarckism (1809)

In **1809** the French botanist **J. B. P. A. de Monet de Lamarck** (1744–1829) published his *Zoological Philosophy* in which he set out his theory of evolution: transformism.

According to Lamarck, the species are transformed with time, under the influence of their surroundings, and through the intermediary of their habits and needs. This theory of mutability had its hour of glory, then fell out of favour. However there has been a revival of interest since the studies carried out by biologists at Harvard Medical School (USA) on mutant *Escherichia coli* bacteria.

Genetic fingerprinting (1985)

Professor **Alec J. Jeffreys**, head of the genetic laboratory at the University of Leicester (UK), has developed a revolutionary and virtually infallible procedure for detecting criminals by analysing their blood, or their sperm in cases of rape: genetic fingerprinting.

This new technique, which continues the work undertaken by the American Professor Southern, the Southern Blot process, will soon replace simple fingerprinting and is used for research into paternity.

The procedure stems from the principle that DNA, which is found in cell nuclei, varies from person to person. These variations, or genetic imprints, can be photographed. The photo, which looks like the bar-codes on packets in supermarkets, allows each person to be genetically identified.

Gametes (5th century BC)

The first person to suspect the existence of male and female reproductive cells was the Greek doctor **Hippocrates** (c.460–c.377 BC): in effect he accepted that the formation of the human embryo involved male and female seeds.

These two seeds are the spermatazoon and the ovum. The former is the characteristic cell of the sperm produced by the testicles. A man produces about 200 million of these per day. The ovum is produced by the ovary. At birth, a girl has about 400 000 ovums whose number then diminishes; by puberty there are only 10 000, of which only 400 achieve maturity.

The gene (1910)

The key person in the saga of the **gene** is a southern gentleman born in Lexington, Virginia (USA) in 1866: **Thomas Hunt Morgan**. In **1910**, while he was working on the fruit-fly *Drosophila melanogaster*, an insect which usually has red eyes, he observed a mutant male with white eyes. Morgan crossed it with a normal red-eyed female, and noted that all the first-generation descendants had red eyes. However among their descendants, he found flies with white eyes again. The charcteristic white eye had been transmitted, according to Mendelian laws, by a fragment of chromosome to which Morgan gave the name *gene*. Today it is reckoned that a person's genetic inheritance consists of around 500 000 genes divided among 46 chromosomes. The essential constituent of genes is DNA.

Synthesis of the first artificial gene (1973)

This was achieved in **1973** by the Indian researcher **Har Gabind Khorama**, at MIT in Cambridge (USA). It involved a double helix of DNA, corresponding to the precursor of ribo-nucleic acid, the gene for tyrosine transfer-RNA. The gene is made up of 199 nucleotides.

WHAT IS GENETIC ENGINEERING?

It was in 1974 that the term genetic engineering made its appearance in the scientific world. The association of the word 'engineering' with 'genetics' – or the science of genes – came about because of the new possibility open to researchers breaking up DNA, separating out the genes and recombining them in another DNA molecule.

Genetic engineering is thus the collection of techniques that enable scientists to break into DNA, isolate genes, identify their structure, modify them if necessary and finally, to introduce them into an organism which might be different from the one from which they were originally taken.

Genetic engineering quickly established itself as a basic technique in all areas of biology. But alongside these fundamental activities, it has also made possible the manufacture of rare substances, or those difficult to extract. Medicine and industry have found new methods of production by using organisms – most often genetically modified bacteria – to work on such projects as the synthesis of insulin, a new vaccination against hepatitis B, or alcohol from biomass for use as a fuel. A new technique, genetic engineering is far from having exhausted all its resources.

Microbiology (1857)

In **1857**, the Frenchman **Louis Pasteur** (1822–95) discovered yeast (microscopic fungi) and explained the fermentation process. Pasteur extended his research to bacteria: this was the beginning of microbiology.

Bacteriology (1870)

From **1870 Louis Pasteur** cultivated and identified *Staphylococcus*, *Streptomyces* and *Streptococcus*. He was thus the inventor of bacteriology.

Pasteur proved that, if the environment was favourable, all cultivated germs could multiply. Since then, nearly a hundred culture environments have been experimented with, and have allowed the isolation of many germs, which these days are known and identified. Each germ carries the name of the person who discovered it.

Virology (1898)

Between 1880 and 1885, in the course of his work on rabies, Pasteur came up against the problem of an illness, then incurable, whose agent he could not cultivate. In fact this was not a bacteria, but a virus. Eventually he was able to develop the vaccine.

In **1898**, the Dutchman **Martinus Willem Beijerinck** (1851–1931) discovered microorganisms even smaller than bacteria: they passed through the finest porcelain filters. Beijerinck also discovered the virus which caused tobacco mosaic disease. This was the birth of virology. In 1935, the American biochemist Wendell Meredith Stanley (1904–71) succeeded for the first time in crystallising a virus.

In 1959, thanks to electronic microscopy, X-ray diffractions and biochemical methods, the Frenchman André Lwoff established a definition of the virus based on the presence of nucleic acids, which made for progress in virology.

Synthesis of a virus (1968)

In **1968** two American researchers, **Arthur Kornberg** and **Robert L. Sinsheimer**, succeeded in completely synthesising a virus, using only two enzymes; a polymerase DNA and a lipase DNA, extracted and purified from colonies of colon bacilli.

Plant propagation in vitro

In **1952** the French researchers **G. Morel** and **J. Martin** of INRA (National Institute of Agronomical Research) obtained from a few cells, placed in an artificial environment, a genuine dwarf plant, that could be transplanted to serve as the basis for producing perfect plants.

The success of Morel and Martin's method enabled diseased plants to be regenerated and to proliferate rapidly in their improved form. It is used now to produce roses, orchids, potatoes, fruit trees, etc.

Mutations (1901)

The first observations of mutations were made by the Dutch botanist **Hugo de Vries** in **1901**. In cultivating and studying plants of different species, he observed some that did not correspond to what would be expected from their original grain. These individual plants were different from their progenitors and their differences were inherited by their offspring. Hugo de Vries gave the name mutations to these hereditary variations. Later it was proved that the modifications sprang from alterations in the genes. Mutations are observed among animals as well as vegetables. Today experimental **mutations** can be obtained by using radiation and chemical products.

Oncology (1981)

Cancer genes or oncogenes were discovered in **1981** by three separate American research teams: that of **Professor Robert Weinberg** at the Centre for Cancer Research, part of the famous Massachusetts Institute of Technology; that of **Dr Geoffrey Cooper** at the Sydney Farber Cancer Institute in Boston, and that of **Professor Michael Wigler**, of Cold Spring Laboratory on Long Island.

Cancer genes are not in themselves generators of cancer. They only become so when they are affected either by carcinogenic substances, ionising radiation or by viruses. Recently various cancer genes have been isolated, for cancer of the colon, the bladder, the kidneys and for a form of leukaemia.

Protein (1953)

A protein can be represented as a chain of amino acids, of which there are 20 different kinds. The English biochemist **Frederick Sanger** was the first to work out the sequence of amino acids in various protein molecules, in **1953**. He determined the sequence for insulin.

Proteins form the main structural components of most animal cells: they constitute connective tissues, skin, hair, ligaments and tendons. Proteins also take part in metabolic processes, when they are called enzymes. Some hormones are also proteins; insulin is an example. Today insulin can be produced by synthesis.

In vitro synthesis of proteins (1954)

In **1954** the American biochemist **Du Vigneaud** achieved the first synthesis of two proteins. This concerned hormones usually secreted by the post-pituitary gland: oxytocin and vasopressin, each of which is made up of nine amino acids. In 1969 the first synthesis of a large molecule was achieved in the USA by Bruce Merrifield and Gutte, of the Rockefeller Institute (later University, New York).

They brought about the attachment of a string of 124 amino acids to each other, anchored to some polystyrene. For this technique, which took over three weeks, they used electronic equipment which enabled 369 chemical reactions to take place automatically, in 11 931 stages.

The Merrifield method is now used to produce numerous syntheses. Moreover, the grafting of two cellular fragments, each belonging to different species, can produce proteins. All these 'miraculous' operations are grouped under the name of genetic engineering.

In 1984 Bruce Merrifield won the Nobel Prize for Chemistry for synthesising chains of amino acids or polypeptides.

THE FIRST PATENTED ANIMAL: A MOUSE

Harvard University (Massachusetts) took out a patent on the genetic transformation worked on mice, the first in the world because no genetic modification of an animal had been patented before. The fathers of this 'invention' (which will be used in cancer research), Philip Leder and Timothy Stewart, have an exclusive right over their creation for 17 years. This is a patent, however, which has given rise to many protests from those who are opposed to genetic experiments. The request was submitted in 1984, and 21 similar requests are being considered by the commission for patents. The *OncoMice*, so-called because they carry human oncogenes, were sold at the beginning of 1989 for $50 each.

Other mice which are much discussed at present, the SCID (Severe Combined Immune Deficiency), are advancing research in immunology thanks to a graft of cells or human tissues which enable the reconstruction in the animal of a part of the immune system of man. Two teams of researchers are working on the SCID: Dr Irving Weissman's, at Stanford University, and Dr Donald Mosier's at the institute of medical biology at La Jolla (California).

And one more: the *transgenic* mice, created by researchers at Integrated Genetics, an American company, in collaboration with a team from Tufts University. With these mice, which have been grafted with a specific gene, they hope to produce certain medicines to alleviate illnesses such as thrombosis.

HLA system (1958)

In **1958**, the Frenchman **Jean Dausset** first described the HLA system (for Human Leucocyte Antigen) in white globules. The HLA system is a series of proteins present at the surface of all the cells of an individual. These proteins are analogous to fingerprints. They vary from one individual to another and in some way may be thought of as a person's identity card. Today it is known that tolerance to grafts depends on the resemblance of the HLA systems of the donor and the recipient.

Immune system (1877)

A little over a hundred years ago, Louis Pasteur came up with the idea that living beings possessed within themselves the means of fighting against sickness. When in **1877** the Russian scientist **Ilya Metchnikoff**, who had discovered that certain cells in living organisms were capable of 'eating' and 'digesting' microbes, passed through Paris, Pasteur asked whether he would join his team. So many stages were involved in the progressive discovery of the immune system, such as 'tissue immunity' described in 1922 by Levaditi and Nicolau, that it was not until the 1950s that a clear idea was formed of the defence system belonging to each organism.

Astronomy

Origins

With mathematics, astronomy is definitely one of the two oldest sciences and the observation of stars probably dates back to prehistory. Born from the needs of daily life as well as from the fears and beliefs of primitive people when faced with unexplained phenomena, it had pride of place as far as the Sumerians and Babylonians were concerned (4th and 5th millennium), and was often confused with astrology, the mystical interpretation of astronomical events to which it owes much. In order to observe the stars the first astronomers used an index bar which was a simple rectilinear rod, as well as a compass, that is to say, a double articulated index bar. They also had the astrolabe – an instrument which allowed one to measure the position of the stars above the horizon – said to have been invented by Hipparchus, the Greek astronomer who lived in the 2nd century BC.

This photograph of the sun setting was taken over a period of five minutes.

Lenses and telescopes

Galileo and the astronomical lens (1609)

It was in **1609** that the Italian scientist, **Galileo Galilei** (1564–1642), first used a magnifying lens to observe the sky. This was his improved version of a lens originally invented in Holland.

Within a short period of time, Galileo made an extraordinary number of discoveries. He discovered the following:

– the marks on the sun which demonstrated a fact which at the time seemed incredible: that the star of the day was an 'imperfect' entity;
– the four large stars of the planet Jupiter: Io, Europa, Ganymede and Callisto, now called the Galilean satellites;
– the phases of the planet Venus, proving that Venus, like the moon, can be lit by the sun from the side. This observation gave substantial support to the theory that the earth revolved around the sun;
– the rings of Saturn, which Galileo could not correctly interpret and which seemed to him to be rather like curious growths resembling ears;
– the myriad of stars that make up the Milky Way;
– the moon's mountains and craters.

In 1633 Galileo was forced by the Roman Catholic Church to renounce the theory of elliptical planetary rotation and was placed under permanent house arrest at his home in Florence where he spent the last nine years of his life.

An eclipse of the sun in Thailand in 1688. Before Galileo perfected the astronomical lens in 1609, the stars had to be studied by the naked eye.

Telescope (1672)

The first telescope was built in **1672** by **Isaac Newton**, the famous British physicist, mathematician and astronomer (1642–1727), who set out in the *Principia* in 1687 the famous laws of universal gravitation. The instrument was perfected in the 17th and 18th centuries, in particular by the British astronomer, William Herschel and the German astronomer, Johannes Hevelius. It was in 1842 that another Englishman, William Parsons, built the first giant telescopes in the garden of his house.

Giant telescopes

Mount Palomar (1948)

The huge telescope on Mount Palomar in California is equipped with a lens that has a 5m *16ft 4.8in* aperture. It was brought into service on **3 June 1948**.

The telescope was initiated by George Hellery Hale, an astronomer and inventor of the spectroheliograph, and built with the aid of a $6 million donation from the Rockefeller Foundation.

Hale's giant telescope has become difficult to use because of the increasing level of pollution coming from the neighbouring city of Los Angeles.

Zelentchouk (1974)

The Zelentchouk observatory telescope, located at an altitude of 2050m *6728ft* in the Caucasus Mountains, Russia, is today the largest in the world. The telescope with its 6m *19ft 8.3in* aperture was designed by Doctor Icannissiani. It went into operation in **1974** after extreme difficulties in manufacturing its enormous 42 tonne lens. This was made by the Lomo centre in Leningrad and is one and a half million times more powerful than the human eye.

Infra-red telescope (1977)

The largest infra-red telescope in the world, equipped with a lens with a 3.8m *12ft 5.7in* aperture, was installed in **1977** by British scientists at an altitude of 4200m *3784ft* on the top of Mount Mauna Kea, an extinct volcano on the island of Hawaii.

Multiple-lens telescope

The world's first multiple-lens telescope was installed at the Mount Hopkins Observatory (Arizona) by the Smithsonian Astrophysical Observatory (SAO) and the University of Arizona. It comprised six lenses with a 1.85m 14ft 9.2in aperture and is the equivalent of a single-lens telescope with a 4.5m 6ft 9in aperture.

This new kind of optical telescope was devised by astronomers in order to reduce the size as well as the cost of the instruments without sacrificing performance.

Keck Observatory (1991)

A sum of $70 million was offered by the **Keck Foundation** at the California Institute of Technology for the construction of the largest optical telescope in the world. Work began on 12 September 1985 on the summit of Mauna Kea, a dead volcano on one of the Hawaiian islands.

The mirror of this enormous telescope which will allow one to observe objects of magnitude 26, that is to say, 200 million times smaller than objects discernible by the naked eye, is to be built using 36 identical hexagonal parts, juxtaposed in order to form a single mirror measuring 10m *32ft 9.9in* in diameter.

It is expected that the telescope will be ready for use by 1991. It will cost some $40 million.

VLT (1997)

The European Southern Observatory (ESO) decided in 1987 to build what will be the largest telescope in the world, the **VLT** or Very Large Telescope. It will take ten years to build but the first part is expected to be ready for use in 1994/5.

The VLT will comprise four astronomic lenses with 8-m diameter mirrors *26ft 3.1in*. A 16m *52ft 6.2in* chamber will extend it further.

It will allow scientists to observe stars inside interstallar clouds, and even to look into the centre of galaxies, and will make it possible to locate potential black holes. The telescope will probably be located in Chile.

The ESO, established in 1962, is a scientific organisation involving eight countries: Belgium, Federal Republic of Germany, Denmark, France, Italy, Netherlands, Sweden and Switzerland.

The VLT, Europe's future telescope, will be composed of four astronomic lenses, each with a 8m 26ft 3.1in *mirror. Together they will form a telescope five times more powerful than any that currently exist.*

FIVE THOUSAND MILLION YEARS AGO ON 9 AUGUST 1988

On **9 August 1988** the European Southern Observatory (ESO) of La Silla in Chile observed the most remote supernova known to date. This supernova exploded in an obscure galaxy, one of a very remote cluster of galaxies situated about five thousand million light years from earth. This means that the explosion observed on 9 August 1988 had in fact taken place five thousand million years before, at a time when the earth and the sun were being formed from a primitive protosolar nebula.

This pile of twisted metal used to be Green Bank radio telescope.

Radioastronomy

First radio telescope (1932)

The very first radio telescope was developed in **1932**, quite accidentally, by an American radioelectrical engineer of Czechoslovak origins, **Karl Jansky**.

A 'continuous whistle'

Employed by Bell Telephone to trace the origins of parasitic signals causing obstructions to radiotelephonic traffic on the North Atlantic, Jansky built a receiver which he had invented himself in Holmdel, New Jersey, USA. With this apparatus he captured the first unfamiliar whistles coming from outer space: 'a continuous whistle' coming from the constellation of Sagittarius, at the centre of our galaxy, 250 000 000 thousand million kilometres *155 350 000 thousand million miles* away.

Although this news caused a sensation, Jansky abandoned radioastronomy in 1938 on account of the indifference which greeted his discovery. However, his name is still used as the terms for the unit of measurement of the radioelectric flux of the stars.

Reber radio telescope (1937)

The first proper radio telescope was built in **1937** by an American engineer of Dutch origin, **Grote Reber**. A keen radio buff, he used his savings to build an instrument in his garden in Wheaton near Chicago (USA). It comprised a parabolic antenna with a diameter of about 3m *9ft 10.2in*, and was azimuthal like all modern radio telescopes. After two years of patience and unsuccessful experimentation, Reber finally received a signal with a wavelength of 1.87m *6ft 1.6in*.

With this success behind him. Reber tried, in 1941 to draw the radioelectric map of our galaxy. This document, published in 1944, marked the beginnings of radio astronomy as a new scientific discipline, officially recognised by astronomers who were eager to use it.

The largest radio telescope in the world (1986)

In July and August **1986** American astronomers from NASA's **Jet Propulsion Laboratory** developed the largest radio telescope ever invented, by electronically linking aerials on earth and on board a satellite. The parabolic aerials measuring 70m *229ft 8.9in* in diameter, located in Japan and Australia, were in contact with one measuring 5m *16ft 4.9in* in diameter belonging to the TDRS communication satellite in a geostationary orbit. The telescope thus created measured 18 000km *11 185 miles* in diameter, that is to say, the largest ever developed. It has made it possible to observe three quasars which astronomers believe to be between 15 and 20 thousand million years old.

Green Bank

The largest radio telescope in the world, the one in Green Bank (Virginia, USA), collapsed during the night of 15 to 16 November 1988. It had been set up in 1962 and comprised mainly a giant parabolic antenna measuring some 100m *328ft 2.4in* in diameter. Thanks to its mobile antenna, this radiotelescope (one of the few to possess such an antenna), was capable of covering the whole northern hemisphere of the sky and, moreover, was situated in an area protected from parabolic radio emissions.

Radio-astronomic observatories (c.1950)

The first radio-astronomic observatories were created in Cambridge, England, and Sydney, Australia, after the Second World War, thanks to the progress made in the areas of radar and electronics during that conflict.

The International Union of Telecommunications (IUT) contributed valuable aid to radio astronomers by deciding, in 1959, to reserve a frequency band of approximately 1420MHz exclusively for the study of signals emitted by cosmic hydrogen over 21cm *8.27in* in wavelength.

GIANT RADIO TELESCOPES

Type	Date	Origin	Characteristics
Jodrell Bank	**1957**	GB	The first large radio telescope to be put into operation. Designed by the British physicist Sir Bernard Lovell.
Nançay	**1958**	France	Installed in Sologne, the Nançay radio telescope is a semi-mobile instrument with an adjustable reflector.
Eifelsberg	**1962**	W. Germany	The largest radio telescope with a mobile antenna. Equipped with a parabolic antenna measuring 100m *328ft 2.4in* in diameter.
Ratan 600	**1974**	USSR	The largest Russian radio telescope.
Arecibo	**1974**	USA	Set up on the island of Puerto Rico, this is a fixed-antenna instrument the reflector of which comprises 38 778 sheets of aluminium. Inaugurated in 1963; put into operation in 1974; under the SETI (Search for Extraterrestrial Intelligence) programme, a 169-second message was sent at a frequency of 2380MHz towards the M13 global star cluster (100 000 stars) in the Hercules constellation. Distance: 25 000 light years. Possible reply: in five hundred centuries' time!

A double rainbow forms over the largest radio-interferometer in the world: the Very Large Array in America.

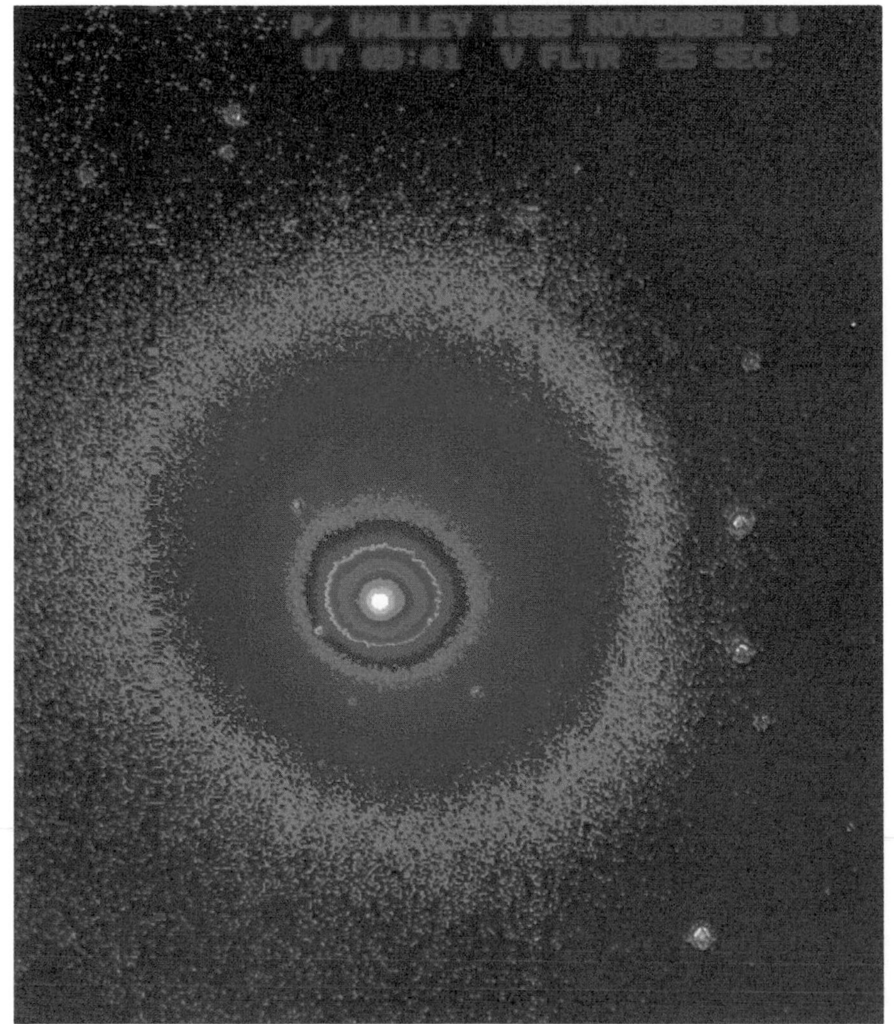

Halley's Comet photographed on 14 November 1985 by scientists at the University of Arizona. The colours represent the different layers of brilliance, the white dot corresponding to the brightest part.

One such transmission, which had been announced by the young Dutch astronomer H. C. van de Hulst in 1944, was indeed detected in 1951 by the American physicists H. I. Ewen and E. M. Purcell, with the help of a spectroscope.

Radio-interferometer (1960)

The synthesised aperture interferometer was developed in **1960** by the British radio astronomer, **Sir Martin Ryle**, who in 1974 received the Nobel Prize for Physics with his colleague Anthony Hewish. Synthesised aperture interferometry is a method which, using two or more antennae, allows one to gather simultaneously with a single receiver several signals from the same source and to make a true chart of the area observed with a resolution similar to what would have been obtained by using a very large instrument.

The first large radio-interferometer was developed by Sir Martin Ryle in 1964 at Cambridge (UK), by using three telescopes 1.6km *a mile* apart.

Radar astronomy (1946)

Radar astronomy is a particular technique in which radio telescopes are no longer used simply as passive receivers but like radars, that is, in an active manner, first as transmitters, then as receivers.

Since **1946**, radar echoes with rather weak transmission power have been obtained from the moon, as well as from other bodies, in order to measure their distances and their movements. This same technique was recently used with the Arecibo radio telescope to establish the first topographical map of the surface of the planet Venus.

Airborne telescopes

First observations (1927)

The first astronomical observations carried out by aeroplane date back to **29 June 1927**, when a twin-engine British Imperial Airways plane was used to photograph a total eclipse of the sun above the London fog. This observation technique has since been used successfully with Concorde.

VLA (1977)

The largest radio-interferometer is the Very Large Array (VLA), built near to Socorro in the New Mexican desert in America.

The VLA comprises 27 metal parabolas measuring 25m *82ft* in diameter, distributed along three bases 25km *15 miles 942yd* long arranged in a Y shape. The whole unit forms the equivalent of a giant dome measuring 27km *16 miles 1379yd* in diameter.

Kuiper Observatory (1965)

Since **1965 NASA** has been equipped with observatory jet aircraft, the Learjet and the Convair 990 which have enabled scientists to discover, in particular, the infra-red emission from the centre of our galaxy. Since 1975 NASA has also used a giant four-engined C 141 Starlifter, specially equipped for astronomy and, in particular, for research into infra-red radiation. This vehicle, named the Kuiper Observatory after the famous American astrophysician of Dutch origin, Gerard P. Kuiper

(1905–73), may be used on more than two hundred nights a year. It is capable of flying for three hours with nine tonnes of observation equipment, at an altitude of more than 14 000m *45 948ft*. The Kuiper Observatory, equipped with a telescope with a 91cm *35.85in* aperture, has already made several important discoveries possible such as, for example, the discovery of the rings of Uranus in 1977.

Space astronomy

Space radio astronomy (1968)

The first radio astronomy satellites were those of the Radio Astronomy Explorer (RAE 1 and 2) from **NASA**, launched in July **1968** and June 1973 respectively.

The first radio telescope in space was put into orbit by the USSR in July 1979 on the orbital station of Salyut 6. This radio telescope, KRT 10, weighing 200kg *441lb*, had an antenna with a diameter of 10m *32ft 9.8in*, carried by a cargo vessel called Progress, and sent into orbit for the purpose of observing centimetre and decimetre waves. The Soviets carried out their interferometry observations of various radio sources in this way in conjunction with an observation station in the Crimea.

Astronomic satellites (1946)

The first observations by means of astronomic satellites were carried out in **1946** by Americans who used V2s salvaged in Germany. However this technique did not allow for long observations. The first astronomic satellites were launched by the United States, (notably Explorer 1 in 1958, under the direction of Joseph Van Allen, enabling the radiation rings circling the earth to be discovered: the Van Allen rings) and by the USSR (the Cosmos satellites, beginning in 1962, which are not exclusively for the purposes of astronomy).

Uhuru (1970)

Launched from Kenya on **12 December 1970**, Uhuru was the first satellite to be dedicated to X-ray astronomy.

After Uhuru, other X-ray satellites were launched. In 1978 the Einstein satellite was launched in commemoration of the centenary of the birth of the great physicist in 1979. Let us not forget the European satellite, Exoset, launched in 1983 by an American Delta rocket whose mission was completed in 1986.

The next X-ray mission, that of NASA's AXAF satellite (Advanced X-Ray Astrophysics Facility), is planned for the 1990s and is expected to last ten years.

COS-B (1975)

Launched in August **1975**, the COS-B satellite, developed by the European Space Agency (ESA), was wholly dedicated to gamma astronomy or high energy astronomy. The mission finished at the end of April 1982.

The COS-B's mission will be continued during the 1990s by NASA's GRO satellite (Gamma Ray Observatory).

IUE (1978)

The International Ultraviolet Explorer was launched on **26 January 1978** as a joint venture between NASA, the European Space Agency and the Engineering Research Council.

Its mission was to observe the stars and, principally, the sun within the wavelengths of the ultraviolet spectrum. Among the data received by this satellite, are some of prime importance with regard to the origin of life on earth.

It is the only ultraviolet explorer in operation and will remain so until work is completed on the Hubble Space Telescope.

IRAS (1983)

Launched on **26 January 1983** by a Thor-Delta rocket, the IRAS (Infra Red Astronomy Satellite) stopped functioning on 22 November of the same year. Built with the cooperation of the United States (NASA), the United Kingdom and Holland, its purpose was to transmit data on bodies in the universe too cold to emit light visible to the human eye.

In the ten months it was in operation, IRAS drew up a cartographical map of the whole sky and found some 200 000 sources of infra-red rays.

The continuation of its mission will be guaranteed from 1992 by the European Space Agency's (ESA's) satellite observatory.

Astro-C (1987)

On **5 February 1987** the Japanese Institute of Space and Astronomy Sciences (ISAS) launched the Astro-C satellite, the purpose of which was to observe neutron stars and perhaps black holes. The satellite, equipped with X-ray and gamma equipment, is presently in orbit at an altitude of about 530km *329.34 miles*.

Space telescope (1990)

Planned for several years, the Hubble Space Telescope (NASA's space telescope in collaboration with the ESA) will be deployed for the purposes of transmitting to earth pictures free from atmospheric impurities, that is to say, exceedingly clear images. Two cameras, two spectrometers and a photometer will allow the telescope to function within the visible, infra-red and ultraviolet spectrum. Hundreds of experiments are planned for the 15 years it is expected to be in operation.

Hipparcos (1989)

The European Space Agency's scientific satellite, Hipparcos, called the 'surveyor of the stars' by astronomers, was launched by the European rocket Ariane-4 in July **1989**. It weighs 1140kg *2513lb*.

Placed in geostationary orbit at an altitude of 36 000km *22 370 miles*, the purpose of this space surveyor is to plot, with a precision one hundred times greater than could be previously achieved, the coordinates and movement of 100 000 stars in our Milky Way of which there are in total 100 000 million.

On board the satellite, two cameras, the optics of which diverge by 58 degrees, will record the images of the stars in our galaxy and enable astronomers to build an exceedingly precise series of three-dimensional charts.

The performance of the optics of these cameras is such that, if the earth were flat, Hipparcos would be able to see, from the summit of the Eiffel Tower, a pigeon's egg placed on the top of the Statue of Liberty in New York!

The USSR plans to launch a giant telescope into space around the year 2000. This telescope will comprise a mirror measuring 10m *32ft 9.8in* in diameter and should be more powerful than the US Hubble Space Telescope which has a mirror measuring 2.4m *7ft 10.5in* in diameter.

Hipparcos will revolutionise astronomy.

Astrophysics

Cepheid stars

The cepheid stars take their name from the first among them: Delta Cephei, discovered in 1784 by the British astronomer John Goodricke. These relatively rare stars (today only 700 of them are known) have a variable luminosity or magnitude.

They are of great importance since they enable scientists to calculate distances within the universe. In fact, there is a linear relationship between the rhythm of the variation of their brightness and their average luminosity. This relationship was proven by the American astronomer, Henrietta Leavitt, in 1912.

The Pole Star is a cepheid star whose period of variation is approximately four days.

White dwarfs

Some stars, whose mass does not exceed that of the sun by more than one and a half times, cave in at the end of their lives giving rise to celestial bodies called white dwarfs.

In 1779

It is not possible to give a date on which a white dwarf was first discovered, but that of the planetary nebula, the Lyre, was sighted by **Antoine Dargulier** in **1779**. The existence of Sirius's companion, Sirius B, was calculated in 1834 by the German Friedrich Bessel, and observed in 1862 by the American optician and astronomer Alvan Graham Clark. Neither of these men understood the exact nature of the subject of their observations.

Eridani B (1910)

It was in **1910** with the discovery of Eridani B that scientists became curious about this object with its peculiar temperature and density.

In 1917 it was established that Sirius B and Van Maanen star had the same characteristics as Eridani B.

A short time afterwards, the quantum theory was to provide answers to questions posed by astronomers: it is a quantum principle, called the Pauli principle which enables the white dwarf stars to be stabilised.

Neutron and pulsating stars (1967)

Although their existence was heralded in 1932 using the quantum theory, it was not confirmed until more than 30 years later.

They were discovered in **1967** by the British astronomer **Jocelyn Bell** when she was working with Anthony Hewish at the Radio Astronomy Observatory at the University of Cambridge, England.

In 1964 she had detected in the sky a very regular radio source which emitted a signal every 1.33730113 seconds. The Cambridge team came to the conclusion that it was a neutron star, the remains of a large star which, after exploding, left a residue the density of which is measured in millions of tonnes per cm^3 with an approximate diameter of 15km *9 miles 565yd.*

The pulsars (or pulsating stars) discovered by Jocelyn Bell are neutron stars which, in emitting radio signals, are like locating points in the universe.

There are now more than three hundred pulsars known, the frequency of which varies between more than 620 times per second to once every 4.3 seconds.

The European satellite Exosat took this X-ray picture of the Milky Way.

Black holes

The existence of black holes comes within Einstein's theory of general relativity, but was predicted by the French astronomer and mathematician Pierre de Laplace in 1796.

These are remains of giant stars the extreme gravity of which would prevent even electromagnetic waves from escaping, including light waves. Their mass would bore a hole in space from which nothing could escape. Such objects could only be detected by the effect produced on the area surrounding them, in particular a nearby star.

Astrophysicists are therefore carrying out research on systems where stars seem to orbit around invisible bodies the mass of which must be superior to that of a neutron star. At present there are three systems which could be considered as serious candidates: Cyg X-1, LMC X-1 and LMC X-3. However, they all lend themselves to different interpretations and cannot at present be considered as undeniable proof of the existence of black holes.

Quasars

During the 1950s and 1960s radio astronomers drew up a chart of celestial radio sources. The University of Cambridge, England, drew up a catalogue of them entitled 3C, considered to be the most important on the subject. By observing, in 1961, one of the radio objects discovered by Cambridge, the astronomer Matthews managed to locate it very precisely. Using a telescope, he discovered a star which was shown to be rather special. In 1963 Hazard, Mackey and Shimming made a second similar discovery.

Distant galaxies

After examination of the results of the objects observed, which were named quasars (quasi-stellar radio source) in reference to their stellar form, were identified as being very remote and very bright galaxies.

More than 20 years after their discovery, it is not known which source of energy feeds the quasars to make them so bright and so active. At present about 3000 quasars have been recognised and recorded by astronomers, one of which, discovered in 1986, is the most distant object ever observed.

The distance separating us from the quasars is such that it cannot be measured either in light years or in parsecs (1 pc = 3.26 light years), because there is no point of reference on which to base a measurement. To express their remoteness one uses the quantity of red in the light we receive from them.

Gravitational mirages (1979)

One day in **1979** the British astronomers **Carswell**, **Walsh** and **Weymann** were extremely surprised to find that two quasars close to each other in the sky were each emitting an absolutely identical light.

It was not long before this phenomenon was explained. It is known that light is not generally propagated in a straight line but that it is sent off course by the field of gravitation of heavy bodies. The astronomers were seeing two images of one and the same quasar, the light of which had reached earth following two different trajectories passing on either side of an enormous galaxy situated between the quasar and us, and acting as a light deflector.

This quasar is now called the double quasar and the phenomenon of the multiplication of images is called gravitational mirage. Many other examples have been discovered since 1979.

Einstein's rings (1987)

A different type of gravitational mirage was discovered in **1987** by an observation team in **Toulouse**. This team made the observation that the arc of light in the star cluster of galaxies Abell 370 come from further afield than does the star cluster itself. It is very likely,

ONE OF THE MOST FASCINATING OBJECTS IN THE UNIVERSE

The PSR 1913+16 binary pulsar (that is its name!) is one of the most fascinating objects in astrophysics. It is a pulsar (or pulsating star) which circles round another star. The latter is invisible but is probably also a neutron star. The pulsar revolves about every 7 hours 45 minutes.

PSR 1913+16 was discovered in September 1974 by American radio astronomers Hulse and Taylor with the help of the large Arecibo telescope. These two men realised immediately that the pulsar and its invisible companion could enable important tests to be carried out regarding the theory of gravity. In fact what was being observed were two neutron stars – stars which create an intense gravitational field – existing in close orbit, with one revolving around the other. Moreover, PSR 1913+16 pulsar itself rotates in a very stable manner and therefore acts as a clock, allowing scientists to measure the time and carry out very accurate tests.

Five years after the discovery of the pulsar, in 1979, Taylor and his collaborators announced that they had observed a regular reduction in its orbital period of 7 hours 45 minutes. At each rotation the period reduces by a tiny amount and it can be demonstrated that this effect is due to the emission of a gravitational wave.

Since 1979, therefore, astrophysicists have held to the theory that gravitational waves certainly exist in nature.

therefore, that this is what is called an Einstein ring, that is to say, a deformed image which looks like the ring of a distant quasar as a result of a gravitational mirage due to Abell 370. An image such as this, in the shape of a ring, very occasionally appears in cases where the quasar, the star cluster and earth are perfectly aligned along the same line of sight.

The most distant galaxy (1988)

The 0902+34 galaxy is the most remote galaxy ever observed. It was discovered in **1988** by **Simon Lilly** of the University of Hawaii using a Canada–France–Hawaii telescope. Only a few quasars are found further away than this galaxy.

The 0902+34 is observed as it was at a time when the universe was only 15 per cent of its present age. The galaxy is, of course, young, and its light seems to indicate that intense star formations occur there.

The existence of such a distant galaxy demonstrates that the universe itself cannot be much less than 15 thousand million years old.

Hidden mass in the universe (1913)

Since the time of the pioneering work by the American astronomer **Fritz Zwicky** in **1913**, it has been very widely believed that 90 per cent of the total matter in the universe is invisible. In fact, all the indications would lead one to believe that, surrounding the galaxies, and in particular our galaxy, there is a giant spherical halo composed of invisible matter. What is the composition of this matter? Black holes? Neutrinos? Numerous theories have been put forward over the last few years, none of which is entirely satisfactory.

From Big Bang to Big Crunch (1965)

The discovery by American radio astronomers **A. Penzias** and **R. Wilson** in **1965** of a cosmic radiation at 3 degrees celsius, indicates that our universe began life about 15 thousand million years ago with an enormous explosion known as Big Bang. If it is true that 90 per cent of matter in the universe is invisible, it means that our universe will probably end within 80 thousand million years with an enormous implosion known as the Big Crunch, the opposite of the Big Bang explosion.

The enigma of solar neutrinos

Since the work of the German-born American Hans Bethe in 1938, it is believed that the nuclear reactions occurring inside the sun, which also serve to supply the latter with light and energy, are well understood.

All the calculations demonstrate that the sun must emit, on account of its nuclear reactions, a certain quantity of very light particles which are called neutrinos. The detection of neutrinos coming from the sun has been undertaken by several laboratories throughout the world, in particular the Davies laboratory in America.

All the experiments show that the sun, mysteriously, emits only a third of the calculated number of neutrinos. Could this be because neutrinos exist in three different forms and that not all these forms are recognised? Astrophysicists are puzzled and some of them have their doubts as to our understanding of the inside of the sun.

Rockets

Rocket (13th century)

The first rockets were, it seems, built in China and India in the distant past. According to Chinese legend, a mandarin once attempted to fly hanging from two paper kites driven by a large number of rockets. What is certain is that some rockets were in operation as early as the 13th century: Tartars used them in 1241 at the Battle of Legnica (lower Silesia, now in Poland).

The only rocket engine used today is the internal combustion rocket engine which uses chemical fuels. Other types of rocket engines, nuclear and electrical, are still only at the planning stage.

Space rocket (1957)

Although the origins of space rockets goes back some 40 years, in particular to the German V2, it was only in **1957** that a rocket escaped from the earth's gravitational pull and sent the famous Sputnik into orbit. Since 1957 more than 2500 rockets have been launched.

First liquid-fuelled rocket (1895)

The exact date of the first liquid-fuelled rocket is not known. Certain people date the invention back to **1895**, when a Peruvian engineer, **Pedro P. Poulet**, built and patented an engine with a 10cm *3.94in* diameter into which was injected a mixture of nitrogen peroxide and petroleum, lit by a candle. Americans readily attribute the first rocket to their compatriot Robert H. Goddard.

The first European to launch a liquid-fuelled rocket, was the German Johannes Winkler whose small rocket HW I (with liquid oxygen and methane) was launched on 21 February 1931 in Germany at Breslau (now Wroclaw, in Poland).

Ariane (1979)

Launched for the first time in December **1979**, the Ariane rocket was invented and perfected for a very precise task: to place satellites in geostationary orbit. For this task, which requires a satellite to be positioned at an altitude of 36000km *22370.4 miles* above the equator, the rocket is assisted by the location of its launching site, Kourou in French Guiana, which is only five degrees from the equator.

Although it has a launching capacity of 1750kg *3858.75lb* for the geostationary orbit (GTO), Ariane 1 was abandoned in February 1986, at the time that Spot was launched, in favour of models 2 and 3 which were capable of launching 2600kg *5733lb* GTO.

It is worth noting that Ariane is the European space rocket, built with the participation of ten western European countries: Belgium, Denmark, France, Great Britain, Italy, Netherlands, Spain, Sweden, Switzerland, West Germany. Members of the European Space Agency (ESA), decided in 1973 to build Ariane, and construction was carried out by about 60 companies from the ten participating countries.

Ariane 4 (1988)

1988 saw the arrival of Ariane 4, the first launcher in the world to provide different options. According to the wishes of the customer and the mass and form of the satellites to be launched, there could in theory be about 40 different versions of the rocket. In practical terms, however, Arianespace will market six different versions, the differences between which will be based on the form and volume of the cap containing the satellites, and on the extra engines which will run on either solid or liquid fuel. Depending on the version, the rocket has a capacity of between 1900 and 4200kg *4188lb and 9262lb* GTO.

Ariane 5

During the last few years of this century, Ariane 5, with its entirely new conception, will be able to devote itself to several missions. Capable of placing 6800kg *14994lb* in geostationary orbit, or 21 tonnes in a lower orbit of 500km *310.7 miles*, the rocket will also be able to launch the space-craft *Hermès*.

Rockets attached to guiding sticks were illustrated in this collection of German water colours.

Space shuttle (1981)

The NASA space shuttle is the first rocket in the history of astronautics that can be recovered and used again. It is a real aerospace vehicle that weighs more than 2000 tonnes and lifts off like a rocket: vertically. The main part, the orbiter, is a kind of delta-winged aircraft that weighs 100 tonnes and is placed into orbit around the earth at low altitude (160 to 1100km *100 to 683.54 miles*). The orbiter re-enters the atmosphere as a glider, and lands on the runway horizontally, like a plane. The shuttle can carry a 39-tonne payload and a crew of between four and seven people, two of whom are the pilots.

Thanks to its jointed arm, the shuttle can place all sorts of satellites into orbit. This revolutionary rocket prefigures the space-ships of the future, which, like all other means of transport, will be reusable.

Decided upon by President Richard Nixon in 1972, the shuttle (as well as its prototype *Enterprise* which was used for test landings) was constructed principally by Rockwell International.

The first shuttle, *Columbia*, had its maiden flight on **12 April 1981**, with **John Young** and **Robert Crippen** at the controls. *Columbia* lifted off from Cape Canaveral, Florida, and landed 54 hours later at Edwards Air Force Base in California.

The second shuttle, *Challenger*, had its first flight in April 1983 but was destroyed in a tragic accident on 28 January 1986. The third shuttle, *Discovery*, had its first flight in August 1984, and the fourth, *Atlantis*, in 1985.

Walking in space

Some highly successful missions have been accomplished by space shuttles and a particularly spectacular one was that by *Challenger* between 4 and 9 April 1983. During this, two astronauts, Story Musgrave and Donald Peterson, tied to the space ship, 'walked' in space. On a *Challenger* trip in February 1984 Bruce McCandless was seen riding a 'scooter' in space.

28 September 1988

At 11.37am on **28 September 1988** the space shuttle *Discovery* was launched, thus marking the United States' renewed interest in manned space flights after an interval of 32 months. A second flight, that of *Atlantis* (the purpose of which was military), followed on 2 December 1988. NASA plans nine space shuttle flights in 1990.

Hermès, the European space shuttle (1997)

The beginning of the preparatory programme for the European space shuttle *Hermès* was originally announced in December 1986. In charge of the shuttle will be the French company Aérospatiale, with Dassault responsible for the aeronautics. The final design for *Hermès* will be completed in 1991. The first automatic space flight is planned for **1997**. The first manned flight is planned for 1998.

Hotol

A project running in competition with *Hermès* is the British rocket-propelled plane, *Hotol*, designed by Alan Bond of the Atomic Energy Authority (AEA).

Hotol would be able to take off from ordi-

Discovery *was the third of America's space shuttles.*

New boosters are tested in the desert.

BOURANE, A SPACE SHUTTLE NICKNAMED SNOW STORM

The Russian space shuttle, *Bourane* or snow storm, was first launched on 15 November 1988 from the cosmodrome of Baykonyr in Kazakhstan. Propelled by a giant rocket called Energia, *Bourane* (which weighs 105 tonnes at take-off) lifted off at 6am Moscow time unmanned, in automatic mode. Forty-seven minutes later, the shuttle was circling in orbit. After going twice round the earth at an altitude of 250km *155 miles 616yd*, *Bourane* switched on its retrorockets at 8.30am and at 9.25am had returned to a runway 12km *7 miles 804yd* from its launch site.

If the landing of the Russian shuttle can be compared to that of its US cousins, *Discovery*, *Atlantis* and *Columbia* (all land without an engine, gliding down), the lift-off is very different. Unlike the US shuttles, at the moment of lift-off, *Bourane* is in a totally passive state: the four engines on the first stage of Energia (800 tonnes of thrust each) ignite at the same time as the rockets on the second stage (200 tonnes of thrust each). The first burn kerosene and oxygen, the second, hydrogen and liquid oxygen.

After 10 minutes 30 seconds of propulsion, *Bourane* separates from the rest of the rocket at an altitude of 110km *68 miles 623yd*. Finally at 160km *99 miles 746yd* altitude, *Bourane*'s principal engine comes into action and propels the shuttle to its circular orbit, at an altitude of 250km *155 miles 616yd*, which it reaches 47 minutes after lift-off.

Capable of carrying a crew of two to four cosmonauts as well as six passengers (scientists and mission experts), *Bourane*, just like the American shuttles, has an enormous cargo hold: 4.7m *15ft 5.1in* is diameter by 18.3m *60ft .73in* long. It can lift off carrying a 30 tonne load and is able to bring back 20 tonnes to earth (maximum landing weight: 82 tonnes).

Bourane's second flight will again take place without a crew and it will not happen before the end of 1990. The shuttle will take passengers only after 1992.

The Russian shuttle project first began in the second half of the 1970s. After long years of secrecy, it was in the spring of 1983, at Bourget, that the Russian cosmonaut, Igor Volk, who will probably be *Bourane*'s first pilot, revealed its existence.

nary airports and will be equipped with a hybrid propulsion system, 'breathing' the surrounding air in the atmospheric phase of the flight. However, at present the project seems to be facing numerous problems.

Sänger (2004)

Parallel to the European project *Hermès*, West Germany plans to have its own space programme which could be implemented around 1995.

The main project in this programme is the space shuttle *Sänger*, named after the German physicist, Eugen Sänger (1905–64), one of the pioneers of the development of liquid-fuelled rockets. Like *Hotol*, the German space shuttle would not need a rocket launcher. Preliminary studies will begin, but the decision whether or not to built *Sänger* will not take place before the year **2004**.

Endeavour (1989)

On 10 May 1989 President George Bush named the fourth US space shuttle *Endeavour* after James Cook's vessel. In August 1768, during his first expedition on board the ship of that name, the British explorer had observed and taken notes on the trajectory of Venus. The new American shuttle is being built at the Rockwell workshops.

Satellites

Transit 1 (1960)

This satellite, launched by the United States in **1960**, was the first to provide navigation aid to vehicles on earth. Using special equipment, a vehicle on earth can ascertain its position through triangulation.

Vostok 1 (1961)

The first manned space-craft, Vostok 1, was also the first artificial satellite with a man on board. On **12 April 1961** the young cosmonaut

Vostok 1, the first manned space-craft, went into space on 12 April 1961.

Yuri A. Gagarin circumnavigated the earth at an altitude of 327km *203.2 miles* in 108 minutes. The first woman in space (and the sixth and last flight of a Vostok) was Valentina Tereschkova in 1963.

Landsat (1972)

In **1972 NASA** launched Landsat 1 under the name of ERTS-1 (Earth Resource Technology Satellite). Observation satellites like Landsat 1 are used for a variety of purposes: cartography; in the search for minerals or water; to follow the movement of swarms of locusts, etc.

Microsatellites (1990)

Known as Lightsats, Piggy bags, Secondary passengers. . . . although others have preferred to call them auxiliary loads. One thing is certain: microsatellites have a future ahead of them since a study carried out by the company Arianespace expects there will be a dozen satellites launched in **1990**, 20 in existence by 1993 and about 40 by 1996.

These microsatellites would however have to adhere to certain rules. They must be electrically 'transparent' – there must be no interference with the launcher – not require any

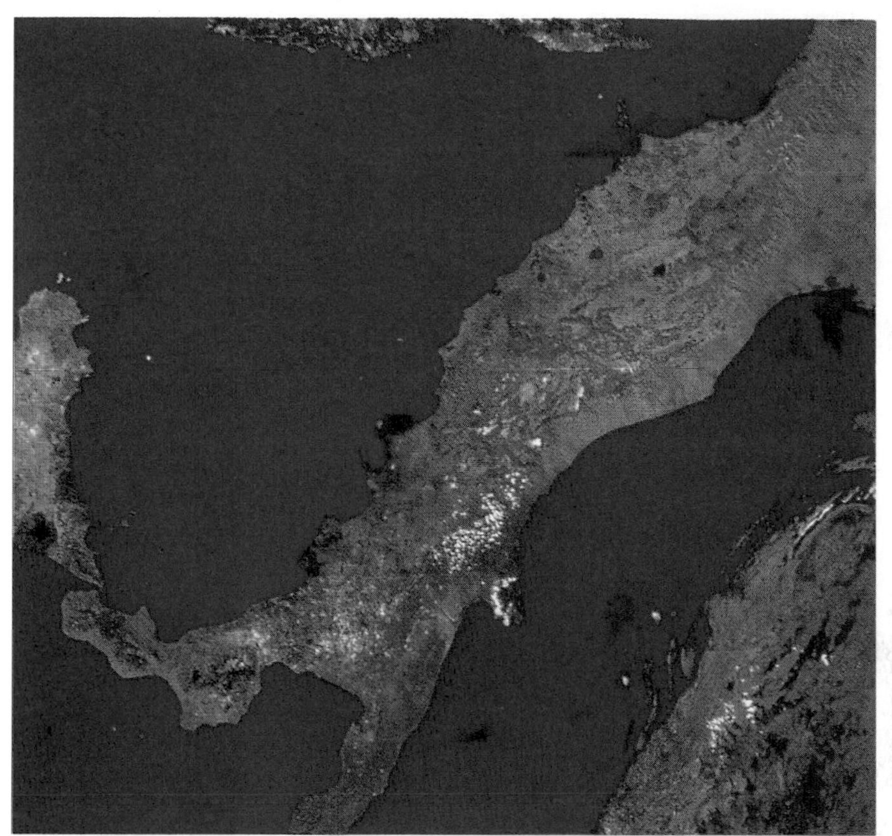

Italy as seen by the American satellite, Landsat.

significant modifications, they must not go above a certain volume (250kg *551lb 4oz* in total per Ariane launcher) and they must not have any impact on the insurance conditions of the main satellite and launcher.

Orbital stations

Salyut (1971)

The USSR was the first country to launch a manned orbital station. Since April **1971** the Russians have put into orbit seven Salyut orbital stations. The last to date, Salyut 7, was launched on 19 April 1982. It was manned by seven Soviet cosmonauts one of whom was the second woman cosmonaut ever.

Although clearly smaller than the American orbital station Skylab, this Russian station is capable of carrying a crew of four cosmonauts for one week. However long flights lasting several months are generally manned with a two-person crew.

Mir orbital station (1986)

The Mir orbital station (*Mir* means peace in Russian), which has been going round the

Vladimir Titov and Moussa Manarov established a new record by remaining in space from 21 December 1987 to the same day a year later. They were joined in their last month by the Frenchman, J.L. Chrétien.

earth at an altitude of between 300 and 400km *186.4 and 248.6 miles* since **29 February 1986**, entered the record books for the longest human stay in space: 366 days passed on board the station by the pair Moussa Manarov and Vladimir Titov. Having left earth on 21 December 1987, the two men returned a year and one day later.

It is worth noting that during this last mission, the Russian station (15m *49ft 3in* in length, 4.2m *13ft 9.4in* in diameter with a useable space of 100m³ *3531.6cu ft*) for the very first time held six cosmonauts at one time, three of whom remained several months on board the Mir before returning to earth.

What is the purpose of the station?

The purpose of the station – which is locked alongside the astrophysical module Kavant – is multiple and includes the following: first, to analyse changes undergone by the human body during a long period in a weightless state (reduction in muscles, weakening of bones etc). The main purpose of this is to prepare for long interplanetary trips such as to Mars in around 2020; second, to carry out observations of earth, as well as of the planets and stars which surround us; finally, to carry out experiments in an atmosphere which is weightless such as the creation of alloys and new crystals, preparation of ultra-pure medicines, experiments on the growth of plants and so on.

New modules (1989)

In order to enhance further still the performance of the orbital station, the Russians have made plans to equip it with new modules. In the spring of 1989, a service module weighing about 20 tonnes where the astronauts will find Manned Manoeuvring Units (or MMU), was placed at one of the six Mir access points.

Subsequently, during the second half of 1989 a module specialising in crystallography and the creation of new substances was put in place. Finally, in 1990, the fourth and last module will be put in place which will be equipped with medical equipment and a laboratory, making it possible to produce ultra-pure medicines.

The orbital station equipped with four modules (astrophysics, amenities, crystallography and medical) will be able to function to its full capacity. We are now awaiting the next stage, when, in the 1990s, the central structure of a new orbital space laboratory weighing 70 tonnes will be sent into orbit.

NEWS OF SEMYORKA

It is Semyorka which launches the manned Soyuz vessels as well as the automatic Mir vessels that take food, water, fuel, air as well as materials for different experiments, to the orbital stations.

The Semyorka rocket (meaning Seventh in Russian) was first launched on 16 November 1963 with the satellite Cosmos 22. Christened R-7 by Sergei Korolev – father of Russian cosmonautics – more than a thousand have been built since then.

Experimental space station (1984)

Christened LDEF (Long Duration Exposure Facility), this experimental space station was jettisoned in 1984. It was the size of a coach, weighing 12 tonnes and was to have been returned to Earth ten months later. However it was found impossible to adhere to this schedule on account of delays which had built up in NASA's programme. It was therefore recovered by a shuttle after spending more than five years in space, exposed to meteorites, cosmic rays and corrosion.

***Freedom* (1999)**

The LDEF is the forerunner to the permanent space station *Freedom* which should be fully operational by about **1999**.

This orbital station will not be solely American, since Europe (the ESA) will be represented by *Columbus*, the European part of the future station, which is to comprise the following: a permanently habitable module laboratory linked to the station; an autonomous module, the MTFF (see below) which will fly in conjunction with it but at different altitudes; a platform in polar orbit containing meteorological and observational instruments.

The American orbital space station was launched in 1984 by President Ronald Reagan and it is an international civil programme. The first part of the programme comprises a 'girder' carrying at its centre of gravity an accommodation module which is permanently occupied and three laboratory modules (the US module, *Columbus* and JEM, the Japanese Experimental Module).

MTFF (2001)

Under the auspices of the ESA, a team of European industrialists (Aérospatiale, Dornier, General Technology System, Casa, Marconi Space Systems) is carrying out research into a series of space station designs which are entirely European. Its first project is the MTFF (Man Tended Free-Flyer).

The space station Freedom *will become operational in 1999. It will comprise living quarters and three laboratories.*

WHAT DO THESE MYSTERIOUS 'COSMOS' SATELLITES DO?

The first Cosmos satellite was sent into orbit by the USSR on 16 March 1962. To date, nearly 2000 Cosmos have made the journey into space, but what do they do?

There are certainly spy satellites, of which there are two types: those that take photographs of sensitive areas, belonging to friends and potential enemies alike; and those that bug telecommunications. With regard to this, it is worth remembering the story of Cosmos 1402 which was launched on 30 August 1982 and which fell back to earth on 24 January 1983 after causing some alarm. This military oceanic satellite was equipped with a nuclear radar system made of enriched uranium (45kg *99lb 3.6oz* radioactive matter!). In the end, Cosmos 1402 disintegrated over the Indian Ocean, burning in the dense strata of the atmosphere.

There are also scientific satellites such as Cosmos 1514 which was launched on 14 December 1983 taking two rhesus monkeys into orbit aged three and four respectively, weighing about 5kg *11lb .4oz.*

However, there are also enormous orbital modules which are called Cosmos to render them commonplace. Such was the case with numbers 1267 and 1443. The first had been moored to the orbital station, Salyut 6, on 19 June 1981 comprising a space train weighing 34 tonnes. The second one was linked in April 1983 to Salyut 7, to form a unit of 40 tonnes in space. These 'carriages', each weighing about 20 tonnes, transported materials needed by future crews of the orbital stations, as well as scientific instruments and provisions. They are already overshadowed by the modular unit built at a later date using the Mir station.

At first the MTFF, which would be circling at an altitude of between 400 and 450km *248.56 to 279.63 miles*, would be linked with a small accommodation module to hold two astronauts for missions lasting three months. This option could be available around **2001**.

To be known as ESS-1 (European Space Station), this mini-station, which is to be sent into orbit by two flights of the future Ariane 5, would have a payload of 2500kg *5512.15lb*. It would need two flights by *Hermès*, one to take the astronauts, another to bring them back to earth. A logistic vehicle, sent into orbit by a conventional rocket, would take charge of the transportation of provisions and fuel during the course of the mission.

SOS orbital stations (1987)

The first space rescue capsule exists on paper: it was presented in **1987** by **British Aerospace**. It is intended that this 'lifeboat', to be based on the permanent space stations, should play the role of rescuer to space ships unable to get their crew rapidly back to earth. Launched by a rocket and then linked to the station, it would be used for emergency evacuations due to, for example, ill-health. A case in point was that of the Soviet cosmonaut Alexander Laveikine who was working on the Mir and was brought back to earth in August 1987 because of cardiac problems which could not be dealt with by the Soyuz crew.

Space vehicles

LEM lunar module

The Lunar Exploration Module (LEM), nicknamed space spider on account of its shape, was developed by the American company **Grumman** to make safer those phases which are without a doubt the most critical in space travel, that is to say, landing and take-off. This 15 tonne 'space lift', built in aluminium and covered with gold leaf, allowed a crew of two astronauts to land on the moon within 12 minutes and to take off within seven. They could then rejoin the Apollo cabin which was revolving round the moon at an altitude of about 100km *62.14 miles*, with the third astronaut involved in the mission on board.

The LEM performed without breaking down throughout all the Apollo flights. On 11 April 1970 it even saved the crew on the Apollo 13 mission after the explosion in mid-flight of an oxygen tank in the cabin.

First walk on the moon (20 July 1969)

It was during the Apollo 11 flight that on **20 July 1969** two Americans, **Neil Armstrong** and **Edwin 'Buzz' Aldrin**, made an old dream come true: walking on the moon. They stayed there for 21 hours 36 minutes.

Abandoned after the last Apollo flight in 1972, the moon could very well become an astronauts' paradise during the next few decades. In the Russian and American files are dozens of projects; there may even be a joint US–Russian mission.

Manned Manoeuvring Unit (MMU) (1984)

In February **1984**, a *Challenger* mission enabled the astronaut **Bruce McCandless**, aged 46, to use the MMU or Manned Manoeuvring Unit for the first time. This $10 million project allows the astronaut complete independence. McCandless was away from the space shuttle for 90 minutes while the latter was more than 270km *167.78 miles* above Hawaii.

It was the first time that a man had 'wandered' in space in this way without being tied to anything. The freedom it gave was extraordinary but it was not without its dangers. If there was a breakdown or defect in the equipment, the astronaut would be completely on his own in space.

Life in space

Space design

Anxious to improve the comfort of the orbital station *Skylab*, NASA called upon the most famous talent in industrial design, **Raymond Loewy**, a Frenchman who has been established in the United States since 1919.

Raymond Loewy was involved with the interior design of the *Skylab* living quarters, that is to say, the arrangement of rooms, choice of colours etc. He also insisted on there being a porthole through which the crew could observe the earth: this was one of the reasons why the mission was so successful.

Edwin Aldrin, seen here, and Neil Armstrong were the first men to set foot on the moon.

WHERE IS THE GEOSTATIONARY ORBIT?

It is above earth's equator at an altitude of 35 900km *21 749 miles*.

The lowest satellites take about 80 minutes to circle the earth. The moon, on the other hand, takes one month. Between the lower orbits and that of the moon, there is an orbit in which the satellite takes exactly one day to circle the planet. This is the geostationary orbit. Any satellite placed there seems to be stationary above a certain point as it takes as much time to circle the earth as that does to make one revolution. This situation obviously has its advantages, particularly for communications satellites towards which one only needs to point antennae without it being necessary to have a tracking system.

The advantages of this orbit mean that it is now already very crowded, in particular with satellites that are no longer operational. The serious nature of the problem means that we need to find solutions; abandoned space craft will have to be collected or placed in a 'parking' orbit higher up if they are worth retaining for future use.

In order to prevent a satellite from falling back to earth, it must travel at a minimum speed of 8km *4.97 miles* per second to tear itself away from the earth's attraction. It must also be placed at an altitude of at least 150km *93 miles 370yd*, so that it is not braked by the earth's atmosphere, which would make it lose speed and fall. In this situation, contrary to a widespread idea, the satellite is not weightless, but feels the attraction of the earth's gravity. The satellite is therefore endlessly falling; that is why its speed is of such importance. As long as this speed is sufficient the satellite will continue to go beyond the earth, and, as it were, to fall alongside it, remaining in its orbit, kept in place by the two-fold action of its speed and the force of gravity.

The new space suit, AX-5, is tested in a water tank. It is designed to be used for routine operations in space.

Some of his ideas were also used in the design of the habitable section of the shuttle. All of Raymond Loewy's preliminary sketches, 3500 of which were drawn up for NASA, were sold at auction by Sotheby's in London in July 1981 when the designer was celebrating his 88th birthday.

AX-5 and ZPS Mk-3 (1996/7)

These are two new types of space suits intended to be used during journeys outside the space station in **1996/7**. Tested since February 1988 by **NASA**, they should allow trips of eight hours; so, with three trips a week this would be a total of between 1000 and 2000 hours of use per year without the need for them to be sent back to earth for maintenance. The ZPS Mk-3 weighs only 68kg *149lb 15oz*, as against 83kg *183lb 2.5oz* for the AX-5. The outdoor space suit weighs 45kg *99lb 3.6oz*.

They are like independent cubicles equipped with all modern conveniences and the astronaut, wearing underwear which automatically regulates temperature, puts on the suit using an opening at the back, and then closes himself in before screwing on the transparent helmet.

Space shower (1990s)

Nothing could be better for the astronauts' morale than to be able to wash in comfort (almost). This is why **NASA** has developed a special shower intended to be used in weightless conditions and which will be fitted on board the orbital station. It is expected to be put into operation in the **1990s**.

93 per cent recycled water

After having tried for three years and spent $400 000, the Johnson Space Centre presented this unusual piece of sanitary equipment: a cell measuring 1.85m *6ft 7in* in height by 1m *3ft 3.4in* in diameter, made of plexiglass preventing any liquid from escaping.

A shower head sprays and the sucks up the water, 93 per cent of which will be recycled, since it is very important to avoid over-consumption of water. Tests have shown that on average it uses 4 litres *7 pints* a minute, compared with a shower used on earth which generally uses up 20 litres *35 pints* a minute! Taking water into space is expensive for NASA: several thousand dollars a litre. Thanks to recycling, each astronaut will be able to have a shower every two days; at present astronauts on space shuttles are allowed only to wash themselves with a sponge!

The NASA engineer, **Rafael Garcia**, who designed the equipment, must find a solution to the problem posed by micro-organisms which sometimes develop in a shower. He will then attempt to develop a space machine for washing dirty clothes.

Exploration of the cosmos

Space probes

Origins (1920s)

The idea of sending space probes to the planets goes back to the beginning of space exploration, that is to say, around **1920**.

In the future, American astronauts will be able to enjoy a real shower in space.

UNKNOWN MOONS

During the course of their long exploration of the universe, Voyagers 1 and 2 have discovered 17 different moons: three belonging to Jupiter; four beginning to Saturn and ten around Uranus. They also revealed the presence of rings around these three planets.

Retransmitted images have also shown us active volcanoes on Io, one of Jupiter's satellites.

SATURN, TITAN AND THE CASSINI PROJECT

In November 1988 the European Space Agency (ESA) took up the American–European Cassini project (named after the astronomer Jean Dominique Cassini, first director of the Paris Observatory founded in 1672 by Louis XIV). The project discovered a division in the rings of the planet Saturn.

Planned since 1982 by the Frenchman Daniel Gautier and the German Wing Ipp, the space probe Cassini should be launched on 9 April 1996 by an American rocket Titan 4/Centaur G. It will reach Saturn in October 2002, after a journey of 1450 million kilometres.

For four years, Cassini will revolve round the system formed by Saturn and its large satellite Titan. The European Huygens probe (named after the Dutch astronomer Christian Huygens who discovered Titan) will then be detached from Cassini and released into Titan's dense nitrogen atmosphere. After being slowed down by an aerodynamic shield, it should reach the unknown surface of Titan and carry out numerous measurements.

The density of the probe has been calculated so that the probe will float if Titan's surface is formed by an ocean composed of ethane, methane and liquid nitrogen.

The German Hermann J. Oberth described in two published works (appearing in 1923 and 1929) the broad principle of interplanetary space flight. During this time, another German called Walter Hohmann became the first person to calculate the conditions of such space flights (flight time, mass of propellants etc) as well as the best orbits for reaching the planets. Some of these orbits are still used today and are known by the name Hohmann orbits as a tribute to their inventor.

Luna 9: towards the moon (1966)

It was on **3 February 1966** with the Russian space probe Luna 9 that an artificial satellite positioned itself on the moon for the first time.

The United States in turn carried out this task on **2 June 1966** with the landing of the automatic probe Surveyor 1 which transmitted 11150 photographs of the surroundings (Sea of Tempests). This was followed in 1967 by the probe Surveyor 5 which carried out the first analyses of lunar soil.

Then in September 1970 the USSR sent the Luna 16 probe which carried out the first core-sampling of lunar soil.

Towards other planets (1970)

The USSR also launched the probe Venera 7 which positioned itself on the planet Venus for the first time on **15 December 1970** and then Venera 9 and 10 which landed on Venus on 22 and 25 October 1975, transmitting the first images from Venus's surface.

The United States sent the probe Mariner 10, which, in March 1974 passed less than 1000km *6214 miles* from Mercury, the nearest planet to the sun.

The United States also successfully carried out their first flights over Jupiter and Saturn, the largest planets in the solar system. First it was Pioneer 10 and 11 which flew above the two planets in December 1973 and December 1974 respectively. Subsequently it was the probes Voyagers 1 and 2 which flew over Jupiter and Saturn, this time getting even closer, as well as over their numerous satellites, before continuing on their flights towards Uranus and Neptune.

Voyager 2 (1977)

It is now 12 years since the American probe Voyager 2 began its long journey into space.

Its three meetings with Jupiter, Saturn and Uranus provided a wealth of information and breathtaking photographs. Launched on **20 August 1977** by an Atlas-Centaur rocket, it has covered some 6000 million kilometres *3728400 million miles*. It brushed against Neptune and her satellite, Triton, on 25 August

PHOBOS NO LONGER RESPONDS

Phobos 1 was sent into orbit on 7 July 1988 followed by Phobos 2 on 12 July. These two Russian probes intended for the exploration of the Mars satellite, Phobos, are lost for ever. The first was lost because of an incorrect manoeuvre. As for the second, all attempts to communicate with it have been abandoned since 27 March 1989.

These two probes formed part of an ambitious Russian and international programmes for the exploration of Mars and its moons, prior to a manned flight towards the red planet around the year 2010.

1989 before exploring the frontiers of the solar system. Voyager 2 should enable astronomers to site more exactly the point at which the sun's influence gives way to the galactic flux of the Milky Way. Then it will sink irrevocably into the universe.

Until 2000 at least

It is expected that the probe will gradually transmit fewer and fewer photographs, but, so long as Voyager 2 has the power, information will continue to arrive on earth. NASA calculates that it will be in contact with Voyager 2 until at least the year 2000. In order to make contact with it, NASA will use its communications network for remote areas in space at Goldstone, California, as well as the 27 antennae of the National Observatory of Radio Astronomy in New Mexico or those of Parkes Radio Observatory in Australia. When it has ceased to send information, Voyager 2 will continue on its journey.

Whales' calls

The probe carries a copper disc on which there has been recorded the sound of a baby crying, train noises, whales 'singing', 60 messages in as many languages and a message of peace from ex-President Carter, all intended for extra-terrestrial beings.

Magellan, Ulysses and Galileo (1989)

The first of the probes, Magellan, was launched towards Venus during Atlantis' mission in May 1989. The European probe Ulysses, and Galileo will be sent towards the sun and Jupiter respectively at the beginning of the 1990s.

Cobe (1990)

A probe for distant exploration, Cobe (Cosmic Background Explorer) will be launched in the United States in 1990 by a Delta rocket. The purpose of this probe is to research Big Bang, the giant explosion which, according to certain scientists, triggered off the expansion of the universe 15 thousand million years ago.

Soho (1993)

In 1993 or 1995, under the ESA 'Soho' programme, five probes will be launched to examine more closely the relationship between the earth and the sun.

Examination techniques

Auscultation (19th century)

With the exception of the medical observations of Hippocrates (460–377BC), the first true auscultation was performed by the French doctor **René Théophile Hyacinthe Laënnec** (1781–1826) using a makeshift stethoscope made from a sheet of paper rolled into a cylinder. The stethoscope in the true sense of the word, from the Greek *stêthos* meaning breast, was invented in 1815 when Laënnec replaced the sheet of paper with a wooden cylinder. The device was further modified and improved by Joseph Skoda of Austria and Cammam of the United States, to become the binaural stethoscope as we know it. The electron stethoscope was invented in 1980 by the Americans Groom and Boone.

Percussion (1761)

Percussion was invented in **1761** by the Austrian physician **L. Auenbrügger**. This clinical mode of exploration allows the condition of certain organs to be deduced by the noise obtained when they are tapped with the fingers. The system was improved by Baron Jean Corvisart (1755–1821), Napoleon's personal physician, and then by the Austrian Skoda, who improved it considerably.

Measuring blood pressure (1819)

Taking the pulse has long been one of the main methods of making a diagnosis. In **1819** the French doctor and physicist **Jean-Louis Poiseuille** (1799–1869) invented the manometer, a mercury gauge for measuring the tension, or what physiologists called the pressure, of the blood. It was succeeded by the sphygmomanometer, from the Greek *sphygmos* meaning pulse, which was a pulsometer developed by an Austrian physician, Siegfried Carl von Basch, in 1881, and by the French physician, Pierre Potain in 1889. This was followed by the broad armband invented in 1896 by the Italian, Scipione Riva-Rocci which exerted a more consistent and reliable pressure. In 1905, the Russian physician, N. S. Korotkov, perfected the method by developing a device which examined the arteries by means of auscultation rather than palpation.

Thermometer (1626)

The first clinical thermometer was a water themometer invented in **1626** by the Italian physician **Santorio**, otherwise known as Sanctorius. The model of the modern clinical thermometer – a graduated glass tube containing mercury – was developed by the English physician, Sir Thomas Allbutt, in 1867.

There are now various types of medical thermometer available, including the disposable oral thermometer invented by an American physician, Louis Weinstein.

Microscope (16th century)

It would seem that the microscope was invented towards the end of the **16th century** by the Dutch optician, **Hans Jansen**, with the help of his son Zacharias. Described by Galileo in

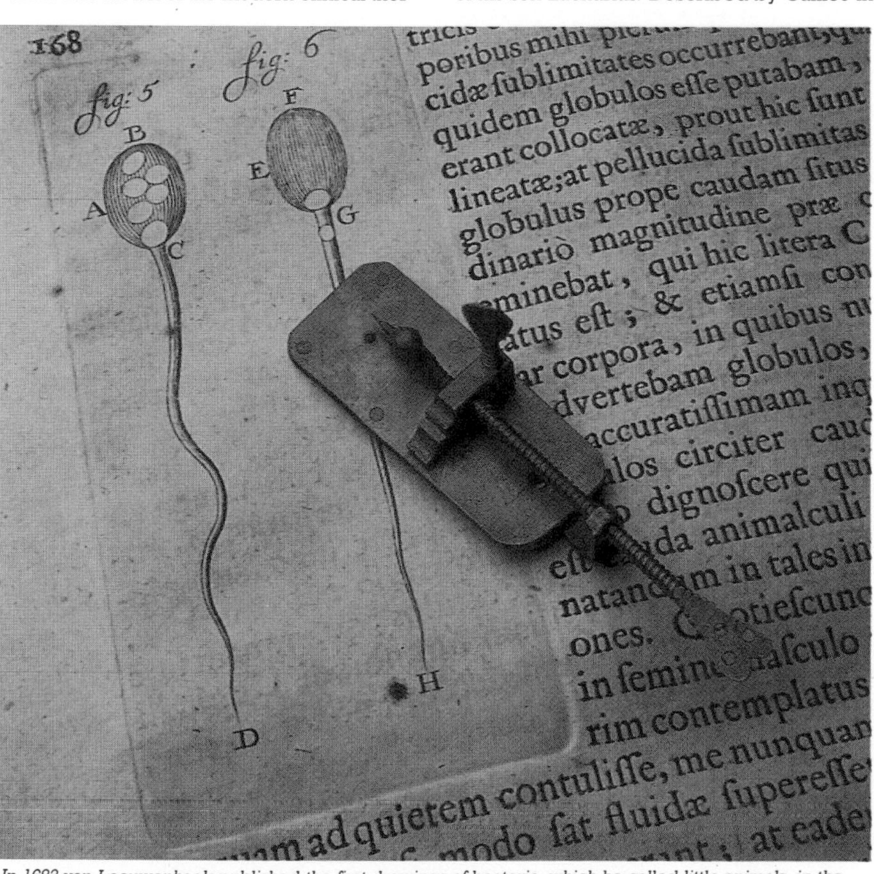

In 1683 van Leeuwenhoek published the first drawings of bacteria, which he called little animals, in the Philosophical Transactions *of the Royal Society in London.*

A spider's head photographed by an electron microscope.

1609, this microscope was quite basic and had minimal powers of magnification.

Another Dutchman, Antonie van Leeuwenhoek (1632–1723) was the first to observe spermatozoa, muscular striation and certain oral bacteria, although magnification was still less than 200 times actual size. The first modern microscopes were constructed after 1880.

Electron microscope (1926)

In **1926**, the German, **Hans Busch**, laid the theoretical foundations for the electron microscope. In 1928 two of his fellow countrymen, Max Knoll and Ernst Ruska, from the Technische Hochschule in Berlin, carried out experiments based on his research which led to the development of the first operational electron microscope in 1933. It was perfected by Ernst Ruska who, with Heinrich Rohrer and Gerd Binning, was awarded the Nobel Prize for Physics in 1986 for the invention of the tunnel effect microscope.

Three-dimensional electron microscope (1985)

Living cells could not be observed using the electron microscope as the beam of electrons directed at the target had to circulate in a high vacuum. In **1985** scientists from the **Massachusetts Institute of Technology** (MIT), directed by Alan Nelson, invented a new process which consisted of placing the sample in a cavity which retained enough air to produce about 1/10th atmospheric pressure and so made it possible to keep the cells alive.

The second advantage of the process was that a three-dimensional image could be projected onto a television screen, by means of a technique based on medical X-ray scanners.

Tunnel effect microscope (1980–86)

In **1980 Gerd Binning** of West Germany and **Heinrich Rohrer** of Switzerland, both working in the IBM research laboratory in Zurich, developed the tunnel effect microscope for which they received the Nobel Prize for Physics in 1986. Based on a principle of quantum mechanics, anticipated at the end of the 1920s, it makes it possible for the surface of a sample to be viewed atom by atom, with a magnification of a hundred million times. The tunnel effect microscope is proving to have an extremely wide range of uses. It operates in a vacuum, in air and in water and its uses range from observing the surface of metals and crystals to the field of biology where it has produced images of DNA and viruses.

Electrocardiograph (1887)

The first human electrocardiogram was recorded in **1887** by **Augustus Désiré Waller** (1856–1922) a physiologist from the University of London, born in Paris.

In 1901, Willem Einthoven, professor of physiology at the University of Leiden in Holland, and former colleague of the French physicist and 1908 Nobel Prize winner, Gabriel Lippmann, developed the loop galvanometer. This made him the true inventor of the electrocardiograph, a piece of equipment that weighed 300kg *661lbs* and required five people to operate it.

By satellite and telephone

It is not difficult to carry out an electrocardiogram but interpreting the result is, on the other hand, a complex medical decision on which the treatment of the patient or accident victim depends.

In **1984 Doctor Christopher Pternitis** developed a long-distance electrocardiograph capable of transmitting the graph via satellite. It is received and interpreted by a specialist who can then communicate the necessary medical treatment by telephone or have the patient admitted to hospital. This piece of equipment, known as Cerfor, is mainly intended for use by sailors making long sea voyages with no doctor on board ship. It came onto the market in 1988.

Electro-encephalogram (EEG) (1929)

In **1929** the spontaneous electrical activity of the brain was recorded for the first time by **Hans Berger**, professor of neuropsychiatry at the German University of Jena. But his recording was met with scepticism because of the weakness of the signal. It was not until the English physiologist and 1932 Nobel Prize winner, Edgar Douglas Adrian, circulated the results in 1934 and defended Berger that the latter received the support of scientific and medical circles.

In 1984 an Englishman, Professor Stores, carried out an experimental continuous recording of an electro-encephalogram for a period of 24 hours.

Radio-immunology

Radio-immunology, invented by the American physicians **Solomon Berson** and **Rosalyn Yalow**, for which the latter won the Nobel Prize for Medicine in 1977, is a combination of two techniques. The first, which is biological, uses the specificity of the immune reaction in order to identify a given organic substance, while the second is physical and marks these substances by introducing radioactive atoms into their molecules.

Test for hepatitis-B (1984)

This test was developed in **1984** by a working party consisting of Diagnostic Pasteur and the French National Blood Transfusion Centre. It was marketed by Diagnostic Pasteur under the name Monolisa, 'mono' for monoclonal antibody and 'lisa' for the elisa technique i.e. enzyme linked immuno sorbent assay. It makes it possible for hepatitis-B to be traced quickly.

Tracing hepatitis-C (1989)

A hepatitis virus which is neither A nor B, and which should therefore logically be referred to as hepatitis-C, has just been identified in the blood by the research team of **Doctor Qui-Lim-Choo** of the Chiron Corporation, a Californian biotechnology company. The hepatitis-C virus is the most common cause of the illnesses which tend to follow blood transfusions such as hepatitis, cirrhosis and cancer of the liver. This is an important discovery which could enable a vaccine to be developed.

The algometer (1988)

Developed by the companies **3M** and **Racia** and based on the work of **Doctor Claude Willer** of the Hôpital Saint-Antoine in Paris, the algometer or instrument for measuring sensitivity to pain measures the threshold above which a stimulus is experienced as pain. It has been tested on hundreds of patients and has been shown to be an objective and accurate method of measurement. It will be used to test the efficiency of certain medicines, prescribe the required dosage of analgesics as well as to measure the level of analgesia during the administration of an anaesthetic.

LipoScan (1989)

Americans can now carry out on-the-spot cholesterol checks by using the LipoScan (TM)-TC, developed by **Home Diagnostics Inc.** and marketed in **1989**.

A few years ago the same laboratories brought out a similar test, the DiaScan (TM)-S, which allowed diabetics to check the daily level of glucose in their blood.

Drug detection device (1989)

ADX is a device invented by Abbott Laboratories which enables drugs and medications such as hashish, cocaine, barbiturates, benzodiazepines and antidepressants present in the bloodstream to be detected within a period of 30 minutes to two hours. This easy-to-use device which came onto the market in 1989 should prove extremely useful in laboratories and hospital emergency units.

Early detection of micro-metastases (1989)

A team from the Munich University clinic led by **Doctor Gunther Schlimok** has developed a method of using monoclonal antibodies to mark the micro-metastases (from the Greek word *metastasis* meaning transition) of breast cancer while the tumor is being removed. This method has made it possible to discover metastases which remained undetected by standard diagnostic techniques such as scintigraphy and tomography. It is currently being tested on stomach cancer and cancer of the rectum and colon. If the initial results are confirmed, it should be possible to give immediate treatment to patients who have a positive reaction to the test.

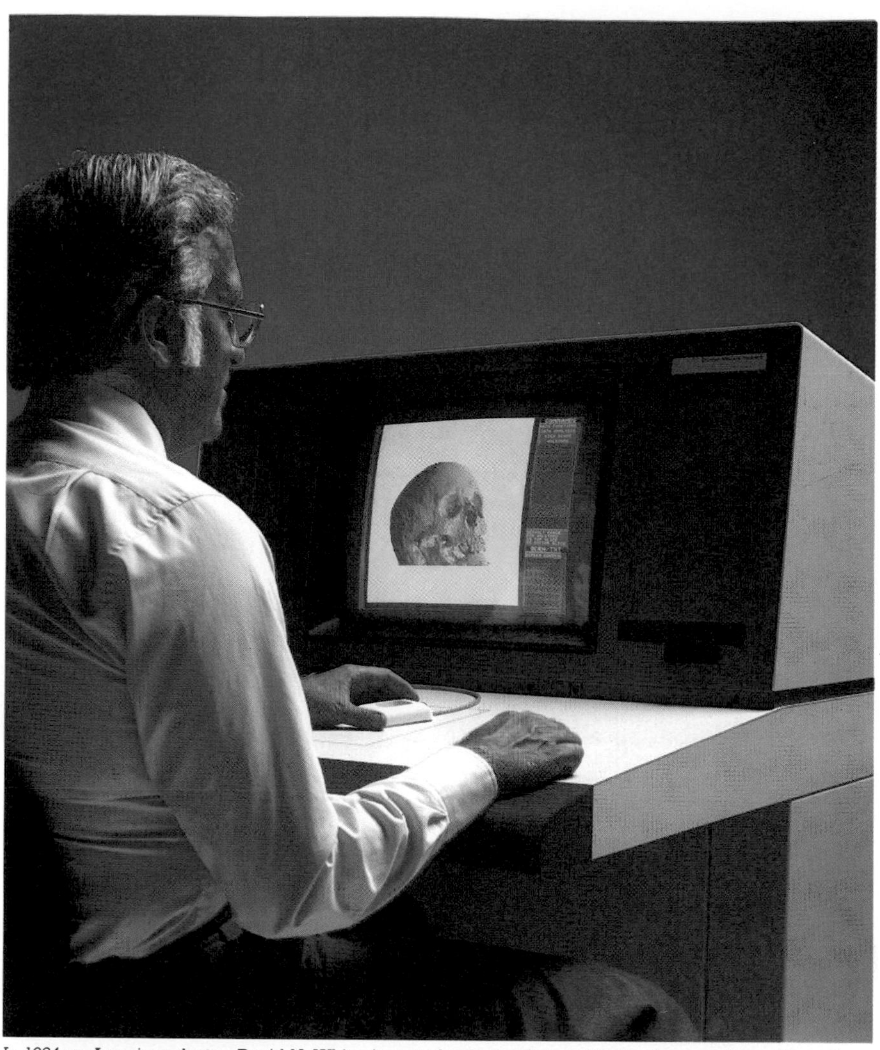

In 1984 an American doctor, David N. White, invented a device that could take X-rays in three dimensions: it is part scanner, part computer.

Medical photography

Radiology (1895)

On **8 November 1895** the German physicist **Wilhelm Conrad Röntgen** discovered X-rays while working in his laboratory. On the 22 December, he X-rayed his wife's hand and was able to see the carpus, the phalanges etc. In 1901 Röntgen received the Nobel Prize for Physics for his discovery.

In 1912 another German physicist, Max von Laue, demonstrated that X-rays resemble visible light in that they are electromagnetic waves, but with a very short wavelength which enables them to pass through opaque matter.

Radiography was initially used for examining the skeleton, but was extended to the other organs by the use of injections of contrasting substances.

Tomography, a process by which a thin layer of an organ is X-rayed to a specific depth, was discovered by the Frenchman André Bocage in 1915. The first tomographies were carried out in 1928.

The first X-ray of the skull was carried out by Walker in 1896; the first encephalogram or X-ray photograph of the brain by the American surgeon Walter Dandy in 1918; the first arteriogram by the Portugese physician Antonio de Egas Moniz and urologist Reynaldo Dos Santos in 1927, and the first X-ray of the pulmonary blood vessels and the cardiac chambers by Ameuille in 1938. Today, traditional radiology accounts for no more than half of medical photography.

Scanner (1972)

The scanner is the result of the combination of the X-ray and the computer. Perfected in **1972** by **Sir Godfrey Newbold Hounsfield**, a British engineer working for EMI, the scanner or tomodensitometer makes it possible to take photographs of cross-sections of tissue in which the detail is a hundred times larger than those produced by the traditional X-ray method.

Echogram (1952)

The principle of the echogram is the application of sonar to the human body. An ultra-sonic source transmits a signal which is reflected by the obstacles in its path. Ultra-sound was first used in the field of medicine in **1952** by the American physician, **Robert Lee Wild**, and then by his fellow American, Leskell, who was the first to observe the heart using ultra-sound. In 1958, the English physician, Ian Donald, carried out the first echogram of the uterus. The method was not widely used until after 1970, but today it can be used to examine any organ. It is used mainly in the fields of gynaecology and cardiology.

Thermal analysis (1950s)

Thermal analysis is another painless and harmless method of examination developed during the 1950s by English and American scientists. It is a method of photographing the tissues using infra-red rays that brings out the differences in temperature due to changes in vascularisation. It is used particularly as a means of checking for cysts and inflammation in the breasts but is replaced by mammography when it is a question of detecting malignant tumours. In 1980 the microwave thermal

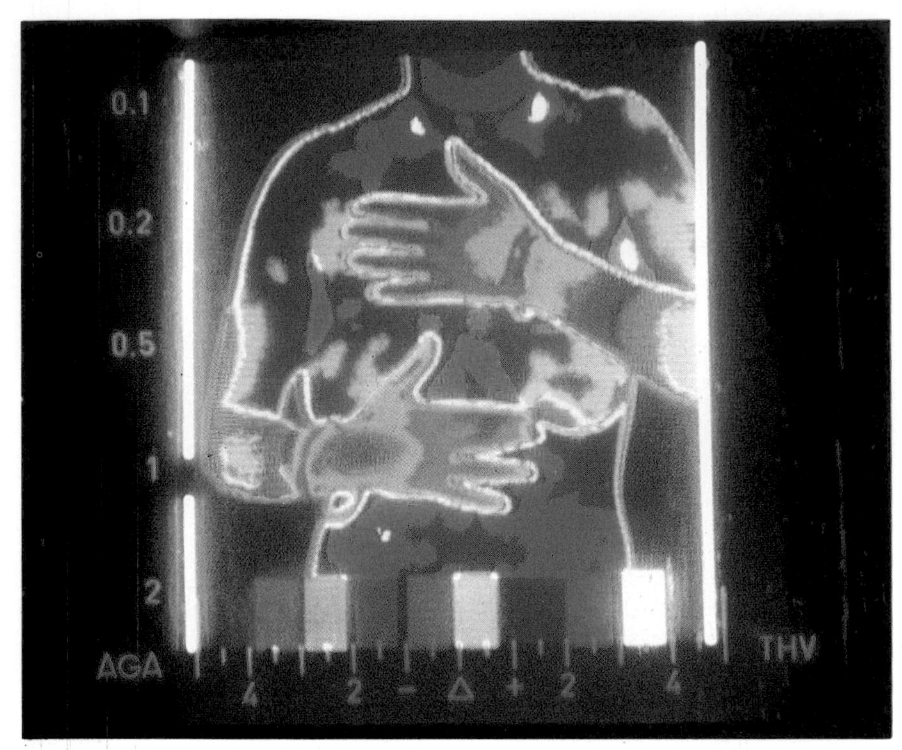

Thermography is a method which allows the differences in temperature around the body to be seen.

analysis process was developed by a French research team led by Professor Y. Leroy of Lille University of Science and Technology.

Scintiscanning (1961)

This technique of medical photography was used for the first time in 1961. It is based on the principle of introducing a radioactive substance such as phosphorus, iodine, thallium etc into the body which then attaches itself to the organ to be examined. Nearly all organs can be scanned in this way, although the technique is currently used mainly for the thyroid gland and the bones. The method has great potential for the future. The use of computers since 1975 has enabled the resultant images to be improved.

Photography by magnetic resonance (1972)

The appearance of this technique in **1972** revolutionised the field of medical photography. It produced clearer and more detailed pictures than the scanner, with the added advantage that it did not use X-rays. The phenomenon of nuclear magnetic resonance (NMR) was discovered in 1948 by the American physicists Felix Bloch and Edward Mills Purcell. In 1972, it was introduced into the medical world by **P. C. Lauterbur**, Professor of Chemistry at the State University of New York at Stony Brook, and the biophysicist, **Raymond Damadian**. The technique uses high frequency electromagnetic radiation which produces changes of energy within the cells which in turn enable the nature of the tissues being studied to be identified. The first pictures of the human body were obtained by Damadian in 1977. At present the technique is mainly used in the field of neurology, but it has other very promising possibilities.

Endoscope (1826)

This is a technique which enables organs to be examined internally, to detect and even treat certain forms of damage. The first instrument was developed by a French doctor, **Pierre Salomon Ségalas** who, in **1826**, performed the first endoscopy of the bladder using a speculum lit by candles. In 1853, another French urologist, Antonin Desormeaux, performed the first rectal endoscopy. The first laryngoscopy was performed in 1829 using an instrument invented by the English physician, Benjamin Babington. The first gastroscopy was performed in 1842; the first electric cystoscopy in 1878 by the German surgeon and urologist Max Nitze; the first tracheobronchoscopy in 1897 and the first arthroscopy in 1951 by Watanabe.

In the 1960s the range of investigations performed by endoscopy was greatly increased and extended due to the development of glass fibre instruments which are both extremely supple and excellent conductors of light.

The Spotlight angioscope (1988)

Developed by **Baxter Laboratories** and brought out in **1988**, the angioscope makes it possible to carry out a particularly efficient internal examination of the arteries. Linked to a colour screen and video recorder this strand, with a diameter of less than a millimetre and a light at the end, is composed of 3000 filaments of which 12 act as both a spotlight and a camera.

Vaccination

Bacteriology (19th century)

The German **Robert Koch** (1843–1910) shares with Louis Pasteur the title of founder of the science of microbiology. Koch became famous in 1882 when he discovered the tuberculosis bacillus, and, in 1883, the cholera vibrion, which he detected in less than a month during an epidemic that raged in Alexandria, Egypt.

Microbes (1762)

The first microbes were discovered by an Austrian doctor, **M. A Plenciz** (1705–86) who published his 'Medico-Physical Studies' in 1762.

Most of the microbes that we know were discovered by the end of the 19th century.

The virus (1892)

The first virus to be characterised was detected in **1892** by the Russian microbiologist **Dmitry Ivanovsky**. It is the cause of tobacco mosaic, an illness which attacks a variety of plants.

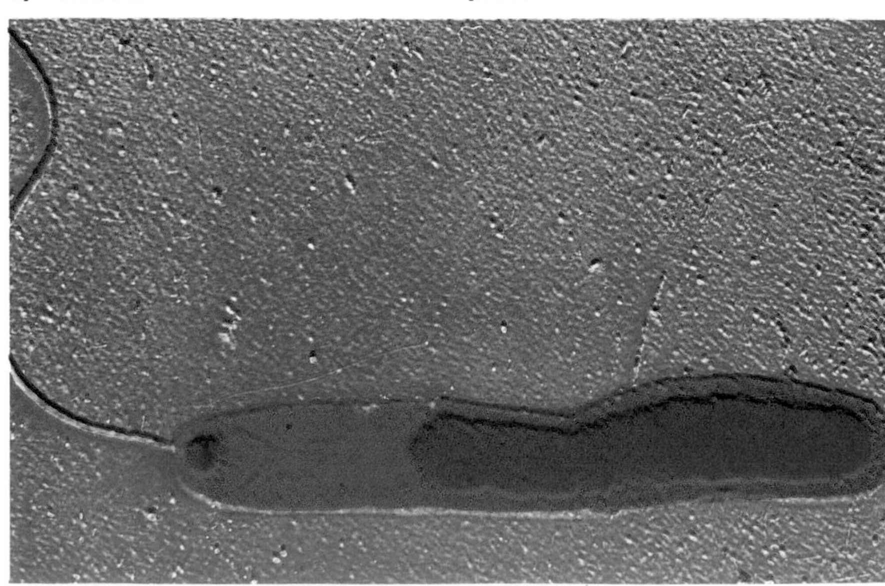
The first microbes to be discovered were found in 1762 by M.A. Plenciz. Here is a cholera vibrium.

VIRUSES AND RETROVIRUSES: WHAT ARE THEY?

A virus is an organism which is composed entirely of a membrane and genetic material, and which measures about one 10 000th of a millimetre. It cannot survive independently, and in order to multiply has to live as a parasite in a living animal or vegetable cell.

When it attacks a cell, the virus attaches itself to it, working its way in and becoming part of the cell's genetic material and forcing the cell to work for it. It diverts the cell from its normal function, causing the illness and ultimate death of the cell it has attacked, and an increase in the number of viruses.

According to their composition, viruses are classified into two main categories, DNA viruses and RNA viruses, which are known as retroviruses.

The function of all living cells, and their hereditary characteristics, depends on the genes situated in their nucleus. These genes consist of complex molecules of deoxyribonucleic acid or DNA. If all is functioning normally, DNA transmits instructions to the various functional elements of the cell. These instructions are carried by a messenger, RNA or ribonucleic acid: a copy of DNA. Normally the formation of RNA from DNA is a one-way process. RNA cannot produce DNA.

In a retrovirus, the reverse reaction does in fact take place thanks to the presence of a special enzyme, reverse transcriptase, discovered in 1970 by the American biologist, Howard Temin, and by Mizutani, which enables the RNA which the retrovirus carries to be converted into DNA. It can then conceal itself in the DNA of the cell which has been attacked and make it work for its own benefit.

Initially, the existence of viruses was only suspected, as they are not visible under the optical microscope, but from 1933 onwards, with the development of the electron microscope, the list of identified viruses kept increasing. Today, certain viruses are identified even before the illnesses which they could cause have chance to appear. In 1956, Werner christened them orphan viruses, i.e. viruses that are looking for an illness.

Culture of hepatitis-B virus (1986)

The hepatitis-B virus, which is responsible for the most serious forms of jaundice, has been cultivated in the laboratory since **1986**. The success is due to a Franco–American team led by **Professor Max Essex** of Harvard University (USA). Tests and vaccines had already been developed, but the above research has now made it possible to test anti-viral medication, which is a major step forward in the fight against this extremely contagious disease.

Other parasites (1880)

Besides bacteria and viruses, a number of other disease-causing microorganisms exist.

In **1880** the Frenchman **A. Laveran** identified the haematozoan, the protozoan responsible for malaria.

In **1881** the Englishman **R. Ross** and the Cuban **C. Finlay** discovered the role of the filariae (parasitic worms found in hot climates) in the transmission of malaria and yellow fever. In **1883** the Englishman **P. Manson** completed his studies by investigating the role of mosquitoes in the transmission of these filariae.

In **1895** the Australian **D. Bruce** investigated the role of tsetse flies in the transmission of sleeping sickness.

Variolation

For quite some time, the only response possible to smallpox epidemics was to flee the area. Nonetheless, variolation, which was the first known of all immunisation methods, was discovered in India or China many centuries ago.

Variolation is based on the fact that smallpox once contracted and cured, prevents a renewed attack of the disease.

Preventive variolation by the application of dried crusts of smallpox lesions to nasal mucous membranes was used regularly in the Far East almost 800 years before Jenner's discovery. It was not known in Europe until 1717. At this time Lady Mary Wortley Montague, the wife of the British ambassador to Constantinople, revealed that she had had her three-year-old son variolated.

The eradication of smallpox

For ten years, there has not been a single case of smallpox throughout the world. In producing the figures which enabled it to claim such a victory, the World Health Organisation (WHO) in Geneva considers that this is one of the most important medical facts of the 20th century. In 1967, it was estimated that two million people would die from the disease which affected between ten and 15 million people annually. The campaign to eradicate smallpox, which took place between 1967 and 1979, cost about $300 million.

Vaccination (1796)

On **14 May 1796 Edward Jenner** (1749–1823), having done considerable work on cowpox (a disease of the cow udder whose French name is *vaccine*), took a sample of the material from a pustule on the hand of a dairymaid, contaminated by the cows, and put it into the arm of a young boy named James Phipps. Ten days later, a pustule appeared on the boy and healed quite normally.

In a second experimental phase, Jenner inoculated the boy with smallpox; there was no harmful effect. The experiment was a complete success, and in 1798 Jenner published his results. In 1799, he perfected his idea and his technique.

The method spread widely in Europe, the East, and the United States. Some 60 years later, Pasteur would make a discovery of still greater general interest in the area of disease prevention. However although the biological principle was different, Pasteur kept the term vaccination, as a posthumous tribute to Jenner.

The work of Pasteur

Assisted by his students, E. Roux and C. Chamberland, **Louis Pasteur** (1822–95) first isolated a number of microbes which cause disease in man.

Pasteur made his first attempt at vaccination to fight the viral disease of rabies. On 6 July 1885 Pasteur injected Joseph Meister, who had been bitten by a rabid dog, with dried spinal marrow taken from rabbits he had inoculated with the virus. The result was conclusive.

In 1922 a French veterinary surgeon, Gaston Ramon (1886–1963), managed to isolate a diphtheria toxin and weakened it in formaldehyde. He thus paved the way to vaccines that cause no ill effects to the recipient.

These Russian peasants, who had been bitten by a rabid wolf, made the journey to Paris to see Louis Pasteur in the hope that they might be cured.

Sera (1890)

Sera, obtained by taking samples of blood serum from a diseased or vaccinated patient (serum which thus contains the desired antibodies), allow either preventive or curative action to be taken against numerous diseases and also against bites and stings from venomous animals, by providing the contaminated individual with protective antibodies.

The principal preventative sera were discovered before 1900:

Anti-diphtheria: discovered in **1890** by the German **E. von Behring**, the Japanese **S. Kitasato**, the Frenchman **E. Roux**, **L. Martin** and **A. Chaillou**.

Anti-tetanus: discovered in **1890** by **Behring**, **Kitasato**, **Roux** and **Vaillard**.

Anti-plague: discovered in **1894** by the Swiss **A. Yersin**.

Anti-anthrax: discovered in **1895** by Italian **A. Sclavo** and the Frenchman **E. Marchoux**.

Anti-cholera: discovered in **1896** by the Frenchman **E. Roux**, the Russian **I. Metchnikoff**, and the Italian **A. Salimbeni**.

This lithograph of the circulatory, nervous and muscular systems was produced in 1823.

The heart and lungs

Circulation of the blood (13th century)

It is generally thought that the British physicist William Harvey discovered the circulation of the blood in 1628. In fact a 13th century Arab physician, Ibn al-Nafis al-Quarashi, had already mentioned the existence of a pulmonary circulation in a work dedicated to the Persian philosopher and scientist, Ibn Sina or Avicenna. This work passed unnoticed until it was referred to in 1552 by the Spanish theologian and physician Miguel Serveto in his theological and medical work, *Restitutio Christianismi*, for which he was burnt at the stake.

From 1550 onwards, several physiologists of the Paduan School, including Matteo Colombo, Carpi and Hieronymus Fabricius of Aquapendente, studied the problem. William Harvey based his work on that of his predecessors and had the inspired idea of considering the heart as a pump which was operated by muscular pressure. Proof of the existence of capillary vessels linking the arterial and venous systems was supplied in 1661 by the Italian anatomist Marcello Malpighi.

Blood transfusion (1667)

There seems to be some question as to who should be given credit for the development and use of blood transfusion techniques. Early work in the field of blood transfusion was carried out in England by R. Lower, in France by J. Denis, in Germany by Mayor, and in Italy by F. Folli. Nevertheless, it is practically certain that Lower was the promoter of experimental transfusion in animals, and that Denis was the first to use it for humans. In **1667**, **Denis** injected one litre of arterial blood taken from a lamb into a young man who had previously been bled. By virtue of its principle and because of the severe dangers inherent in its use, the method was immediately condemned and forbidden.

In 1821 the study of transfusion in animals was again taken up. The method was defined in 1875, but interhuman transfusion was developed only as of 1900, when the work of the Austrian K. Landsteiner demonstrated the existence of four large blood groups. In 1910, the Czech serologist Jansky designated these groups by the letters, A, B, AB and O.

In 1940 Landsteiner crowned his achievements by identifying, with Wiener and Levine, the Rhesus factor. It is named after the kind of monkey in which it was first identified. The discovery provided an explanation for the haemolytic reaction in newborn babies.

Artificial blood (1979)

In **February 1979**, a Japanese doctor, **Ryochi Naito**, injected himself for the first time with 200ml of artificial blood, Fluosol DA, which is a totally synthetic derivative of petroleum and milky white in colour.

In 1966 two Americans from the University of Cincinnati, Professors Glark and Gollan, had demonstrated that a mouse submerged in liquid perfluocarbons could survive because there was enough oxygen in the liquid to prevent it dying of suffocation. But these fluocarbons did not mix with the blood.

In 1967 an American professor, Henry A. Slaviter from the University of Pennsylvania, succeeded in emulsifying the perfluocarbons by adding albumin. But in spite of this step forward, the emulsion ran the risk of agglomerating and blocking some of the capillary vessels.

The first success

Ryochi Naito was to first to experiment successfully. In April 1979, during an operation at the Fukushima Centre, an emergency injection of Fluosol DA was given to a man with such a rare blood group that a transfusion would have been impossible.

Subsequent research

At the end of 1985 an American scientist, Professor A. Hunt from the University of California, announced that he had perfected artificial red corpuscles, and at about the same time a report by Doctor L. Djordjevich of the Rush-Presbyterian-St-Luke Medical Centre in Chicago gave an account of similar research

One of the most important uses for artificial blood would be to reduce considerably the spread of blood-transmitted diseases such as AIDS.

Discussion are currently being held about the marketing of Fluosol DA, developed by the Midori Juji Laboratories.

TPA: the enzyme that prevents heart attacks (1984)

TPA or Tissue Plasminogen Activator is an enzyme that has a thrombolytic effect i.e. it dissolves blood clots. It was produced in **1984** by the American company **Genentech**, based on the research of Doctor Collen of Louvain in Belgium. Its commercial production was undertaken in 1986 by Carl Thomas GmbH, a West German pharmaceutical company.

A hormone for the red corpuscles (1988)

For the first time a hormone which plays an

The experiment shows that a mouse can breath in a fluorocarbon emulsion which contains 20 times more oxygen than water does.

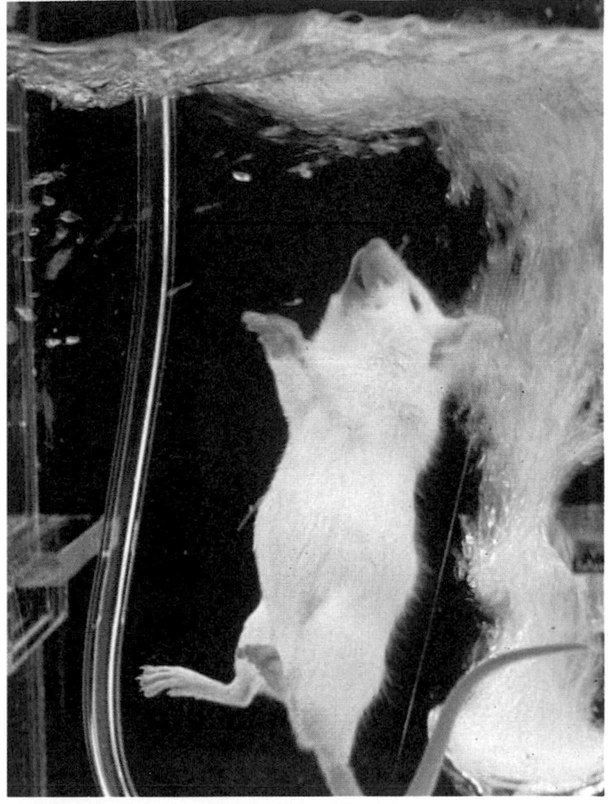

essential part in the physiology of the blood – erythropoietin – is going to be produced commercially as a result of the genetic engineering carried out by the Ortho-Cilag Laboratory. It is a medication which marks a major step forward in the treatment of patients suffering from renal insufficiency who are treated by dialysis and who are often unable to produce enough of this hormone. This result is anaemia which requires continual blood transfusions.

Artificial heart (1957)

The artificial heart was invented by **Willem Kolff** who began his research in **1957** at the Cleveland Clinic in the United States. In 1970 he joined scientists from other countries at the University of Utah, to develop models for the artificial heart.

The Jarvik

In **1976** one of his colleagues, Doctor Robert Jarvik, invented the Jarvik 7, an artificial pneumatic heart which operated on compressed air and was connected, at the time, to a compressor which weighed 150kg *330lb*.

It was declared suitable for use on humans by the University of Utah in 1981, and in 1982 the Food and Drug Administration gave permission for it to be implanted in a patient in a situation where there was no alternative form of treatment. On 2 December 1982 the Jarvik 7 was implanted for the first time by Doctor William Dr Vries into a voluntary patient, Mr Barney Clarke.

Since then, 91 Jarviks (the Jarvik 8 came out in 1986) have been implanted throughout the world. However, it is a process which is being increasingly disputed as it has only a 37 per cent success rate, and the use of the artificial heart is being restricted. It is not so much a permanent solution as an interim measure for patients awaiting a heart transplant.

Other artificial hearts

In addition to the Jarvik, other artificial hearts include the Pen State, developed by Doctor William Pierce, and the Buecherl System, developed by Professor Emil S. Buecherl of Berlin.

The heart pump (1989)

The Hemopump, developed by Doctor Richard K. Wampler, is a miniature turbine which is designed to take over temporarily from the cardiac muscle in a situation where the latter is receiving an insufficient supply of blood. This temporary heart, 6cm *2.3in* long and 6cm *2.3in* in diameter, is inserted percutaneously and directed along the femoral artery to the left ventricle which is responsible for pushing the blood into the aorta. Once in place the pump, which is operated by an electric motor outside the body, rotates at a rate of 28 000 rpm to ensure a supply of blood. The Hemopump is extremely valuable in that it provides support at such critical moments as restarting the heart after a heart attack and recuperation before an operation. Its use is limited in that it can only be left in place for a maximum of six days.

A TALKING POINT FOR TOMORROW: A DRILL FOR CLEARING ARTERIES

For many years, specialists have been looking for a method of clearing arteries obstructed by atheroma, or fatty deposits, in order to prevent an infarct of the muscular tissue of the heart. Among the instruments suggested for the purpose, the Rotablator, developed by an American, David Auth, seems the most efficient. It is a miniature rotary drill which revolves at a speed of 200 000rpm and which is introduced into the arteries percutaneously. The drill has the advantage of destroying the atheroma while at the same time smoothing the walls of the artery. The 'drilling' operation requires only a local anaesthetic and three days in hospital. It is possible that the method could replace the heart bypass operation and transluminal angioplasty which are both more traumatic and, in 20 to 30 per cent of cases, need further treatment.

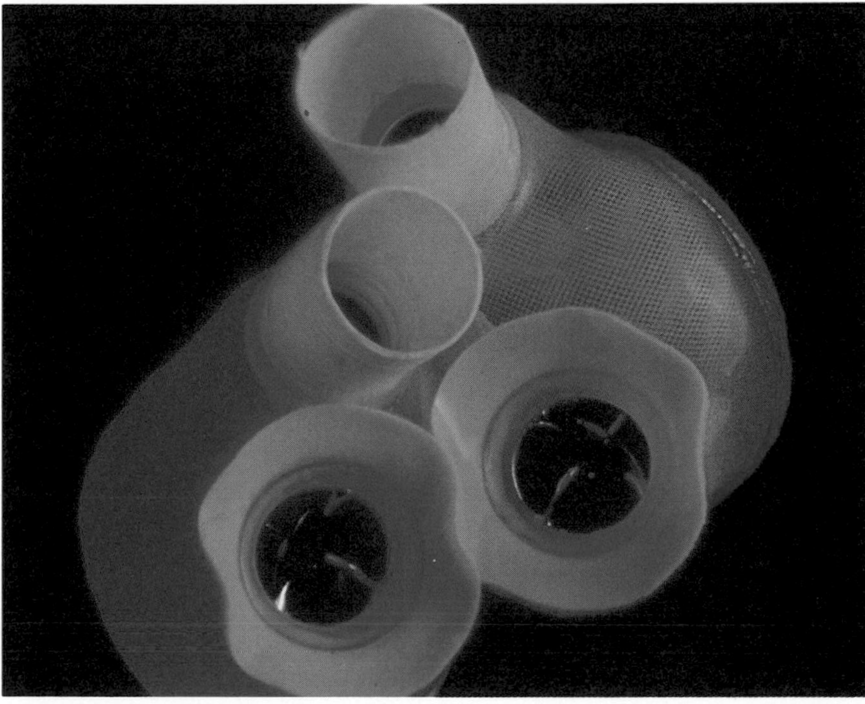

The success rate of the Jarvik has not been outstanding.

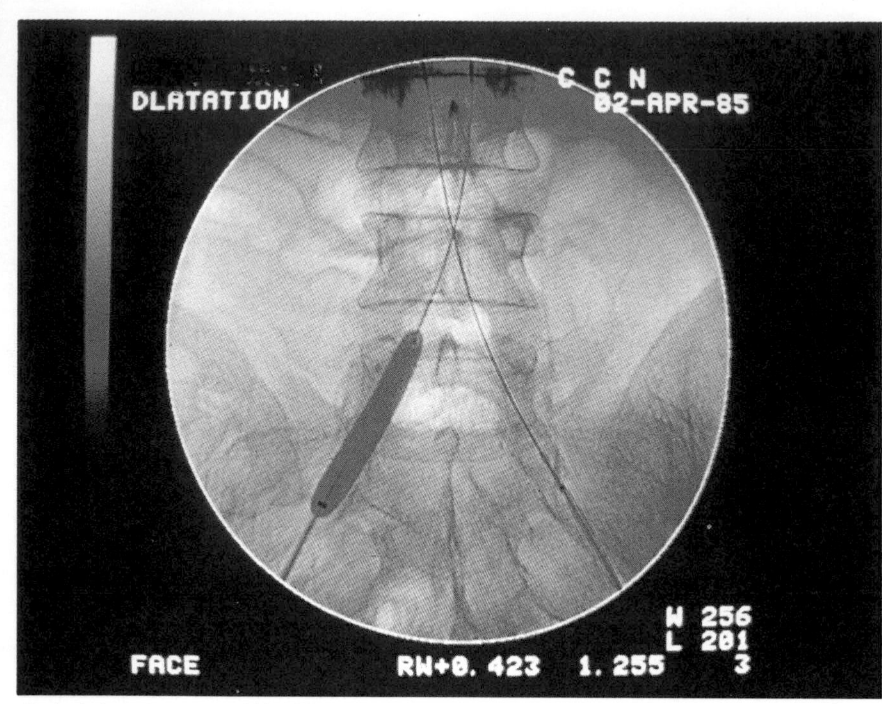

Angioplasty restores normal blood flow: a balloon is inserted into the artery and inflated to dilate it.

The pacemaker (1958)

The cardiac pacemaker was invented in **1958** by Doctor **Ake Senning** of Sweden. The first implants took place in the early 1960s. The pacemaker is capable of stimulating other organs as well as the heart.

In October 1986 a baby was born in Manchester suffering from a congenital malformation of the heart. It was given a cardiac pacemaker when it was only three days old. This was the first operation to be carried out in Europe on such a young child.

The programmed pacemaker (1986)

In **1986** a German company, **Biotronik**, developed a programmed cardiac pacemaker based on blood temperature, which is a good indicator of the level of activity of the patient. For example, as the patient climbs the stairs, the blood temperature rises and the device increases the heart rate accordingly.

The pacemaker pill (1986)

In 1986 an American company, **Arzco Medicals Electronics**, brought out the first pacemaker 'pill': an electrode contained in a gelatine capsule.

The nuclear pacemaker (1988)

In November **1988** a 47-year-old man made history by being the first person ever to receive a double pulse nuclear cardiac pacemaker. The Pulsar N-1, which operates by using a tiny pastille of plutonium coated with titanium, was implanted under the skin next to the chest. It should last for between 20 and 40 years, whereas single pulse nuclear pacemakers, which only act on a single cardiac function, last for 16 years. Today, between 3000 and 4000 people are fitted with a single pulse pacemaker.

Angioplasty (1964)

In 1954 an American doctor, Charles Daughter, had the idea of inserting a tiny balloon inside an artery and inflating it at the point at which the artery had become constricted. The idea was put into practice for the first time in Zurich, by the Swiss doctor **Andreas Grüntzig** in **1964**, since when the technique has been greatly developed.

The defibrillator (1970)

Certain irregularities in the muscular contractions of the heart cause what is known as a ventricular fibrillation, i.e. uncontrolled contractions of the cardiac muscle. The defibrillator, developed in **1970** in the United States by **Professor Mirovski**, is implanted directly onto the heart, making it possible to identify any irregularity in the contractions and for immediate action to be taken. In this way the risks entailed in an emergency admission to hospital can be eliminated and the patient's chances of survival increased.

The device was not marketed until the beginning of the 1980s.

Iron lung (1927)

Philip Drinken, an American professor at Harvard University, designed the iron lung in **1927**. It was tested on a young girl at Boston Hospital on 12 October 1928.

The first model was made from bizarre objects: two vacuum cleaners alternately produced a positive and a negative pressure on the patient's thorax.

Surgery

Neolithic origins

The first recorded surgical operation, an amputation, dates back to Neolithic times, between 5000 and 2500 BC. But the skeleton of a Neanderthal Man, about 45000 years old, found in the Zagros Mountains, in Iraq, seems also to have undergone an amputation. Its missing right arm was due neither to chance nor to an accident.

The sorcerers, doctors and surgeons of the time were also the first to perform trepanation on living patients, some of whom appear to have survived this terrifying operation. In fact, some of the trepanned skulls show evidence of healing.

Autopsy (3rd century BC)

The study of anatomy began to develop as a result of the autopsy. Ancient religions prohibited the mutilation of the body. It was not until the 3rd century BC and during the reigns of Ptolemy I Soter and Ptolemy II Philadelphus that the first examinations of a corpse were carried out in Alexandria by the famous Greek anatomists and physicians, Herophilus and Erasistratus. For a long time after this the practice was prohibited.

Papal authorisation

It was not until the medical renaissance of the 13th and 14th centuries in Bologna and Padua, and then only with papal authorisation, that the Italian physician, Mondino dei Liucci (c.1270–c.1362) was able to publish accounts of the dissections he had performed, in his *Anatomia*.

Anaesthesia

We know from very early documents that certain methods were used to suppress pain during medical operations. The Assyrians, for example, cut off the circulation to the brain by pressing on the carotid arteries while performing a circumcision.

During the 1st century AD, Pliny the Elder refered to a method of making painless incisions and injections by the use of the mandrake. But before the discovery of anaesthetics, the main obstacle to the surgeon's work was the pain suffered by the patient.

Anaesthetic gases (1799)

In **1799** the English chemist **Sir Humphry Davy** (1778–1829) described the analgesic and laughter-provoking effect of nitrous oxide (laughing gas). To demonstrate these effects, he inhaled the gas to ease the pain brought on by an abscess on his tooth.

Some dentists, notably the American Horace Wells in 1844, used this chemical compound when extracting teeth. Wells said 'A new era is beginning in dental surgery. It hurts no more than a pin prick.' Wells died in 1848 from a wound to the femoral artery and as a final recourse inhaled the gas while dying.

General anaesthesia (1842)

General anaesthesia with ether was first used by the American **C W. Long** in **1842**. Americans William Morton and John Collins Warren operated on a neck tumour at Massachusetts General Hospital on 14 October 1846 after placing the patient under general anaesthesia with ether. After their success, the use of ether spread throughout the United States, and Robert Liston, a Scottish surgeon, introduced it to Europe.

Chloroform became popular after ether. Doctor Simpson, a professor of obstetrics in

This 1st century AD *illuminated manuscript shows the treatment of dislocations.*

Edinburgh, used the gas, which had been available since 1831, in a pure form in 1834. After Queen Victoria had been given chloroform during the delivery of her seventh child, anaesthesia was adopted in all hospitals.

Local anaesthetic (1884)

Local anaesthetic was developed in **1884** using cocaine, by the Austrian ophthalmic surgeon **Carl Köller**. The effect of cocaine was subsequently improved by the addition of adrenalin in 1902, and then it was replaced in 1904 by novocaine.

However, the earliest written record of local anaesthetic dates back to Pliny the Elder who, in his *Natural History*, gives the recipe for an anaesthetic poultice made from crushed mandrake leaves mixed with polenta.

Epidural anaesthetic (1885)

The epidural anesthetic was described for the first time in **1885** by an American neurologist, **J. Leonard Corning**. This regional anaesthetic is carried out by injecting an analgesic into the epidural space, i.e. the space surrounding the spinal cord, between the eleventh dorsal and the fourth lumbar vertebrae. This deadens the pelvic organs (the uterus, kidneys, prostate etc). It was rediscovered in France in 1901 at the Hôpital Tenon by the French surgeon, Fernand Cathelin and physician, Jean Athanase Sicard, and then neglected until 1970 when it began to be widely used, particularly in obstetrics.

Cryoanaesthesia (1978)

According to 17th century medical texts, doctors in Finland used a technique whereby, in order to reduce fractures and to replace joints, limbs were bathed in iced water. The process made the operation virtually painless. More recently, in **1978**, the **Spembly Llord Method** has enabled an English manufacturer to market ST-2000 Neurostat equipment which extends the range of this technique. These specially shaped cryosounds are able to reach deep-seated nerves and relieve very different types of pain such as backache, post-operative pain or pain resulting from a trapped nerve.

Asepsis (1844)

Asepsis or medical and surgical hygiene was invented by a Hungarian physicist, **Ignaz Phillip Semmelweis**, in **1844**. Asepsis by boiling and by dry heat autoclave was first achieved by the French surgeons Octave Terrillon and Louis-Félix Terrier in 1883. In 1889 the American surgeon William Stewart Halsted introduced the use of rubber gloves.

Antisepsis (1867)

The English surgeon **Joseph Lister** (1827) concentrated his efforts on antiseptics. Prior to all operations, he sprayed carbolic acid in the room, and disinfected the instruments and the area of the patient's skin where the incision was to be made. He published the results of his work in *The Lancet* in **1867**.

One of the first antiseptics was honey, which the Egyptians used on wounds.

Dressings and bandages (7th century BC)

The first evidence is recorded on slate tablets discovered during the digs in Assur and Nineveh in Assyria and written by one of the best known medical practitioners of the 7th century BC, **Arad-Manai**. In 1825 the French surgeon, Antoine Labarraque, introduced the chemical disinfection of wounds. In 1840, the English physicians Sir Astley Cooper, Robert Liston, Syme and Macartnay introduced a new form of dressing, a piece of cotton cloth which had been moistened and covered with sticking plaster. In 1864 another English surgeon, Joseph Lister, used dressings moistened with diluted carbolic acid which acted as an antiseptic.

Soluble dressings (1947)

The soluble dressing was invented simultaneously in **1947** by **Jenkins** in **United States** and **Robert Monod** in **France**. It consists of a small gelatin sponge that can absorb 20 to 50 times its weight in blood, before gradually dissolving in the body.

Sutures (1820)

The French surgeon **Pierre-François Percy** (1754–1825) invented wire sutures in **1820**. Catgut sutures appeared in 1920. It is interesting that the famous Hispano–Moorish surgeon, Abulcasis, was already using catgut in the 10th century as he liked its suppleness, its strength and the fact that it is resorbent.

Terylene sutures were invented in 1950, and in 1964 the American company 3M invented the Steristrip, a type of dressing which knits the wound together but avoids stitch marks.

Treatment for fractures (3000 BC)

The Egyptians invented the earliest form of support for fractures. In **3000 BC**, **Athotis** recommended strips of cloth soaked in mud.

Plaster (18th–19th century)

In 1798 William Eton, a member of the British Consulate in Persia, noticed that plaster was used there to support fractures. In 1850 a Dutch military doctor from the Royal Hospital, Doctor Antonius Mathijsen, developed a method whereby strips of linen sprinkled with dry plaster were prepared in advance and soaked when required.

Glass fibre and resin plaster (1982)

In **1982** the American company **3M** developed the Scotchcast, a glass fibre strip impregnated with a polyurethane-based resin. This plaster-resin, which is extremely resistant, is waterproof and is only a third of the weight of a traditional plaster cast.

Appendectomy (1735)

The English military surgeon **Claudius Amyan** performed the first successful appendectomy in **1735**.

On 27 April 1887 in Philadelphia George Thomas Morton (the son of William Morton, one of the pioneers of anaesthesia), operated on a young man who was suffering from acute appendicitis thereby saving his life.

Plastic surgery (3rd century BC)

The great progress made in the field of anatomy during the 3rd century BC in Alexandria encouraged surgeons to attempt the first operations in facial plastic surgery. **Amynthas of Alexandria** performed the first operation on a nose.

Plastic surgery as we know it appeared around the same time on both sides of the Atlantic. In **1891**, **Roe** of the United States and in 1898, Joseph of Germany, invented rhinoplastic surgery i.e. the alteration of the shape of the nose by surgery for purely aesthetic reasons. In 1907 the French surgeon Hippolyte Morestin described his method for the resection of enlarged breasts. In Vienna, in 1928 and 1930 respectively, H. Biesenberger and E. Schwarzmann made important contributions to mammary plastic surgery.

The facelift

In **1925** a surgeon, **Suzanne Noël**, performed facelift operations under local anaesthetic in the patient's home.

Neurosurgery (1918)

Modern neurosurgery was first introduced in the United States in **1918** by **Harvey Williams Cushing**. It was introduced in France in 1936 by Thierry de Martel and Clovis Vincent who invented the technique of carrying out brain surgery while the patient remained seated. In the same year, the Portuguese physician Antonio de Egas Moniz invented arteriography.

Some important dates

In 1937 Fiambert invented the lobotomy which has now been virtually abandoned.

In 1950 Talairach of France invented the sterotaxy. One of the first applications of this process was in surgery for Parkinson's disease performed by the French surgeon Fenelon.

In 1960, another French surgeon, Guiot, operated on a tumour of the pituitary gland through the nose.

In 1962 a radiologist, Djindjian, and the neurosurgeon, Hudart, performed the first arteriography of the spinal chord, which enabled operations to be carried out on angiomas of the marrow.

Since 1970, Serbedinko of the USSR has been avoiding operations by inserting a probe into the artery. At the end of a probe is a tiny balloon which the surgeon releases at the point chosen to block or embolise the artery.

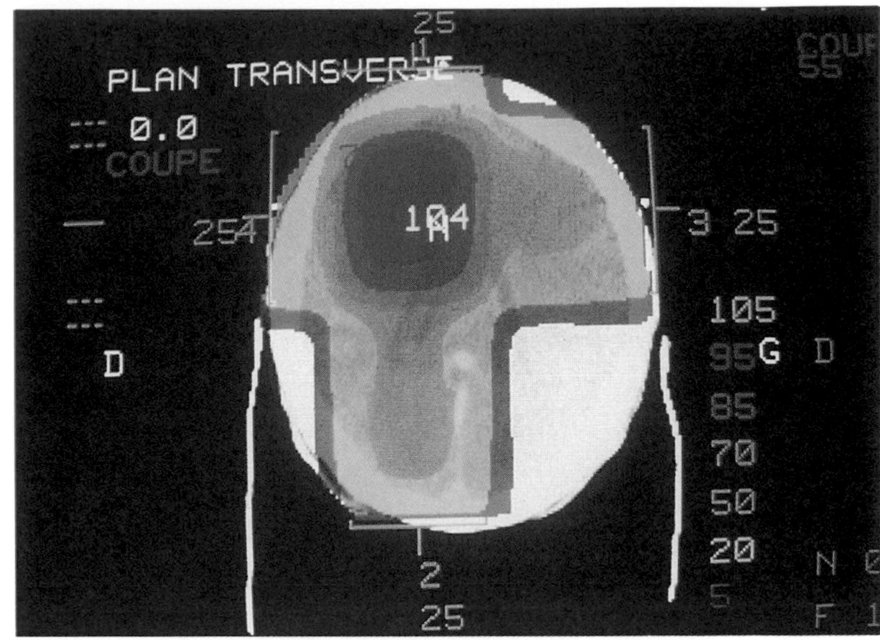

Radiotherapy of a brain tumour is controlled by a computer.

Cerebral radiotherapy (1982)

The multi-beam convergent irradiation unit is a unique piece of equipment. It makes it possible to treat brain lesions which were hitherto inaccessible. The first version was invented in **1982** by the Argentinian neurosurgeon, **Oswald Betti**.

Transplants

Origins

In the 17th century the Boiani, a family of surgeons from southern Italy who practised empirical medicine, developed a method of re-building the face based on the method of layering vines carried on in their region. They were the first surgeons to transplant skin tissue by a method still in use today, the skin graft.

But grafting was not developed further until 1958 with the discoveries relating to the HLA system made by the French haematologist and immunologist, Professor Jean-Baptiste Dausset, Nobel Prize winner for Physiology and Medicine in 1980.

Cyclosporin-A (1972)

The second important stage occurred in **1972** when Doctor **J.-F. Borel** of the **Sandoz Laboratories** in Basel, Switzerland, discovered the immunosuppressive properties of cyclosporin-A. This is a substance found in a mushroom which grows on the Hardangervidda, a high plateau in southern Norway. The first tests on humans took place in 1978 and cyclosporin-A came into general use for bone marrow and organ transplants in 1983.

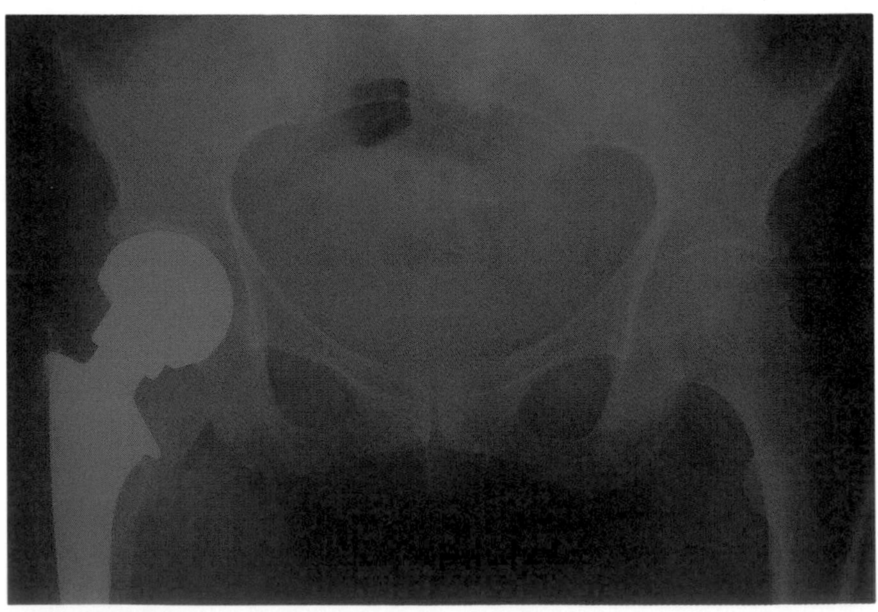

In 1960 an English surgeon, John Charnley, began fitting hip replacements at his hospital in Wigan.

Major successes

Hair: N. Orentreich, New York, 1959; the technique invented by Doctor Yamada of Japan in 1985 according to which artificial hairs are implanted individually; the technique of implanting into a cushion of air, invented in 1986 by Doctor Gilbert Ozun of the Hôpital Foch in Paris.
Heart: The first heart transplant was performed on a chimpanzee by J. D. Hary of Chicago in 1964. The first human heart transplant was performed by Dr Christiaan Barnard in Cape Town, South Africa in 1967. The first successful heart transplant was performed by Professors Cabrol and Giraudon on Emmanuel Vitria on 27 November 1968. Vitria died in 1987, 18 years after the transplant. The first infant heart transplant was performed in London in 1984 by Professor Magdi Yacoub who also carried out a heart and lung transplant on a baby in 1986.
Cornea: Elschwig, Prague in 1914.
Arm: Ronald A. Malt and J. McKhann, in Boston, USA, in 1962.
Liver: In 1988 a double transplant of half a liver was performed on two women suffering from acute hepatic insufficiency by Professor H. Bismuth of the Hôpital Paul Brousse in Paris.
Bone marrow: Thomas, USA, 1957.
Bone: R. L. and Jean Judet, Paris, 1950.
Skin: Jacques-Louis Reverdin, Geneva, 1969.
Kidney: Performed by R. Lawler of Chicago in 1950, the patient survived for five years. The first completely successful transplant was carried out by the American physician John Putnam Merrill of Boston, in 1954.
Testicle: In 1977 S. Silver of Sacramento, USA, transplanted a testicle from one twin to another. In 1984 Wang Linglong of Hubei, China, performed transplants between brothers and parents.
Aortic valve: The French surgeons, J. P Binet and A. Carpentier of the Hôpital Marie-Lannelongue had the first success in 1965. The same surgeons inserted a synthetic valve in 1970.

Skin culture (1950)

In **1950 Professor Howard Green** of the Massachusetts Institute of Technology (MIT), discovered that the fibroblasts of human skin multiplied extremely well when a sort of 'fertiliser' made from cancerous 3T3 cells was added. The cells were irradiated to prevent them from reproducing while at the same time allowing them to produce the required nutrient, with the result that cell culture was achieved.

As a result of this discovery, the research team led by Howard Green developed a technique which enabled 60cm² of skin to be obtained in 20 days from 1mm² of skin taken from a newborn baby.

Synthetic skin (1986)

In **1986** two American scientists, **Professor John Burke**, a surgeon at the Massachusetts General Hospital, and **Professor Ioanis Yannas**, of the Polymer Laboratory at the MIT, saved several hopeless cases by performing synthetic skin grafts. Their work was based on the patent held by two Americans, Howard Green and Eugene Bell, who created a second artificial layer of skin from beef collagen, which is well received by the human body, and from silicone plastic. This 'skin' is then sterilised and frozen in alcohol. The technique enables epidermis to be recreated.

Cryosurgery (c.1960)

Cryosurgery is a method by which pathological tissue is destroyed by the use of extreme cold, i.e. below −40°C. It has the advantages of being painless and causing no risk of haemorrhage. In about **1960** the American neurosurgeon **Irving S. Cooper** (sponsored by Union Carbide), developed a cryosound using liquid nitrogen which reached a temperature of −180°C. Other instruments were used for cataract surgery and the treatment of tumours of the larynx, and haemorrhoids. In 1964, M. J. Gonder and W. A Soanes used cryosurgery to treat adenomas of the prostate gland. In 1975 a German surgeon, A. J. Keller, invented cryocautery.

Cryogenic probes (1986)

In **1986**, **Patrick Lepivert**, a French doctor from Nice, in collaboration with the Centre for Nuclear Research in Grenoble, developed tiny cryogenic probes which could be used in the treatment of varicose veins. Those probes, which are adapted to the size of the veins, make it possible to operate without using an anaesthetic, and to avoid post-operative complications and scars. The development of these micro-probes should enable them to be used in the same way as laser beams, and at a much lower cost, to destroy inoperable tumours in the colon and digestive tract.

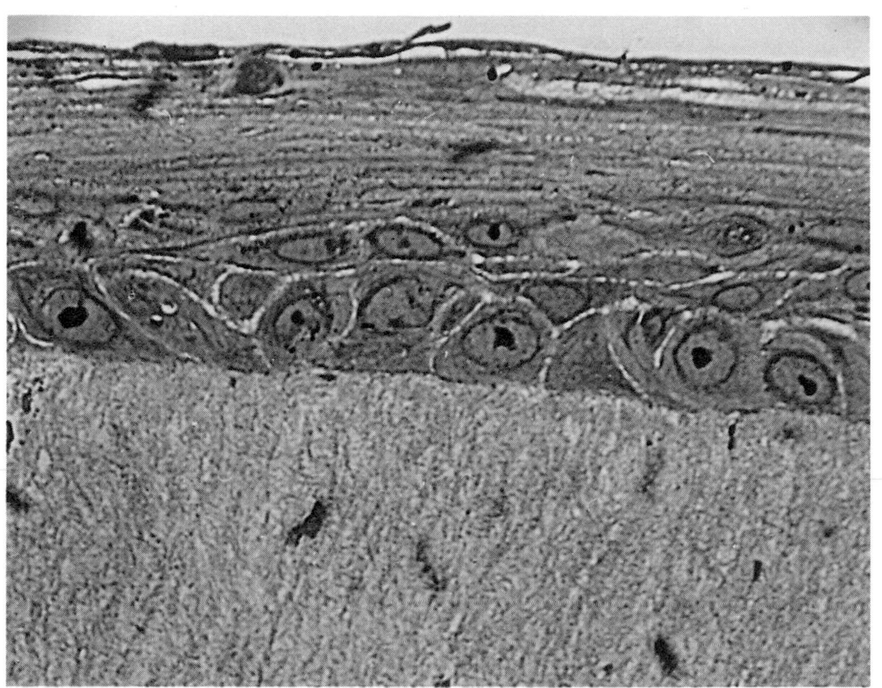
The epidermis of an artificial section of skin.

Shock waves against kidney stones (1982)

In **1982** three German professors in Munich, **Christian Chaussy**, **Egbert Schmied** and **Walter Brendel**, working with the Dornier company (manufacturers of fighter planes), developed a device that can disintegrate kidney stones by shock waves. The stones are broken down into tiny grains of about 1.5mm *.0585in* which can then be eliminated in the urine.

Treatment of gall stones by laser (1986)

In **1986** a German doctor, **Ludwig Demling** from the University Hospital of Nuremberg in West Germany, obtained extremely interesting results by the internal use of the laser. An endoscope is inserted into the gall-bladder where it is directed at the stone or stones and the surgeon fires a shot with the laser which destroys them without damaging the surrounding tissue.

Pencil-laser (1987)

This is a surgical laser which, according to its inventors, **Professor Jean Lemaire** and his research team at the French University of Science and Technology in Lille, is as easy to use as a pencil. This CO_2 type laser, known as the Optro 20, was presented in **1987** as the first in Europe to combine the most recent developments in the fields of optics and electronics.

Transparent X-ray table (1989)

Orthopaedic surgery, which is used for example in the treatment of road accident victims, often uses X-ray treatment before, during and after operations. An invention has been available since **1989** which eliminates the visual obstacle of the operating table. It is a table made of composite materials that are completely transparent to X-rays. Developed jointly by two French surgeons **Gindrey** and **Letourneur**, the table, known as the T 3000 and manufactured by Tasserit, has two positive advantages. By completely eliminating the image of the metal parts of the table, the intensity of the rays can be reduced by 50 per cent. Also, the operation time is shorter so the patient does not require such a strong anaesthetic.

GREAT FEARS, GREAT DISEASES, GREAT DOCTORS

Since the beginning of time, people have been terrified of contagious diseases. Today, it is AIDS, previously it was the plague, cholera, syphilis, smallpox, tuberculosis. There were even diseases such as the 'sweating sickness' or miliary fever in the 15th and 16th centuries which disappeared suddenly after claiming many victims. Nowadays in tropical countries sleeping sickness, and especially malaria, kill millions of people every year.

A plague victim is treated by a doctor (covering his mouth), as the wife makes magic signs.

Plague

The word plague was the name given to any epidemic disease which claimed large numbers of victims. The first recognised plague epidemic occurred in AD 542.

25 million dead

The second pandemic occurred between 1346 and 1353. It started in India, spreading to the Middle East and then Italy. In five years it killed 25 million people throughout Europe: a quarter of the population. In Britain the population fell from 3700000 in 1350 to 2000000 in 1377. The disease gradually died out, although it persisted within Europe, reappearing in certain centres: Venice in 1487, Milan at the beginning of the 16th century where it claimed the lives of 190000 of the 250000 inhabitants, London in 1655, Marseille in 1720.

Doctors were powerless to stop it, although they continued to carry out their duty at the risk of their own lives.

Hong Kong, 1894

During the third pandemic in 1894, the plague bacillus of the rat, as well as the human bacillus, was identified in Hong Kong by the Swiss-born French bacteriologist, Alexandre Yersin, who had been sent by the Institut Pasteur to study, in situ, the disease which had just broken out in China. In 1898, P. L. Simond identified the role played by the flea in the transmission of the disease. In 1897, Waldemar Haffkinse developed the first anti-plague vaccine.

50 years of pandemic

During the third pandemic which lasted for 50 years and killed 12 million people in India between 1898 and 1948, there were only 91 cases recorded in Europe. It was followed by minor epidemics in Algeria in 1930 and 1944. Today, the plague is still endemic in several regions throughout the world. The early use of antibiotics as soon as the disease is diagnosed has affected the outcome of the disease which is no longer necessarily fatal.

Leprosy

Leprosy is a legendary disease. It is one of the oldest diseases in the world. The name comes from the Greek *lepsi* meaning scales, and *lepein* meaning to scrape or chafe.

The earliest descriptions of the disease were imprecise and passed on by word of mouth. It appears to have existed endemically in India and along the banks of the Nile well before the 1st century AD. In Egypt, the examination of the skeleton of a Coptic mummy of the 5th century AD constitutes the earliest osteological proof of the existence of leprosy. It appeared much later in Europe. It is thought that it reached Greece from the East, following the campaigns of Darius, King of Persia. Subsequently, the legions of Alexander the Great and Julius Caesar helped to spread the disease.

Lazar houses

In the parable, Lazarus, covered in sores, lay at the gate of the rich man. This is the origin of the term lazar house. With the return of the crusades, there was a new outbreak of leprosy. During the 13th century, there were as many as

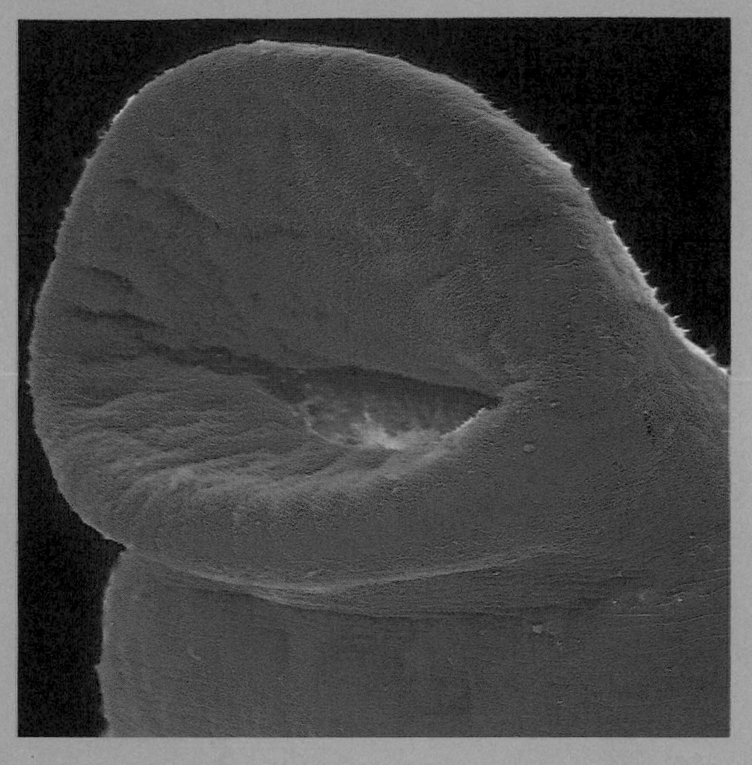

This is the mouth of a schistosome, the blood fluke that causes bilharzia. A vaccine has now been developed to counter this appalling disease which affects some 200 million people.

20 000 lazar houses in Europe. The first to be recorded was at Saint-Claude in the Jura Mountains around AD 460.

The disease was spread northwards by the barbarian invasions to places such as Iceland and Norway. It was in fact a Norwegian doctor, Armauer Gerhard Hansen, who discovered the leprosy vaccine in 1873, while working among lepers in his country. This was a major step forward.

Today

Leprosy is still rife in regions such as Central and Southern America, the West Indies, Black Africa and Asia. It persists in Europe in the Balkans and Portugal.

Cholera

Cholera, which has always been endemic in India, particularly in Bengal, for a long time spread no further than South East Asia. Suddenly, with the development of navigation and trade which took place during the 19th century, it spread throughout the world in a matter of 20 or so years. 1817 saw the first of six pandemics which affected the whole of the Eastern World. It reached Europe in 1823 and by 1837 had claimed more than a million victims.

The comma bacillus

During the fourth pandemic in 1883, which was particularly virulent in the Mediterranean and Egypt, the German doctor, Robert Koch, also famous for his work on tuberculosis, went to Alexandria to study the disease. He was responsible for discovering the cholera vibrio or comma bacillus.

Today

The disease is still active in India and South East Asia although there is a very effective vaccine which is compulsory for travellers visiting tropical countries. However, there is no guarantee that there will not be a new outbreak of the disease in countries with a more temperate climate.

Syphilis

The exact origin of the disease is not known, and there are many theories as to how it first appeared. An engraving by Albert Dürer, dated 1484, suggests that syphilis was in existence at the time. Syphilis is also supposed to have been identified on skeletons from central Russia dating back to 2000 BC.

The New World and war

There was an epidemic of syphilis throughout Europe after the sailors with Christopher Columbus returned from the New World in 1493. Wars, with their accompanying migrations of men (and prostitutes), were dangerous propagators of this disease.

The unhappy shepherd

In 1530 when the disease broke out again in Venice, a city with 11 654 known prostitutes, the Italian doctor, Frascator, named the disease syphilis after the unhappy hero of his poem *Syphilis sive Morbus Gallicus* (Syphilis or the Disease of France). The shepherd, Syphilis, has offended Apollo who afflicts him with a terrible disease which has the most awful symptoms. The skin falls away from his limbs, leaving his bones exposed, his teeth fall out, his breath become fetid. But he is saved by the miracle remedy, lignum vitae, which was recommended at the time. Treatments using lignum vitae and mercury were in fact carried out until 1907! It is difficult to say how many victims were claimed by syphilis during the course of the centuries. But it is possible to say that in the absence of any really effective cure, tens of millions of people died from the disease.

A major step forward

From the 19th century onwards, iodine and iodide were prescribed. After 1910, an arsenic-based preparation was used which had been developed by the German doctor, Paul Ehrlich (1854–1915). This changed the outcome of the illness.

In 1905, a major step had been made with the discovery of the agent responsible for the disease, *Treponema pallidum*, by the German microbiologist, Fritz R. Schaudinn (1871–1906), and the syphilis specialist, Erich Hoffman.

The real cure for syphilis only came with the discovery of penicillin. This does not mean, however, that the disease has disappeared. Far from it! There are about 65 million people throughout the world who currently have syphilis; but at least it can now be treated effectively, particularly if it is diagnosed in its early stages.

Malaria

Malaria has been in existence since ancient times, and it is likely that Alexander the Great suffered from it during his campaign in Mesopotamia during the 6th century BC.

The disease is endemic in all hot regions throughout the world, and particularly in swampy areas.

A voluntary scientist

The pathogenic agent *Plasmodium* was discovered by the French doctor, Alphonse Laveran, in 1881. In 1907 this remarkable man, who for 25 years carried out voluntary research at the Institut Pasteur, was awarded the Nobel Prize, half of which was devoted to his research in tropical pathology.

Resistant to quinine

After the discovery of quinine in 1820 by the French chemists Joseph Pelletier and Joseph Bienaimé Caventou, it became possible to treat patients suffering from the disease and to protect those who were not affected.

Today the anopheles mosquito has become resistant to insecticides, and *Plasmodium* has become immune to quinine.

A possible vaccine

Malaria is the most widespread and the most serious of the transmitted diseases. It threatens half the world's population and is one of the main fatal illnesses. About a hundred million people contract the disease each year of which some two million die. In the UK there were 2000 cases of imported malaria in 1984 and three deaths. The Institut Pasteur and several other laboratories worldwide are trying to develop a vaccine, a task which is made more difficult by the many transformations undergone by the pathogenic agent during the contamination process. However, it appears that the Colombian doctor, Manuel Elkin Patarroyo, is about to make a major breakthrough.

Gynaecology

Sanitary towels (1921)

The American company **Kimberley-Clark** of Wisconsin marketed the first commercial sanitary towels under the name of Kotex in **1921**.

Ernest Mahler, a German chemist working in the United States, had invented a cotton substitute made from wood pulp to compensate for the lack of dressings in hospitals. Nurses acquired the habit of using these cellulose-padded dressings as hygienic menstrual towels.

When the Kimberley-Clark company, which was already manufacturing cotton-wool bandages, learned of this, they began marketing them.

Tampons (1937)

In the 1930s the American **Earl Hass** thought of a way to modify the surgical tampon. He wanted to eliminate the inconvenience and the embarrassment caused by use of the sanitary towel. In **1937** he applied for a patent and founded the Tampax Company. After some improvements, the use of the tampon spread throughout the world after the Second World War.

Contraception

The vaginal tampon was already being used in Egypt in the time of the Pharoahs. Other Mediterranean peoples such as the Syrians used small sponges soaked in liquids such as spiced vinegar water which were supposed to have spermicidal properties. In 1984 there was a return to these origins when a sponge impregnated with spermicide came on to the market.

Modern spermicides, brought out in the 1970s in pessary or cream form, are 97 per cent reliable. One of the most reliable of these products is Nonoxynol-9 which is also noted for its protection against AIDS.

Sheath

The invention of the sheath is attributed to the Italian **Gabriele Fallopia** (1523–62), who was professor of anatomy at the University of Padua from 1551 to 1562. The sheath was made of cloth and was intended primarily as a means of combatting veneral disease; its contraceptive value was only secondary. Condom, a physician residing at the court of Charles II, was the inventor of the modern contraceptive which was named after him. Nowadays the contraceptive is widely available and has reverted to its original use as a protection against infection, and particularly against the spread of AIDS.

Diaphram (1881)

In **1881** in Holland the first birth control clinic was opened under the direction of Doctors Rutgers and Aletta Jacobs. They advised women to use the diaphram developed by the German **Mensiga**.

Coil (1928)

The first effective intra-uterine device was the silver ring designed by the German **Ernst**

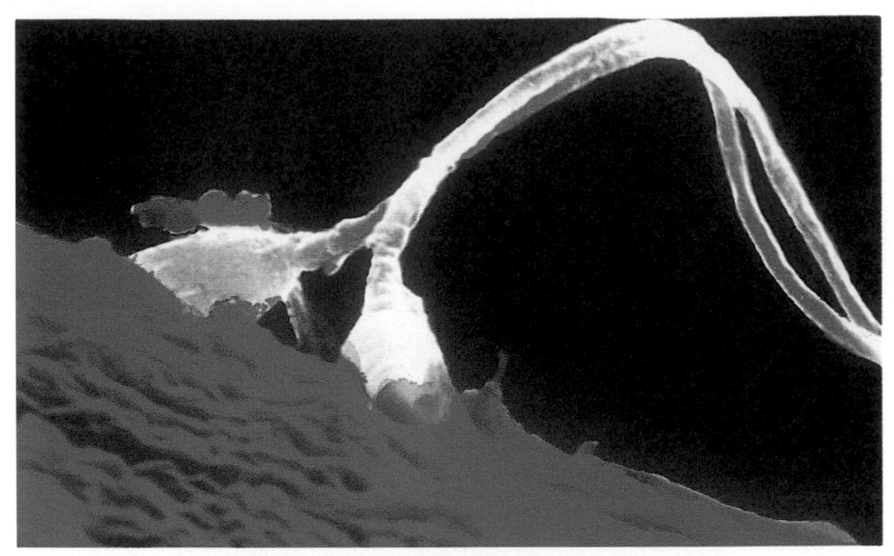

Human fecundation observed by a scanning electron microscope.

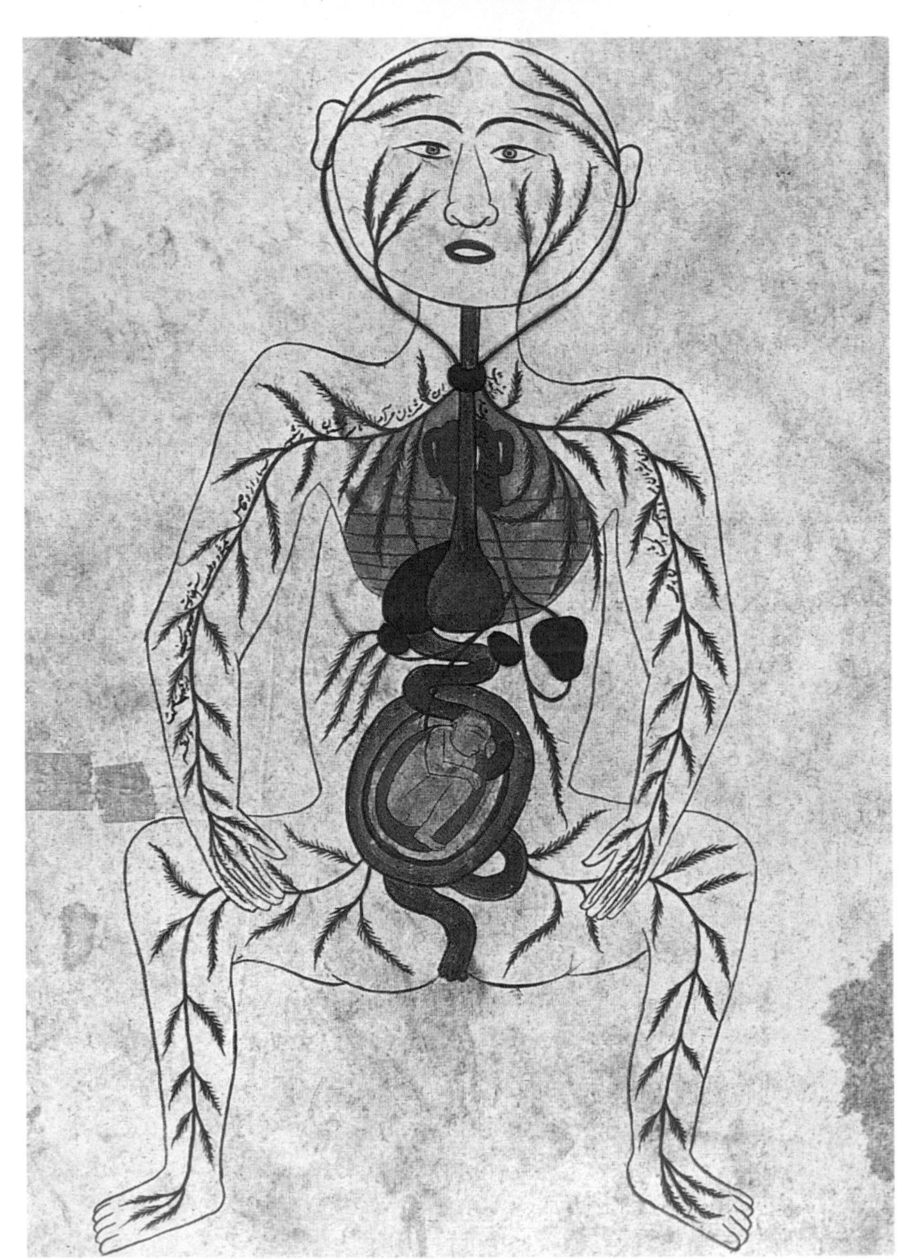

Artists have long been fascinated with fertility and birth.

Grafenberg in **1928**. Measuring 1.5mm .0585in in diameter, it was made of silver thread rolled into a spiral.

The pill (1954)

The pill was invented in **1954** by the American **Gregory Pincus** of the Worcester Foundation for Experimental Biology, Massachusetts, and **John Rock**. These doctors worked for five years to develop a definitive contraceptive that would be 'without danger, sure, simple, practical, suitable for all women, and ethically acceptable for the couple.' The initial clinical tests were performed in 1954, and the first large-scale experimentation took place in 1956 in San Juan, Puerto Rico with 1308 female volunteers.

The first pill to be marketed was Enovid 10, manufactured by G. D. Searle Inc. of Illinois, USA.

Anti-pregnancy vaccination (1990)

The World Health Organisation (WHO) in Geneva is currently sponsoring the first tests on a contraceptive vaccine for which 30 women have volunteered. The testing began in February 1986, at the Flinders Medical Centre in Adelaide, Australia. The vaccine has been developed over the past ten years by Ohio State University, USA, in collaboration with the WHO and various pharmaceutical laboratories, and could be on the market by the middle of the 1990s.

Caesarean section (7th century BC)

The caesarean is an operation that was first mentioned in a law set out by **Numa Pompilius** (715–672 BC), the legendary second king of Rome. According to his law, no woman who had died during her confinement could be buried until her infant had been removed via an abdominal incision. The name of the first of the Roman emperors, Caesar, is derived from the Latin word for cut *caedere*, because one of his forebears had given birth by caesarean section.

The first modern caesarean seems to have been performed in 1610 by the German surgeon, Trautmann.

Obstetric forceps (17th century)

Obstetric forceps were invented in the **17th century** by the English surgeon, **Peter Chamberlen**. Until then, a kind of hook, known as a head-hook, had been used. The French surgeon and obstetrician, André Levret improved the technique of the instrument, particularly the curve of the blades, in the 18th century. In 1838, there were 144 different types of forceps. After further improvements, the forceps of the French obstetrician, Stéphane Tarnier, were generally adopted.

Feeding bottle (antiquity)

Feeding bottles have been used since antiquity. At first they were in the form of jugs with two openings, one to fill the container, and the other, in the shape of a beak, to feed the baby.

Until the end of the 18th century, the teat was made from a small piece of rolled linen, one end of which soaked in the container while the other was sucked by the child. Teats were also made of sponge, softened leather or dried cow udders.

When rubber was discovered in the second half of the 19th century teats made from this new, hygenic material became popular.

This 15th century miniature shows a caesarian delivery. The name of the first Roman emperors derives from the operation.

Incubators (1880)

The incubator was invented by **Budin** of France in **1880**. These first wooden incubators were heated by saucepans of hot water placed beneath them. In 1894, **Lion**, a French doctor from Nice, invented the first incubator for premature babies.

The babies of science

Test-tube babies (1978)

The first test-tube baby was 12 years old on 25 July 1990: Louise Brown was born on **25 July 1978** at Oldham Hospital. Louise's scientific 'fathers' were **Doctors Patrick Steptoe** and **Robert Edwards**. They were the first to perform this scientific exploit. An ovum taken from the mother was fertilised in a test tube by spermatozoa from the father and the resulting embryo was re-implanted in the mother's womb. Today, thousands of test tube babies have been born throughout the world.

Frozen embryos (1984)

On **11 April 1984** Zoe, the first 'frozen' baby, was born in Melbourne, Australia. This was made possible by **Doctors Linda Mohr** and **Alan Frounson**, biologists at the Queen Victoria Hospital. Zoe was born from an embryo formed in a test tube and preserved for two months in liquid nitrogen at a temperature of −196°C *−320.8°F*.

In April 1987 it was announced that the first English test-tube twins would be born . . . 18 months apart. Amy and Elisabeth were conceived on the same day, but one embryo was re-implanted straight away and the other 18 months later.

A mother without ovaries (1988)

In December **1988**, Europe witnessed a scientific miracle. A 37 year old woman without ovaries gave birth to a daughter. A little over a year before, the woman had had to have her ovaries removed and she and her husband decided on a test-tube fertilisation. The two resulting embryos were re-implanted in the mother's womb after the removal of her ovaries. One of them embedded itself and, with the help of hormone treatment, the pregnancy was successful.

Therapeutics

Acupuncture (2000 BC)

The **Chinese** have been practising acupuncture, one of the branches of their traditional medicine, since about 2000 BC. The theory is that the cause of a disease can be explained by a disruption in the flow of energy, which can be remedied by action taken on one or more of the points situated along the meridians: pathways along which energy is transmitted around the body by oscillation and vibration. This action usually consists of the insertion of needles, but also of the application

IVF, GIFT, ZIFT, FIP. . . WHAT DOES IT ALL MEAN?

They are all methods of assisted reproduction, the most famous of which is IVF (In vitro Fertilisation), still referred to as IVFETE (Fertilisation in vitro and Transfer of the Embryo). It was initially used only in cases of sterility of the fallopian tube, i.e. when the tubes were blocked or missing. The cases for the use of IVF have been extended to all forms of sterility with no obvious cause. GIFT (Gamete Intra-Fallopian Transfer) was described in 1984 by Ricardo Asch as an alternative to IVF for sterility with no known cause as GIFT requires at least one of the tubes to be healthy. Fertilisation takes place in vitro as GIFT cannot be carried out through natural channels. It is of particular interest where there is no obvious cause for sterility.

ZIFT (Zygote Intra-Fallopian Transfer) is a method described by Braekmans in 1986. The fusion of the spermatozoa and the oocytes takes place in a test tube and not in the womb. The replacement is done under coelioscopy. The advantage of ZIFT is that it is possible to tell whether or not fertilisation has taken place. ZIFT is especially recommended in cases of male hypofertility. According to Doctor Devroey, a Belgian doctor who is one of the initiators of the method, the success rate is higher than that of IVF, about 30 per cent. But the method can be carried out only if one of the tubes is healthy.

IPF (Intra-Peritoneal Fertilisation) was described by Forrer, Dellenbach and Nizand in 1986. The method consists of introducing into the peritoneal cavity spermatozoa which have been prepared the day after ovulation has started in the woman (as a result of stimulation). The idea is to encourage fertilisation by placing the spermatozoa near the tube which is bathed in peritoneal fluid. IPF is therefore suitable for sterility with no obvious cause, and also male sterility.

of heat (moxas) or massage. Today, even electric currents and laser beams are used.

Antibiotics (1889)

In the course of his work, Louis Pasteur noted the vital competition that makes some bacteria fight against each other. This fact was repeatedly confirmed and was attributed to the action of an *antibiote* by the French scientist **Vuillemin** in **1889**. Penicillin was the first antibiotic discovered, and it remains the most important because of its curative effects and its almost complete absence of toxicity.

Penicillin (1928)

It was by chance that the Scottish bacteriologist **Alexander Fleming** (1881–1955), working at St Mary's Hospital in London, discovered penicillin. While working on staphyloccoci in **1928**, he discovered that they were destroyed by the mould which had contaminated them.

A team of researchers at Sir William Dunn School of Pathology at Oxford, Howard Florey, Ernst Chain and N. G. Heatley, continued Fleming's work and concentrated extracts of *Penicillium notatum* in order to obtain purified penicillin, which is more active. Mass production of penicillin began in 1943.

It is interesting to note that during the 9th and 10th centuries, Arab doctors of the Baghdad School were already using the curative properties of a mould that appeared in farinaceous foods. Vegetable dust blown into a patient's nose and mouth was used as a cure for ailments such as catarrh.

Monoclonal antibodies (1975)

The first research on monoclonal antibodies was published in May **1975** by two immunologists, **Georges Köhler** of Germany and **César Milstein** of Argentina, both working in Cambridge, England. In 1984 they were awarded the Nobel Prize for their research which consisted of combining, in a test tube, lymphocytes from mice which produce antibodies, with myeloma (cancer) cells. In this way they obtained hybrid cells or hybridomas that were able to survive indefinitely, continuously producing a single antibody specific to the illness against which the animal had been immunised. In therapeutics, monoclonal antibodies are able to direct a chemical substance against a given target located by the antibody which acts as a vector.

Antihistamine (1937)

An antihistamine is a drug that combats the effects of histamine and is used especially to fight allergies, a type of condition that was discovered in 1906 by the Austrian C. von Picquet. The first active antihistamine was discovered in **1937** by **Bouet** and **Staub**. The first synthesised antihistamine was created by the Frenchman, Halpern, in 1942.

Antipyretic agent (5th–4th century BC)

The first antipyretic agent (a remedy that lowers the temperature) was discovered by **Hippocrates** and was based on camomile.

Aspirin (1853)

Aspirin or acetylsalicylic acid was synthesised by the Frenchman **Charles Gerhardt** at the University of Montpellier in **1853** but he was not particularly interested in its practical use.

In 1893 a chemist working at Bayer rediscovered aspirin to treat his rheumatic father. Bayer began to market the drug in 1899 under the name Aspirin, formed from *a*cetyl + *spir*aeic + *in* which was a popular ending for the names of medicines at the time. In the Treaty of Versailles in 1919 Germany surrendered the brand name to the Allies as part of her war reparations.

Recent studies have shown what had been suspected for a long time: that taken in moderation, aspirin is good for the heart.

The world-wide consumption of aspirins: a hundred thousand million tablets are swallowed each year.

Chewing gum for smokers (1986)

Nicoret was conceived by the Swiss company **Ciba Geigy** as an aid to fight the desire for tobacco. It is a product based on nicotine which is released as the gum is chewed and absorbed directly via the mucus in the mouth.

Chemotherapy (1964)

Chemotherapy is the treatment of illnesses,

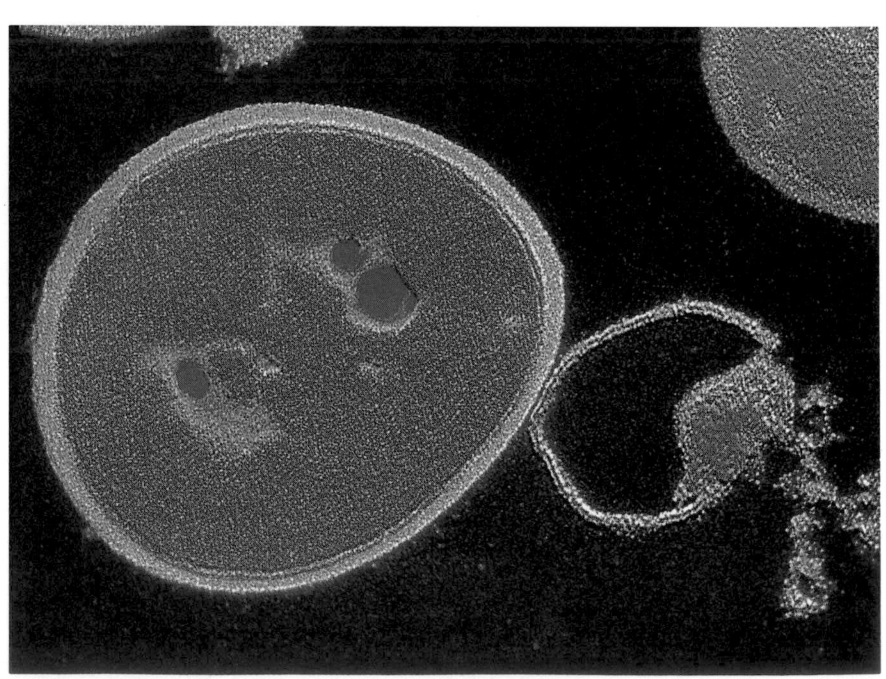

Germs may build up a resistance to one type of antibiotic but can still be destroyed by another.

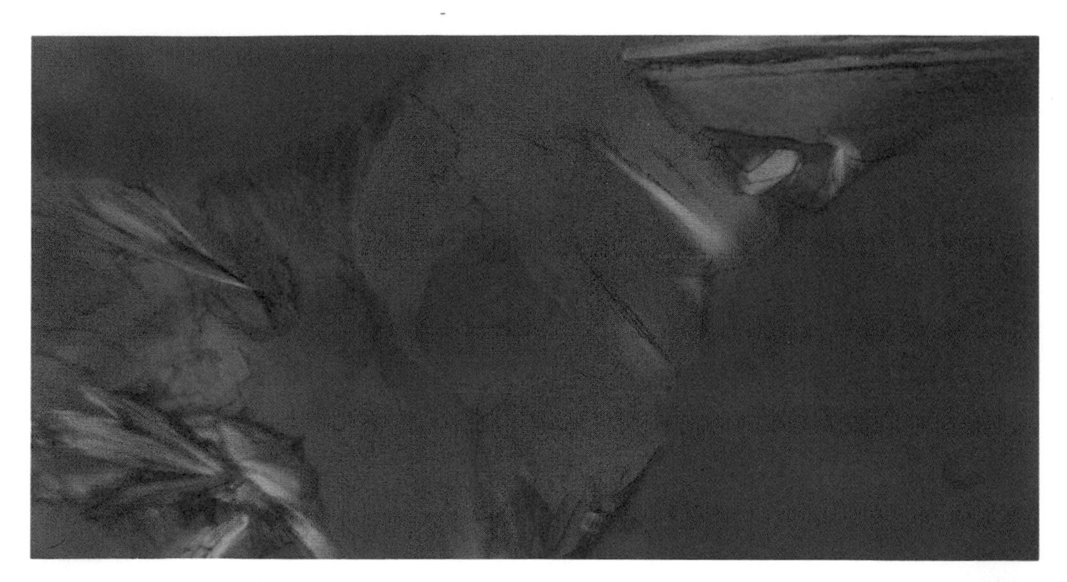

Bacteriologist Professor Alexander Fleming discovered penicillin accidentally while working at St Mary's hospital, London, in 1928. This discovery was to revolutionise medicine.

particularly cancer, by chemical substances or drugs. In **1964** an American professor, **G. Rosen**, used chemotherapy for the first time, and before trying any other form of treatment, on two types of cancer. His work was based on experiments carried out on the rat by Professor Brooke.

The German surgeon working in Boston, Emil Frey, referred to the new technique as initial or neo-adjuvant chemotherapy.

Homeopathy (1796)

The German **C. S. Hahnemann** (1755–1843) created homeopathy in **1796**. In 1790 he had been struck by descriptions of the properties of the cinchona, and by the incoherence of the explanations given for them, and decided to test its action on himself. He took large doses of cinchona over several days and indeed suffered the symptoms of an intermittent febrile state, identical to the fevers that cinchona cured.

Hahnemann then extended his experiments to belladonna, Digitalis, and mercury and verified the law of similitude: any substance capable of inducing certain symptoms in a healthy person is also able to make the analogous symptoms disappear in a sick person.

Hahnemann based his theory on extensive experimentation with healthy subjects, and this led him to pronounce the second fundamental principle of homeopathy: the remedy acts not by virtue of its quantity, but in proportion to its dilution.

Mechanical knee (1987)

The knee plays an important part in pathology in the field of sport. The mechanical knee, which is a perfect reproduction of the human knee, will enable progress to be made in areas such as the use of orthopaedic appliances which maintain joint articulation, designs for bindings and new protective devices for skis – and skiers. The knee took **Doctors L. Paulos**, **P. France**, **G. Jayaraman** and **T. Rosenberg** (orthopaedic specialists and biomechanics in Salt Lake City, USA), two years to perfect. The different sections were moulded from the muscles and ligaments of a dead athlete.

Immunotherapy

Immunotherapy is a recent science. It is a method of treatment based on the reactions of the natural defences of the organism against illness.

Interferon (1957)

Interferon was first discovered in **1957** by a British virologist, **Alick Isaacs**, and **J. Lindenmann** of Switzerland. It is one of the substances produced by the organism to combat viruses. In 1969, these substances, active in the immune system which is connected to the lymphatic system, were called lymphokines. Earlier, in 1966, the Americans J. David and V. Blum had identified in a culture of T-lymphocytes (a type of white corpuscle), a factor which inhibited the migration of macrophages (another type of white corpuscle). This was the first in a series of discoveries which were extremely promising for the future of therapeutics, a field in which 148 companies and 248 research centres throughout the world are currently involved.

Genetic engineering

Interferon was subsequently subdivided into categories and was not really developed further until it became possible to produce it by genetic engineering towards the end of the 1970s. Interferon 2A was marketed in 1987, after being tested for two years on tricholeucocyte leukaemia and Kaposi's sarcoma.

Interleukins (1979)

Other very well known substances which act as a defence are the interleukins, so called in 1979 because they act between lymphocytes (or lympholeukocytes), either causing the production of more lymphocytes or provoking the latter to produce other substances.

Interleukins 2 and 3 (1985 and 1986)

The most recently discovered of the interleukins, which are also subdivided into categories, is interleukin 3. It was discovered in **1986** by **Doctors Steven Clark** and **Yu Chang Yang**, of the Genetic Institute in Cambridge, USA. The anti-cancer effect of interleukin 2 was demonstrated in **1985** by the American doctor, **Steven Rosenberg**. Although they provide an excellent defence against viruses,

DEADLY CELLS USED TO COMBAT CANCER

After radiotherapy, chemotherapy and surgery, immunotherapy could become a fourth weapon against cancer. The method, which is still in its early stages, and which has had its failures, was presented in November 1988 by Professor Steven Rosenberg, a pioneer of immunotherapy who, among other things, has carried out research on LAK lymphocytes, deadly cells which are naturally present in the human immune system.

In the United States, Doctor Rosenberg has carried out research in collaboration with the company, French Anderson, in order to try to understand the reasons for the many failures of his treatment. During 1989 ten patients, in an advanced stage of the illness, received some of their own cells which had been genetically marked, so that their progress within the organism could be monitored. This is the first time that a genetic manipulation test has been carried out on human beings. The first results should be known in a year.

SHAKESPEARE'S HAND

Designed by a Californian, Scott Shakespeare, who was born without a left arm, the Simplistic Hand is made of lightweight plastic, is pliable, capable of grasping and looks like a real hand. In fact it could be mistaken for one, with its colour, tracing of the veins and fingerprints. It is less expensive, not as heavy and just as efficient as the electronic limbs currently used which are both unaesthetic and a source of embarrassment.

Scott Shakespeare, the founder of the company, TAPS, The Advancement of Prosthetic Services, presented his invention at the 1988 Inventions Convention in California.

SOME IMPORTANT DATES IN THE HISTORY OF AIDS

The following dates are important in the fight against AIDS.

1983: The team of Professor Luc Montaignier of the Institut Pasteur in Paris isolated LAV. In the same year, a research team from the National Cancer Institute, led by Robert Gallo, discovered the retrovirus HTLV 3 which is identical to LAV.

1986: A second AIDS virus was isolated by the team of Professor Montaignier in collaboration with Portuguese doctors. This virus, HIV 2, is particularly virulent in Africa.

At the same time, the Franco–American team of Doctor Max Essex of the Dana Farber Institute of Boston and Doctor Francis Barin of Tours, described another virus, HTLV 4. But it seems likely that HIV 2 and HTLV 4 are one and the same virus.

1987: The Institut Pasteur published the sequel to the virus HIV 2 and brought out a second diagnostic test, Elavia. The first test, Elisa, was developed by the same team in 1985.

1988: Research was carried out in the United States by scientists of the Frederick Center, Maryland and a team from the Center for Disease Control in Atlanta on a second generation of (rapid) detection tests which will make it possible to detect the virus as soon as it enters the organism.

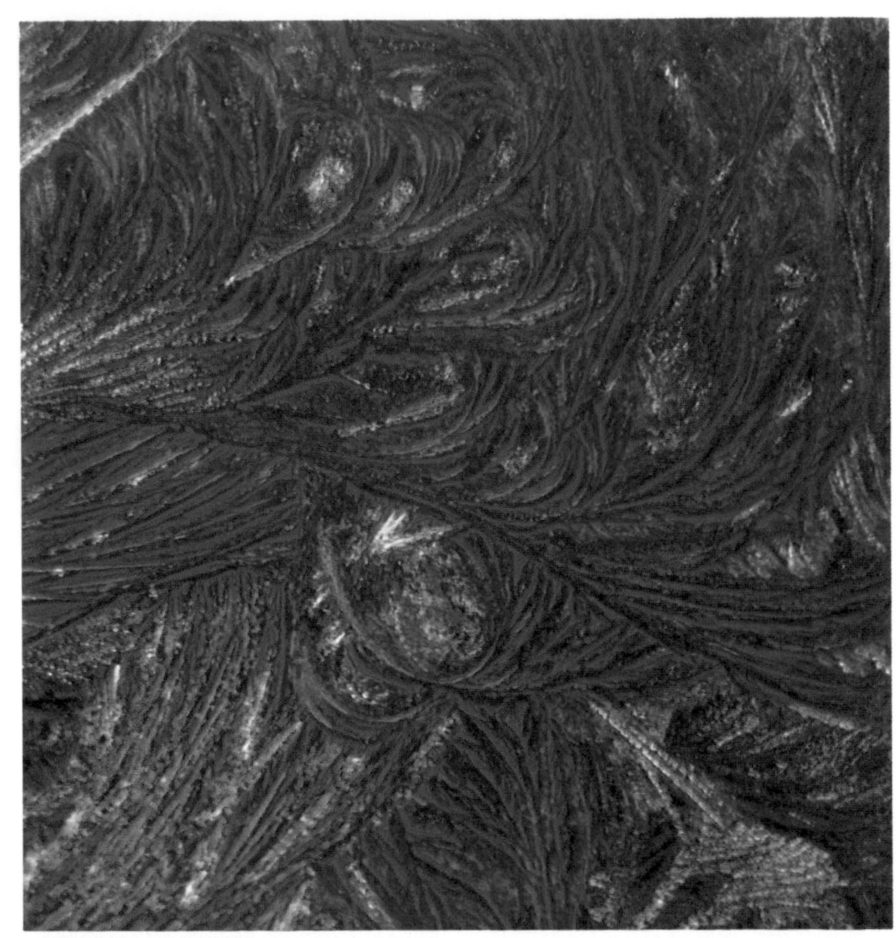

AZT has given hope to AIDS sufferers, of whom there are over 9600 in the UK. Some 50 per cent of victims die within a year of diagnosis.

interleukins can have a toxic effect if they are administered in large quantities.

TNF (1975)

TNF or Tumor Necrosis Factor was discovered in **1975** by **Carswell** and his colleagues at the Sloan Kettering Memorial Institute in New York. It is a protein produced by certain white corpuscles which selects and attacks tumorous cells. The production, study and classification of all these proteins has led to a review of knowledge in the field of immunology. These substances have potentially far-reaching effects and implications, not only in the fields of cancer research and virology, but also in the areas of parasitic, inflammatory, infectious and allergic diseases, and in vaccinations and transplants.

Intravenous injection (16th century)

Elshots was the first to inject medicinal products into human veins in the middle of the **16th century**. However, we know that 'intravenous infusions' had already been tested on animals.

In 1655 Schmidt treated syphilis by intravenous injection.

Syringe (1657)

The principle of the syringe was established by the Italian Gattinara in the 15th century, but it was not until the 17th century that practical trials were carried out by the Englishman **C. Wren** and the Irish scientist **Robert Boyle** in **1657**.

The Englishman Fergusson was the first to use glass, whose transparency allowed the injection to be monitored. However, it was the Frenchman Luer who produced the first all-glass syringe in 1869 and risks of infection fell as a result of its use.

Finally it is to the Irishman Rynd in 1845 and the Scot Wood in 1853 that we owe the method of subcutaneous injection.

Safety hypodermic (1987)

In **1987** the French doctor **Jean-Lous Brunet** of the Hôpital de la Croix-Rousse in Lyon, patented a safety device which can be attached to hypodermic needles or vacutainers for taking blood samples. It eliminates any risk of contamination from the blood in illnesses such as AIDS and hepatitis.

The system enables the used needle to be automatically re-capped as soon as it is withdrawn from the vein or muscle.

A similar type of invention in the same area is the disposable hypodermic which it is impossible to re-use. It was invented by four Danish medical assistants working with drug-users in Aarhus on the east coast of Denmark.

Jetnet (1988)

The handling of cutting instruments and hypodermics used in even minor operations is a source of contamination which, particularly in the case of AIDS, represents a very real danger for medical staff. A new system has been developed by **3M** which ensures the complete isolation of used instruments and eliminates the need for direct contact. Presented in **1988**, Jetnet consists of a hermetically sealed plastic receptacle with a cover in which there is a clamp-operated mechanism. Scalpel blades, needles and other instruments can be separated, kept in sealed receptacles and then incinerated without ever being touched by the person using them. The cover can be sterilised and re-used. The level of prevention achieved in this way is considered to be complete.

Insulin (1921)

Discovered by the Romanian, **Paulesco**, in **1921**, insulin was isolated a few months later by the 1923 Nobel Prize winners the Canadian physiologists John James Macleod and Sir Frederick Grant Banting, of the University of Toronto, and the American physiologist, Charles Herbert Best .

Synthetic insulin was obtained in 1964 by Panatotis and Kastoyannis of the University of Pittsburgh, USA.

In 1978 the Eli Lilly Laboratories successfully synthesised the human insulin gene, which was a major step in the production of insulin by genetic engineering.

In 1982, the first human insulin obtained by genetic engineering appeared on the market.

Anti-cholesterol machine (1986)

This process was invented in **1986** by a German biochemist, **Wilhelm Stoffel**. It is based

These rather attractive crystals belong to a modern bogeyman: cholesterol. A machine that filters out the 'bad' cholesterol was developed by German biochemist, Wilhelm Stoffel.

THE LATEST NEWS IN THE FIGHT AGAINST AIDS

In 1989 there was no decisive progress made in the fight against AIDS. New anti-viral drugs were researched but AZT is still the most effective medication. According to the most recent statistics, it can prolong the life of AIDS sufferers by 200 per cent.

At the Fifth International Conference on AIDS, held in Montreal, in June 1989, Professor Jonas Salk and his team announced that they had successfully vaccinated a healthy (seronegative) chimpanzee against the AIDS virus.

The vaccine they used, developed by Doctor Clarence Gibbs of the National Institutes of Health, consists of the complete HIV virus, released from its membrane and inactive with an adjuvant which enhances the antigenicity of the vaccine.

The vaccinated chimpanzee was then given the AIDS virus. Six months later, there was no trace of the virus in its system. Two other (seropositive) chimpanzees were also vaccinated, with the same result: six months later there was no trace of the virus in their systems.

A second type of experiment using the same vaccine on humans instead of chimpanzees, was carried out by Doctor Alexandra Levine of the University of South Carolina. Nineteen seropositive patients in a pre-AIDS state were given the vaccine. The aim was to prove that the vaccine was harmless, and the result was categorical. The Salk vaccine was not toxic, and what was even better, 12 out of the 19 patients saw an improvement in their immune system. Widespread therapeutic tests will now take place on thousands of seropositive patients in both the United States and France, run jointly by Professor Jonas Salk, the American company, Immune Response Corporation and Pasteur-Vaccin, a branch of the Institut Mérieux.

on haemodialysis, a system of filtering the blood, and is used for people whose kidneys do not function. It enables the particles of so-called bad cholesterol (Low Density Lipoprotein, LDL) to be eliminated, while the good cholesterol is completely restored to the patient after the filtering process. Such a technique can be applied to patients who have suffered from an excess of cholesterol since birth and, regardless of diet, will give them a quality of life which would have been impossible until now.

Anti-sugar medication (1985)

The **Servier Laboratories** have developed a medication which enables weight to be lost naturally by reducing the need for sugar. This anti-sugar medication has already been successfully tested in the United States by Professor Wurtman of the Massachusetts Institute of Technology (MIT), and has been available, subject to medical advice, since **1985**.

Medication to treat river blindness (1987)

In **1987** the American company **Merck and Co.** developed a medication to treat river blindness. The disease is spread by a fly and there are 18 million suffers throughout the world, mostly in Africa. The new medication, Mectizan or Ivermectine, will be offered free of charge to various health programmes in the Third World.

Patch (1952)

Soon there will be no need to take pills or drops; medication will be administered in the form of subcutaneous mini-implants, self-adhesive bandages and self-adjusting miniature pumps. The first invention using this type of technique was the Spansule, created by the American company **Smith, Kline and French** in **1952**.

In 1975 the Alza Corp. brought out Ocursert which when placed behind the eyeball releases pilocarpine for the treatment of glaucoma.

Self-adhesive anti-smoking device (1988)

After their self-adhesive device to combat sea sickness, the Swiss laboratories **Ciba-Geigy** have brought out a new patch to help people

A MEDICATION TO INCREASE DESIRE: A SECRET RECIPE

LY 163-502 is the code name of the mysterious molecule produced by the American pharmaceutical laboratories, Elli Lilly, which is supposed to have the power to increase sexual desire. The molecule, which is still being tested, could be the first truly effective medication to combat the lack of sexual drive. But you will have to be patient as the tests, which began in May 1988, are still going on.

This panel constantly reminds passers-by of the number of deaths from lung cancer in the US.

give up smoking. The TNS or Trans Nicotine Patch System is a stamp soaked in nicotine which is stuck onto the skin. The stamp, like the chewing-gum, also brought out by Ciba-Geigy, helps the smoker to give up smoking without suffering from withdrawal symptoms. It is currently being tested in Switzerland.

The time pill (1986)

Benzodiazepines, or tranquillizers, were discovered by the American doctors **Fred W. Turck** and **Susan Losee Olsen**, and scientists of the Northwestern University of Evanston, Illinois. Better known as Valium and Librium etc, they are used in the treatment of insomnia and enable biological rhythms to be reset when they have been disturbed by a time change, for example. According to the time they are administered, the pills can either bring forward or put back our biological clock which, among other things, synchronises our sleeping/waking rhythm. One of the first people to have identified this biological cycle of approximately 25 hours, known as the circadian rhythm, was the spelaeologist, Michel Siffre, during one of his underground expeditions in 1972.

Psychoanalysis (1885)

Psychoanalysis is at once a method of understanding psychological and psychopathological phenomena, and of treating mental illness. It was developed around **1885** by the Austrian doctor **Sigmund Freud** (1856–1939) in Vienna. Freud did not invent the idea of the psychic unconscious, but he undertook a systematic exploration of it.

At first received with considerable scepticism by the medical community due to its novel propositions regarding sexuality, psychoanalysis has acquired a more and more important place in medicine and in psychology.

Quinine (1820)

In the 17th century the Indians of Peru entrusted the secret of the 'sacred bark', or Cinchona bark, to the Jesuit missionaries who were both numerous and powerful in the country at the time. In **1820** the French pharmacists **Joseph Pelletier** and **Joseph Bienaimé Caventou** isolated the alkaloid contained in the cinchona bark: quinine. It was partially synthesised in 1931 by Rabe, but it was not until 1944 that total synthesis was achieved by Doerning.

Radiotherapy (1934)

Radiotherapy, the use of rays for therapeutic purposes, was developed after the discovery of X-rays in 1895, and of radioactivity. It became an area of specialisation independent of radiodiagnosis in **1934** when the French chemist **Irène Joliot-Curie** (1897–1956) and her husband **Jean-Frédéric Joliot** (1900–58) discovered artificial radioactivity. The first apparatus capable of transmitting radiation which could reach relatively deep-seated tumours was fed by a current of 250 000 volts. This method was gradually abandoned and replaced in 1956 by the cobalt bomb.

Sleeping pill (1st century BC)

The first sleeping pill was invented during early Roman times by the medical writer, **Celsus**, who gave patients suffering from insomnia a pill made from the mandrake and henbane. Today it is known that both these plants are narcotics.

The treatment of pain

Bracelet for the treatment of neuralgia (18th century)

During the 18th century, bracelets and necklaces for the treatment of neuralgia were extremely popular. An Austrian physician Franz Anton Mesmer (1734–1815), who lived in Paris, made a considerable amount of money by selling 'magnets which would draw out the pain'. The doctrine of mesmerism, based on 'animal magnetism' was named after him.

Vitamins (1910)

Although the influence of vitamins on the body was only demonstrated in 1910, their effect had long been suspected, particularly in connection with deficiency diseases such as

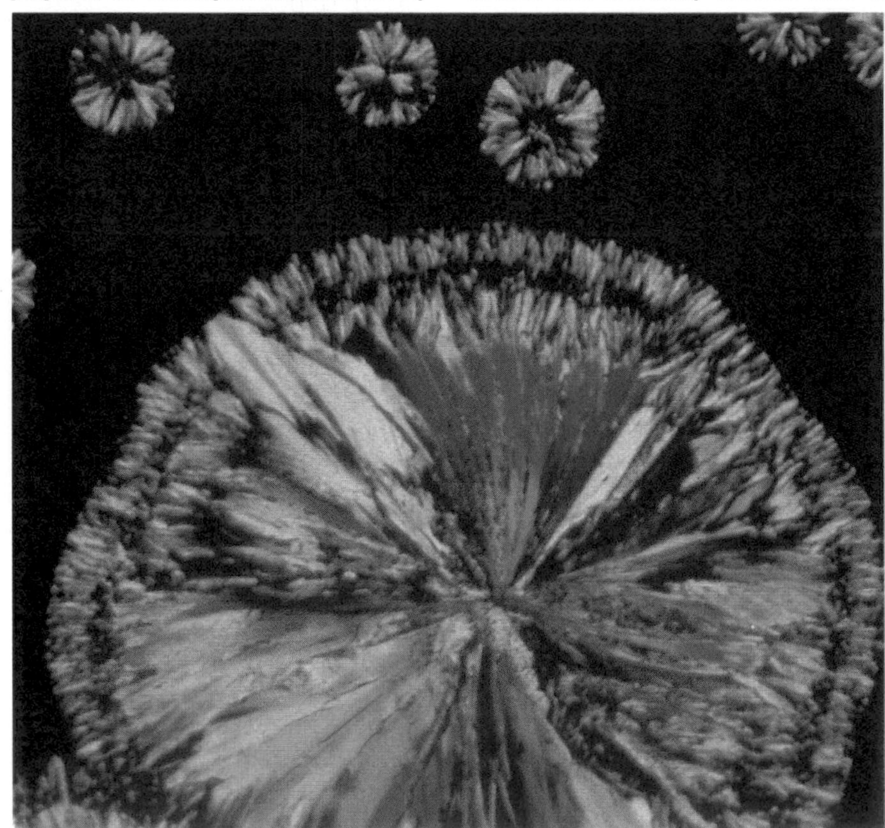
Small quantities of vitamins are essential if our bodies are to remain healthy. Above is B1.

Alka-Seltzer was developed by the Miles laboratories in Indiana (USA) in 1931. Here advertised as a cold remedy, it is now better known as a hangover cure.

scurvy and beriberi. The English navigator Captain James Cook had been one of the first to mention them in a letter dated 7 March 1777 and addressed to the Royal Society of London.

In **1910** the Polish chemist **Casimir Funk** isolated vitamin B1 in unpolished rice and gave the name *vitamine* to this 'amine' which was so essential to life.

In 1936 the American chemist Robert R. Williams synthesised vitamin B1, and in 1959 a German physicist, Dieter Muting, invented an anti-mosquito pill based on vitamin B1. Vitamin C was isolated in 1928 by Saint Györgyi from the juice of the capsicum, and synthesised in 1933 by the Swiss chemist, Tadeus Reichstein.

Eyes and ears

The cataract operation (1748)

The first description of a cataract operation by 'lowering the cornea' was given by the surgeon, Anthyllus, in the 2nd century AD.

In **1748** the French opthalmologist **Jacques Daviel** performed the first cataract operation by extraction of the crystalline lens. At the time, the reputation of Daviel was equalled only by that of the British surgeon William Cheselden who had restored the sight of a patient who had been blind since birth.

Operation by ultra-sound (1976)

In **1976** the American **Charles Kelman** invented an instrument which made it possible to remove the cataract through a tiny incision after fragmenting it with ultra-sound. The operation was performed under local anaesthetic.

In 1986 Doctor Kelman also invented a completely harmless, flexible ocular implant, which can be inserted through an making an incision in of 3mm *.12in.*

Operation by laser (1979)

In **1979 Professor Danièle Aron-Rosa** developed an operation technique using a super-quick-acting Yag laser which made it possible to operate without making an incision in the patient's eye.

Artificial crystalline lens (1952)

Invented in **1952** by the English doctor **Harold Ridley**, this polymethylmethacrylate (PMMA) or plexiglass lens is placed behind the iris.

Multifocal crystalline lens (1989)

A new ocular implant has revolutionised cataract surgery by making it possible for the eye to accommodate without the patient having to wear spectacles.

Previously the artificial crystalline lens which replaces the opaque natural lens has restored the patient's vision without being able to accommodate. The lens is usually adjusted for long vision so that the patient has had to wear spectacles for close work.

The multifocal intra-ocular implant developed by the American company **3M** diffracts the light in such a way that there are two focal points, one for long and one for short vision. This device, and other similar ones, are currently being studied in many centres throughout Europe and by the Food and Drugs Administration (FDA).

The corneal graft (1949)

This form of treatment for certain types of myopia and severe hypermetropia is known as refractive lamellar keratoplasty and rectifies the curve of the cornea. The idea was first conceived in **1949** by the opthalmic surgeon, **Professor Barraquer** of Bogota, Colombia

Since 1983, keratoplasy has been performed at the Rothschild Foundation in the departments run by the American surgeons, **Doctor Ganem** and **Professor P. Couderc**.

Incision of the cornea (1955)

This method of treatment, radial keratotomy, consists of making radial incisions on the inner surface of the cornea in order to correct myopia. It is intended to replace the use of spectacles and contact lenses. It was developed in **1955** by **T. Sato** of Japan only to be abandoned and then resumed in 1979 by the Russian practitioner Sviatoslav Fiodorov who, with 50 surgeons, created a sort of production line in his hospital in Moscow which carried out 22000 operations in one year. This system enables the most able specialists to concentrate on the important part of the operation, while the initial and final stages are supervised and dealt with by other practitioners.

Spectacles (1280)

Very early on in history, various attempts were made to remedy sight defects. However, the magnifying glass was not invented until the 11th century.

In **1280** the Florentine physicist **Salvino degli Armati** (1245–1317) developed two eye glasses which, at a certain degree of thickness and with a certain curve, magnified objects. He can therefore be said to have invented spectacles. He told his secret to his friend Alessandro della Spina, a Dominican friar from the Monastery of Saint Catherine of Pisa, who subsequently revealed it. At this point the spectacles had convex lenses for long-sighted people.

Concave lenses for short-sighted people appeared at the end of the 15th century.

The first spectacles consisted of two lenses made of beryl, a sort of crystal, set in a circle of wood or horn. They were later joined by a stud.

Bifocals (1780)

The American **Benjamin Franklin** invented bifocals for long-sighted people in **1780**; in those days, bifocals were simply two lenses bound together by a metal frame. Bifocals made from one piece of glass were developed by Bentron and Emerson for the Carl Zeiss company in 1910. In 1908 J. L. Borsch had invented a process that allowed the two lenses to be soldered together.

Varilux Pilote (1985)

The Varilux Pilote spectacles were invented in 1983 by **Essilor Svenska**, sponsored by the Swedish airline SAS, and have been on the market since **1985**. They make it possible to see at close range not only by lowering the eyes, but also by raising them. This makes them extremely suitable for the professional activites of those who, like airline pilots, need to be able to see at close range above as well as below eye-level.

Light-sensitive sunglasses (1938)

Doctor Edwin Land, the founder of the Polaroid company, invented polarising sunglasses

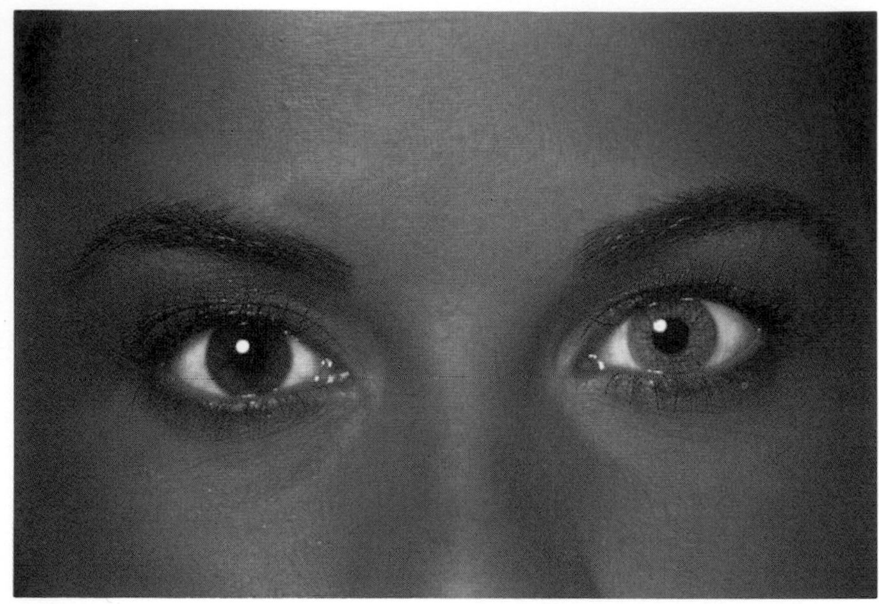

These contact lenses, patented by the American company Wesley-Jessen, turn brown eyes blue.

which cancel reflections in **1938**. As early as 1939 a chemist at Corning Glass, Doctor R. H Dalton, had begun to develop glasses sensitive to variations in light. His invention, patented by Corning in 1964, was put on the market in 1967 as Photogray.

Anti-rain glasses (1989)

Anyone who wears glasses and plays a sport knows only too well that when it is raining or snowing, before long it becomes virtually impossible to see anything.

The unmarkable, anti-rain lenses invented by the Japanese researchers for Nikon, are a real innovation. Water slides over the surface of the glass without clinging or leaving marks. They have been on the market in England since 1989.

Contact lenses

Origins

The first person to think of contact lenses was **Leonardo da Vinci**. In his *Codex on the Eye* he described an optical method for correcting refraction defects by immersing the eye in a water-filled tube that was sealed with a lens.

In 1686 the Frenchman René Descartes performed the experiment and it was performed again at the end of the 18th century by two Englishmen Thomas Young and John Herschel. Herschel's experiment involved applying a layer of gelatin to the eye, held in place by a lens. This was to correct his astigmatism.

1887: glass contact lenses made their appearance due to the efforts of August Müller in Germany, Fick and Sulzer in Switzerland, and Kalt in France.

1936: the German company I. G. Farben made the first plexiglass contact lenses. This same material is used to day for hard lenses.

1945: an American Tuohy invented the corneal lens which covers only the cornea

1964: the Czech Wichterle developed flexible hydrophilic lenses.

Holographic lenses (1987)

These lenses are for people who are both short- and long-sighted. They are called Diffrax, use a technique based on holography, and their entire surface corrects the vision. Even in half light they adapt to the vision of the wearer. They were invented by the English company **Pilkington**, which is well-known for its range of inventions in the field of glassware, and came onto the market at the beginning of 1987.

Disposable lenses (1989)

The first disposable contact lenses, developed by the American laboratories **Vistakon**, have been brought onto the market. They are designed to be worn continuously for a week at a time and then discarded and replaced by a new pair. They are suitable for mildly short-sighted people.

Electric hearing-aid (1901)

The American **Miller Reese Hutchinson** was 26 years old when he invented the first electro-acoustic apparatus designed to amplify sounds for the deaf. One of the first users of the invention was Queen Alexandra who wore her Acousticon during the coronation ceremony. She presented a medal to the young inventor as a mark of her gratitude.

Implants for the severely deaf (1961)

In **1961 Georg von Békésy**, a Hungarian physicist and physiologist living in the United States, was awarded the Nobel Prize for Physiology and Medicine for his discoveries concerning the stimulation mechanism of the inner ear or cochlea.

In 1973 the company 3M developed a cochlear implant which made it possible for severely deaf people to hear and interpret most sounds.

Scientists throughout the world have pursued similar lines of research with considerable success. In 1977 the Bertin implant was developed by the French doctors Pialoux, Claude Chouard and MacLeod, and in 1981 an Australian, Professor Graeme Clark, produced the 'bionic ear'. In 1988 the most recent model of the device was implanted in a five-year-old girl who had been completely deaf since birth. It was the first time that such an implant had been performed on a child of that age.

Dental surgery

Dentures (16th century)

The earliest complete denture on record was found in a grave in Switzerland that dated from the early **16th century**. The dentures appear to have been made from the femur of an ox which was cut and then carved. The upper and lower sections were joined by a metal wire.

Dentures were more widely used when it became possible to place the false teeth on a base which could then rest on the jawbone and adapt to its shape. In 1864, the American company Goodyear brought out a vulcanised rubber which made it possible to produce this type of base.

False teeth (1788)

Until the 19th century, false teeth were made from the bones of familiar or exotic animals; hippopotamus bones were the most widely used as they were the strongest. But these teeth became worn, turning brown and giving off a nauseating smell and had to be changed every 18 months to two years. So, different materials were sought.

As early as 1770 the French apothecary, A. Duchâteau had tried to produce a complete denture made from a mineral paste.

In **1788 Dubois de Chemant**, a Parisian dentist, produced hardwearing dentures by taking a wax impression from which he made a plaster model.

In 1817 the first porcelain teeth appeared in the United States, and in 1825 S. W. Stockton of Philadelphia commercialised what had previously been a rather unsophisticated process.

Crowns and bridges (3rd century BC)

A piece of a dental plate made in gold was found in Tanagra in Greece. The plate, which consisted of four elements, had been made to replace the two lateral incisors. It appears that the tomb in which it was found could date back to the **3rd century BC**. The two central incisors had bands round them and acted as a support for the extension of the false teeth which they held in place. This is the earliest existing evidence of a bridge.

In his *Traité des Dents* (Treatise on Teeth) the French dentist, Pierre Fauchard (1678–1761) discusses different ways of achieving fixed dental plates, the bridges promoted by W. H. Dwinelle almost a century later in 1856.

Crowns made from porcelain or industrial resin, known as jacket crowns, appeared around 1895.

Solid ceramic (1979)

This new type of ceramic developed by **Doctor Sozio** of the United States makes it possible to construct artificial teeth that are superior to the original. The material is so solid that it

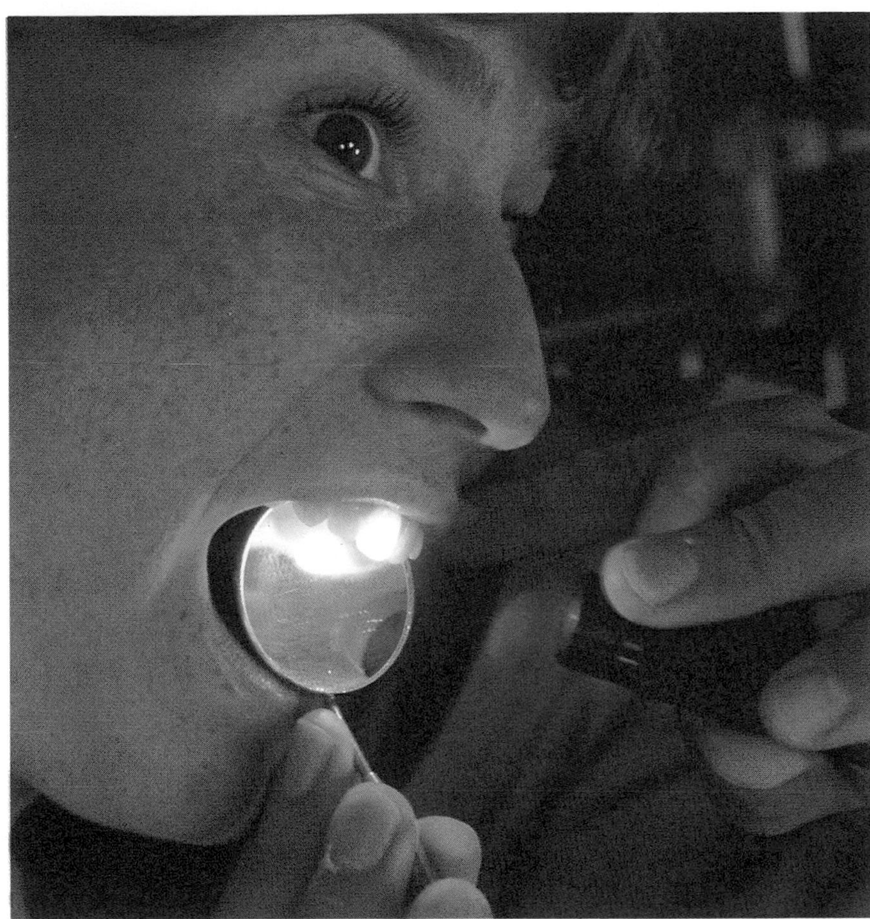

Lasers have found an application in dental surgery.

does not require any reinforced metallic support.

Dental implants (1965)

Dental implant as a method of treatment for people who have had all or some of their teeth removed has existed for the last 30 years. Pegs are inserted into the jaw which act as a support for false teeth. But it is a method which, over the years, has produced many failures. In **1965** positive progress was made as a result of a method of implantation developed by a Swedish professor, **Per Ingvar Branemark**, a biologist at the University of Gothenburg. The method, which has a success rate of 81 per cent for the upper jaw and 91 per cent for the lower jaw, is the only one recognised by the American Dental Association.

Fillings (9th century)

The method of filling dental cavities resulting from dental caries is attributed to the famous **9th century** Muslim physician, **Abu Zakariya Yuhanna Ibn Masawaih** (776–855), also known as Mesuë Major, who used gold for his fillings. In the 15th century, important progress was made by the use of gold leaf which made it possible to fill the cavities completely, a technique developed by the Italian surgeon, Giovanni Arcolani. In 1853 Makins replaced gold foil with porous gold. This was replaced by soft or non-cohesive gold, and then by cohesive or adhesive gold used by Arthur in 1855.

Compounds (1963)

In **1963 Bowen** of the United States discovered a formula for a synthetic resin, for rebuilding and restoring teeth.

Since then, many improvements have been made in the fields of resistance to erosion and adhesive quality. Nowadays, adhesives are being used which prevent the recurrence of dental caries.

The use of ultraviolet rays, blue light and halogen lamps gives a better polymerisation of the product. This also gives the dentist more time to concentrate on the delicate operation of putting the composite in place.

Toothpaste

The earliest formula for toothpaste was given at the end of the **1st century AD** by a Roman doctor, **Scribonius Largus**. It consisted of a mixture of vinegar, honey, salt and glass frit!

Pliny the Elder recommended urine as a mouthwash, and this use of urine, particularly as treatment for dental caries, persisted until the 19th century. The explanation given was that urine, which was warm and acidic, neutralised the decaying action of the cold, damp secretions from the pituitary gland which flowed from the brain into the mouth.

Extraction of teeth

In China the instruments preferred for pulling teeth were the fingers. For five or six hours a day, tooth-pullers practised pulling out nails hammered into thick planks.

The invention of the first pair of dental pliers is traditionally attributed to the famous doctor and anatomist, Erasistrates of Alexandria, during the 2nd century BC.

Celsus, the great Roman medical theoretician of the 1st century AD, suggested placing a piece of flock soaked in the sap of tithymalis, a variety of euphorbia, in the painful cavity, a method which broke the tooth into pieces.

In the early 16th century, an instrument appeared which made it possible to extract teeth without the risk of crushing them or fracturing the crown. The invention was known as the pelican and was attributed to the Italian surgeon, Giovanni Arcolani. The first so-called anatomic forceps were designed by the London-based French surgeon, J.-M. Everard (1800–82) in collaboration with Sir John Tomes (1815–59)

Oral anaesthetic (14th century)

The first evidence of an oral anaesthetic is provided by a surgeon from Padua, grandfather of the famous Italian religious and political reformer, **Girolamo Savonarola**. He made his patients chew tiny cloth sachets filled with henbane, poppy seeds and mandrake. The sap from these plants produced an insensitivity of the mucous membrane which made it possible to make a painless incision.

The mouthwash (15th century)

In the **15th century** the Italian surgeon from the Salerno School **Giovanni Plateario** was the first to recommend a mouthwash to accelerate post-operative healing. This was a major step forward at a time when the risk of oral sepsis, or infection, was extremely high.

Dental transplants (11th century)

The Muslim physician and surgeon Abulcasis of Cordoba (Abul Kasim or Abu al-Qasim Khalaf ibn Abbas as-Zahrawi) advised that teeth that were knocked out accidentally should be reimplanted. He had them held in place for several weeks with ligatures which enabled them to re-root.

Towards the end of the 13th century Nicolas Falucci developed the transplantation technique whereby teeth were taken from a donor, living or dead. During the 18th century, transplantations became very popular.

Dental equipment

In his treatise on medicine and surgery Abulcasis, the Muslim physician from Cordoba (see above), describes an impressive dental arsenal consisting of pliers, elevators and ligatures which has long been the prerogative of the dental surgeon.

The dentist's chair (16th century)

In the early **16th century Giovanni Plateario**, abandoned the standard operating position whereby the patient's head was held tightly between the surgeon's knees. He adopted a low chair with a shorter back rest which gave the dentist easier access to the patient's mouth.

The following dates represent the main stages in the development of the dentist's chair: *1810:* chair with a folding stool for the dentist; *1848:* the headrest; *1855:* the jack-operated chair invented by Ball; *1871:* the swivel chair invented by Harris; *1872:* the iron chair operated by a pneumatic jack and which could be tipped backwards, invented by Alexander Morrison; *1877:* the chair operated by hydraulic pump, invented by the Johnston brothers; *1950:* the electrically operated chair.

The drill

The first surgical drill dates back to the 1st century AD. The Roman surgeon Archigenes developed an instrument which was set in motion by a rope and which drove a drill by rotation.

The mechanical drill (1864)

In **1864 Harrington** of England had the idea of activating a drill using a clock mechanism in which the spring could be wound up and then released.

In 1868 the American, Green, introduced a drill driven by pedal-operated bellows.

On 7 February 1871 the American, J. B. Morrisson, introduced a model which could achieve a speed of between 600 and 800rpm. In 1874 Green constructed a system which was activated by an electric wheel and reached a rotational speed of between 1200 and 4000rpm.

Since 1958, air-driven turbine-operated drills can reach a rotational speed of between 300 000 and 400 000rpm.

Dental scanner (1987)

The dental scanner or T-Scan was invented by two Americans, an information technologist called Rob Golden and a dentist called William Maness, who consider their invention the most important since the discovery of X-rays. The scanner is a small device shaped like a jawbone. As the patient 'bites' the scanner, an image is simultaneously projected onto a screen which gives the dentist an immediate view of the teeth and enables him to see what treatment may be necessary. It provides a quick method of examination which means treatment can be carried out without delay.

The first machines

The abacus (3000 BC)

The abacus, ancestor of the calculator and of the computer, is of Babylonian origin and dates from around 3000 BC. The word abacus is derived from the Semitic term for dust. In its old form the abacus was in fact a slab of wood covered in fine sand on which figures were written with a stylus. The abacus later took the form of a bead frame. Hindus, Greeks, Romans and Europeans (in the Middle Ages) all used the abacus in one form or another. Nowadays, the bead frame is still used in India, China *(suan pan)*, Japan *(soroban)*, and Russia *tschoty)*. An abacus performs the four mathematical operations: addition, subtraction, multiplication and division.

The abacus, in the form of a bead frame, is still used in China, India, Japan and the USSR. This forerunner of the calculator and the computer is 5000 years old.

Pascaline (1642)

In **1642 Blaise Pascal** (1623–62) made the first calculating machine, which was the true ancestor of our modern pocket calculator. Pascal was a highly gifted young man, for he invented this machine at the age of 19, while working on conic sections, to help his father collect taxes in the central region of France. Counting whole numbers cogwheels in a mechanical gear system performed additions or subtractions which could involve up to eight columns of figures at a time. This machine, which Pascal named the Pascaline in 1645, worked in the same way as a car milometer.

Arithmetical machine (1624)

The German **Wilhelm Schickard**, a professor at the University of Heidelberg, built, in **1624**, the first arithmetical machine capable of performing the four basic operations. He called it a calculator clock.

Stepped reckoner (1671)

In **1671 Gottfried Wilhelm von Leibnitz** (1646–1716), the German philosopher and mathe-

Blaise Pascal was only 19 years old when he invented the mechanical adding machine to help his father.

Babbage's concept for a programmable digital computer was too ambitious for the time.

matician, invented a mechanical calculator which was similar to the Pascaline, but more refined. Pascal's machine could only count. Leibnitz's could multiply, divide and calculate square roots. However both these two calculators were based on the same mechanical technique, 'single step' calculation. They repeated the same operation, such as, for example, a series of additions. Many modern computer programs also work in this way.

Analytical engine (1835)

In **1835** the Englishman **Charles Babbage** (1792–1871), who was professor of mathematics at Cambridge University, presented the concept of an 'analytical engine'. This machine, which was completely new, was in fact the world's first digital computer.

The analytical engine combined arithmetical operations with decisions based on its own calculations. It used a system of 50 cogwheels, and data was entered by means of punched cards.

The first ever programs were written for this machine by Ada, Countess Lovelace. Unfortunately 19th century technology was not sufficiently advanced to put most of Lovelace and Babbage's brilliant discoveries into practice. Thus only a rudimentary version of the analytical engine was built.

Numerical control (1805)

Without realising the importance of his invention, the Frenchman **Joseph-Marie Jacquard** (1752–1834) used numerical control in the operation of a mechanical loom.

The Jacquard loom was, in **1805**, the first machine to use a punched hole in a card to represent a number and thus control the pattern of its weave.

Electric totaliser (1886)

Hermann Hollerith (1860–1929), an American statistician, found fame by combining Babbage and Jacquard's system of punched cards with electromagnetic inventions. He thus made a major contribution to the development of computer technology. In **1886**, when he was working on the American population census, he tried placing punched cards over little bowls of mercury. He then dropped metal pins through the holes into the mercury, to complete an electrical circuit. This system of electromechanical detection enabled Hollerith's totaliser to classify data and enter it into a ledger. His systems were put to good use during the 1890 census.

Hollerith went on to develop punching and sorting machines, which were precursors of today's computer peripherals. In 1911 he helped to set up the Computing Tabulating Recording Company, which became the International Business Machines Corporation (IBM) in 1957.

First attempt to build a computer: the Z1 (1931)

Early in the 1930s the German engineer **Konrad Zuse** made computers that operated in binary mode. The Z1 was followed by the mechanical Z2, then by the Z3, a relay computer which could perform a multiplication in three to four seconds.

Zuse was hindered by the slowness of his machines and in 1940 he suggested to the German government that electromechanical relays should be replaced by electronic tubes. But Hitler could not see the point of this and the project was shelved.

Binary computer (1939)

The first binary computer was made in **1939** by the American mathematician **George R. Stibitz** at the Bell Laboratories and was called the Model 1 Relay Computer or Complex Number Calculator.

It consisted of a logical mechanism in which the data output consisted of the sum of the data entered. Stibitz used telephone relays in his computer, which functioned in the binary 'all or nothing' mode (in other words they only used the digits 1 and 0), with the aim of developing a universal computer. He assembled the computer in one weekend, using a few discarded relays, two lightbulbs and fragments of a tobacco jar.

The first computers

Colossus (1943)

Right from the start of the Second World War, British numbers theorists tried to find a way of decoding German messages. They put **Doctor Alan Mathison Turing** at the head of a team charged with solving this problem.

Before the war Alan Turing had clarified the notion of calculability and adapted the notion of algorithms to calculate certain functions. He had thus postulated the Turing machine, which was theoretically capable of calculating any calculable function. In **1943** at Bletchley Park, the first electromechanical computer, Colossus, went into operation. It was formulated by **Professor Max H. A. Newman** and built by **T. H. Flowers**. This computer contained more than 2000 electronic tubes and could process 5000 characters a second.

It was a specialised machine, which did its job very well: right until the end of the war, the British government was kept informed about German plans.

Harvard Mark 1 (1944)

The first fully automatic calculator in the world was Harvard Mark 1, at that time called the IBM Automatic Sequence Controlled Calculator. The machine was presented by **Howard**

Aiken of Harvard University, and its development had been encouraged and financed by T. J. Watson, then president of IBM. It weighed five tonnes and contained 800km *500 miles* of tape.

This calculator improved on Babbage's dream and included two innovations: a clock intended to syncronise the diverse sequences of operations, and the use of registers – an idea that was picked up by every other manufacturer.

A register is a device used by the computer to store information for high-speed access. The bits of data stored in the register could represent a binary number, an alphabetic character, or a computer instruction.

Universal electronic computer (1946)

After signing a contract in 1943 with the Ballistic Research Laboratory, **John W. Mauchly** and **J. Presper Eckert**, two American scientists from the University of Pennsylvania, set to work. In **1946** they presented ENIAC (the acronym for Electronic Numerical Integrator and Calculator), the first universal electronic computer. It weighed 30 tonnes, occupied a surface area of $160m^2$ *191.36sq yd* and contained 18000 electronic tubes. By means of electronics, it brought speed to the world of computers. It was used to calculate ballistic trajectories.

Stored program computer (1948)

John von Neumann joined the team at the Institute of Advanced Study in Princeton, New Jersey, and it was here that the idea of a machine with a stored program was conceived. In 1946 Neumann, with Arthur W. Burks and H. H. Goldstine, published 'Preliminary Discussion of the Logical Design of an Electronic Computing Instrument'. This was a crucial document in the history of computer science: the program became a sequence of numbers stored in the computer's memory. EDVAC (Electronic Discrete Variable Automatic Computer) was capable of operating on and therefore changing the stored instructions and was thus able to alter its own program.

The first machine to incorporate von Neumann's principles was built at the University of Manchester, England, in **1948**. In 1950 the first computer intended for business use came onto the market: the Universal Automatic Computer (UNIVAC I).

Concepts

Mechanised calculation (1617)

In **1617** the Scotsman **John Napier** (1550–1617) found a way of expressing division by a series of subtractions and multiplication by a series of additions. He thus became the inventor of logarithms.

Napier's technique, which made it possible to perform any calculation simply by repeating the same operation several times, opened the way to calculation using mechanical means.

Algorithm (18th century BC)

In the 18th century BC, Babylonian mathematicians of the time of Hammurabi formulated algorithms in order to solve certain numerical problems. An algorithm is a series of elementary actions designed to solve a problem.

The idea of mechanising algorithms goes back to the year 1000, particularly to the work of the Frenchman Gerbert d'Aurilliac (c.938–1003), who became Pope Sylvester II.

A computer program is the translation of an algorithm into a well-defined language.

Computer program (1835)

While Charles Babbage was designing his analytic engine, **Ada, Countess Lovelace** (1815–52) was writing programs for it.

In fact it has recently been revealed that without the help of the Countess (who was the daughter of the famous poet Lord Byron), Babbage's machine would certainly not have been built.

The Countess of Lovelace was thus the first programmer. Her work foreshadowed such things as subroutines and automatic programming.

Coding (19th century)

The invention of coding can be attributed to the American **Hermann Hollerith**. The coding of punched cards is often called the Hollerith code, in memory of this scientific forerunner. Later notable contributions were made by the Frenchman Emile Baudot, who invented the telegraphic code which was patented in 1874. However, coding goes back to the first mechanisms to use punched cards: Jacquard's looms and Babbage's machine.

Binary logic (1859)

George Boole (1815–64) an English logician and mathematician, developed symbolic logic and, specifically, binary logic operators such as AND, OR etc. Boole's rigorous system makes it possible to mechanise logic, operating with 0 and 1 only: 0 meaning off and 1 meaning on. This is how electronic logic circuits work in computers today.

Ada, Countess Lovelace wrote the first computer programs for Babbage's machine.

Cynbernetics (1940)

Cybernetics, as a science, was invented by **Norbert Wiener** in **1940**, but the word was not coined until 1948 by Wiener and A. Rosenblueth. It comes from the Greek word *kybernētés* meaning a steersman or pilot.

Cybernetics is the study of automatic communication and control mechanisms in machines as well as in humans.

Bit (1946)

The term 'bit' was created by **John Tukey** in **1946**, while he was working on the ENIAC.

It is a contraction of the two words 'binary digit'. A bit is a binary unit of information. It refers to either (or 'one') of the two elements, 0 and 1, by means of which information is coded in a computer.

The computer scientist's bit is in fact the same as the thing that telegraphists had for years been calling a 'moment', using five-moment codes (Baudot code) and eight-moment codes (bytes).

Byte (1961)

A byte is a group of eight bits, or eight binary elements. The byte appeared as a basic unit of information on the Stretch, a high-powered, transistorised computer built by **IBM** in **1961**. Today the byte is universally used to represent a character (a letter or figure).

Computer capacity is usually given in kilobytes (one kb equals 1024 bytes), megabytes (one Mb equals 1024kb), gigabytes (one Gb equals 1024Mb) or terabytes (one Tb equals 1024Gb).

Real time working (1946)

A system is said to work in real time when it provides an immediate result. A highly accurate computer, the BINAC (Binary Automatic Computer), was developed in **1946** by the Americans **J. P. Eckert** and **J. W. Mauchly**. It consisted of two computers which simultaneously carried out the same calculations, whose results were then compared. The BINAC was the first computer to work in real time. Its accuracy was phenomenal for its day: in 1949 one of the two processors which made up BINAC worked faultlessly for 44 hours non-stop.

Time sharing system (1961)

A team headed by **F. Corbato** at Massachusetts Institute of Technology (MIT) designed, in **1961**, the Compatible Time Sharing System (CTSS) for the exploitation of IBMs 700 and 7090. The first time sharing system to be marketed was on the PDP1 in 1962.

Work stations (1980)

Work stations are in fact simply computers, misleadingly named by the American firm Apollo which invented them. Current models are very powerful, more so than minicomputers. A work station is a machine in the form of a microcomputer, designed to carry out a precise function – calculation or drawing (which requires a powerful processing capacity) – and linked to a central system and shared peripherals (file storage, flatbed plotters).

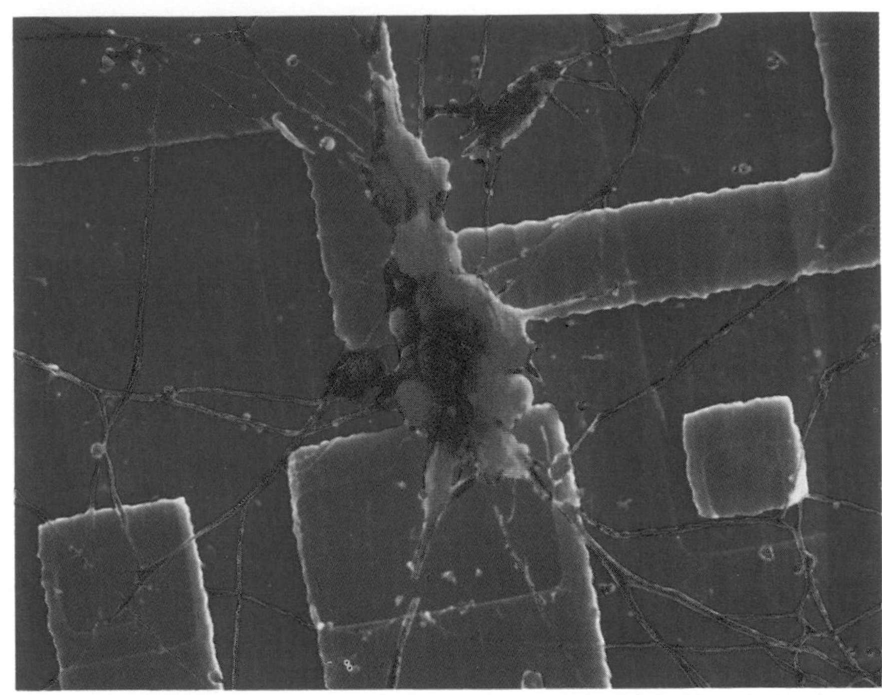

This is a microphotograph of a neurone developing on an integrated circuit. Scientists are looking for ways of modelling computer architecture on the human brain.

NEUROCOMPUTER SCIENCE ON THE AGENDA

Neurocomputer science, or the study of computers whose architecture is modelled on that of the grey matter of the brain, was abandoned after being very fashionable during the 1950s and 1960s. But the limited progress in 'classical' artificial intelligence and expert systems has led some scientists to take up this type of research again.

In 1987 neuronic software was developed in Baltimore (USA) and a neuron network in Finland. The importance of these neuronic machines is that they are capable of learning and of recognition.

In the face of these new perspectives, Europe has decided to launch an ambitious programme: Brain. This involves 28 European laboratories and about a hundred scientists working on six projects. An amusing detail: the scientists at the Bell Laboratories (USA) carry out their research on the brains of slugs. Slug brains are very simple and their nerve cells (or neurons) are the right size and easy to reach.

Software

Operating systems

Origins (1954)

An operating system is software (or a program) which is used not by the person using the computer, but by the computer itself. It comes between the machine's electronic circuits and the application software (word processing, accountancy) with which the user is in contact.

It acts like the conductor of an orchestra, co-ordinating the functioning of the computer's different elements. In microcomputing it is also a tool for standardisation, since it enables any software to function (with some modifications) on several different computers, provided they use the same operating system. Each one is designed for a particular type of microprocessor. The first operating system dates back to large computers: **Gene Amdahl** designed it in **1954** on an IBM 704.

CP/M (1976)

In **1976 Gary Kildall** invented the first operating system for microcomputers. He was working on assembling systems from components of different origins and found it a tedious chore having to rewrite all his programs every time he tried to read different floppy disks on the same central unit. The adaptor software that he wrote at the time was the core for the CP/M operating system. For some reason, Gary Kildall chose to give his company the grandiose name of Intergalactic Digital Research! The fact remains that CP/M rose to the rank of standard operating system for eight-bit professional microcomputers equipped with a Zilog Z 80 microprocessor. Today, apart from some home computers, CP/M is hardly used and its proprietor has adopted the more modest name of Digital Research.

MS-DOS (1981)

In **1981** IBM asked **Microsoft** (a small company which has since expanded greatly) to provide them with an operating system for their future microcomputer, the PC. Bill Gates, who was Microsoft's young owner at the time,

Buddhist monk Phra Kru Sripiriyatkal has created software based on ancient Chinese and Japanese buddhist texts.

then bought Seattle Computer Products' Tim Patterson's 16 bit operating system SCP-DOS. Having adapted it, he christened it MS-DOS and delivered it to IBM (who call it PC-DOS). MS-DOS has since been considerably improved and is today the most widely used for professional microcomputers, since it is used by all IBM compatibles.

The irony of the story is that Bill Gates first advised IBM to go to his competitor, Digital Research, the producers of CP/M. Digital Research refused to sign the promise of secrecy required by IBM, thus losing the market and at the same time taking the first step on their path to decline.

Unix (1970)

The Unix operating system grew out of research done at Bell Laboratories and at the University of California at Berkeley. It was originally designed for minicomputers by **Ken Thompson** and **Dennis Ritchie**. Today there are versions of it for almost every sort of machine, from portables to supercomputers. The name Unix itself dates from **1970**, and the first version was marketed in 1975.

The main advantage of this rather complex operating system on a microcomputer is that it is multi-task and multi-user, which explains its growing share of the market. The two main versions for micros are called Unix System V and Xenix. AT&T (to which Bell Laboratories belong) have invested considerable sums in the development of the Unix operating system. At the time when its Unix System V Version 3 seemed set to become the world standard, seven major builders representing 40 per cent of world computer science announced that they were setting up a foundation which would develop its own Unix standard. The Open Software Foundation was born in May 1988. Formed of IBM, HP, DEC, Bull, Siemens, Nixdorf and Apollo, it seems all set to fight the monopoly that AT&T was building up.

OS/2 (1987)

In 1987 Microsoft and IBM launched a new operating system which will make it possible for the first time to get the full use from microcomputers which use the Intel 80286 processor. The particular feature of OS/2, which was developed jointly by the two companies, is that it makes it possible to break through the 640k memory barrier imposed on IBM PCs and compatibles by the MS-DOS system. It also enables the writing of applications programs which can be run conjointly on the same computer. It has been available since the beginning of 1988 and will be used on IBM machines equipped with an 80286 or 80386 processor.

Although OS/2 was developed collectively – more than 120 programmers worked on it – it is another success for Microsoft's founder Bill Gates. His is the archetypal computer science success story. Co-author of the first Basic and supplier of MS-DOS and now of OS/2 to IBM, he combines a genius for programming with a talent for management. He is one of the rare men who has been able to influence the future of the micro all by himself. For a day in 1987 he was a dollar billionaire.

Interface software

Integrating (or interface) software consists of utility programs situated between operating systems and applications software. They make for easier movement from one application to the next, with each taking place in a different window displayed on the screen. They are also intended to facilitate the adaptation of applications software from one type of computer to another.

EVERYONE'S TALKING ABOUT: COMPUTER VIRUSES

A virus is a particular sort of program which is introduced into a system to perform various functions, such as espionage by recording information (in particular passwords), and the destruction of programs, data and backups.

Logical bombs, which act in the same way, are different from viruses in that they are of internal origin: they are directly incorporated into a program and can be activated in two ways: either at a predetermined time, or by means of a signal. Logical bombs are particularly used to conceal traces of fraudulent operations (for example, embezzlement).

The first virus to become widely known (for they have existed unmentioned for a long time) appeared in the United States, at Stanford University in California. The word 'cookie' was constantly written on campus screens.

Visi-On (1982)

The American company **Visicorp** was the first to have the idea of integrated software. Visi-on, which was first launched in November **1982**, was made up of an integrating module on to which the required applications modules, such as word processing, spreadsheet, graphics, or file management applications, could be grafted. Visi-On could only use applications specifically designed for it and ultimately failed.

Windows (1985)

This was launched for the first time in November **1985** by **Microsoft**. It required more than 50 man-years of work. In April 1987 it was adopted as the standard integrator by IBM. Windows is multi-task (in some conditions) and will to a certain extent accept applications modules not specifically designed for it. The Windows interface may make it possible for IBM compatibles to become more user-friendly like the Macintosh.

Postscript (1985)

It was the American company **Adobe Systems** that developed the page description language Postcript, which was first used in microcomputing in Apple's Laserwrite printer, launched in January **1985**.

The function of this kind of language is to program the typographic and graphic styles used in the composition and layout of a document. It is used by programmers writing software used in these fields and functions as an interface between the software and the laser printer, greatly improving printing quality. Its adoption by Apple, and then in March 1987 by IBM, granted it the status of standard.

Languages

All software of whatever type is written using a programming language, which has its own vocabulary (its instructions list) and syntax. The most basic language (from which all others are ultimately derived) is machine language, a series of binary or hexadecimal numbers which are directly comprehensible to the computer but hard for the programmer to manipulate.

Assembler (1950)

Assemblers are types of languages close to machine language, but at a higher level, therefore easier to use. They are specific to a particular computer or microprocessor. Signs of the first assembler can be found in **1950**, in the EDSAC, one of the ancestors of the large computers of today, which was developed in Cambridge (UK) by **M. V. Wilkes'** team. The first commercial assembler was the SAP (Symbolic Assembly Program), which was developed by the United Aircraft Corporation and installed on an IBM 704.

Basic (1965)

Invented in **1965** at Dartmouth College (USA) by **Thomas E. Kurtz** and **John G. Kemeny**, Basic was originally designed to help students to learn programming. Basic stands for '*B*eginners *A*ll purpose *S*ymbolic *I*nstruction *C*ode'. It was developed at a time when microcomputers did not exist and was adapted for use on them in 1974 by Bill Gates and Paul Allen, the founders of Microsoft. Today Basic is the standard language of all home and office microcomputing and its reputation as a beginner's language with little power is becoming increasingly unjustified.

Pascal (1969)

Named after the French mathematician, Pascal was developed in **1969** in the United States by **Niklaus Wirth**. The idea governing its design was to give programming students their own language, which was well structured and would instil good writing habits. It is more powerful than Basic and is often used in microcomputing, in universities and to write wide-circulation professional software.

Cobol (1959)

Cobol (*Co*mmon *B*usiness *O*rientated *L*anguage), the prime language for business management applications, was launched in **1959** under the aegis of the Conference on Data Systems Languages in the United States. It belongs to the world of the large computers, but although it now seems very dated it is still widely used.

Artificial intelligence programming languages (1956)

Logical reasoning and formal calculus have led to entirely new programming methods. Today more than a hundred languages are used in the field of artificial intelligence (AI).

The first AI programming language IPL (Information Processing Language) was invented in **1956** by American scientists **A. Newell**, **D. Shaw** and **F. Simon**. It was developed specifically to write the LT (Logic Theorist) program, capable of resolving mathematical logic problems.

Prolog (1973)

One of the main languages adapted to artificial intelligence and chosen by the Japanese for their project to build a fifth generation computer is Prolog. This was designed in **1973** by **Alain Colmerauer** and his team at the University of Luminy-Marseille, France. Its basic principle, which draws on mathematical logic, was revolutionary: instead of telling the computer how to solve a given problem – as happens with traditional languages, which are called procedural – the programmer simply sets out the data for the problem. Prolog then gets on with finding the solution. Although Prolog can function on supercomputers (as is the case in the fifth generation project), it has recently come into use on microcomputers of the IBM PC type and even on some home computers!

LISP (1958)

Although old by computer standards, Lisp is the most commonly used language in artificial intelligence. As early as **1958 J. McCarthy** had developed the particular concept of list processing that forms the basis of this language whose name is a contraction of 'list' and 'processing'. Since Lisp needs a lot of memory it generally requires fairly large computers to run properly, but microcomputer based versions have been written.

C (1972)

This language is rapidly gaining popularity in the development of wide distribution software applications. C was created in **1972** at the Bell Laboratories in the USA by **Dennis Ritchie** – one of the inventors of the Unix operating systems – specifically to help the development of Unix. C has the innovative feature of mixing high level code (which is easy to manipulate) with low level code (which is efficient). It is based on BCPL, a language developed at Massachusetts Institute of Technology and at Cambridge University (UK).

Ada (1979)

Ada was developed in **1979** after five years of effort by a team from CIL-Honeywell-Bull, headed by the Frenchman **Jean Ichbiach**, in response to a invitation to tender by the American Defence Department in 1974.

This language takes its name from Ada, Countess Lovelace, who may be regarded as the earliest programmer.

Ada has become a world standard. It is in general use for all American military applications and by NATO.

Alsys, the company set up by Jean Ichbiach in 1980, is involved in microcomputers too, and in 1987 produced Ada compilers for them. In 1988 the global market for products and services related to Ada use was assessed at $2.74 billion.

Applications software

The term applications software refers to programs which deliver a specific service directly to the user, such as word processing, printing payslips, or playing games.

Visicalc (1979)

Visicalc was invented by two Americans, **Dan Bricklin** and **Bob Frankston**, and launched in **1979** on Apple computers. It is the world's first spreadsheet and an excellent example of how lively the minds of microcomputing inventors are. Certainly before Visicalc no programmer working on large computers had thought of such a program, which combines a table of figures with a table of formulae which determine the relations between the figures. As a result, it is possible to construct very complex models (for example the provisional budget of a company), and to see in an instant what would happen if one of the hypotheses were changed: the spreadsheet automatically recalculates all the figures. Today the spreadsheet programs are among the most widely used software on microcomputers, and there are hundreds of them. Similar tools have been developed for large computers.

Wordstar (1979)

The first major word processing software for microcomputers was launched in **1979** by the American company **Micropro**. It is still on the market today and has given rise to a whole host of programs of the same type.

1-2-3 (1982)

This software with the strange name has given rise to a whole new generation of integrated software. It was designed by one man, **Jonathan Sachs**, and launched in October **1982** by the American company Lotus. Its particular feature is that it combines three functions in one microcomputer program. It is above all a very high power spreadsheet, but is also combined with a small file management facility and most importantly with a graphics module which makes it possible to visualise in curve form any group of figures presented in the spreadsheet. 1-2-3 is one of the most widely used software products on IBM microcomputers and has made its producer, Lotus, a world leader.

dBase II (1980)

Launched in **1980** by the American firm **Ashton-Tate**, dBase II (pronounced dee-base two) is the archtypal database management software for microcomputers. It makes it possible to classify, sort and select information according to numerous criteria. It was developed in October 1979 by **C. Wayne Ratcliffe**, who had marketed it under the name Vulcan before selling it to Ashton-Tate.

Its successor, the higher performance dBase III Plus, was the world's bestselling software in 1986. dBase IV is now also available.

Mac Paint (1984)

Written by **Bill Atkinson** in **1984**, Mac Paint led to thousands of beginners becoming interested in computers. This graphics software specially designed for the Apple Macintosh works on an instinctive basis: the user draws by moving the mouse on the desk after selecting one of the symbols 'pencil', 'paintbush', 'spray' and even 'rubber' which are displayed on the screen and give lines of different appearance and thickness. The user simply selects a paint pot to colour in a shape, the lasso to grasp part of the drawing and move it elsewhere, and so on. Since its creation, Mac Paint has had countless imitators.

Mac Write (1984)

Mac Write was presented in **1984** at the same time as the Mac Paint software, and was designed for use on the Macintosh. It was the first word processing software intended to be used without training. **Randy Wigginton**, **Ed Ruder** and **Don Breuner** from Encore Systems designed it for Apple.

Page Maker (1985)

On **15 July 1985** the American company Aldus perfected revolutionary software designed for the Apple Macintosh and called Page Maker. It was the first software to enable a single individual to write, lay out, paginate and print a newspaper or book, including illustrations, using only a microcomputer and a laser printer.

Agenda (1988)

A simple but brilliant idea for everyday software. Invented by the American company **Lotus** in **1988**, Agenda classifies a jumble of information that it is given. So 'call Mr Smith', 'plan the Book of Inventions and Discoveries', 'find such and such an inventor's name' and other such scattered and dissimilar notes which the user has not classified are regrouped by the Agenda software to make for better organisation.

Artificial intelligence

AI is a scientific discipline consisting of writing computer programs that attempt to model human intelligence. AI formalises human knowledge and reasoning, whereas data processing merely manipulates information.

According to this definition, AI could date back to 450 BC when the Greek philsopher Socrates envisaged reducing all reasoning to a simple calculation in the form of something like geometry.

Fundamental papers relating to AI were published as early as 1953 in America. However, the term was used officially for the first time at the International Joint Conference on Artificial Intelligence held at Washington (USA) in 1969.

Today, the most popular application of AI is in building expert systems.

Expert systems

An expert system (ES) is a software program characterised by its ability to reason by logical inference starting from a problem set by the user. The system uses a base of knowledge and a set of rules called production rules, drawn up by a human expert.

In the early 1960s, some researchers were already putting forward the idea that the laws of reason, combined with the power of a computer, could produce systems that go beyond the capacity of human experts. However, there are severe theoretical reservations.

The first operational ES appeared in the early 1970s. There are very many of them today, in different domains.

Oldest expert system (1961)

In **1961 J. R. Slagle** produced a thesis at the Massachusetts Institute of Technology (MIT) on a heuristic program for solving problems related to symbolic mathematics. This was the beginning of SAINT (*S*ymbolic *A*utomatic *In*tegrator), which culminated in MACSYMA, presented in 1971 by two MIT researchers, W. A. Martin and R. J. Fateman. Today an improved version of MACSYMA surpasses most human experts in performing symbolic differential and integral calculus.

The DENDRAL project (1967)

The DENDRAL project grew out of research by scientists at Stanford University in California and has produced two expert systems, DENDRAL and METADENDRAL.

DENDRAL's origins, attributed to the Americans **B. G. Buchanan**, **E. A. Feigenbaum** and **J. Lederberg**, the 1958 Nobel Prizewinner for Medicine or Physiology, go back to **1967**. DENDRAL analyses the results of chemical experiments and deduces from them the structures of unknown compounds. METADENDRAL is an extension of DENDRAL. It sets out and selects the fragmentation rules of organic structures. It was developed by M. Mitchell, B. G. Buchanan and E. A. Feigenbaum.

Components

Transistor (1947)

The transistor was invented at the end of **1947** by three physicists at the Bell Laboratories, USA: **William B. Shockley**, **John Bardeen** and **Walter H. Brattain**. The transistor is a true semiconductor triode. It is the electronic component which characterised the second generation of computers. It is a solid-state component, and rapidly became widespread. It was originally made with germanium but after 1960 transistors used silicon, which is more stable.

Transistors can detect, amplify and correct currents; they can also break them. They can produce very high frequency electromagnetic waves and open or close circuits in the space of a millionth of a second. They have allowed advances to be made in information technology, communications, aeronautics and have also made electronic watches and pacemakers possible.

Quantum effect transistor (1989)

Presented as the transistor of the year 2000, a quantum effect transistor has been developed by the research centre of Texas Instruments in Dallas (USA). It works on the basis of the wave behaviour of electrons and on the tunnel effect, which in particular requires a control of nano-electronics (electronics on the scale of a thousand millionth of a metre).

A hundred times smaller than today's transistors, it reaches execution speeds a thousand times faster, which means many changes can be envisaged in the electronic devices and computers whose construction it will make possible.

This new transistor, which has been studied since 1982, was developed for the US Army.

Microprocessor (1971)

In **1971 Marcian E. Hoff**, then working for Intel, developed the first microprocessor which he baptised the 4004. Hoff brought together the elementary functions of a computer on a single electronic component (an integrated circuit). It contained the equivalent of 230 transistors and was a four-bit processor.

Z 80 (1976)

In **1976** Zilog launched one of the micro-

The first integrated circuit (originally known as a single-crystal circuit) was invented by Jack Kilby of Texas Instruments in 1958.

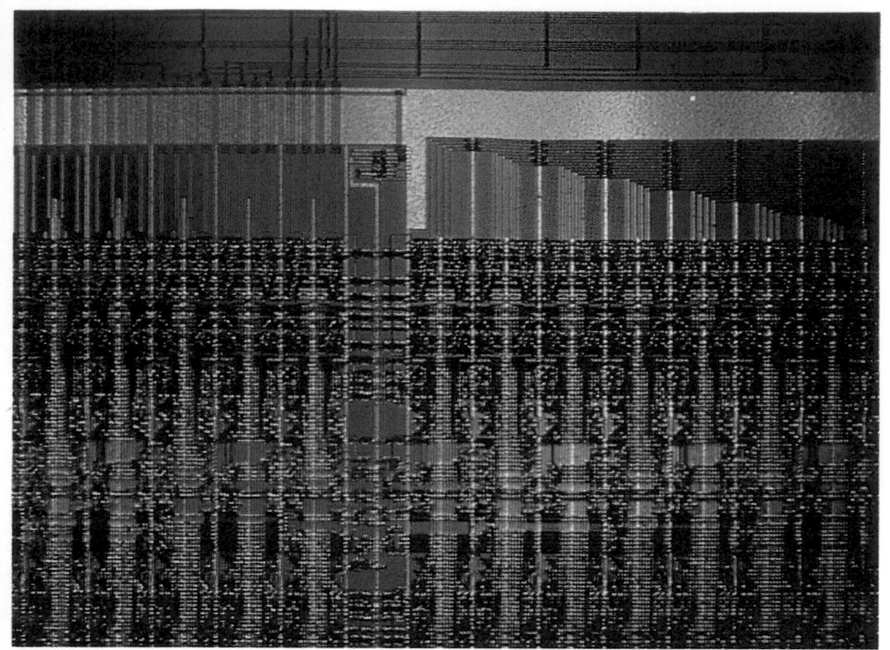

The first microprocessor was created in 1971. This tiny wafer of silicon had a tremendous impact on industrial production processes.

processors that did most for the microcomputing boom, the Z 80. The operating system CP/M and leading software (Wordstar, dBase II) were written for this 8 bit processor.

6502 (1976)

In **1976 Chuck Peddle**, one of microcomputing's great pioneers, developed the 6502, an 8 bit microprocessor marketed by the American company MOS Technologies. It was chosen by Stephen Wozniak and Steve Jobs to equip their first Apple II, and has had a fairytale career: it is still a core element in the latest versions of the Apple II, and an extraordinary software library has been designed around it.

8086 (1978)

In June **1978 Intel** launched the first 16 bit microprocessor to have commercial success, the 8086. It was developed by a team headed by Bill Pohlman. It has considerable strategic importance, since one of its versions, the 8088, an 8–16 bit processor, was chosen by IBM to equip their first microcomputer, the PC. Since then, IBM has bought part of Intel's capital, and all their microcomputers use Intel microprocessors.

80386 (1985)

In **1985 Intel** launched their 32 bit microprocessor, the 80386, which offers microcomputers performance at the level of yesterday's minicomputers (it is capable of carrying out three to four million instructions a second), while being compatible with the software designed for its predecessors (8086 and 80286). In particular it can run several programs designed for an 8086 simultaneously each working as though it had a real 8086 entirely at its

Jack Kilby surrounded by some of the devices that his invention made possible.

THE CHIP IS 32 YEARS OLD!

Thirty-two years ago, on 12 September 1958, Jack St Clair Kilby, a young American engineer from the University of Illinois who had recently been taken on at Texas Instruments, showed the results of his work to some of his co-workers. He had assembled a few transistors and capacitors on a single support. This discovery was to revolutionise the electronics world. A patent for this first integrated circuit was applied for in 1959 and granted in 1964.

An integrated circuit is an electronic mechanism in which different components (transistors, resistances, capacitors, etc.) are diffused or implanted, then connected up within a thin layer of semiconductor, such as silicon, enabling the formation of complex electronic circuits which carry out complete functions.

The first commercial application of integrated circuits was in 1964, in a hearing aid. We should not forget that the term 'chip' applies to the active element in an integrated circuit.

Today the integrated circuit industry represents a market of some $500 million and forecasts are that it will be the world's number one industry at the beginning of the next century.

disposal. The first microcomputers using the 80386 appeared at the end of 1986.

68000 (1979)

In the United States in **1979**, **Motorola** launched the 68000, first of a series of microprocessors competing with those of Intel. This 16–32 bit processor is known above all for having been chosen by Apple for their Macintosh, and, secondarily by Atari for their ST series of personal microcomputers. It gave rise to several other versions, notably the 68020 used by the Macintosh II.

Graphics processors (1986)

At first the main function of microprocessors was calculation. With the growing stress on user-friendliness, the manipulation of graphics has come to take up an increasing proportion of microcomputer power. This is why in **1986** **Texas Instruments** and **Intel** each separately launched graphics processors designed to ease the burden of the main processor. Texas' 34010 is particularly used in extension cards for IBM PC compatibles and Intel's 82786 is its direct competitor.

Integration

Because the integration of electronic components grew out of many different technologies, its beginnings are hard to trace. The problem was to produce increasingly complex and small integrated circuits. In the case of a computer, the time taken by a signal to go from one circuit to the next limits its performance. Increasing the micro-electronic density inside a package is then of prime importance.

In the early 1970s, electronic engineers began talking about large scale integration (LSI). This meant a single 'chip' could hold around 500 components. Some five years later very large scale integration (VLSI) appeared, with up to 10 000 active elements.

Today super-large scale integration (SLSI) makes it possible to build supercomputers such as the Cray. Its chips contain more than 50 000 transistors in a space the size of a pinhead.

Transputer (1983)

According to some specialists, this is the most important invention in computer science since the microprocessor. It was designed by **Ian Barron** for the **Inmos** company (UK) around a very high integration microprocessor featuring a new microscopic architecture which enabled it to carry out parallel processing of information. Classical computers, by contrast, work 'serially', one bit at a time.

The first transputer, Inmos' T 414, was highly regarded but did not sell well. But the performance of the T 800, which went into mass production in November 1987, is impressive. Its architecture gives it 12 times more power than Intel's 80386 microprocessor. Its originality lies in the way that it enables design of multiprocessor systems whose architecture can be modified by software, depending on the problems to be solved.

The European program to build a supercomputer is based on the transputer.

Silicon chips are at the base of integrated circuits and microprocessors. Silicon is an insulator at low temperatures and becomes a conductor under the effects of light and heat.

Internal memory (1947)

The internal memory of a computer is the mechanism which enables it provisionally to store information just before, during and after it is processed. In **1947** at Manchester University, the Englishman **F. C. Williams** experimented with electrostatic tubes used as a memory. In 1949 at Cambridge University, UK, EDSAC, one of the computer's first ancestors, used delay lines. These were tubes filled with water or mercury, blocked off at both ends with crystals which transformed electronic signals into sonic vibrations. Because sound travels more slowly than electricity, it was possible to store some information in them. Legend has it that Alan Turing advised putting gin in the delay lines. Then, in 1949, Jay Forrester used ferrite cores for the memory of the Whirlwind. Cores continued to be used until 1964, when they gave way to semiconductors.

Random access memory (1970)

The first component of dynamic *R*andom *A*ccess *M*emory, or RAM, was produced in **1970** by **Inter**. This circuit, developed by **Bob Abbott** under the direction of **Les Vadasz**, was called 1103 by the American company and had only a 1 kbit capacity (1024 bits or units of information). Random access memory, which can be both written to and read, forms the usable working memory of computers.

Semi-conductor memories (1987)

The capacity of random access memory, or RAM circuits, is continually increasing. Although in 1986 chips of 256 kbits were still regarded as pioneers, April of that year saw IBM incorporate one million bit (1 Mbit) chips into its new large computers. And in February 1987, when IBM were announcing that they had succeeded in industrially manufacturing a component of 4 Mbits, NTT (the Japanese telecommunications company) was already presenting a 16 Mbit prototype.

Static RAM (1988)

The first static memory capable of storing 4 Mbits (or 512 000 bytes or characters) was constructed by **Sony**, using a submicronic

ECOLOGY AND COMPUTER SCIENCE

A new product to clean semi-conductors (used in transistors) was developed in 1989 by the Japanese company Daikin, who specialise in the manufacture of 'microchip cleaner' fluid. This is a fluorised alcohol, 5-PP, which is not dangerous to the ozone layer.

STRONGER AND STRONGER, SMALLER AND SMALLER

Ten million transistors in 25cm^2 *3.875sq in*: the SCAMP (*S*ingle *C*hip *A* *M*ainframe *P*rocessor) from Unisys contains as many circuits as a minicomputer. The machine built with this processor, the Microframe, measures 17 × 58 × 43cm *6.7 × 22.8 × 16.9in* and weighs 23kg *50lb 11.5oz*. It was designed as a microcomputer and in fact is very like one.

In addition, there are several advantages to having all a computer's central circuits on the same chip. The first is that the elementary electric signals exchanged between the circuit's different modules have less distance to travel and thus move more quickly. The result is less production of heat, which makes cooling the system easier and leads to an equal reduction in consumption of computer power.

Furthermore, the reduction in the number of components means that construction tends to be more accurate, thus leading to a reduction in maintenance costs. This also facilitates assembly of the system, leading to reduced manufacturing costs. The Microframe costs half the price of an equivalent minicomputer.

technique (the individual elements of circuits are only 0.5 micrometres *.0000195in* wide). SRAM (*S*tatic *R*andom *A*ccess *M*emory) works more quickly than dynamic RAM and enables information to be preserved when there is no electric current passing through it.

Microcomputers

Electronic pocket calculator (1972)

The first electronic pocket calculator was developed by the Americans **J. S. Kilby**, **J. D. Merryman** and **J. H. van Tassel** of **Texas Instruments**. The patent was applied for in **1972** and granted in 1978. The calculator is preserved at the Smithsonian Institute in Washington.

In 1973 Hewlett-Packard brought out pocket calculators that were programmed to suit the needs of a particular domain (for example finance, or economics). In 1976 the same American producer marketed the first programmable calculators, which were true pocket computers.

Micral (1971)

The first microcomputer in the world was French. At the end of **1971**, **François Gernelle**, an engineer with R2E, designed it to respond to the French agricultural research institute's need for automatic regulation. R2E's boss, André Truong Trong Thi, a Frenchman of Vietnamese origins, was won over by the invention and decided to manufacture computers built around a single microprocessor. This very first microcomputer was called Micral. When R2E was bought by Bull, all that remained of Micral was the brand name.

Altair 8800 (1974)

Many producers who have now disappeared argued over who launched the first American microcomputer. But the one which really set things in motion was the Altair, produced by the company MITS which was set up by **H. Edward Roberts**. In December **1974** the magazine *Popular Electronics* published a bombshell of an article: readers were invited to buy through the post a kit with which they could build themselves a real computer, based on Intel's 8080 microprocessor. It cost $397 and was extremely basic compared to today's microcomputers. In place of a keyboard it had 25 switches which had to be moved in a cumbersome sequence in order to start the machine, and it had a 256 byte random access memory (2000 times less than many of today's PCs). Nonetheless, 200 orders were received on the first day.

IBM 5100 (1975)

Six years before their famous PC, IBM had already built a microcomputer, the 5100. However, its distribution remained confidential.

Apple II (1977)

The Apple II, brought out for the first time in the United States in May **1977**, was the first commercial product of the legendary company founded by Steve Jobs and Stephen Wozniak. It was an improvement on the first Apple of which a hundred or so had been manufactured the previous year and which was simply a kit sold directly to members of the Homebrew Club, the first computing club in the United States. The Apple offered what seemed revolutionary features at the time: a keyboard, a Basic language provided and ready to use, and a simple connection for a screen or disk drive.

IBM PC (1981)

It was on **12 August 1981** that the number one company in world computer science announced its entry into the microcomputing market with the IBM PC. The arrival of this microcomputer, designed by a team of young computer scientists headed by **Philip Estridge**, created a real standard throughout the world. Not only did it reassure company heads, for whom microcomputing had up till then been synonymous with do-it-yourself constructions by amateurs, but it was copied, with improvements, by all the major producers. IBM let this happen. Thus was born the IBM compatibles industry.

4000 PAGES ON ONE CHIP

A chip that can store eight million characters, or 4000 typed pages: that is the capacity of the D-RAM (*D*ynamic *R*andom *A*ccess *M*emory), a semi-conductor memory produced by Fujitsu. This memory is contained on a chip with an area of 100mm^2 *.155sq in*. Dynamic memories keep information as long as they have an electric current passing through them.

TRANSPUTERS AND MICROCOMPUTERS

The transputer is an ultra-fast microprocessor. It is connected to a microcomputer using an extension card and enables the computer to work ten or 20 times faster. These cards are already available: Atari has presented a prototype which connects to the Atari ST. The American company Levco has announced extension cards for the Macintosh. The German company Parsytec is bringing out the Megaframe/IBM card for PCs and compatibles.

Apple Macintosh (1984)

At the beginning of **1984** Apple presented the Macintosh, the most original microcomputer and the one most likely to revolutionise the way in which non-specialist users interact with their machines. The Apple team which developed the Macintosh, personally led by Steve Jobs, took up ideas that had been used in the same company's Lisa in 1983; a mouse, a high definition screen and graphic symbols representing programs and data.

These techniques had been outlined several years earlier at Rank Xerox's PARC Laboratory. They saved the user from having to be concerned with the internal workings of the computer.

Macintosh II and Macintosh SE (1987)

When on **2 March 1987** Apple launched two microcomputers which could accept software designed for Macintosh and also that written for the IBM PC, they brought together two worlds which had hitherto been thought incompatible.

Amstrad PC 1512 (1986)

Amstrad created a real stir in September **1986** when they announced their PC 1512: this microcomputer, which was compatible with the IBM PC but twice as fast and easier to use, cost three times less. In Europe this machine opened up IBM standard to home users for the first time.

Alan Sugar, Amstrad's Managing Director, is happy to admit that he understands nothing of computer science. All the same he is a formidable negotiator. He begins by setting the price for his future products – always much lower than those of his competitors – then he argues every step of the way with his suppliers in order to hold to those prices.

In this way in 1984 Amstrad, a British producer of hi-fi systems, launched a cheap high performance home computer, the Amstrad 464, and the company joined the club of microcomputer producers.

A POCKET PC

It weighs 450g *1lb*, measures 20cm × 11cm × 3cm *7.8 × 4.3 × 1.1in* and is battery operated: nevertheless Atari's pocket PC is a true computer, and it is compatible. It differs from other models, not only in its size and weight, but also in its mass memory, which uses not discs or diskettes but cards similar to credit cards.

PC–Macintosh interface (1987)

It was originally impossible to copy information from a Macintosh disc on to a PC. With the MatchMaker card developed by **Micro Solutions** (Illinois, USA) in April **1987**, this became simple. The card is inserted into the PC and enables it directly to access data on Macintosh discs, and vice versa. On the other hand Macintosh programs do not work on the PC.

The IBM Personal System/2 (1987)

On **2 April 1987**, IBM changed the rules of the microcomputing game by launching a new range of microcomputers called Personal System/2, or PS/2, with a different design from that of the previous PC and PC-AT.

The four models in the PS/2 series are faster, more compact and easier to use, their screens are easier to read, they have greater RAM and mass memory and are designed to run several programs at once. They seem almost impossible to copy.

The marketing of this new range unleashed war once again in the microcomputing world. Will IBM succeed in setting a new standard and will they make all the competition look outdated? Or will an alliance be formed between customers tired of constantly having to renew their hardware and software and the IBM PC compatible producers that will stop the world's number one producer?

University microcomputer (1989)

Steve Jobs' (Apple's supplanted founder) lastborn is a computer aimed at academics and researchers. This machine looks like a black cube and is produced by his new company, **Next**. It uses a Motorola 68030 microprocessor and four coprocessors and has the first erasable optical disk as its mass memory, with a capacity of 256MB (millions of bytes or characters).

Supercomputers

Cray 1 (1976)

The first commercial supercomputer (or supercalculator), the Cray 1, was created under the supervision of **Seymour Cray** in **1976**. It contained 200000 integrated circuits, was freon-cooled and could perform 150 million operations per second.

NASA's Cray 2 seen behind the reservoirs of fluorocarbons needed to keep the computer cool.

Bestseller: the Cray X-MP (1982)

In **1982 Steve Chen**, a Taiwanese immigrant to the United States where he did his engineering studies, designed the world's first computer featuring parallel architecture, the Cray X-MP.

The Crays were overtaken by Control Data's Cyber 205, which performs 700 million operations a second.

Seymour Cray regained the advantage with the Cray 2, presented in 1986. It has a two gigabyte central memory and performs 1.6 thousand million operations a second. Its architecture is very compact to facilitate the movement of information, and to stop it overheating it is submerged in a fluorocarbon fluid.

Cray Y-MP (1988)

At the beginning of **1988** Cray, who still dominate the market with 60 per cent of supercomputers in service (since Control Data left the field open to Cray Research by abandoning the ETA 10 and the supercomputer market), presented the Cray Y-MP. It is capable of performing two thousand million operations per second and costs $20 million. It is used by NASA, for instance, but should also be of interest to the biotechnological, aerodynamic and chemical industries, since it enables the generation of complex three-dimensional simulations.

The SSI-BM project (1995)

In December **1987 Steve Chen**, the brilliant inventor of the Cray X-MP, decided to leave Cray and to set up his own company, Supercomputer Systems Inc. (SSI), which got the backing of IBM at the beginning of 1988. The two firms have joined forces to develop a supercomputer with parallel architecture that is a hundred times more powerful than today's machines. Based on techniques which have not yet been fully measured, this project is as ambitious as it is risky. We shall know the answer around **1995**.

The European project Supernode (1995)

At the end of 1987 the British and French presented the prototype of a supercomputer built as part of the European ESPRIT programme, called Supernode. It consists of an assembly of transputers. It is therefore an MIMD (Multiple Instructions Multiple Data) machine, with highly parallel architecture.

The announced advantage of Supernode is its cost, which is considerably lower than that of supercalculators of comparable power. Its problem is that the design of a highly parallel operating system and applications software is proving very difficult.

Fifth generation computers (1982)

Researchers worldwide are working on computer architecture with a view to developing it for artificial intelligence. The Japanese have launched an ambitious programme with the help of the Ministry of International Trade and Industry. This led to the creation of a special institute in **1982**, the **Institute for New Generation Computer Technology** (ICOT), which is to be given $450 million over ten years. The work is directed by the highly respected expert Kazohiro Fuchi.

Computer peripherals

The terminal (1940)

The first experiment that involved a terminal connected to a remote computer was conducted in **1940** by Bell Laboratories. The computer was in New York and the terminal at Dartmouth College, New Hampshire.

Light pen (1963)

Light pens are accessories with which users can draw on their screens, as they would do with a real pen. A light pen enables them to move part of a drawing, to 'take' a colour from a 'palette' and use it to 'paint' the surface they touch with the tip, and to command different functions. The first light pen was presented in **1963** at the Massachusetts Institute of Technology (MIT) by **I. E. Sutherland**, who was associated with the conversational graphics system Sketchapad. Today this accessory is chiefly used on computer-aided design (CAD) consoles.

The mouse (1965)

The mouse is a small device that slides in all directions on a desk and which makes it possible to interact naturally with the computer. Its used was popularised by Apple with the Lisa and the Macintosh models in 1983. However, it was the little-known American inventor **Douglas Engelbart** who conceived and designed the mouse at the Stanford Research Institute in the mid-1960s. His brilliant idea was to have the computer operator place his or her hand on a small box or mouse. A sphere on the underside of the mouse is used to measure movements which are then transmitted to the computer via a lead – the tail of the mouse. These movements are translated to the cursor on the screen: if the mouse is pushed to the right the cursor goes to the right; if the mouse is pushed away from the user the cursor moves up, and so on. This revolutionary input device, originally found only on Apple computers, was adopted by IBM in 1987.

AFTER THE MOUSE, THE GLOVE

After three years of research, two engineers from a small Californian company, VPL Research, Thomas Zimmerman and L. Harvill have developed the Dataglove. The glove consists of fibre optic strands sandwiched between the layers of fabric along each finger. The glove allows the user to 'dive' into a display unit and enables him or her to manipulate in three dimensions the objects represented on the screen. According to VPL Research, the glove can be used in robotics, nuclear physics, space travel, and even cartoon animation. In the future this device could well change human–computer interaction.

Tactile screen (1985)

In **1985** Zenith (USA) presented the first tactile screen system, based on surface acoustic wave technology, which is even easier to use than the mouse or light pen: all the user has to do to give a command is touch a section of the screen.

Speech recognition (1950)

The first machine to recognise ten numbers pronounced by a human voice as a series of sonic signals was built at the Bell Laboratories in the United States by **K. H. Davies**. That was in **1950**. Since then, speech recognition has given rise to promises rather than to spectacular demonstrations. The keyboardless typewriter or computer which responds to the sound of the human voice, is the dream of many scientists.

TODAY'S CD-ROM

The first CD-ROM, called Perinorm, which stores 80 per cent of all European and international technical standards and regulations, was presented in March 1989. The documents archived in this way represent more than 100000 standards. This CD-ROM is sold by annual subscription (with a monthly update) and has been produced as part of the future single European market, due to come into being in 1992, to help industrialists cross the barrier of international regulations. The project's designers are DIN (West Germany), AFNOR (France) and BSI (UK).

Speech synthesis (1978)

The first electric speech machine, the Voder (Voice Demonstrator), was built in 1933 by the American **H. W. Dudley**. It was followed in 1939 by the Vocoder (Voice Coder). Speech synthesis is based on the theory of visible speech, formulated in 1948 by the Americans **R. K. Potter**, **G. A. Kopp** and **H. C. Green**, who showed how phonemes (vocal sounds) correspond to graphic traces.

Speech synthesis, which poses far fewer theoretical problems than speech recognition, is used in many domains, such as industry, cars and games.

Mass memory (1805)

The first mass memory, in other words a medium enabling the permanent storage of data, was invented long before computer science by the Frenchman **Joseph-Marie Jacquard in 1805**. This was the punched card, and it was designed for his loom. Hermann Hollerith used it again in 1890 on his machine designed for the American census. Then came magnetic tape, tested for the first time on an EDVAC in 1949, and removable discs, which were first marketed by IBM in 1962.

Floppy discs (1950)

Floppy discs, universally used on microcomputers, were invented in **1950** at the Imperial University in Tokyo by **Doctor Yoshiro Nakamats**, an inventor who boasts of having 2360 patents which include golf clubs and loudspeakers. He granted the sales licence for the disc to IBM.

Nondegradable diskettes (1987)

The 5¼in discs which make up a large proportion of diskettes on the market are fairly vulnerable to damage by scratching, heat, humidity, ash, etc. The Verbatim disc developed by **Kodak** in **1987** is covered with a thin film of Teflon, which protects the user against nasty surprises. Two types of this disc were originally developed, then at the end of 1989 Kodak brought out another two.

Hard disc card (1985)

In 1985 the American company **Plus Development Corporation** revolutionised mass memory technology by launching an extension card designed for IBM PCs equipped with an extra-flat hard disk of 10MB. Before this, hard discs (large capacity fixed magnetic media) were awkward and fragile. The Hardcard, as it is called, is only 2.5cm *.985in* thick. Many other producers have followed the lead.

CD-ROM (1985)

The CD-ROM, invented by **Philips** and promoted throughout the world in collaboration with **Sony**, is simply a laser-read compact disc, similar to those used in hi-fi systems, but adapted to computing uses. It has the advantage of containing a thousand times more data than a diskette. Its disadvantage is that the data on it can be read, but new data cannot be written on to it.

The CD-ROM began to take off in 1988. It

The Dataglove permits the operator to manipulate an image in three dimensions on the screen.

The inventor of the floppy disc, Yoshiro Nakamats, shows off one of his creations.

found professional outlets: Renault decided to put all the technical documentation for its network (12 000 graphics plates and 80 000 pages of text) on to CD-ROM; training courses for Airbus pilots are now on CD-ROM, etc. It should soon reach a broader public, since Atari and Apple have announced CD-ROM readers compatible with their microcomputers.

Interactive compact disc or CD-I (1989)

Put forward by **Philips** and **Sony** at the first CD-ROM conference in Seattle organised by Microsoft in March 1986, the CD-I arrived on the professional market in 1989 and should reach a wider public in 1990.

Derived from CD-ROM technology, it stores interactive programes, combining high quality sound, animated images and texts. The disc is read by a reader resembling a hi-fi laser deck connected to a television screen and stero system, and which is in reality a powerful computer.

Write once optical disc (1987)

Write once optical discs, called WORM (*W*rite *O*nce *R*ead *M*any times) have been the object of much research but are not yet on the market. IBM's announcement in April **1987** of a disc reader of this type, with a 200MB capacity, that can be connected to all its microcomputers has not yet been followed by results.

Digital optical disc (1984)

In **1984** the French firm **ATG** marketed its numerical optical disc system, called Gigadisc. It is designed for large scale storage of all sorts of documents: texts, photos, sound, computer data. The information is read using a laser beam. The Gigadisc can store up to two thousand million bytes, which is the equivalent of 600 000 pages of typed text.

Printer (1953)

The printer enables data provided by the computer to be printed on paper. The first fast printer worthy of the name was developed in **1953** by **Remington Rand** (USA). It printed 600 lines of 120 characters a minute.

The most widely used technologies fall into two types: impact printers (using dot matrix or daisy wheels) and non-impact printers (thermal or laser transfer).

Dot matrix printer (1957)

Dot matrix printing is an invention which has had the greatest of success: the majority of printers use this principle. The print head has a vertical row of needles which are propelled forward electromagnetically as the head runs over each line. The first dot matrix printers were marketed by **IBM** in **1957**.

Thermal printer (1966)

Invented in **1966** by **Texas Instruments**, thermal printing was first used for microcomputers. The print head is made of resistant needles which are heated when an electric current is passed through them. This technique requires special paper and gave way to 'thermal transfer', where the needle does not directly heat the paper but an inked ribbon. IBM retained this technology on its recent typewriter-printers.

Daisy wheel printer (1978)

The American company **Diablo**, since bought by Rank Xerox, invented this procedure which was inspired by techniques used on typewriters.

Laser printer (1985)

We owe the development of laser printing to **Canon**: the Japanese producer sold the machine to their competitors, which enabled Canon (and the others) to keep their prices very low. The market opened up and other manufacturers developed similar technology such as diode systems. For large machines laser printing's chief competitor is ionography (invented in 1983 by the Delphax company) and magnetography (exclusively from Bull). All these techniques are based on the use of a drum which draws in powdered ink, using either an electrostatic or a magnetic effect, and deposits it on the paper to be printed, like a photocopier.

Renov-Ink (1987)

This procedure was invented by the Frenchman **Krikor Hovelian** and it was patented in **1987**. It makes it possible to re-ink the printer ribbon in just a few minutes and without manipulation. It is particularly useful in the case of printers whose ribbon cassettes are no longer manufactured. In 1989 the Herbin-Sueur company, ink specialists for 300 years, began marketing and manufacturing Renov-Ink.

Applications

Computer graphics (1950)

The art of computer graphics can be traced back to the graphics made for wallpaper by Burnett in California from 1937 onwards. These graphics were based on Lissajous figures. But it was **Ben F. Laposky** who, in **1950**, really founded the art of computer graphics.

These pure products of computer technology are beginning to invade our universe. Some are made up from elements that already exist in the computer's memory, but others are derived using mathematical formulae or random procedures.

The work of the companies Robert Abel Associates, Digital Equipment Corporation (USA) and Sogitec (France) has now become famous in this field.

Automatic translation (1950)

In 1946, W. Weaver and A. D. Booth thought of using a computer to help with translation. But the techniques had not been perfected, and it was not until **1950** that **Weaver** and **Booth** could try out their idea.

In 1970 Doctor P. Thomas developed a universal translation system, SYSTRAN. It was first put to spectacular use during the meeting of Apollo and Soyuz in 1975. It was adopted by the European Community in 1981.

There are some systems that can translate in particular fields. One example is the Canadian TAUM METEO, which was presented in 1976 by a working group from Montreal University. But none of these systems can work without human intervention and experts do not expect that there will be any decisive progress in this field for some time.

Music (1956)

Composers were the first to use computers for artistic creation. The first of these were the Americans **M. Hiller** and **M. Isaacson** in their work on the 'Illiac Suite' in **1956**. Also important are the works of M. Phillipot and I. Xenakis, C. Risset and M. Matthews from the Bell Laboratories (USA), and lastly those of the Vincennes Group.

At the end of the 1970s, computers became widely used in all fields of music (from pop music shows to teaching).

Cinema (1964)

The film industry was quick to grasp the possibilities of computer graphics. One interesting example is the pioneering work of **Peter Foldes** in his film *The Hunger*, but Steven Lisberger was the first to use all the computer's possibilities in the shooting of *Tron* in 1982.

Image animation (1951)

Computerised image animation was first experimented on at the **Massachusetts Institute of Technology** (MIT) in **1951**. But it was not until the early 1960s that the potential of the technique was fully understood. Today it is used in the fields of medicine, architecture (with models in three dimensions), space exploration, and chemistry.

Flight simulators (1970)

Around **1970 General Electric** supplied NASA with the first flight simulation programs. The power of today's machines provides amazing possibilities and all pilots are now trained on simulators. The advantages of this are obvious: pilots can try out difficult manoeuvres without risk of losing lives or destroying a plane.

The world of simulation (1988)

Simulation is now present in almost every domain: such phenomena as nuclear power plant accidents, a pollution cloud over the Alps, molecules that do not exist, Formula 1 cars that are faster than real ones and chariot races can all be simulated. From high-tech industry to electronic games, computer graphics recreate reality. Things have reached the point where Californian scientists have made an astounding wager: the aim of the Biosphere 2 program is to simulate all the ecosystems of the planet.

Digital control (1956)

The digital control of machine-tools first appeared in **1956**, the year that the *A*utomatically *P*rogrammed *T*ools (APT) language was created for the US Air Force.

Eye control (1986)

In **1986** the American company **Analytics Inc.** developed a prototype computer which obeys sounds and the eye. It uses an infra-red beam to record eye movements. The user stares at a point on the screen, then gives a command to a micro and the machine carries it out immediately. This system could facilitate the control of robots or the selection of components on an assembly line.

Computer-aided design (CAD) (1960)

Computer-aided design was born in the **1960s** in the context of the major American military aeronautics programmes.

The term refers to a set of techniques which can be used to create data that describe an object to be designed, to manipulate that data in a conversational mode and to arrive at a finished form of the design.

After its adoption by the military, CAD penetrated civil aeronautics and the motor and computing industries. It enables an object (for example a car) to be drawn in three dimensions and to be examined in a great number of theoretical circumstances, even before the building has begun. Today CAD plays an essential role in almost all fields of industry.

Computer-aided drawing

Computer-aided design is sometimes simply computer-aided drawing: the image is then seen in two dimensions, which is sufficient for the needs of architects or electronic circuit designers.

Robots

Automata

The first automata are to be found in antiquity at the time of of Hero of Alexandria (1st century AD). The Arabs kept the tradition going. In 809 Sultan Harun ar-Rashid gave an animated clock to Charlemagne.

But it was not until the work of Jacques de Vaucanson (1709–82) that a machine was built that perfectly imitated natural animation. In Paris in 1738 Vaucanson exhibited a duck which astounded everyone: it flapped its wings, swam, smoothed its feathers with its beak, drank and pecked; furthermore, after a certain time it evacuated the food it had taken in, in the form of a soft substance.

The torch was then taken up by two Swiss: Pierre and Henry-Louis Droz. In 1733 they constructed a drawing machine. Their creation was so perfect that they were put on trial for witchcraft. Fortunately for them, there was growing public interest in scientific methods and the verdict was in their favour.

Androids

In our imagination, robots are destined to replace people in certain tasks and should therefore look like them. But industrial robots, whose numbers are ever-increasing, look just like machines, whereas androids look like us. The most famous is certainly Disneyland's Lincoln, although it is a little out of date technically speaking.

In Mirapolis in 1988, a young Frenchman, Pascal Pinteau, presented Leonardo da Vinci, the most complete show android in the world. It weighs 100kg *220lb 8oz* and real hair has been implanted into its supple, translucent skin.

SUPER-COCKPITS AND SUPER-COPTERS

It was in the laboratory of the Wright-Patterson airbase (Ohio, USA) that Dr Tom Furness and his team conceived the super-cockpit. It is a helmet with three-dimensional vision to be used not only by American army pilots but also by surgeons. The helmet transmits all types of information to the wearer – information from the real world and information calculated by computers.

By wearing a super-cockpit that has been linked to a probe inserted into a patient's veins, the surgeon would be transported, through vision and sound, into the body of the patient.

A pilot would see on the screen in his helmet the terrain necessary for his mission – not the real world.

In France the engineers at Thomson-CSF have worked on their super-cockpit for four years. It is being developed for the pilot of the future 'super-copter', a combat helicopter scheduled to appear at the end of this century. The helmet will have a visor upon which all the information will appear. The pilot will talk to the computer which will answer him, outlining the manoeuvres necessary to escape from or destroy an enemy.

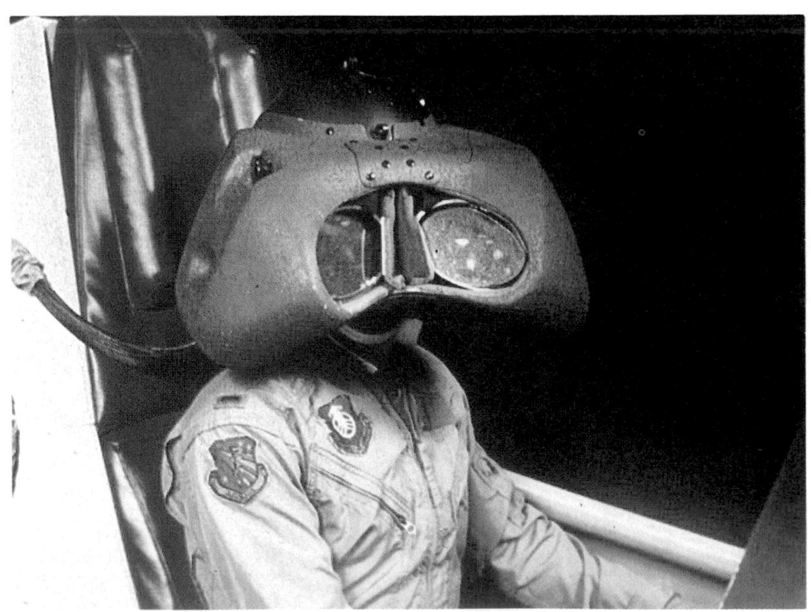

All the pieces of information a pilot needs are projected onto the interior of his helmet: the kind of terrain, the position of the enemy, their weapons, etc.

The word 'robot' (1920)

The word 'robot' was coined in **1920** by Czech dramatist **Karel Capek** for his play *R.U.R. (Rossum's Universal Robots)*. The term comes from the Slavonic root *rab*, meaning 'slave', which became the Czech word *robota* or forced labourer.

Police robots (1983)

Denning Mobile Robotics (Massachusetts, USA) has been working on security robots since **1983**. In 1987 it signed a contract with one of America's largest prisons, to supply about 600 robot police wardens. At first they will work as part of a team with humans but it is envisaged that they will be able to replace people in the execution of certain duties.

The Denning robots are also capable of guarding factories and banks from robbers.

Perspiring robot (1988)

This robot, called Manny, was invented by American engineers from the **Battelle Institute** in Columbus (Ohio). It cost the tidy sum of $2 million and is designed to test military clothing.

Besides the perspiration which is 'secretes' by means of a series of minuscule tubes which carry the water to the surface of its 'skin', Manny can also walk, squat, get down on all fours and crawl using a set of 40 joints. It differs from perspiring people in one fundamental way: it gives off no smell.

Robot shearer (1991)

In Australia there are 165 million sheep to shear. Hence the need to invent a mechanised shearing procedure. After ten years of research the Australian company **Merino Wool Harvesting** has developed prototype robot shearers, capable of shearing a sheep in 100 seconds instead of the three minutes taken by a human. These robots are due to come on to the market in **1991**.

Robots in medicine

Robot patient (1980)

The first of these was the Japanese company **Koken**'s robot patient in **1980**. The following year in California Professor W. P. Harvey created a robotic heart.

Robot nurse (1983)

Melkong (Medical Electric King Kong) was created by **Professor Hiroyasu Funakubo** of Japan. It can hold a patient in its arms, wash him, put him to bed, and tuck in the sheets!

Diagnostic robots (1984)

Thanks to the funds from the Japanese Cancer Institute, a palpating robot is being studied at Waseda University, Tokyo. Equipped with fingers covered with artificial skin, and connected to a computer, this robot palpates breasts to detect tumours.

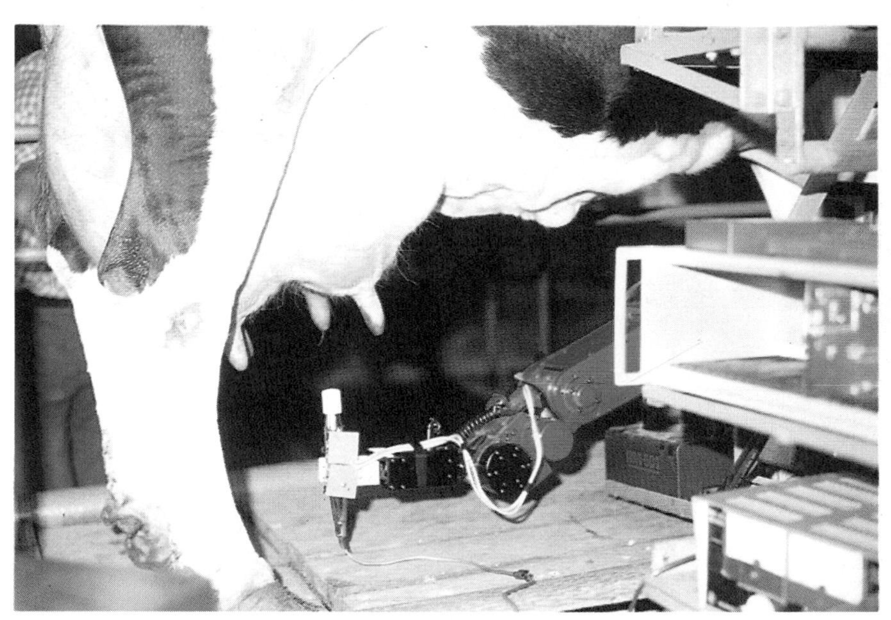

This robot milks cows and records the thermal profile of the udders thanks to an infra-red sensor.

Walking robots (1983)

One of the major areas of research in robotics is that of giving robots the means of moving, in particular so that they can work in hostile environments where people cannot go.

Bipeds

The aim of the inventors of the Chiba Walker 1, Tsutomu Mita, Toru Yamaguchi, Toshio Kashiwase and Taro Kawase from Chiba University in Japan, is to produce an exact simulation of human leg movements. The Chiba Walker has two legs equipped with several joints which give it a very human gait. All it needs now is a body, arms and a head!

WHL-II also imitates human walking and slowly waddles on flat ground – at no more than 150m *492ft 3.6in* per hour. Its feet are equipped with sensors which transmit information to the 'brain' (the computer), which controls movement. The main problem for WHL-II is keeping its balance. This robot was designed by Professor Ichiro Kato and built by the company Hitachi.

Quadrupeds

The Japanese Titan III can move over uneven ground and also climb stairs. Each of its jointed limbs can move in three directions and has a sensor at the end.

Jean-Pascal Herbin from Valenciennes University is the father of the EMA 4, which can reach a speed of 120m *393ft 10in* an hour.

Multi-legged robots

The most developed of these is undoubtedly Odex 1, from the American company Odetics, which can overcome obstacles whilst carrying a load of 500kg *1102lb 8oz*.

Hexapode 1, the brainchild of Jean-Jacques Kessis (University of Paris VII), imitates the walk of insects. It has many variations of this at its disposal, but cannot choose the best one itself.

Robert MacGhee at the University of Ohio has created the ASV, which is 5m *16ft 5in* long and can cross ditches 3m *9ft 10in* wide; it was designed for the US Army.

Domestic robots (1986)

Hero 2000 was born in **1986**. Its big brother, Hero 1, was designed in 1982 by the American company **Heathkit**. The aim of both was educational: they were designed to initiate young people into robotics, but they can also teach languages. They can also carry out a number of useful tasks, such as carrying packages. Many, less ambitious personal robots can be programmed to perform a certain number of tasks, but fundamentally these are just modern automata.

A robot for nuclear power plants (1988)

Using its two jointed caterpillar tracks, Centaur II, developed by the **French Commission for Atomic Energy** jointly with the **Cyberg** company, can move around in all sectors of nuclear installations in the case of incidents giving rise to dangerous radiation levels. It is autonomous and waterproof and can perform a certain number of simple operations. It should be on the market soon.

Agriculture

Agricultural machines

Ploughing and sowing

Origins (4000–3000 BC)

The history of the invention of agricultural machines truly begins with the industrial age. However, the 18th and 19th century inventors were heirs to a tradition that was many thousands of years old.

The earliest evidence we have of agricultural activity – which doubtless began long before – appears in pictograms representing the swing plough, an implement with a symmetrical ploughshare that ploughed a shallow furrow (4000 BC in Mesopotamia, 3000 BC in Egypt and, in Europe, the rock engravings of the Valley of Wonders, 3500 BC). The plough proper, a tool with an asymmetrical ploughshare that makes deep ploughing possible, is more recent (1st century AD).

Industrial plough (1730)

In **1730** the Dutchman **Joseph Foljambe** developed the Rotherham plough (so-called after an important manufacturer in South Yorkshire), which marked the beginnings of industrial manufacture.

The first plough made entirely of cast iron was made by the Englishman Robert Ransome in 1785. In France in 1825, Fondeur, a blacksmith from the department of Aisne, invented the first metal swivel plough with a joined front axle.

In 1837 the American John Deere patented the first steel plough, with an all-in-one ploughshare and mouldboard, and F.S. Davenport patented the two-wheeled Sulky in 1864, which was supplanted by the three-wheeled Sulky, known as the Flying Dutchman, in 1884. The disc plough first appears in the United States in 1847 and the version built by John Shearer and Sons in 1877 was particularly successful in Australia.

Tractor-drawn plough (19th century)

In the second half of the 19th century steam engines sometimes replaced animals to draw the plough, machines such as those built by the Englishmen John Heathcoat in 1832 and John Fowler in 1852; the steam cultivator by Halkett in 1860; steam tractors from the American firm Case, from 1829, and the steam locotractor built by the Frenchman Albaret in 1856.

It was not until the 20th century that the tractor-mounted plough first appeared.

Seeders and dibbles (prehistory)

If we regard the digging stick as the ancestor of this family of tools, then it is fair to say that seeders and dibbles go back into prehistory.

In 1660 the Italian Taddeo Calvani invented a seeder consisting of a box which was mounted on two wheels.

In the industrial age there were a number of inventions in the field of seeders: the Englishman Jethro Tull's seed drill in 1700; the ploughshare seeder of another Englishman, James Smyth in 1800, the Frenchman François Dombasle's cylinder seeder in 1805, etc.

Harrow (AD c.800)

The harrow in the form of thorny branches or bushes, was already known to the Egyptians, and the Romans used logs stuck with wooden spikes.

However the harrow that we know today – a rectangular or triangular implement which is pulled along by animal or machine – dates from around **800 AD**.

In the 19th century the harrow entered the industrial age. In England in 1840 James Smith invented the flexible harrow with metal chains; in 1854 the American H. Johnson patented the disc harrow; in 1869 another American, Davis L. Garver, developed the vibrating harrow and in 1872 the Frenchman Emile Puzenat invented the zigzag harrow.

Of all agricultural machines the harrow is perhaps the one whose form has changed the least, since today's tractor-drawn harrows are triangular or rectangular frames, like those of the Middle Ages.

Clod-crusher (1841)

The use of the roller to break up clods of earth dates back to the end of the 16th century, but the type used today is the crosskil, named after its inventor, the Englishman **William Crosskil**, who patented it in **1841**.

Haymaking, reaping and harvesting

Reaper (1822)

The American **Jeremiah Bailey** was the true inventor of the mechanical reaper. Patented in **1822**, his rotary reaper used a cutting disc which turned horizontally a few centimetres from the ground.

The cutting arm, which is specifically for haymaking, was invented by the American William Manning in 1831, but it was not manufactured industrially until 1850.

Tedder (1820)

In **1820** Englishman **Robert Salmon** designed a machine which lifted cut grass and turned it over so that it could be completely dried by the action of the sun and wind. In one day the machine could do the work of 15 women, and even more if the horse kept up a good pace.

However this implement was too brutal and spread the hay around, and these disadvantages prevented its development. Twenty years later, fork tedders were invented. These were used up until the 1930s, when there were replaced by machines that combine raking, tedding and draining.

This Egyptian painting shows the ploughing, sowing and harvesting processes in beautiful detail.

Jethro Tull pioneered a drill that prevented waste and allowed farmers to sow seeds in straight lines, making it easier to weed between them.

Harvester (1st century AD)

The harvester already existed at the beginning of the Christian era. It was widely used in Gaul and throughout the Roman world, and Pliny the Elder (AD 23–79) gives a very precise description of it. Strangely, the harvester disappeared into oblivion after this and only reappeared in mechanised form in the industrial age.

After 1800 many different types of harvester were produced, among them notably the combine harvester, invented by the American Lane in 1818 and put into operation in 1838 as a result of improvements by Moore and Hascall; Cyrus McCormick's tractor-drawn reaper in 1831; the self-binders from the Americans John F. Appleby (1858) and Walter Wood (1871), and the self-propelling combine harvester, whose prototype was put forward in 1888 by the American Best and which did not come into full operation until 1944, when it was launched by the American firm Massey-Harris.

The world leader in the field is now the American company New Holland, with 27.8 per cent of the market.

Axial threshing (1975)

The American companies **International Harvester** and **New Holland** began to research into an axial threshing combine harvester in 1962, but the first prototype was not built until **1975**. With its central rotor and system of separation, this machine represents the greatest change since motorisation.

Pollen aspirator (1988)

Among other things, pollen is used in medical research on allergies, the development of hybrid seeds for reafforestation and the genetic improvement of some plants, such as larch. In order to increase and speed up pollen collection, **Patrick Baldet** and the Seeds and Forest Plants division of CEMAGREF in France developed a self-propelling machine which encloses the tree in a sort of 'cage'. Collection is carried out by vacuum suction within this cage, when injections of compressed air drive the ripe pollen out of the male flowers. The pollen in suspension inside the cage is then filtered. It takes 1 minute and 30 seconds a time to harvest a tree, in other words 12 grammes of pollen can be collected in three operations. It took three years to develop this pollen harvest machine, which was first used in **1988**.

According to their publicity, Massey-Harris' harvester was the best in the world.

Radar detection of underground water (1988)

Radar capable of detecting water in the substrata was developed in **1988** by scientists at ***Agriculture Canada***. This modern version of the famous diviner's rod emits signals into the ground by means of two metal antennae, and picks up the echo. According to its inventors, it will enable farmers who use irrigation systems to rationalise their use of water.

Agronomy

Hydroponics (1860)

The growing of plants without soil, or hydroponics, has become a part of daily life. Tomatoes, cucumbers, lettuces, lemons, oranges and avocados, coming from Holland or Israel, are all now grown without the medium of soil, which has been replaced by renewable nutrient solutions.

This is up-to-the-minute technology, but the invention is an old one. Indeed it was in **1860** that two German scientists, **Ferdinand Gustav Sachs** and **Knopp**, succeeded in growing plants on a simple mineral solution.

In 1914, a Frenchman called Mazé, drew attention to the role that rare elements, which are present as impurities in ordinary water, could have in hydroponics. However it was not until 1940 that molybdenum, a fertilising element, completed the list of elements thought to be indispensable to the constitution of nutrient solutions.

From then on, the use of hydroponic cultures in the field of agriculture could seriously be envisaged. Relatively large-scale production made it possible to feed American troops

This machine gathers 12g 0.4oz of pollen from a tree in a minute and a half. The pollen is used for research into allergies.

in the Pacific islands during the Second World War.

The growing of plants without soil, using both hydroponics and aeroponics (when the water and mineral salts are sprayed directly on to the roots), has been extensively developed over the last ten years and will make it possible in the near future to plant real gardens, orchards or cottage gardens to feed the space towns of the future.

The first tomatoes officially labelled 'hydroponic' were marketed in Canada in 1988.

Insecticides (1st century AD)

As far back as 2000 years ago, the Chinese were using powdered Oriental Chrysanthemum (Pyrethrum) to kill fleas. In the Middle Ages and during the Renaissance, entomologists prescribed arsenic against insect infestations, following the example of the Borgias, who used it to rid themselves of their enemies. In 1681 the first arsenic compounds for treating plants were recorded.

Many tropical plants were used against insects, such as Tumbo from Brazil and Ecuador, Nikoe from Surinam, etc. Their roots contain a powder called rotenone, which affects the nervous system of insects. Its properties were discovered in 1920, but DDT, which was first synthesised in 1874 and was used as an insecticide from 1939, supplanted rotenone. DDT is a highly polluting product and is now partially or entirely banned in many countries.

Fastac (1979)

In **1979**, taking their inspiration from chemical structures similar to those of the pyrethrum, the American company **Shell** developed a new synthesised insecticide: Fastac. Fastac is a high performance insecticide, harmless to some insects, notably bees, which gather nectar from treated plants.

THE WONDERFUL TOMATO TREE

The tomato tree is a spectacular example of hydroponic culture. It was presented in 1986 at the Tsukuba exhibition in Japan and can produce 10 000 tomatoes in six months, without soil or sun.

THE FLY THAT ATTACKS VIPER'S BUGLOSS

Every year in Australia viper's bugloss, a plant that originated in the Mediterranean, causes $31 thousand million worth of damage by poisoning sheep and cattle. Until now there has been no way of stopping it. But in 1988 Len Foster, a specialist in biological pest control, found the answer in France with the *Dialectica scalariella*, a minute fly whose larvae devour viper's bugloss leaves and kill the plant. Tests are now being carried out in two Australian states but it will be five years before the first effects are felt.

THE PLANTS OF THE FUTURE WILL BE TRANSGENIC

In 1987 the Belgian company Plant Genetic System took out a major option on the future in the field of genetic engineering by developing a number of so-called transgenic plants.

As a result of the manipulation of their genes, these new plants have proved resistant to attack from some parasites (in the case of tobacco, for example), or can themselves select the action of predetermined herbicides (in the case of tomatoes, potatoes and sugar beet).

When the genes of a bacterium, *Bacillus thuringiensis*, long known for its effectiveness in controlling various caterpillars, are grafted on to tobacco plants, the plants become resistant to attack from *Manduca sexta*, a parasite which was previously very damaging to them. *Bacillus thuringiensis* acts by producing little protein crystals which dissolve in the stomach of the insect and poison it. Scientists at Plant Genetic System have managed to introduce the gene which governs the synthesis of this toxic protein into the cell of a tobacco plant.

Using the technique of multiplication in vitro of this non-differentiated cell, it has been possible to produce an entire plant bearing the genetic insecticide. If early results prove promising, this little revolution in the plant world is destined to develop rapidly, enabling many transgenic plants to come on to the market in the future, with considerable economic effects.

Fungicides (c.1850)

About **1850** farmers began to use sulphur, first as a powder and then in the form of a paste, because its fungicidal properties had been recognised.

Fungicides are used to fight illnesses caused by microscopic fungi. However some fungi can have beneficial effects. In 1985 scientists from INRA, the French national centre for agronomic research, isolated a fungus microbe which makes it possible to accelerate the process of degradation of lignite (wood): an invention which will be of great service to the economic development of the biomass.

Parasite farming (1984)

Since **1984** the first industrial farming of a useful insect (*Encarsia formosa*) has been in

RAPESEED AND PHARMACOLOGY

The University of Ghent and the Belgian company Plant Genetic System have developed a strain of rape which has been specially programmed so that its seed produces therapeutic molecules. Until now rapeseed was used entirely for oil; now a particular type of protein (neuropeptides) can be extracted from this rapeseed and soon, perhaps, blood and growth factors. This is a world first in the field of an entirely new biological concept: molecular farming.

The tobacco leaf on the left is affected by tobacco mosaic virus (TMV); the one on the right has been inoculated. The same leaves are shown in the box under ultraviolet radiation.

progress. This insect is intended to control the populations of another insect which attacks greenhouse tomatoes and cucumbers. The project took four years of research and development by INRA scientists.

Self-destructing virus (1987)

Scientists from the Institute of Virology in Oxford have developed a new weapon against insects: a self-destructing killer virus. Through genetic engineering, this virus has been stripped of its natural defences. Once it is swallowed by the insect, the micro-organism dies all by itself. The first experiments took place in Scotland on a caterpillar that attacks pine trees. This procedure would be less dangerous than the usual chemical pesticides which kill indiscriminately. Another method of fighting viruses is to 'vaccinate' plants. The first 'vaccinated' plant is a form of tobacco, which came into being as a result of the work of the CNRS and the IBMP in Strasbourg.

Ray Ely pictured with one of his eight miniature horses, created by selection.

Animal husbandry

New races

Over recent years research on animal biology has made it possible to create new races, such as the Renitelo, a new race of very robust bovines which was created in 1962 by the Madagascans from three strains of zebu. The smallest horse in the world (40 to 60cm *15.76 to 23.64in* at the withers), the Falabella, was the successful outcome of work by a vet of Irish origin, Patrick Falabella. Falabella moved to Argentina in the last century and began his research in 1868. It was continued by his descendants, who also experimented unsuccessfully on increasing the size of horses.

In 1984 an Israeli scientist, Dan Rattner, invented the gobex, a cross between goat and ibex.

In 1986 there was a zoological event of double importance at the Nantes Veterinary School: the natural birth of the first female chabin. It is quite exceptional for a male chabin, a cross between a sheep and a goat, not to be sterile. This baby chabin is the offspring of a female chabin and a ram.

At the end of 1987 a lamb was born to a mother who was entirely 'manufactured' (a chimera, half goat, half sheep) from a few cells from a goat embryo which were inserted into a sheep embryo. This genetic engineering was carried out by a team of reproduction physiologists at the University of California, headed by Gary Anderson.

Dainty was born in South Africa and is the smallest Yorkshire terrier in the world. At birth she weighed 4g 0.1oz *and now hits the scales at 500g* 1lb 1½oz.

Barbed wire (1874)

In 1874 **Joseph Gliden** from Illinois (USA) built the first machine capable of producing barbed wire in large quantities. This cheap mode of fencing quickly spread throughout vast regions of western America. However, disputes soon broke out between farmers who were installing fences and big stock breeders who were accustomed to herding their cattle cross country as they pleased.

Artificial insemination (1780)

While the first and perhaps mythical instance of artificial insemination applied to horses in 1322 is attributed to the Arabs, the true forerunner was the Italian **Lazzaro Spallanzani** (1729–99), who carried out the first recognised experiment of insemination on a bitch in 1780.

Despite the success of this experiment, it was not until the 20th century that new research was carried out, notably in Russia with the work of Ivanov, in 1907, and Milanov in 1933.

Artificial insemination was mainly developed after the Second World War and became a common practice, particularly for cattle.

British Telecom's Prestel Farmlink service makes it easier for farmers to find bulls for their cows. Here the search is on to find suitable matches for Victoria and Maydream.

To fight against so-called killer bees, engineers in Tennessee (USA) have developed a transmitter that allows them to follow the queen bees' movements.

Embryo transplantation (1975)

The first genetic experiments on embryo transplantation in cattle were tried out in **1975**, in the laboratories of British **doctors Rawson** and **Polch**.

Distance alarm (1987)

Denis Carrier, a Canadian inventor from Notre-Dame-du-Nord, Quebec, developed this distance alarm (patented in the United States in **1987**), which is intended to be used by farmers who have an animal that is about to give birth.

Milking machines (19th century)

The invention of the milking machine dates from the 19th century. The first model was made by the American **L.O. Colvin** in 1862. After this, in 1889, the Scotsman William Murchland developed a constant suction machine.

In 1895 another Scotsman, Doctor Alexander Shields, had the idea of using the principle of the pulser, which gave rise to intermittent suction. This was less painful for the cow than Murchland's machine, which caused inflammation. The process was perfected by Hulbert and Park in 1902 and by Gillies in 1903.

THE CHICKEN IS 8000 YEARS OLD

Or, more precisely, the domestication of the chicken. According to recent discoveries by Doctor Barbara West of the British Natural History Museum and Professor Ben Xiong Zhou of the Archaeological Institute in Beijing, chickens were domesticated in Indochina as early as 6000 BC and had reached the Yellow River basin in North China by 5900 to 5400 BC. They are thought to have reached Europe between 800 and 300 BC.

Automatic milking system (1977)

On **23 March 1977** the Swedish company **Alfa-Laval** patented an entirely automatic milking system. The only manual operation remaining was to attach the cups to the udders. A single farmer can now milk large numbers of cows. The Atlas 400 is an important invention as milking represents more than 50 per cent of a dairy farmer's work.

Milk refrigeration (1850)

The first system for refrigerating milk was invented in **1850** by the American Lawrence. His invention represented a great step forward. Until this point, the temperature of milk as it came from the udder (38.5°C *101.3°F*, the cow's body temperature) was extremely favourable to the proliferation of bacteria. Fast refrigeration was thus one of the fundamental stages in the preservation of milk for distribution, and permitted the development of the dairy industry.

Cream separator (1878)

The separator-creamer, based on a system using centrifugal force, was patented in **1878** by the Swede **Carl Gustaf de Laval**. Before Laval's separator, milk was left to stand and separation occurred as a result of the different weights of buttermilk and cream.

The centrifuge principle was afterwards put to use in many industrial fields and in biological laboratories.

Food bank for dairy cows (1972)

In **1972** the Swedish firm **Alfa-Laval** presented farmers with an automatic system for individual feeding, intended for dairy cattle: the Alfa Feed. By means of a computer-run distributor, this system allows fully automatic controlled individual distribution of concentrated foodstuffs for dairy cows raised in free range sheds.

Beekeeping (3000BC)

The Greeks attributed the invention of beekeeping to Aristaeus, the son of Apollo, and the Egyptians had a myth to explain the origins of bees and honey, according to which the latter was the tears of the god Ra. Egyptian beekeepers started under the Old Kingdom (between 2780 and 2280BC). They made great use of honey, since they did not know about sugar.

Fishfarming

Origins (antiquity)

It was the Romans who first had the idea of farming marine animals, fish or crustaceans, in enclosed waters (fishponds). The Chinese also used artificial spawning grounds.

Aquaculture (1933)

In **1933** the Japanese **Professor Fujinaga** first solved the problem of getting shrimps (*Penaeus japonicus*) to reproduce in captivity.

Between 1890 and 1910 billions of cod and lobster larvae had been released into the ocean by American biologists to multiply, in the hope of improving productivity: in vain. Professor Fujinaga's experiments at last opened the way to modern aquaculture, whose production (in fresh and salt water) is today more than 25 million tonnes per annum.

Seaweed farming (1972)

In **1972** the American **Howard A. Wilcox** put forward the idea of using seaweed to produce methane as an energy source. His calculations showed that if a 'field' 856km *531 miles 1616yd* long of seaweed of the species *Macrocystis pyrifera* could be cultivated in the ocean, the entire annual gas consumption of the United States could be produced. The idea is to harvest the seaweed, which the action of the sun's rays causes to grow very fast, and then to use a bacterium to digest it, with a resulting emission of methane.

Mussel beds (1235)

After being shipwrecked on the French coast in **1235**, an Irishman, **Patrick Walton**, settled there. Between tall wooden stakes planted in the sea he stretched nets to catch birds. He soon noticed that these stakes became covered in mussels, which seemed to grow remarkably well there. He had the idea of planting many more stakes, close together and linked by racks. He called these strange barriers 'bout choat', which became *bouchot*, the French term for these mussel beds.

Oyster farming (2nd century BC)

The first oyster farmer recorded by history was a Roman named **Sergius Orata**. He built an oyster bed on his property, in Lake Lucrin, near Naples, at the beginning of the 2nd century BC.

Orata built fishponds that connected with the sea but protected the oyster brood from waves. The young oysters were provided with posts to which they could cling and grow in proper conditions of temperature and light. Sergius Orata's know-how was such that he made a fortune selling his oysters. His contemporaries said of him, 'He could grow oysters on a roof.' Lake Lucrin disappeared in 1583 after an earthquake and a volcanic eruption.

Currently, Japan is the number one producer of oysters in the world.

Cultured pearls (about 1899)

At the end of the last century (the patent came into the public domain in 1921), **Kokichi Mikimoto** from Japan invented a procedure which made it possible, with pearl oysters, to obtain the first pearls called cultured pearls. Natural pearls are those formed without any deliberate outside action.

Natural pearls were more sought after (and thus more expensive) than diamonds until the invention of cultured pearls. They were mainly gathered in the tropical seas by divers who went to a depth of 40m *131ft 3in* without breathing apparatus.

In the hands of sometimes unscrupulous dealers, the intensive exploitation of pearl oysters was slowly but surely exhausting the natural beds in Ceylon, Japan, the Red Sea, Polynesia and the Persian Gulf when the invention of cultured pearls opened up new horizons.

Pearls have since been cultivated in specialised 'sea farms' and have become a product almost like any other. It takes three years for a pearl to form around a core of nacre which is implanted in the pearl oyster. One oyster can take two to three grafts during its life, but no more. In 1985 about 68 000 pearls were produced in this way by Japanese pearl farms, which, since 1968, have faced competition from the pearl farms of French Polynesia, whose pearls are highly prized. A connoisseur may pay up to 10 000 dollars for a single cultured pearl, if it is of the first water and perfectly formed.

Domestic animals

Cat (2100 BC)

Nearly all of the proof we have on the domestication of the cat comes from ancient Egypt. The first record of the cat's existence dates from approximately **2100 BC**. Any previous traces are probably those of a wild species. It was the Romans who introduced the cat to Europe.

It was only about a hundred years ago that cats began to be selected for reproduction. Towards the middle of the 19th century, breeding and exhibiting began in Britain.

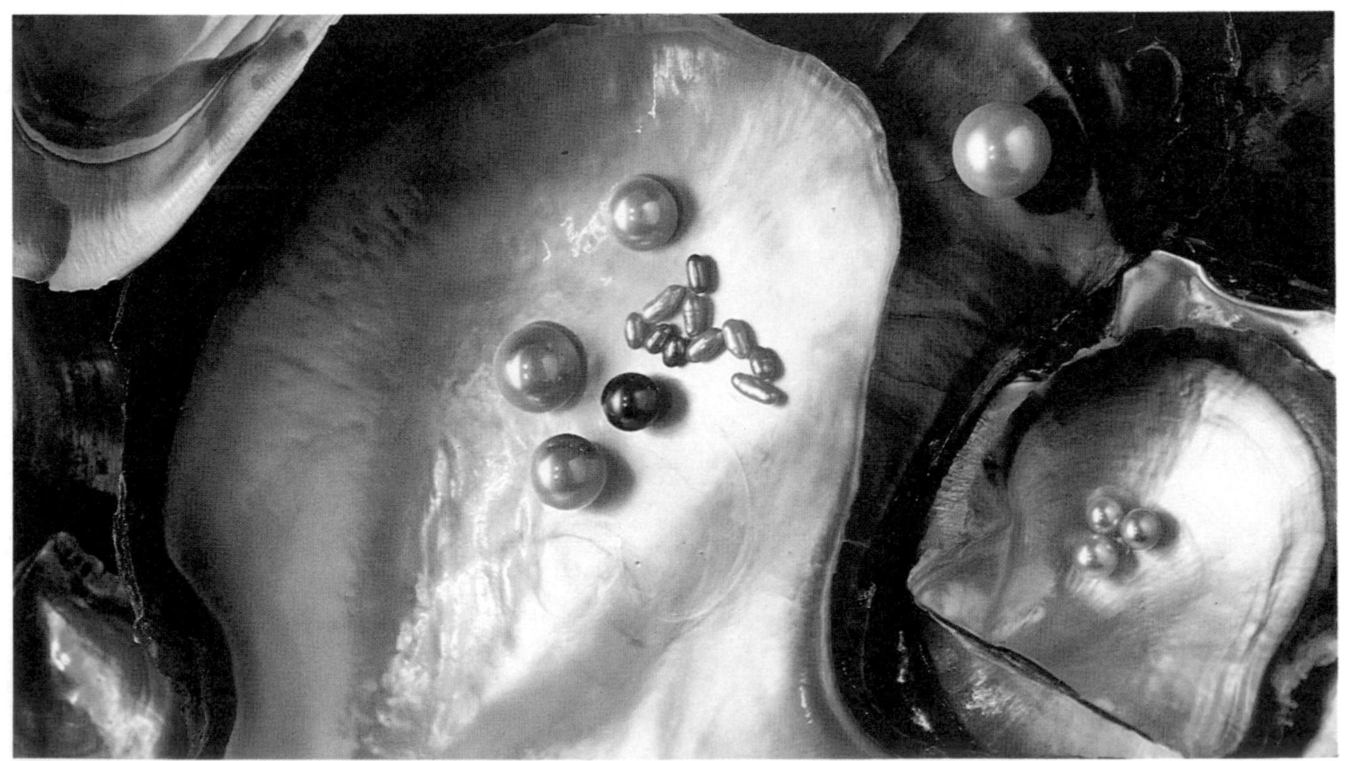

At the end of the last century cultured pearls were developed. In this photo are Japanese fresh water pearls, a Tahitian black pearl, and one from the South Seas with a diameter of 17mm ⅝in.

Dog (Neolithic period)

The dog has accompanied man ever since the latter began a more settled existence; that is, around **5000 BC**.

Breeds of dog such as the mastiff and the greyhound are ancient, featuring in Assyrian sculpture and Egyptian art of the Old Empire (between 2780 and 2280 BC).

The German shepherd first appeared around 2000 BC and is the descendant of the Persian shepherd. However the standards for this breed were not established until the end of the 19th century by Max von Stephaniz.

Other breeds are more recent: the boxer appeared at the end of the 19th century and the characteristics of the cocker spaniel were established in 1893. The Pekinese was originally the exclusive property of the Imperial Family of China and the Empress Tseu-Hi (1835–1908) set out the breeding regulations.

Tinned food for dogs and cats (1865)

When did 'Whiskas', 'Chum' and other tinned foods for cats and dogs first come on the market? The first industrialised food for domestic animals was a cake for dogs, manufactured in the United States around **1865** by **James Spratt**. In 1907 the Bennett Biscuit Company began to manufacture dog biscuits in the shape of bones.

Gardening

Wheelbarrow (Middle Ages)

The origins of the wheelbarrow are obscure. In Europe one of the oldest representations of a wheelbarrow, used on cathedral building sites, dates from the **12th century**. But the wheelbarrow must already have been in use in Asia, and more specifically in China. Moreover the latter has an astonishing variety of vehicles on the wheelbarrow theme, the common feature of all of them being that, like all other wheelbarrows, they have a single wheel.

Water grabber (1984)

The water grabber, invented by the nurseryman **Léon Beck** and marketed in **1984**, looks like a grain of salt or a fragment of a grain of rice; but when it is immersed in water it will absorb up to 700 times its volume in a few hours. Using this invention, flowers and plants can water themselves, with a real independent reservoir at their disposal.

The water grabber has had worldwide success and has been marketed in the United States since 1986.

Plantophon (1987)

Presented at the Brussels Trade Fair in **1987**, this 'telephone' for plants was invented by a Belgian, **M. Heusquin**. It is shaped like a probe and penetrates into the heart of the plant. When the plant is thirsty, a little red light flashes, accompanied by an audible signal which repeats every five seconds. Once the plant has drunk its fill, the device gives a little bleep and a green light comes on when the desired level of water has been reached.

Lawn mower (1831)

The invention of the lawn mower in **1831** was due to two Englishmen: **Edwin Budding** for the design and **Ferrabee** for the manufacture. The horses' hooves were covered by rubber boots so that they would not damage the lawn while pulling the cylindrical blades of the machine.

At the beginning of the second half of the 19th century, the steam mower was produced, followed by the motor-driven lawn mower. Black and Decker produced the first successful lightweight model in 1969.

Flymo (1963)

A mower based on the hovercraft principle was produced by **Flymo** in Britain in **1963**. This mower with a vertical blade has a fan which creates an air cushion under its plastic skirt. The machine is therefore very light to use, even on slopes, and is practical for negotiating obstacles.

A German, Heinz Zipfel, updated the air cushion system, creating the Fremo GMS in 1977. This machine mows however long the grass is, collects the cut grass, is not damaged by damp (and can therefore mow in all weathers), and makes it possible to mow even quite steep slopes, etc. The system has been patented in five countries, including France, Germany and the United States.

Garden hose (1850)

Imitating rain so as to provide plants with water is a procedure that was described in AD 800 by a Benedictine monk. At that time water was poured from a jug and allowed to run between the fingers onto the ground.

Around **1850** the first gutta-percha hoses began to replace watering cans and the watering cart drawn by a donkey.

Automatic watering kit (1968)

A real transformation of garden watering techniques grew out of an invention by two Germans, Kress and Kestner, from the **Gardena** company. Using moulded plastic components they developed a system of automatic connections which could be simply clicked together without any tools (or leaks) to assemble all the equipment necessary for watering (hoses, nozzles, taps, etc.). In 20 years 200 million Gardena kits were sold.

Tree replanter (1985)

It is very difficult to transplant an adult tree without damaging it or ruining its chances of settling in its new location. The Hydra-Brute Tree Replanter, an American invention from

Richard Hinchcliffe invented this rather eccentric protection from hayfever.

Lakeshore Industrial brought out in **1985**, is very easy to handle and currently has the best performance of any implement of its kind. It does everything – digging up the tree, transporting and replanting it – and it can be operated by one person.

Trees and flowers

Bonsai (3rd century BC)

The Chinese started cultivating 'pun-sai' trees in the Qin dynasty (**3rd century BC**). These were almost certainly trees naturally dwarfed by bad weather or soil, which were then replanted in decorated pots by connoisseurs to recreate miniature landscapes.

Bonsai culture became an art during the Tang dynasty (between the 7th and 10th centuries AD). Long the preserve of the nobility, it gradually reached all strata of the population after the beginning of the 17th century.

It was not until the 12th century that bonsai trees began to arouse the interest and then the passion of the Japanese, who then developed the techniques which make miniaturisation possible.

The bonsai tree in the centre of the picture cost its owner, Mr Akao, $880 000. It is 600 years old, 4m 13ft *high and 8m* 26ft *wide.*

Dahlia (1789)

The first dahlia was sent from Mexico to Madrid in 1789. It was given its name by the Spanish botanist **Cavanilles** in honour of the Swedish botanist Andreas Dahl, himself a student of Carolus Linnaeus, the Swedish founder of the modern classification of plants.

Orchid (1898)

In **1898** the Frenchman **Noël Bernard** discovered that orchid seeds, in order to germinate and guarantee the first stage of development of the embryo, need to be in contact with a microscopic fungus that lives in symbiosis in the plant's roots. Bernard invented a way to allow the seed to germinate independent of the plant.

In 1922 an American, Doctor Lewis Knudson, developed an environment which allowed germination without the presence of the fungus.

Tulip (1554)

Augier de Busbecq, Austrian ambassador at the court of Suliman the Magnificent, was also an amateur botanist. In **1554** he discovered the new flower in Persia and sent seeds and bulbs to Vienna.

A French botanist sent the first tulips to a trial garden in Leyden, Holland, some years later and the reign of the Dutch tulip began.

However, it was not until the 1730s that tulip mania began and fortunes were gained and lost because of the flower as it spread to gardens throughout Europe and then to the rest of the world.

The black tulip was created in 1986 by crossing two dark varieties.

THE ASTOUNDING LACQUER TREE

Lacquer was the first industrial plastic that human beings invented, around the 13th century BC. Like rubber, it is obtained by tapping the bark of a tree which is very common in central China, *Rhus verniciflua*. Each tree is tapped every five to seven years and can produce 50g *1.7635oz* of lacquer.

Lacquer is a long-lasting, plastic varnish which is not damaged by acids or alkalis, is resistant to water, solvents, bacteria and temperatures of −250°C *−418°F*. It is also a good electrical insulator.

In China cooking utensils have been lacquered for thousands of years to provide a cheap alternative to the bronze dishes that were the preserve of rich families.

An artisan begins to lacquer an object with a poorer quality lacquer and finishes with one of the best quality, which is obtained by tapping 15-year-old trees. Each layer has to dry before the next can be applied, and on precious objects, which are often later engraved, a hundred layers may be applied.

The traditional colours are black, red, brown, yellow, gold, green and a specific shade called 'pear skin' (Japanese *nashiji*), made from gold dust and gamboge resin.

In the 2nd century BC, the Chinese improved their technology by discovering, somewhat by chance, a procedure for preventing lacquer from drying by evaporation. This was to throw crabs into their lacquer pots, as the cellular tissue of these crustaceans contains powerful chemical agents which inhibit the enzyme that causes hardening.

Black tulip (1986)

After its long existence as a legend, the black tulip is now a reality. In 1979 the Dutch horticulturalist **Geert Hageman** put pollen from the 'Wienerwald' variety on the stamen of a 'Queen of the Night', creating the first black tulip. It was not until **1986** that the bulb he thus obtained and raised produced its first flower. The demands of reproduction meant that only three bulbs existed in 1987 and a dozen in 1988. Fields of black tulips are still some way off.

Fruit and vegetables

Apricot (300 BC)

The apricot tree came from China with the caravans of the silk route, before becoming established in the Middle East, then gradually spreading to the rest of Europe. History tells us that it was brought to the Graeco-Roman world three centuries before Christ by Alexander the Great.

It is thought the name comes from the Arabic *al-birqūq*, itself derived from the Latin *praecoquum*, meaning 'early ripener', since the tree is one of the first to flower in spring.

Pineapple (1493)

It seems that either Christopher Columbus or his companions were the first Europeans ever to taste a pineapple in Guadeloupe. The Indians of the Carribean had long been growing this fruit, which they called Nana, which means 'flavour'. Its exceptional qualities soon made it the Nana Nana, or flavour of flavours, from which its Latin name *Ananas comosus* is derived. However the English were more struck by its appearance than its taste and called it the pineapple.

In Europe the first attempts to grow pineapples under glass took place in Italy at the beginning of the 17th century, and then in the greenhouses of the Botanical Gardens of Leyden in Holland.

Over the last 20 years, and as a result of fast transportation, the pineapple has become an ordinary product. France is Europe's number one pineapple consumer.

Asparagus (antiquity)

Asparagus seems to have first appeared on the Baltic shore (or in the Mediterranean Basin, according to some). It was greatly appreciated by the ancient Egyptians, then by the Greeks. The Romans began to grow it seriously; however, it remained a luxury for rich gourmets.

It made its way on to our tables during the Renaissance, but did not become truly democratised until 1875.

Carrot (16th century)

The wild carrot came from Afghanistan long before the time of Christ. The Greeks and Romans did not set much store by it and it was used only for limited medicinal purposes before the **16th century**, when the **Italians** brought it to our attention. Although it was tasty, the carrot of the time had almost nothing in common with the vegetable of today. It was a thin white or yellow root and fairly tough. It took centuries of selection and hybridisation to give it the orange colour it has had since the early 19th century.

Lemon (1000 BC)

Some say the lemon tree first appeared in the foothills of the Himalayas, others say it was in the Malayan archipelago. Whatever the case, it was cultivated by the Chinese a good thousand years before the time of Christ.

The Romans were the first Westerners to introduce it into their gardens. They called the fruit the 'apple of the Mede', in reference to a people who lived in what is now Iran. The fruit was mainly used for medicinal purposes, as an antidote to poisons and venom and as an insect repellent.

It was not until the Crusades that the lemon tree arrived in the Mediterranean countries, even though the Arabs had already done much to spread it during their conquests.

The lemon was one of the few gifts from the Old to the New World, thanks to Spanish and Portuguese navigators.

One of the lemon's greatest glories was that it made possible the prevention of the terrible scurvy which had been decimating ships' crews at sea. In the mid-18th century **James Lindt**, a Royal Navy surgeon, discovered the remarkable anti-scurvy properties of this fruit.

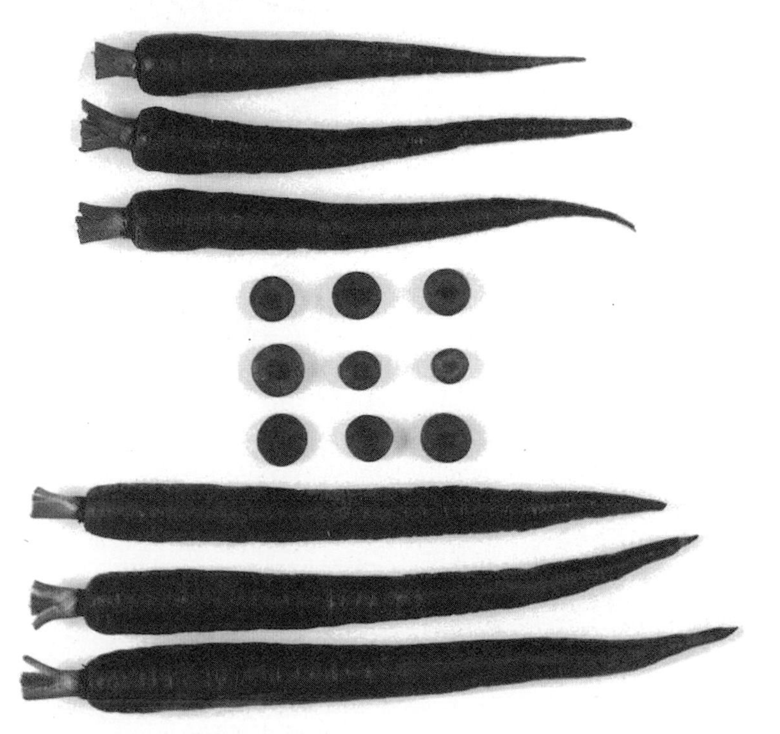

It was not until the 16th century that the carrot reached European kitchens. These 'super carrots' were developed at the University of Wisconsin. They contain four times more carotene – the substance that our bodies transform into vitamin A – than ordinary carrots.

It was not until 1932, a century and a half later, that the lemon's anti-scurvy properties could be attributed to its high vitamin C content.

Watercress (c.1550)

This fast-growing salad vegetable has always grown in ditches and on the banks of streams in countries with a damp climate. But it was not until the middle of the **16th century** that a German, **Nicholas Messinger**, developed the first watercress beds at Erfurt. For several centuries these remained a German speciality.

Chicory (1850)

In the underground passages of the Brussels Botanical Gardens, where he was Head Gardener, **M. Bréziers** had the idea of forcing a few roots of the chicory used in coffee and then eating the young shoots. The original strain has been extensively hybridised and selected so that today we can enjoy this gently bitter salad vegetable all through the winter.

Tomato (1596)

Of South American origin, the tomato was brought to Europe in **1596**. At first cultivated as a curiosity, it was considered to be a violent poison.

This vegetable-fruit had to wait more than two centuries before its alimentary qualities were recognised. The President of the United States, Thomas Jefferson, cultivated the tomato in his garden but didn't eat the fruit.

The first harvest of tomatoes was sold in 1812 in New Orleans, America.

Strawberry (1714)

It is to **François Frézier**, an officer of the French Navy and author of a *Treatise on fireworks*, that we owe the pleasure of eating today's fat juicy strawberries. Frézier had been sent to the South American coast in **1714** to study the fortifications of Chile and Peru. The five plants of *Fragaria chiloensis* (Chilean strawberry) that he brought back to Britanny in France did not fruit, but they pollinated other species which had been brought from Virginia, to give the first of our modern strawberries.

Haricot bean (1529)

The haricot bean was imported from America in **1529** by **Piero Valeriano** for Pope Clement VII. It was called *phaseolus*, then *fayol*, finally acquiring the name haricot when it replaced broad beans in the traditional mutton stew or 'haricot of mutton'.

Orange (15th century)

For a long time it was believed that oranges were the famous 'golden fruit' of the garden of the Hesperides, the three nymphs of the setting sun, which was later located as being at the foot of the Atlas Mountains. But this was just a legend, since neither the Greeks nor the Romans knew about oranges.

The bitter Seville orange was no doubt brought to the Mediterranean basin by Arab navigators and was rediscovered by the Crusaders in Palestine. But the sweet orange that we enjoy today is a mutation of a fruit which originated in India or China and did not appear in Europe until the beginning of the **15th century**.

Williams' pear (1770)

It was in the garden of **Mr Stair**, a schoolmaster at Aldermaston in Berkshire, that the Williams' pear, one of the finest varieties, first appeared in **1770**. Was it the result of deliberate hybridisation or of a mutation in his orchard? Nobody knows, and the inventor of this tasty fruit did not even leave it his name. It was a neighbouring nurseryman, Mr Williams, who bred, marketed and gave his name to the new fruit.

Pomato (1978)

This is a hybrid plant, artificially formed in **1978** by the German **Professor Georg Melchers** by combining protoplasts from a potato and a tomato. The fusion of protoplasts from different plant species is a technique destined to have a great future in agriculture.

Grapefruit (1809)

Count Odette Phillippe, a surgeon in Napoleon's army captured at the battle of

VANILLA: THE FRUIT OF AN ORCHID

When they discovered the New World, the Spanish conquistadors learned about vanilla, a precious product used by the highest Aztec dignitaries to flavour their chocolate. It was a pod, the fruit of a creeper-like climbing orchid, which botanists were soon trying to cultivate in very different places.

But the vanilla plants transplanted to Spain and then, early in the 19th century, to Reunion Island, flowered without producing any pods. It was discovered that a particular insect was missing: a bee with a long proboscis which fertilised the plant when it came to collect nectar from the flowers.

In 1836, in the greenhouses of the French Natural History Museum, the botanist Neumann perfected the first artificial fertilisation technique which made the development of vanilla cultivation possible.

A few years later a black slave called Albius working for a plantation-owner on Reunion Island had the idea of using a bamboo spike to tear the little tongue that separated the orchid's reproductive organs and then pressing on the anther so that the pollen came in contact with the stigma. This technique, as simple as it was clever, represented a great step forward for vanilla cultivation and earned Albius his freedom. Vanilla production became very important for Reunion Island and Madagascar, which now provide more than three-quarters of the 1500 tonnes produced worldwide.

The development of synthetic vanillin made from lignin has taken up only a modest share of the market (about four per cent) and seems to be losing ground in relation to the natural product.

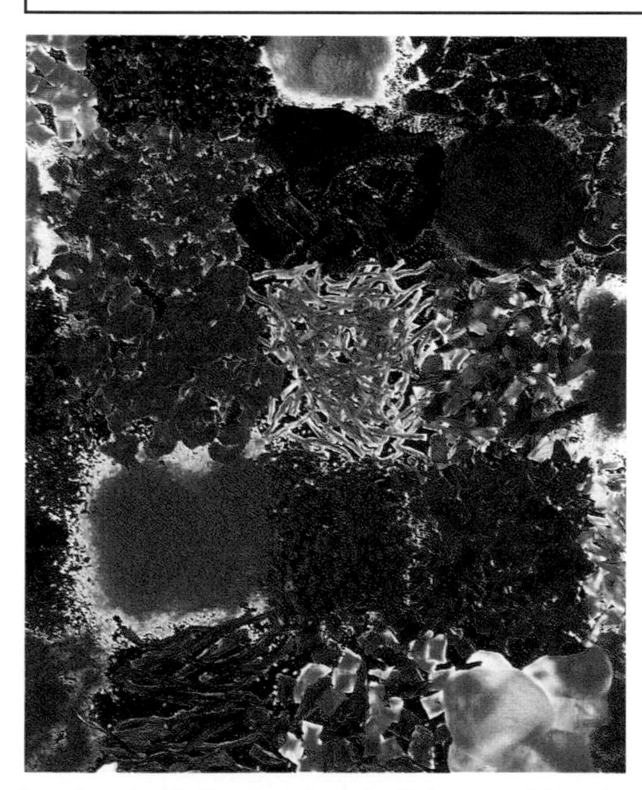

The American E.W. Flosdorff demonstrated that lyophilisation (freeze drying) could keep certain foodstuffs in good condition.

Trafalgar, was imprisoned by the British in the Bahamas. It was there that he discovered the *Citrus paradisi* or grapefruit, which grew there in abundance. When he was freed two years later, he set himself up as a doctor in Charleston, South Carolina, before moving to Florida, where, in **1809**, he established a citrus fruit plantation using seeds and plants from the Bahamas. Sixty years later, the grapefruit was a major source of revenue for Texas, Arizona and California.

It was not until 1914 that the fruit crossed the Atlantic and arrived in Palestine, where large orchards were planted.

Green grapefruit (1985)

This new fruit obtained in **1985** by Israeli scientists from **Jaffa** is the result of hybridisation between the white grapefruit and the shaddock. It has a sweet taste and was named Sweetie by its inventors. Its skin is a fine spring green with a tinge of yellow.

Potato (1554)

The potato was first introduced into Spain in **1554** by the conquistador **Pizarro**, who brought it from South America where it had long been cultivated by the Indians of the Andes Mountains in Chile and Peru. From Spain the potato spread to the British Isles in the 17th century, then to Belgium at the end of the 17th century and at the beginning of the 18th century to Austria, Germany and then to all of Europe except France. The French regarded the potato as unfit for human consumption.

Kiwi (1959)

Chinese currant and Yang Tao are just two of the names given to *Actinidia chinensis*, a plant discovered on the banks of the Yang-Tse river in 1845 and brought to the West at the end of the last century.

In 1906 this prolific creeper arrived in New Zealand where it aroused the interests of the local farmers who were the first in the world to plant orchards and export this new fruit.

The New Zealanders wanted to mark the originality of their product and in **1959** chose to give this fruit the name of the symbolic bird of their country: kiwi.

Pollution, oil slicks and 'green' inventions

Rivers and oceans

A clean-up operation on the Thames (1889)

In **1889** the first sewage treatment plant was built on the Thames. The river's 336 kilometres *208 miles 1391yd* were so polluted by the middle of the 19th century, that only eels could survive in it. As a result of the construction of the first plant, which was followed by others, the river became repopulated with many other species of fish. But in 1945, the population of London had reached eight million and the Thames was again becoming a dead river. Once more, extensive measures were taken to control pollution and give the river a new lease of life. Today the Thames contains salmon!

An automatic warning station (1988)

Two years after the disaster which seriously polluted the Rhine, an automatic warning station, the first link in a chain which is to be set up along the entire length of the river, was opened on **18 November 1988** at Huningue in the Haut-Rhin, France. The station is equipped with a total organic carbon detector, a dissolved hydrocarbon detector, a heavy metal detector, a fluorometer for detecting dye molecules, and a pesticide detector.

The equipment, each element of which is controlled by an independent microprocessor, is completed by a selection–rejection mechanism (the memory bank of the river) and a co-ordinator equipped with a printer as well as a visual display unit. The station also has a tele-data transmission system linked to stations controlling hydraulic structures such as sluices, locks and dams which have to be adjusted in order to limit the effects of pollution.

The state of the river Lea in east London demonstrates the effects of irresponsible dumping. Many species of freshwater fish, waterfowl and other wildlife are endangered by our carelessness.

Hyacinths to treat pollution (1981)

The water hyacinth (*Eichhornia crassipes*) which grows on rivers in Africa and South America spreads at an alarming rate and is amazingly prolific. However, if it is controlled it provides a remarkable means of filtering and recycling sewage as it thrives on waste.

In **1981** the town of San Diego in California (USA) introduced a pilot scheme, based on research carried out by **NASA**. They built a plant for the treatment of sewage using water hyacinths with a capacity of 100000 litres *22000 gallons* a day. The plants flourish on the phosphates and nitrates and absorb heavy metals. Some of the hyacinths are also harvested – five tonnes during the summer and two during the winter – for use as compost and fertiliser.

A sponge to control oil slicks (1975–87)

At first it looks like an ordinary white powder, but it completely absorbs petroleum on contact, becoming plasticised and forming a sort of solid, rubbery pancake which can then be removed.

The product, which has the added advantage that it does not harm animal or plant life, has been used in Japan since 1983 to control oil slicks and pollution in the ports. The Japanese are used to seeing these giant sponges saturated with petroleum hydrocarbons being removed by the cleaning squads. But contrary to popular belief, polynorbornene which is marketed in Japan under the brand names Fixol and TFN-2, is in fact a French invention.

In **1975 Claude Stein** and **André Marbach**, researchers for the French industrial group **CdF Chimie**, developed the product which was originally intended to be used in the manufacture of a special type of shock absorbent rubber. It proved to be highly absorbent of petroleum products and the idea of using it in the fight against pollution suggested itself automatically. But for various economic and legal reasons, this use of polynorbornene was abandoned in Europe, although it proved itself at the time of the *Amoco-Cadiz* disaster in 1978.

However, there are outlets for the sponge in Japan and the United States where the pollution laws are different, although things are beginning to change in Europe. One of the main areas in which it will be used is in the fight against pyralin pollution, a product used as an electric insulator in transformers, which is non-biodegradable and filters into the ground, polluting cultivated land via the groundwater table.

Unfortunately, in the event of an ecological disaster on the scale experienced in Alaska in March 1989 following the accident of the petrol tanker *Exxon Valdez*, this method of soaking up oil does not appear to provide the ultimate solution.

Inipol 90 (1989)

This new biodegradable dispersant was developed by the French laboratories, **Elf-Aquitaine**. Dispersants are used after an oil spillage at sea when it is important to prevent the slick from being driven towards the coast by the wind. The dispersant breaks up the slick, disperses it in the water and breaks down the hydrocarbons. **Inipol** was tested for the first time outside the laboratory at the time of the Alaskan oil slick resulting from the *Exxon Valdez* accident in March **1989**.

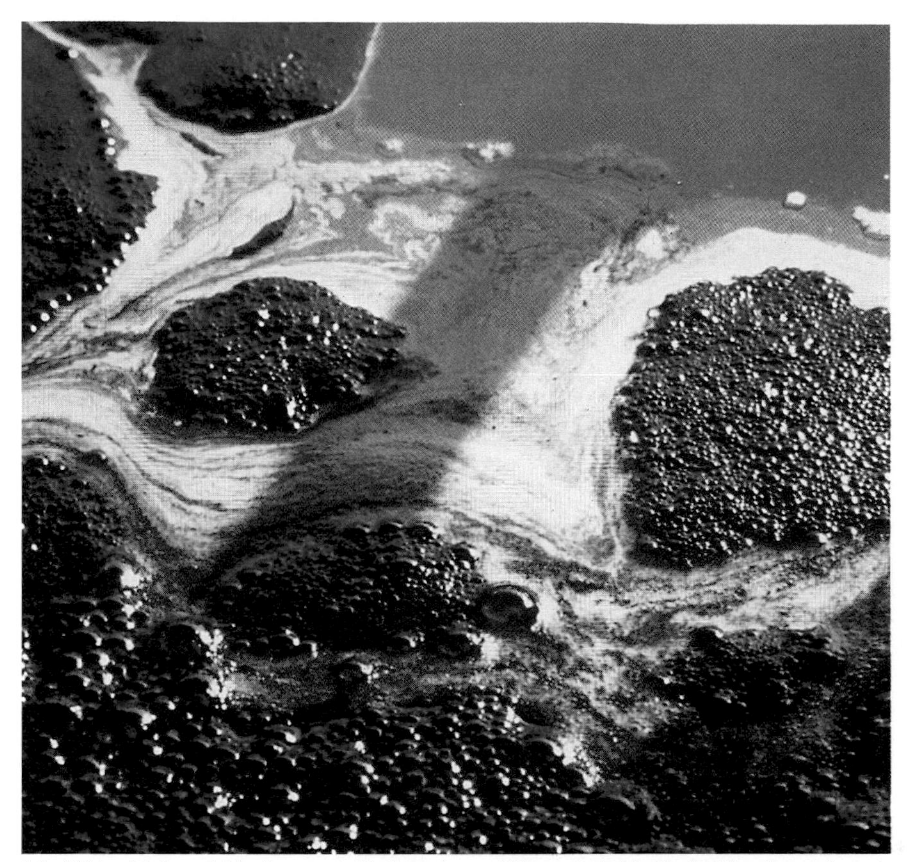

Inipol 90 is a biodegradable dispersant of hydrocarbons which was tested for the first time in a real accident in 1989. In the North Sea there are many spillages each year.

EROS-2000 (1988)

This is not, as its name might suggest, a method of protection against sexually transmitted diseases for the 21st century. It is a programme, the **European River Ocean System**, set up in **1988** by the European Communities Commission to study the pollution of European estuaries, coastlines and rivers. The first workshop of EROS-2000 was devoted to studying the pollution of the Mediterranean coast using the oceanographic ship *Discovery*. The results showed that the body of water off the coast was in a generally good state, due in particular to the rapid replacement by water entering the Mediterranean from the Atlantic via the Straits of Gibraltar. However, the concentration of hydrocarbons off the

WATER: A PROBLEM FOR THE END OF THE CENTURY

It is impossible to survive without water. Between 50 per cent and 70 per cent of the total body weight of a normal adult consists of water. The percentage varies as it is inversely proportional to the amount of fat. Water has been proved to be indispensable, not only to the life of human beings, but to life on earth in general. On planets where there is no water, no form of biological activity has so far been observed.

Since 1886 water intended for human consumption has been sterilised before being distributed, initially using the oxidising properties of ozone, discovered by the Dutch physicist Martinus van Marum. Since then, many other methods have been used, but ozone still occupies a significant place in the standardisation process of drinking water. Today, it is no longer enough simply to kill the bacteria contained in water in order to make it drinkable, since the pollution produced by man has altered the very chemical composition of the natural water supplies used by water authorities and distribution companies.

Today, drinking water is a commercial product and its composition is strictly regulated by standards (laid down in the European Community by the Council of Europe in 1980), which must be observed. For example, there must not be more than 0.5mg of nitrates and 50mg of magnesium per litre. Twenty-four components are regulated in this way.

If drinking water does not present a serious problem in Western countries at the moment, it will surely become one of the main world issues by the end of the century.

MOSES TO THE RESCUE OF VENICE

Since November 1988 Moses, or *Mose* in Italian, has been in place at the mouth of the Lido. It is an electro-mechanical experimental module, the first element of an enormous gate designed to protect the lagoon, and Venice, from high tides. The rectangular metal frame, 32m *104ft 11.5in* by 25m *82ft* weighs 1200 tonnes. Attached to its base by two hinges is a huge steel tank filled with water which, when emptied, is raised at an angle of 45 degrees and prevents water from entering the lagoon. If Moses proves successful after a year's trial, between 60 and 70 similar modules will be installed on the seabed at each entrance to the lagoon between now and the year 2000. They will operate as soon as the level of the tide exceeds 1.10m *3ft 7in.*

There is nothing new in Saint Mark's Square being submerged by the *aqua alta* or high tide, but it is a phenomenon which has become increasingly frequent in recent years. Whereas between 1920 and 1929 it happened on average six times a year, between 1980 and 1986 the average rose to 44 times per year! This is due to several factors, mainly the result of the destructive influence of human activity such as excessive pumping of the groundwater table, the diverting of rivers etc.

The Moses project is the most spectacular and one of the most costly operations launched by the *Venezia Nova* consortium, formed by 26 of the largest Italian companies for the protection of the City of the Doges.

This is coal slurry entering the North Sea. Several hundred thousand gallons of pit waste are dumped everyday off the coast.

coast was between two and ten times that found in the North Atlantic.

Europe's dirtiest river basin is that of the Mersey and its tributaries. A cocktail of industrial and agricultural waste and acid rain has left many of Britain's waterways completely fishless.

Biodegradable plastic (1989)

In July **1989** the Italian company **Ferruzzi** announced that they had produced the first biodegradable plastic. This is an extremely important event within the context of the modern problem, particularly in ecological terms, of reducing the amount of waste.

This new material is produced mainly from maize starch combined with fossil-based products chosen in order to maximise the biodegradable properties and compatibility of the plastic with the environment. It is therefore a plastic which has natural origins.

The first article to be made from this revolutionary new material was a Mickey Mouse watch, produced in collaboration with the Italian weekly of the same name and intended to familiarise the general public with thermoplastic starch. It was indeed a large scale campaign as 750 000 watches were produced.

Beaches

In 1979 Britain was required by an EC Directive to identify 'beaches that are used by a large number of bathers' and to test the seawater for pollution. Initially, Britain nominated just 27 beaches compared with 3000 each by France and Italy! Eventually Britain named 392 bathing areas; only 51 per cent passed EC standards in 1986. In 1990 33 per cent of bathing beaches still fail to meet the European Community's maximum counts of 10 000 coliforms (intestine bacteria) per 100ml of seawater.

Everyday Britain's coastal waters receive 300 million gallons of mostly untreated sewage; as a result ours are the most polluted waters of any country conforming to EC standards.

Shopping trolleys on the beach at Whitehaven in Cumbria.

Protection of the environment

First anti-pollution edict (1382)

In **1382 King Charles VI** of France published an edict that outlawed the emission of foul smelling gases in Paris.

In England in the 18th century, a decree forbad the lighting of fires when parliamentary sessions were taking place at Westminster.

Nearer our own time, in 1967, the Council of Europe first defined atmospheric pollution.

Modern legislation (1864)

The first legislation on atmospheric pollution during the industrial age came into force in the United States on **1 January 1864**. Known as the **Alkali Act**, it was a response to complaints from people living in the vicinity of factories using the Leblanc method to produce alkaline carbonate, a substance which had been used since 1823 and which released large quantities of hydrochloric acid into the atmosphere.

Nature conservation (1864)

The modern concept of nature conservation was invented in **1864** by the American diplomat **George Perkins Marsh** (1801–82). In a short work entitled *Man and Nature*, he explained how it was possible to preserve nature from the effects of the human race by using popular organisations to protect it on a national level.

The fight against smog (1952)

On **8 September 1952** London was in the grip of the worst smog on record. Cars drove into each other, aircraft were unable to find runways and pedestrians fell into the Thames. Some 4000 people died of respiratory disorders in the weeks that followed. As a result, the British Government decided to introduce measures which formed the legislation known as the **Clean Air Act**.

Within 30 years or so, the number of hours of sunshine in London have more than doubled, and over 150 different species of bird have returned to the capital compared with about 60 at the time of the smog.

Biodegradable washing powder (1964)

Because standard detergents did not break down naturally, in 1956 a professional organisation in Britain asked manufacturers to develop biodegradable washing powders. In **1964**, Germany was the first country to make the use of washing powders containing biodegradable products compulsory.

First ecological movement (1968)

The first ecological movement was founded in **1968** by **Cliff Humphrey**, an American student from the University of Berkeley in California.

The 'Refuse Pyramid' (2005)

This strange construction consisting entirely of rubbish will actually exceed the height of the Pyramid of Kheops which is 146.6m *480ft 10in* high. It is to be built on the Fresh Kills rubbish tip on Staten Island, south of Manhattan in New York, where since 1948 some one hundred million tonnes of waste have been dumped. The New York authorities have decided to build a monument from their refuse, a rather special mountain which, at a height of 156m *511ft 8in* by the year 2005, the estimated date of its completion, will be one of the highest on the East Coast of America.

WHAT HAPPENED TO THE DINOSAURS?

At the end of the secondary era, 66 million years ago, the dinosaurs suddenly disappeared from the face of the earth, as did 70 per cent of all other living creatures.

Since the American physicist, L. Alvarez, and his collaborators published their famous article in *Science* in 1980, scientists tend to think that the cause of this sudden disappearance was of an astrophysical nature. It is possible that it was the result of an impact caused by a collision between the earth and a body about ten kilometres *6 miles* in diameter, probably an asteroid. It is a well-known fact that many asteroids cross the earth's orbit. Such an impact would have made a huge crater and produced many side effects such as the destruction of the ozone layer, acid rain, the greenhouse effect, the clouding of the atmosphere by dust, etc. Any one of or a combination of these effects could have been the cause of the disappearance of the species.

According to another theory, an enormous volcanic explosion was the cause of their disappearance. Whatever the reason, it would appear to be generally accepted that the extinction of such a large number of the creatures living on the earth 66 million years ago was caused by a natural disaster of catastrophic dimensions.

Erts-1 (1972)

The satellite Erts-1 (Earth Resource Technology Satellite), launched on **23 July 1972** by **NASA**, can be considered the first ecological satellite. Renamed Landsat 1, and weighing 930kg *421lb*, this observer of the earth was placed at an altitude of 930km *577miles* on a circular orbit. It passed over the same point every 18 days and its cameras and infra-red sensors were able to trace the changes and development of different types of pollution: tree diseases, swarms of insects, etc.

Erts-1 was followed by the launch of a second Erts in 1973, then by numerous other American, Russian and French satellites: Landsat, Meteor, Priroda and Spot.

The desert campaign (1988)

The Japanese had the idea of combating the advance of the desert with the highly absorbent resins which had been used successfully in the nappy industry.

The experimental stage of the project, known as **Green Earth**, was launched in the spring of **1988** by the Japanese Minister for International Trade and Industry in collaboration with the Egyptian Minister for Land Improvement.

The experiment will last for five years and include the construction in Egypt of a 'factory for stopping the advance of the desert'. There is a lot at stake for the African and Asian countries where the desert is encroaching at a rate of 600 000km² *231 600sq. miles* per year, more than the entire surface area of France!

From 1990 the Egyptian factory is scheduled to produce a resin-based acrylic fibre compound, capable of retaining between 500 and 1000 times its own weight in water. Japan is the main world producer of these artificial fibres which are far less expensive than the conventional means of irrigation used in the desert.

WASTE IN THE FIERY FURNACE

An experimental plant for the high temperature incineration (4000°C *7232°F*) of solid chemical waste using an electron torch, is being constructed at Pont-de-Chaix in France. It is the first of its kind in Europe. The French electricity company, EDF, has collaborated in the construction of the plant which could solve the sensitive problem of the destruction of chemical waste. The electron torch, which operates on the basis of a fluid composed of electrically neutral gas molecules, positive ions and negative electrons, was so named in 1928 by the American chemist, Irving Langmuir, winner of the Nobel Prize for Physics in 1932. Generally speaking, and looking towards the future, it is hoped that it will one day be possible to use electron torches (at temperatures of several million degrees . . .) to destroy nuclear waste.

Environment-friendly refrigerators (1989)

The first refrigerators have been produced which contain half the amount of freon gas in their foam insulators. Freon belongs to the chlorofluorocarbon (CFC) family held responsible for the deterioration of the stratospheric ozone layer.

The initiative in the decision to reduce the production and use of CFCs was taken by the Swedish group, **Electrolux**, the current world leaders in the production of household electrical appliances.

An environment-friendly chainsaw (1989)

This piece of equipment has been on the market since **1989**. It was developed by the German company **Stihl**, the second largest

Ozone: we cannot live without it

An altitude of 25km

A layer of ozone, produced by the action of the sun's short ultraviolet rays (0.12 to 0.20μ) on oxygen, surrounds the earth at an altitude of 25km *15 miles 942yd.* Because of its absorbent properties, ozone is able to stop ultraviolet rays with a wave length of less than 0.20μ from entering the earth's atmosphere. Without this protective layer it would be impossible to survive because of the powerful thermic effect of the part of the solar spectrum below 0.20μ.

Depletion continues

There is a considerable amount of evidence which proves that this protective layer is being destroyed. Between 1979 and 1984 it was reduced by 20 per cent above the South Pole and continues to be depleted since it was reduced by a further 16 per cent in 1984/5. Southern Argentina is now under direct threat from the harmful rays of the sun, as is Australia where there has been a marked increase in the number of cases of skin cancer.

Causes

The precise causes of this depletion of the ozone layer are not known, but it is certain that chlorofluorocarbons or CFCs, the gases used in aerosols, refrigerator cooling systems and the manufacture of certain plastics, are a very strong contributory factor. These gases are non-biodegradable and on reaching the upper atmosphere they are destroyed by the ultra-violet rays, giving off chlorine which in turn reacts with the ozone to produce a chloric compound and oxygen.

A solution must be found

Whatever uncertainty there may be surrounding the causes, scientists are certain about one thing: a solution must be found. Until 1987 the annual production of CFCs was 1.1 million tonnes. At the end of the year, leading manufacturers throughout the world met in Montreal to decide on a research programme with a view to discovering a gas which would not harm the ozone layer but which had the same industrial properties as CFCs.

In March 1989, an international conference was held in London to discuss ways of saving the ozone layer and with the particular aim of ending the production and use of CFCs before the end of the century.

Time is running out

The fact of the matter is that time is running out. A second hole in the ozone layer has been discovered above the North Pole and only an immediate reduction of 85 per cent of the world production of CFCs would make it possible, at best, to stabilise their concentration in the earth's atmosphere. But even this would not prevent the ozone layer from continuing to be depleted for many years to come, since the life expectancy of CFCs in the atmosphere is between 70 and 120 years.

Discovered in 1781

Ozone (O_3), a gaseous element formed from three oxygen atoms, was discovered in 1781 by the Dutch physicist, Martinus van Marum, who noticed its smell in air through which electric sparks had passed. In 1840 the German chemist Christian Friedrich Schönbein gave it the name ozone from the Greek *ozein*, meaning having a smell, and its formula was established by Soret.

Surveillance of the ozone layer

On 22 March 1989 a research programme was implemented at the University of Wuppertal in West Germany to develop a system for monitoring changes in the ozone layer by satellite. The system, known as CRISTA (a telescope for observing the atmosphere and a cryogenic spectroradiometer), will be on board the West German satellite ASTROSPAS, which is due to be launched in 1993.

In January 1989 the state of the ozone layer was tested above Kiruna in Sweden. At last governments around the world are waking up to the dangers of the greenhouse effect.

HFA 22

Atochem, the only French company producing CFCs, has developed a CFC substitute for aerosols, HFA 22, which is non-flammable and has a very low level of toxicity. Its adverse effect on the ozone layer is 95 per cent less than that of CFC-11 and 12 and its contribution to the greenhouse effect is minimal. About a third of the CFCs produced are used in the aerosol industry, except in the United States where aerosols containing CFCs were banned ten years ago. Switzerland is to follow suit in 1991. The largest Japanese manufacturer of glass, the Asahi Glass Company, has developed two substitutes for freon gas, used in aerosols and the refrigerator industry: HCFC-225ca and HCFC-225. The Japanese have undertaken to reduce their production of freon by 50 per cent before 1998.

A new type of freon

Scientists from the Russian Institute of Applied Chemistry have succeeded in synthesising two new types of freon gas which do not destroy the ozone layer on decomposition. However a further four or five years' research is necessary before they can be brought onto the market. The USSR has undertaken to introduce the use of freon substitutes in aerosols by 1993.

chainsaw manufacturer in the world, and is fitted with a catalyst which reduces the amount of exhaust fumes. This is the first time that such a device has been developed for a two-stroke engine.

An ecological peace-keeping force (1989)

In March **1989 Federico Mayor**, the head of UNESCO, suggested introducing patrols of scientists as a sort of ecological peace-keeping force responsible for the environment and the enforcement of protective measures.

Engines and fuel

Unleaded petrol (1990)

The petrol obtained by refining crude oil is lead free. Chemists decided to add alkyl leads in the 1920s in order to increase the performance of the internal combustion engine. The American company, General Motors, was responsible for the widespread introduction of the process.

However, the exhaust fumes have proved to have an extremely high level of pollution and in view of the damage they have caused, particularly in the destruction of forests, it was decided to produce lead-free petrol. It has been on sale in America since 1975 and in Japan since 1977. In the European Community, all new cars manufactured after **1 October 1990** will run on lead-free four-star petrol.

Acid rain

Britain is now recognised as one of Europe's worst polluters with nearly two-thirds of her sulphur emissions being blown towards Scandinavia. Nonetheless a 1987 survey showed that the UK has the highest percentage of damaged forests in Europe.

In 1984 the 30 Per Cent Club was formed by 21 members who include Belgium, France and Italy. The Club's aim is to lower their sulphur dioxide emissions by 30 per cent by 1993, and six countries have already reached their target.

In 1989 Britain decided to cut sulphur dioxide emissions by 60 per cent by the year 2000.

ACID RAIN: IS IT DESTROYING OUR FORESTS?

Human activity – and in particular the combustion of fuel in electric power stations, industry and transport – is responsible for the pollution of the atmosphere. For several years, another product of this activity, acid rain, has been blamed for destroying forests. Various recent studies show, however, that things are not quite so straightforward. In the first place, it is obvious that rain is not entirely responsible for the spread of acid pollution and that dry pollution, caused by the wind, is a contributory factor. It is also known that this type of pollution is carried via the upper atmosphere and extends far beyond its point of origin with the result that there is nowhere in the world which is not affected by it. According to readings taken by the American Agency for the Protection of the Environment, the effects are worldwide. It is responsible in particular for causing a mist effect which occurs even in very remote places and which interferes with the climate by creating a screen to the sun's rays, which in turn causes cooling of the land and warming of the upper atmosphere. This type of pollution is the result of the discharge of sulphur dioxide and nitric oxide which combine with the molecules of water in the atmosphere to form sulphuric and nitric acid.

Catalytic exhaust (1909)

The need for lead-free petrol has raised the issue of an invention which was first patented on **17 April 1909** by **Michel Frenkel** of France: the 'method of deodorising exhaust fumes'. The catalytic exhaust was developed by General Motors in 1974 and has become compulsory in the United States.

In October 1987, Switzerland also made the catalytic exhaust compulsory in an attempt to combat the harmful effects of pollution, in particular the destruction of her forests which have been reduced by 40 per cent.

At the end of 1988 the American company Ford produced a new platinum-free catalytic exhaust which complies with the required standards but is much less expensive, and which has been fitted on some of their cars since 1989. The catalytic exhaust converts the polluting exhaust fumes into harmless substances by means of a series of chemical reactions. In the final stages of these reactions, a precious metal, usually platinum, is used.

Anti-pollution engine (1988)

This engine with its low pollution 'weak mixture' was developed by the Japanese company **Toyota** and presented at the Geneva Motor Show in March **1988**. It conforms to the

A sensor analyses acid rain above the Vosges mountains in France.

GREEN ENERGY

Current research is directed towards the use of 'green energy', in other words of the biomass provided by vegetation on land and in the oceans. Ethanol, which is already used as a car fuel in many countries although it is considered rather sluggish, will probably be given a new lease of life by the development of specific types of engine.

In another area, water hyacinths and seaweed provide the main hope as a replacement for fossil hydrocarbons.

Experimental biophotoreactors are also being tested in several laboratories. These are devices which use the photosynthetic properties of seaweed to make them produce certain types of molecules from light and a nutrient liquid. Some seaweeds produce hydrogen by this method which will provide an excellent fuel for the future and, under good conditions, the *Botryococcus* can produce 40 per cent of its own weight of hydrocarbons. Research is also being carried out to find ways of growing fresh water algae in order to give the best possible yield, for example *Scenedesmus* which can provide more than 20 tonnes per hectare of dry combustible material per year.

THE HURRICANE FIGHTERS

This is the name given to an American shock squadron which consists of 12 Air Force Lockheed C-130s used for meteorological observation, based in Mississippi, and two sea reconnaissance P3 Orion aircraft based in Miami. This observation force, formed 31 years ago, is a key source of information for the meteorological services, providing details of wind speed at the centre of a hurricane, its atmospheric pressure, direction and the speed at which it is travelling.

most demanding anti-pollution standards and does not need a catalytic exhaust.

Environment-friendly diesel (1989)

The German manufacturers **Volkswagen** have developed a new diesel engine for their Golf which is less harmful to the environment. It is a 60-horsepower engine fitted with a turbocompressor which, by improving fuel combustion, reduces the harmful effect of the exhaust fumes. It is also designed to take an oxidising catalyst.

Diesel engines are currently under scrutiny in West Germany where they are suspected of dispersing carcinogenic soot particles into the atmosphere.

Meteorology

Origins

One of the first readings of meteorological observations on record dates back to the Yin Dynasty, c.1300 BC. But it can be said that the scientific study of meteorology began in about 340 BC with the publication of Aristotle's *Meteorologica*, which constitutes the first treatise on the atmosphere. However in 278 BC Aratus did recommend the use of frogs as a means of forecasting the weather!

From 1855 the French astronomer, Urbain Le Verrier (1811–77), was responsible for setting up a network of meteorological data between the various European observatories.

Rain gauge (4th century BC)

This very ancient apparatus was used in **India** in the **4th century BC**. It was then found around 1440 in Korea.

In the West, the use of the rain gauge became common after 1639, thanks to the work of the Italian Benedetto Castelli.

Barometer (1643)

The mercury barometer was the invention of the Italian physicist and geometer **Evangelista Torricelli** (1608–47), a student of Galileo.

Using this apparatus, Torricelli succeeded in demonstrating that air had its own weight, which varied depending on the circumstances, and that the weight variations (atmospheric pressure) could be measured by studying changes in the height of the mercury column in the tube.

Ship's barometer (18th century)

The first ship's mercury barometers made their appearance during the **18th century** although they were not immediately popular with sailors who preferred the old empirical methods of predicting the weather.

In 1858 an English admiral, called Fitzroy, had the idea of providing all fishing ports with a barometer.

Weather radar (1949)

In 1924 the Englishmen E.V. Appleton and M.A.F. Barnett demonstrated the existence of the ionosphere (the area of the atmosphere in which the air is highly ionised), using the reflection of continuous waves.

However, the first campaign to use radar for weather study was carried out in **1949** in the United States for the Thunderstorm Project, under the direction of **M.H. Byers** and **R.R. Braham**.

Weather radar allows scientists to track weather balloons equipped with reflectors, helps the detection of large storm clouds and

A RADAR TO DETECT WIND SHEERS

Wind sheers are a meteorological phenomenon which are extremely dangerous for aircraft and difficult to detect. A mass of cold, heavy air moves suddenly downwards towards the ground from a cloud mass or even out of a clear sky. It hits the ground and rebounds in horizontal jets with incredible force. These wind sheers can prove fatal if they take an aircraft by surprise at low altitude or during take-off or landing. They are violent, concentrated and suddenly reduce the lift of an aircraft which does not then have the height or the speed to re-establish itself and falls like a stone. Until now there has been no instrument reliable enough to detect them with any degree of certainty.

American airports near the Rockies or on the Great Plains, two places which are particularly affected by turbulent air conditions, are extremely interested in the major step forward represented by the development of a Doppler effect radar with a margin of error of only two per cent. However, it will not be brought into service until 1992.

The Doppler sodar can improve air safety by identifying wind sheers.

SKY BLUE

In 1852 the German physiologist Ernst von Brücke (1819–92), made the first attempt to give a scientific explanation of the colour of the sky. His theory that the sky blue colour was due to the diffusion of sunlight by the very fine particles suspended in the atmosphere was taken up in 1869 by the Irish physicist, John Tyndall.

In 1897 the British scientist Lord Rayleigh developed the explanation by demonstrating that the light was not diffused by dust particles or droplets of water, but actually by the air molecules. This explanation was developed even further when, in 1928, the Indian physicist Chandrasekhara Raman, winner of the Nobel Prize for Physics in 1930, discovered the effect which has been named after him and which proves that the blue colour of the sky is the result of an interreaction between the air molecules and the sunlight. The blue of the sky in fact contains light frequencies which are not present in the solar spectrum and have been added by the air molecules.

rain, and enables the internal structure of some cloud masses to be examined.

Sodar (1968/9)

The sodar (sound detection and ranging) is a radar that, using sound waves that are emitted vertically and broadcast back by turbulence caused by changes in temperature, allows the three components of wind to be measured. The principle behind this device was described in **1968/9** by the Americans **L.G. McAllister** and **G.G. Little**.

Doppler sodar (1979)

In **1979** the French company Bertin patented the Doppler sodar as a result of research carried out by **J.-M. Fage**. Sodar was developed in order to improve the safety of air traffic by providing a continuous measurement of wind by means of teledetection. The thermal structure of the atmosphere at high altitudes – its head – is also measured. Wind sheer and the movement of atmospheric pollution can be monitored.

The Doppler effect was described by Christian Doppler, an Austrian mathematician and physicist who lived from 1803 to 1853. The effect concerns changes in wavelength that occur when the source of a vibration moves closer or further away.

Lidar (1976)

Lidar is a colidar (coherent light detection and ranging). The transmitter is a laser and the receiver a telescope supported by a photocell detector. It appears that the American electrical engineers, **Louis D. Smullin** and **G. Fiocco** were the first to use the Lidar for meteorological purposes in **1976**, in order to detect aerosols at a distance of up to 140km *87 miles*.

The French company Crouzet is working in collaboration with the American company Spectron to develop the Lidar so that it can be incorporated into the control panel and make it possible for air crews to detect wind sheers.

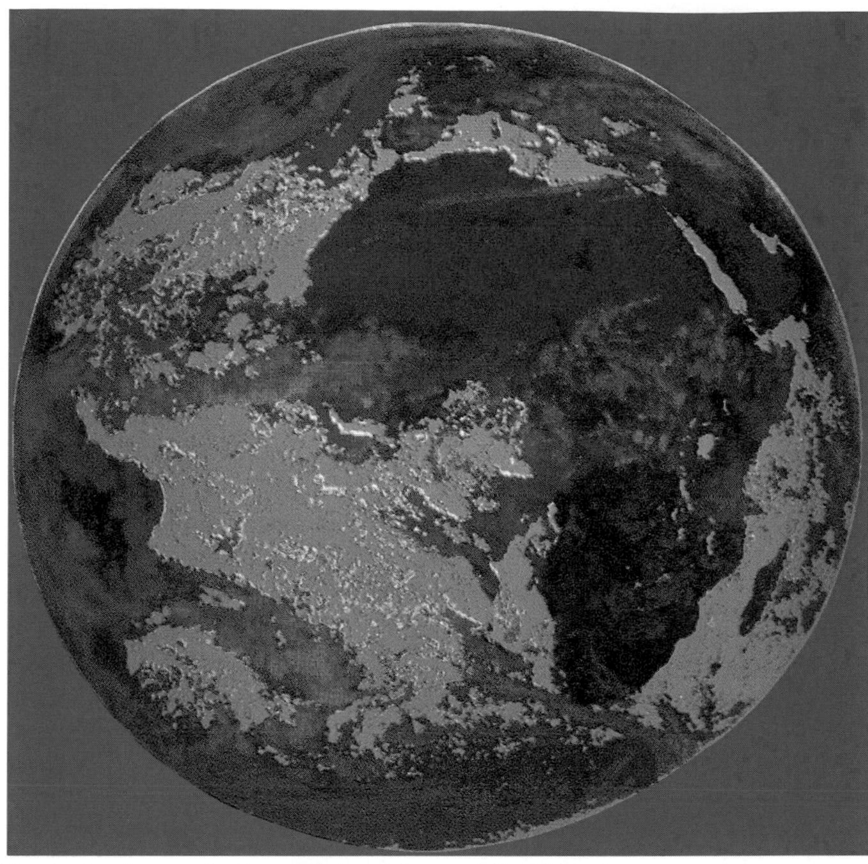

The planet earth as seen by Europe's first geostationary satellite, Meteosat.

Weather satellite (1960)

The first weather satellite was launched by the United States on **1 April 1960**. The **Tiros 1** satellite was built by RCA.

On 24 August 1964 Nimbus 1 was launched; this satellite produced the first good quality night photographs.

The first geostationary satellite was launched over the Pacific Ocean on 6 December 1966.

Meteosat (1977)

On the **23 November 1977** Meteosat, the first European geostationary satellite, was launched in the United States. Initially designed to have a lifespan of three years, Meteosat-1 in fact kept working for eight years, providing meteorologists with the maps and satellite pictures seen daily on our television screens. It has not been in use since the end of 1985 and is at present drifting in orbit, having been replaced by Meteosat-2 which was brought into service in June 1981.

The radiometer of Meteosat-2 has scanned the earth's surface more than 110000 times, which is twice as often as any other device of this type installed on board a geostationary satellite.

In 1989, Meteosat-4 (MOP-1) was launched. It was built by SNIAS, the French aerospace company, under the auspices of the European Space Agency (ESA).

Weather cube (1988)

The weather cube was invented by **Toshiba**. It is a small box which can be placed on a desk and gives a weather forecast for the next eight hours within a radius of 20km *12 miles 753yd*. A screen registers four possibilities – fine, changeable, cloudy or rainy – by means of semiconductors which measure variations in pressure.

Isotherms

The first maps showing the average annual isotherms (lines linking places of equal temperature) for the Northern Hemisphere, and which used the word 'isotherm' for the first time, were published by the German explorer, Alexander von Humboldt (1769–1859).

Geophysics

Origins (6th century BC)

In the **6th century BC** the Greek scholars **Thales** and **Anaximander of Miletus** were the first to collate the various pieces of information about the properties of the earth. They can justly be considered the founders of geophysics.

In the 4th century BC the Greek geographer Pytheas, born at Massilia (modern Marseilles), put forward the idea that the earth was round. He was a skilled astronomer and mathematician and undertook long voyages to prove his theory. He reached the ancient Kingdom of Thule (probably Iceland or the Faroe Islands) and explored the Germanic coastline of the North Sea. He was also the first to explain the effect of the moon upon the tides.

The brilliant Greek philosopher, Aristotle (384–322 BC), set down the geophysical

THE PLANET AMAZON SEEN FROM A RAFT

The raft in question is, of course, the famous treetop raft designed by Gilles Ebersolt. This airborne raft, powered by a huge hot-air balloon, is able to land gently on the forest canopy which it can then study from its unique and privileged vantage point.

The first of these observations of the equatorial forest was carried out in French Guyana in 1986. From 15 August to 30 September 1989, the third generation of (steerable) treetop rafts carried out an exploration of the Brazilian and Peruvian Amazon.

knowledge of his time in two of his works, *On the Heavens* and *Meteorologica*, which continued to be influential works in the field until the 17th century.

The Greek mathematician, astronomer, geographer and philosopher, Eratosthenes (284–192 BC), continued the teachings of Aristotle and was the first to produce a fairly exact calculation of the circumference of the earth.

A subsequently incorrect interpretation of the measurements of Eratosthenes was to have surprising consequences. In the 2nd century AD, the Greek astronomer and geographer, Claudius Ptolemaeus of Alexandria (Ptolemy), made an erroneous calculation of the circumference of the earth based on the work of one of Eratosthenes' successors, Posidonios of Rhodes (AD 135–150). His calculation, preserved by the Arabs and rediscovered in Europe during the 14th century, provided the cartographic information used by Christopher Columbus when he discovered America by chance.

Geographical map (575 BC)

There is a strong possibility that it was **Anaximander of Miletus** who compiled the first map of the world in **575 BC**.

The Greek astronomer, mathematician and geographer, Ptolemaeus of Alexandria (AD 90–168), compiled 26 maps which were used until the 16th century.

In the Middle Ages, the only valid maps were the *portulans* and maps compiled on the basis of information supplied by sailors, used for navigation between ports and giving detailed descriptions of coastlines. The earliest of these to be preserved is the Pisan map compiled in Genoa in 1285.

During the Renaissance, sailors developed the science of measurement. In 1569 the Flemish cartographer Gerhard Kremer, known as Mercator (1512–94), produced maps of the whole of the known world on 18 sheets, and taking into account the relief, or sphericity, of the earth.

The first maps of European countries appeared around 1530.

The first map of North America was produced in Boston in 1677 by John Foster.

An underwater atlas

Between 1983 and 1988 the *Jean Charcot*,

This map of India dates from 1519.

THE CONTINENTS ARE MOVING. . . .

This was the inspired idea that occurred to the German meteorologist Alfred Wegener in 1912 as a result of the realisation that South America fitted into the Gulf of Guinea. Therefore, it appeared that the Atlantic Ocean had been created by the separation of Europe and America. Alfred Wegener's theory of continental drift was not acknowledged by the scientific circles of his time. It was not until the beginning of the 1960s that his idea was considered seriously.

During a Franco–American scientific expedition in the summer of 1974, the rift in the seabed which separated the two continents was observed for the first time.

Wegener's theory is now accepted by most scientists under the name of plate tectonics. They believe that the continents are in fact moving in relation to each other on ten or so plates. Sometimes the plates slide one above the other or collide, as happens in Japan where the collision of three plates off the coast of Tokyo creates incredible tension which produces volcanic eruptions and earthquakes. During the summer of 1985, Operation Kaïko studied the phenomenon in detail when a Franco–Japanese team of scientists on board the *Nautilus* dived to a depth of 6000m *19680ft* to observe at close quarters the tension created at the point of contact of the three plates.

flagship of IFREMER, the French institute for oceanic research, made a round-the-world oceanological voyage which lasted for 1400 days. As a result, an atlas consisting of 60 maps of the oceans' beds was published in 1989. It reproduces ridges and deeps as well as underwater volcanoes with unprecedented precision.

Oceanography

Underwater exploration

Diving equipment

The diving bell (4th century BC)

The diving bell, which already existed in ancient times, was described in detail by **Aristotle** (384–322 BC). There is no trace of it during the Middle Ages, but it reappears in Italy and Spain during the 16th century. In 1583, to the amazement of several thousand onlookers, including Charles V, two Greeks went to the bottom of the River Tagus in Toledo, Spain, and resurfaced without getting wet and without their lamp being blown out.

In 1552 Venetian fishermen carried out similar experiments in the Adriatic in the presence of the doge and senators. About the same time, the Venetians also invented a 'diver's hood' known as the bagpipes.

Halley's bell (1721)

In **1721** the British scientist and astronomer **Edmond Halley** developed the first diving bell really worthy of the name. It was perfected in 1786 by the English civil engineer, John Smeaton.

The diving barrel (1721)

The progression from the diving bell to the early versions of the diving suit happened almost imperceptibly, so that it is impossible to give a precise indication of the origins of the first equipment which allowed the diver freedom of movement. However, it is generally thought that in **1721 John Lethbridge** designed a piece of equipment shaped like a barrel with two holes for the arms and a glass peephole so that the diver could see underwater. The equipment was not particularly practical as the diver was forced to live on his stomach and to return repeatedly to the surface in order to breathe.

A basic diving suit (1796)

The first diving suit in the true sense of the word was invented in **1796** by the German, **Klingert**, from Breslau (now Wroclaw in Poland). It consisted of a domed cylinder made of thick tinplate which completely covered the head and torso of the diver, leaving his arms free. A short-sleeved bodice and a pair of leather trunks protected his limbs from the water pressure, and the whole thing was completely watertight. Two glass-covered holes at eye-level enabled the diver to see. A tube, one end of which was above the surface, was fitted to a hole level with the diver's nose, while a second tube, placed next to it, was intended to evacuate exhaled air. Two lead weights were attached to the diver's waist to provide ballast.

The first wind-powered petrol tanker was launched in 1980 by the Japanese. It has proved itself over the years, particularly as an 'environment friendly' marine vessel.

Diving suit with air pump (1829)

The first really effective diving equipment was invented in **1829** by the Englishman, **Siebe**, who was commissioned to supply the French Navy until 1857.

Diving suit with automatic air regulator (1865)

The piece of equipment, which played an essential part in the development of the diving suit, was the result of collaboration between a mining engineer, **Benoît Rouquayrol**, and a ship's lieutenant, **Auguste Denayrousse**, in **1865**. The air regulator fulfilled the function of an artificial lung in the sense that the diver's lungs actually regulated the intake of air by acting directly on a distribution valve. This forerunner of the air 'bottle' was developed in 1936 by the French naval officer and inventor, Yves Le Prieur. Rouquayrol and Denayrousse were also responsible for inventing what can be considered the forerunner of the diving mask and breathing tube in about 1870. The mask was further developed by another naval officer, de Corlieu, who also invented flippers.

The aqualung (1867)

This version of the diving suit was presented at the Paris Universal Exhibition in **1867** by the New York Underwater Company. It consisted of a metal helmet and waterproof suit, but the diver also carried a tank of air on his back. The air was compressed at 17 atmospheres, and was enough to last one man for three hours at a depth of 20m *65ft*.

Air bottle (1936)

In **1936** the French inventor **Yves Le Prieur**, developed equipment which consisted of a bottle filled with air compressed at 150kg/cm² *2134lb per sq. in* and a pressure-reducing valve linked to the bottle which distributed air at a suitable pressure to a waterproof mask covering the face.

aqualung developed by the French underwater explorer and engineer, **Jacques-Yves Cousteau** in collaboration with E. Gagnan. The aqualung had a pressure-reducing valve invented by Gagnan during the war to enable cars to run on gas, and was the first modern version of the aqualung as we know it.

Newtsuit

The Canadian company **International Hard Suits** of Vancouver has developed a revolutionary suit which enables the diver to reach a depth of 300m *984ft* and work at normal atmospheric pressure while breathing oxygen. The main innovation of the suit, which has been named the Newtsuit, lies in the 20 watertight aluminium and titanium joints. A system of automatic hydraulic valves maintains the correct pressure inside the joints. The suit displaces its own weight of water i.e. 275kg *606lb 6oz* when on the ground.

This new type of suit has eliminated the need for divers to be subjected to long periods of decompression, and the use of extremely costly gas mixtures.

The *Nautilus* (1985)

The *Nautilus*, weighing only 18 tonnes and able to carry three men, was developed in **1985** by **IFREMER**, the French institute for oceanic research. It was developed to allow access to 97 per cent of the seabed with maximum efficiency and manoeuvrability. It consists of a titanium sphere, 2m *6ft 6.72in* in diameter and 15cm *5.91in* thick, and is equipped with remote handling equipment in the form of two carbon fibre arms.

It proved particularly efficient during the exploration of the wreck of the *Titanic* during the summer of 1987. The *Nautilus* with its support ship the *Nadir* has taken part in subsequent expeditions, including Hydronaut, its most important expedition of 1987, which was a biological study trip to investigate hydrothermal sources off the coast of Mexico. At the end of 1988, the *Nautilus* made 21 descents during a 22-day study trip to the Nazca Fault near the Galápagos Islands. It then went to an area known in French as the *permis*, an authorised site reserved for exploitation by France which lies between Mexico and the Hawaiian Islands, where it was used to study the distribution and collection methods of polymetallic nodules which have formed there during the course of thousands of years.

Shinkai 6500 (1989)

The *Shinkai 6500* is the new Japanese oceanographic research submarine which is capable of descending to 6500 metres *21 320ft*, a depth which no other submarine has reached. Constructed by **Mitsubishi Heavy Industries Ltd**, the *Shinkai 6500* carries three people and can explore 98 per cent of the seabed, compared with 97 per cent for the *Nautilus*.

Robot diver (1986)

Jason Jnr was the name given to the robot used in America in **1986** by the **Woods Hole Oceanographic Institute** in the exploration of the seabed and in particular of the wreck of the *Titanic*. Attached by a cable to the submarine *Alvin* at a depth of 3800m *12 464ft*, Jason Jnr was lowered to the wreck which lay at a depth of 4000m *13 120ft*.

The Jason project has been fully operational since 1988. It consists of two interconnected remote-controlled vehicles, equipped with cameras and a control arm. The two robots, controlled from a ship on the surface, can reach a depth of 6000m *19 680ft*.

Sea helicopter (1986)

The first submarine helicopter, known as *Deep Rover*, was constructed in **1986** by the Californian company, the **Deep Ocean Engineering Company**. *Deep Rover* is a completely independent unit which can take a diver to a depth of 1000m *3280ft* for a period of up to eight hours. Its main advantage is that the occupant breathes air at normal atmospheric pressure which eliminates the need for decompression stages. From behind the panoramic window of his cabin, the pilot controls the movement of two remote controlled arms which can perform such tasks as welding, bolting, sawing etc. It is designed to carry out inspection and repair work on pipelines and oil rigs.

Scubaphone (1986)

In **1986** a longstanding dream came true for **Thomas Murdoch**, **Steen Soloman** and **Bent Larsen**, the founders of Orcatron Manufacturing in Vancouver, Canada. They developed the Scubaphone, a radio system which makes it possible for divers to communicate clearly with each other underwater at a distance of up to 1200m *3936ft*, or with the surface at a depth of up to 80m *262ft 5in*. The system is incorporated into a special helmet containing batteries, with a microphone fitted into the mask. Microelectronic circuits eliminate distortion caused by variations in temperature and pressure and filter any radio interference. The signal is converted into an acoustic wave and then reconverted on reception.

Underwater computer (1986)

In 1986 a team of research scientists from Deakins University in Australia, led by **Bob Pedan**, developed a personal underwater computer which could be carried on a diver's back with the miniature keyboard attached to the wrist. Both the computer and keyboard are enclosed in a watertight casing which is pressure-resistant at a depth of more than 100m *328ft*. It enables the diver to record cartographic and archaeological information on the spot.

Mir-1 and Mir-2 (1987)

The Russians began testing their first two research submarines in the Atlantic at the end of **1987**. Mir-1 and Mir-2, capable of operating at a depth of more than 6000m *19 680ft*, were constructed by the Finnish shipyards, Rauma-Repola under the scientific and technical supervision of the Russian Academy of Sciences. They have several advantages over foreign submarines of the same type. They have a greater travelling speed and operating range, a large energy reserve, the ability to change depth without discharging ballast, two manipulator arms and underwater drilling equipment. Both the Mir-1 and Mir-2 carry a crew of three people and have reached a depth of 6170m *20 237ft 7in* and 6120m *20 073ft 7in* respectively.

Aquarius (1987)

At the end of **1987** a highly technically advanced underwater station known as Aquarius was brought into service for an unspecified period. It is located on the floor of the Carribean, off the coast of the island of St Croix in the Virgin Islands. The station consists of three compartments and can hold up to six people. It is 13m *42ft 7.7in* long by 4m *13ft 1.4in* wide and is 5m *16ft 5in* high. It replaces the Hydrolab station which provided a base for 200 missions between 1966 and 1985. The missions, carried out under the auspices of the American Administration for Oceans and the Atmosphere, were concerned with ichthyology (the study of fish), oceanography and underwater engineering.

Volcanoes and earthquakes

Origins (5th century BC)

Empedocles, the Sicilian-born Greek philosopher and naturalist, was responsible for the earliest studies in vulcanology in the **5th century BC**. His observations, which were carried out on site (Mount Etna in Sicily is still the most important active volcano in Europe), led to the realisation that there was molten igneous matter at the centre of the earth.

Seismograph (132BC)

The earliest known seismograph was invented in China in **132 BC** by **Chang Heng**. The device, which he called a seismoscope, unlike modern seismographs did not record all the features of an earthquake but simply registered the direction of the main tremor.

Evaluation of intensity (1857)

In **1857** the British seismologist and engineer **Robert Mallet** was the first to try to evaluate the intensity of earthquakes.

In 1880 Rossi of Italy and Forel of Switzerland developed a scale, graded from one to ten, to evaluate the extent of the damage caused by earthquakes in inhabited areas. In 1902 the scale was extended to 12 degrees by the Italian vulcanologist, Mercalli.

In 1931, Wadati of Japan suggested a scale which was applied to both inhabited and uninhabited regions. This new scale was improved by the American, Charles Richter.

The Richter scale (1935)

The Richter scale, developed in **1935** by the American **Charles Richter** in collaboration with the German-born American geophysicist **Beno Gutenberg**, is based on the conversion of the intensity of seismic waves recorded by seismographs into logarithmic calculations. The scale made it possible for earthquakes to be classified on the basis of energy rather than superficial effects.

Forecasting volcanic eruptions

As a result of research made public in **1987 Professor Bruce Crow** and his team from the Los Alamos Centre for Volcanic Research in New Mexico consider that it will one day be possible to forecast volcanic eruptions by the analysis of gases and particles given off by the volcano. Basic variations in the amount of gold, platinum and iridium contained in the gases and particles could indicate an imminent eruption. The sudden appearance of precious metals in minute quantities before certain eruptions is a phenomenon which has intrigued vulcanologists for a long time.

Forecasting earthquakes

Since **1971** an American team from the **Lamont Doherty Geological Observatory** has also been working with the seismologists from the California Institute of Technology, CalTech, to develop methods which will make it possible to forecast earthquakes by establishing the characteristic conditions which precede an earthquake. Specialists from CalTech use lasers to observe the two edges of the famous San Andreas Fault in California, an area which is very prone to earthquakes.

In 1988 a researcher from the University of California, Leon Knopoff, and two members of the Russian Academy of Science, Vladimir Keilis-Borok and I.M. Rotwain, announced that they had developed a new method of forecasting involving a complicated mathematical formula based on the seismic activity preceding 14 tremors which occurred in the United States between 1938 and 1984 and which registered more than 6.4 on the Richter scale. The formula was subsequently applied to 20 earthquakes that had occurred in other parts of the world since 1963 with the result that it was possible to 'forecast' 16 of them.

The VAN method (1983)

The VAN method of forecasting earthquakes was developed in January **1983** by three engineers from the Institute of Physics of the University of Athens, **P. Varostos**, **K. Alexopoulos** and **K. Nomicos**. Although still in the experi-

It was at the site of Etna on Sicily that Empedocles carried out the first studies of volcanoes.

San Francisco was again devastated by an earthquake in October 1989.

DISCOVERY OF A 350 MILLION YEAR OLD MOUNTAIN RANGE

Since 1989 the Dutch Institute of Applied Research has been taking part in a sounding operation being carried out in Bavaria to a depth of 15km *9 miles 567yd* to find the remains of a 350 million year old mountain range which stretched from France to Czechoslovakia. The Dutch Institute of Applied Research was commissioned by the West German Ministry of Research to develop a highly perfected sound probe able to withstand temperatures of at least 300°C *572°F* and pressures of at least 1000 bars, a bar being the unit of measurement of atmospheric pressure and equivalent to 10^5 pascals and the pascal being the SI unit of pressure. The depth of 15000m *49200ft* should be reached after about ten years.

The deepest geological sounding is currently being carried out in the USSR on the Kola Peninsula where a depth of 12066m *39576ft* was reached during the first stage. The sounding was temporarily interrupted and then resumed in 1988. It is hoped to reach a depth of 14000m *45920ft* during the second stage which is due to last until 1990. The Kola sounding is the only one of its kind in the world to be carried out in crystalline rock. It has already provided invaluable information on the deeper structure of the earth and made it possible to resolve some fundamental geological problems.

mental stage, it has produced some interesting results and can be considered as an important step towards an improved method of forecasting earthquakes.

The disastrous earthquake which shook Armenia on 7 December 1988 again raised the question of the extension of the VAN sensor network.

A computer for volcanoes (1985)

Professor Kosuke Kamo from the University of Kyoto in Japan developed a computer-linked recording system capable of forecasting volcanic eruptions. The system, which is placed near the crater, can detect very slight changes in volcanic activity. It was this system which, by recording an increase of one millimetre on a crater, made it possible to forecast the eruption of a volcano on the Sakurajima Island on **3 December 1985**.

FOBES (1986)

Fred Holmuth of the Los Alamos Laboratories in America has developed an extra-sensitive optical fibre device for the detection of the very slight variations in tension which occur in the substratum prior to an earthquake. The device, the Fiber Optic Borehole Earth Strainmeter known as FOBES, is so sensitive that it can detect the slightest changes in formation of the order of one ten billionth of what is normally measured. It can therefore detect movements of half a millimetre over the distance between Los Angeles and New York. By implanting his sensors in different territories, Fred Holmuth hopes to produce a map of the movements of the substratum with a view to forecasting the quite considerable earthquakes which are well known in the area of the San Andreas Fault in California.

Lava pours down a road in Kilauea, Hawaii, burning the asphalt surface.

Power

Hydraulics

Archimedes' screw (3rd century BC)

One of the greatest scientists of antiquity, a Greek from Syracuse, **Archimedes** (287–212 BC), invented the hydraulic screw. This device is made of an inclined cylinder that encases a broad-threaded screw. It is used to raise water, to serve whatever purpose one wishes. The device is introduced into a body of water, then the screw is rapidly turned, so that the water rises from whorl to whorl.

Hydraulic ram (1796)

In **1796**, the Frenchmen **Joseph** (1740–1810) and **Etienne** (1745–99) **de Montgolfier** had the idea of utilising the kinetic energy of running water in a pipe to force a portion of the liquid mass to a higher level than its source. The energy was transferred through what they called the ram effect. The principle was to be improved by Amédée Bollée (1844–1917), another builder of hydraulic rams.

Steam engines

Connecting-rod system (14th–15th century)

One of the most important inventions of the Middle Ages was the connecting-rod system, developed towards the end of the 14th and beginning of the 15th century.

The system enabled a continuous circular movement to be converted into a rectilinear up-and-down movement, or vice versa. The absence of such a technique had, until then, seriously limited technical development. The system spread rapidly and its uses were many and varied e.g. saws then pumps and wheels. Two centuries later, it contributed to the development of the steam engine.

Steam engine

Origin (1st century AD)

Steam power dates back to ancient times when the Greek engineer and mathematician **Hero of Alexandria (1st century AD)** described an *aeolipile*, a steam operated wheel which at the time remained a project of purely technical interest without there being any thought of finding other uses for it. Steam power did not really develop until the concept of atmospheric pressure was better understood.

Atmospheric engine (1661)

The aim of the first steam engines was quite simply to put atmospheric pressure to work. It was the Italian physicist, Evangelista Torricelli who, in 1643, first discovered and demonstrated its existence. A little later, in 1654, the German physicist and mayor of Magdeburg, **Otto von Guericke**, carried out a spectacular experiment which demonstrated the force of this pressure (*see* Science). In **1661** he invented a machine which consisted of a metal cylinder inside which a piston moved up and down. Using his pump he created a partial vacuum inside the cylinder and, quite naturally, the piston plunged downwards within the cylinder, at the same time lifting a weight by means of a system of ropes and pulleys. This was an important date in history since, for the first time, air pressure was seen to be performing a function.

Denis Papin's engine (1679)

It appears that it was the French inventor **Denis Papin** who first had the idea of creating a void behind the piston by using the evaporation and condensation properties of water. In **1679**, he invented a safety valve for his steam digester, the prototype of the pressure cooker, which gave him the idea of developing a cylindrical machine in which the piston would be steam operated. In 1679, he was also responsible for the voyage of the first steam ship on the River Fulda in Germany. But he did not have the money necessary to exploit and therefore to reap the benefits of his ideas and inventions. He died in poverty, forgotten, in London in 1714.

Newcomen's engine (1712)

In the early 18th century there was a problem in the mines caused by water filling the galleries which had to be pumped out quickly. In 1698, the English engineer, Thomas Savery, registered a patent for a steam pump which was capable of extracting water from great depths. But the machine, which had no piston or valve, proved dangerous and was abandoned. However, the patent protected the rights for the use of 'power produced by fire

THE BIGGEST DAM IN THE WORLD

If it is put into practice, the biggest hydroelectric project in the world will be in China. According to a study carried out by a group of five Canadian companies, the project to construct a giant dam on the Yangtze river in China is theoretically both economically and ecologically viable. However, an environmental organisation which is also Canadian has estimated that a dam of the production capacity envisaged, i.e. 17000MW, would flood the Yangtze Valley for a distance of 600km *372.8 miles*, causing the loss of more than 40000 hectares *154.4sq miles* of agricultural land, the disappearance of about 20 towns and villages and the enforced migration of about a million people.

The steam engine freed ships from the vagaries of the wind and tides. J.M.W. Turner's painting shows the old and the new, side by side.

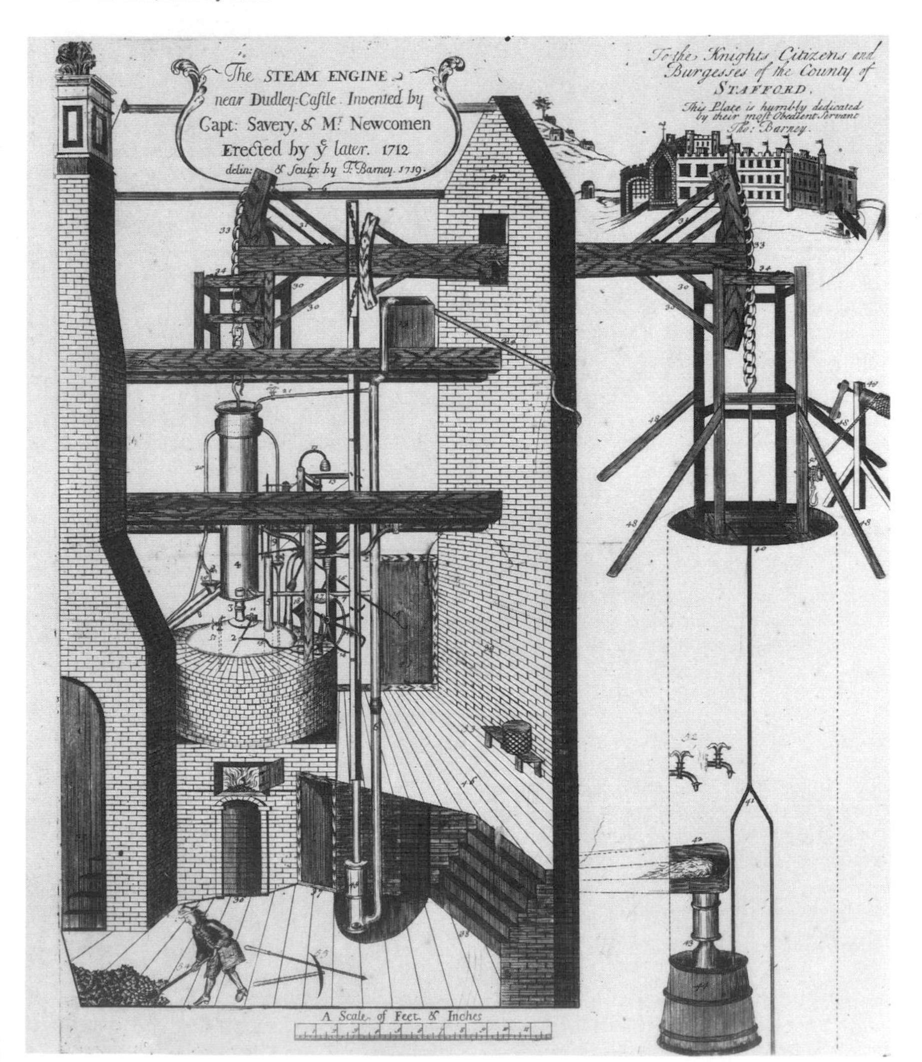

This print shows Thomas Newcomen's atmospheric engine at the colliery at Dudley Castle.

for all types of machine'. In order to circumvent the difficulty, **Thomas Newcomen** joined forces with Savery in 1705 and in **1712** constructed the first real steam engine. It was manufactured and sold in large quantities and was the prototype of the machine developed by the Scotsman James Watt at the end of the 18th century.

Watt's steam engine (1765)

James Watt (1736–1819) had the brilliant idea of converting the steam engine into an actual steam motor capable of driving the machinery of rapidly expanding industry.

In **1765** Watt constructed one of Newcomen's machines, fitted with a condenser which was essential for high efficiency, and then, in 1783, he converted Newcomen's machine into a double-effect machine.

Between 1776 and 1800 Watt joined an important Birmingham manufacturer, Matthew Boulton, and produced about five hundred engines, thus making the steam engine the true instrument of the Industrial Revolution.

In 1881, physicists gave the Scotsman's name to the international unit of power: the watt. He was also responsible for the measurement of horse-power as a result of experiments carried out in 1783.

Multiple expansion engine (1803)

In **1803** the Cornish mining engineer **Arthur Woolf** (1776–1837) patented the first multiple expansion engine. This was a machine with two or three cylinders, in which the steam expanded consecutively. Woolf's engine was used for high pressure machinery.

High-pressure engine (1805)

Around **1805** the Cornishman **Richard Trevithick** (1771–1833) perfected the first high-pressure steam engine. As a mining engineer, Trevithick was well aware of the problems of pumping water from the shafts. He realised that by increasing the pressure on the steam from an engine, he could do away with its condenser, which was cumbersome and heavy.

Trevithick is also the inventor of the first steam locomotive.

Metal ring piston (1816)

In **1816** the Englishman **John Barton** patented a metal ring piston which was subsequently widely adopted for use on cars.

Water-tube boiler (1825)

The first really efficient water-tube boilers were developed from **1825** onwards by the Cornishman **Sir Goldsworthy Gurney** (1793–1875) who wanted to manufacture steam carriages and therefore needed a lightweight boiler. For legal reasons, these inventions were unfortunately never marketed.

During the 1860s Gurney's system reappeared, first in the form of a limited circulation and then a free circulation boiler. The most widely-used examples were developed by the Englishmen Babcock and Wilcox in 1867, and the Frenchmen Belleville and Serpollet in 1886. The boilers of fossil fuel power stations are of the same type.

Francis turbine (1855)

In **1855** *Lowell Hydraulic Experiments* was published by an American of English origin, **James Francis**. In it he describes the invention of a reaction turbine intended for medium to small falls of water. Today it is the most used reaction turbine.

Pelton turbine (1870)

In **1870** the American engineer **Lester Allen Pelton** perfected a turbine which was inspired by the horizontal bucket wheel used in the mountainous regions of California. The turbine is still used for high heads of water which have a low rate of flow, and is the only action turbine in use today.

Kaplan turbine (1912)

In **1912** the Austrian **Viktor Kaplan** (1876–1934) had the idea of developing the propeller type turbine by making it possible to adjust the blade angle, although it was not produced commercially until 1924. The turbine was designed to harness the energy from low heads and wide, slow-flowing rivers, and is used by tidal power stations.

The dam (3000 BC)

The oldest dam on record was built in Egypt about 3000 BC. Initially, dams were used to create reservoirs to supply water to towns and for irrigation, but later they were used to produce energy.

It was the French engineer, François Zola, father of the author, who in 1843–59 constructed the first modern arch dam near Aix-en-Provence, but it was a technique which was not widely adopted. There are two other types of dam: the gravity dam and the coffer-dam. The first of these, after being studied mathematically, was constructed on the River Furan at St-Etienne in 1861–6, and the first big dam of the second type was the Panama Dam built in 1912.

One of the most famous dams is the Itaipu Dam in Brazil, constructed in 1982.

WATER-WHEELS

Type	Date	Origins	Characteristics
Vertical water-wheel	**Between 5th and 3rd century BC**	Middle East	An elevating but not a driving wheel. Probably the first water wheel invented.
Undershot water-wheel	**1st century BC**	Middle East	The first hydraulic engine in history.
Horizontal water-wheel	**1st–2nd century AD**	Middle East	Paddle-wheel
Overshot water-wheel	**4th–5th century AD**	Europe (Roman Empire)	Wooden or iron bowls (buckets) replace the blades. The driving wheel used throughout industry in 18th century.
Breast wheel	**12th–13th century**	Unknown	The blades are on the side of the wheel.
Short bladed undershot paddle-wheel	**1827**	France	Also called the Poncelet Wheel after its inventor Jean-Victor Poncelet (1788–1867). Forerunner of the turbine.
Angle bladed paddle-wheel	**1851**	France	Invented and patented by Sagebien, it was a development of the breast wheel. Also used as an elevating wheel.

Hydro-electric power (1869)

On **28 September 1869** the French paper-maker **Aristide Bergès** (1833–1904) was the first to convert the mechanical energy of a waterfall in the Alps into electrical energy which could operate the machines in his factory. He achieved this by the use of forced conduits.

In 1886–7 he organised an extremely dangerous operation. At a depth of 25m *82ft* below the bottom of Lake Crozet, which stands at an altitude of 1968m *6459ft*, he excavated an overflow gallery, to collect surplus water.

Electricity

Thermoelectric effect (1823)

In **1823**, the German physicist **Thomas Johann Seebeck** (1770–1831) observed that if heat was applied to the junction of two different metals which were joined in a closed circuit, it had the effect of making the needle of a compass deviate. An electric current had therefore been created within the wiring. The Seebeck or thermoelectric effect found a practical application about a century and a half later with semi-conductor techniques.

Lead–acid storage battery (1859)

In **1859** a French physicist **Gaston Planté** (1834–89) invented the first electric storage cell: the lead–acid battery.

The German physicist Johann Wilhelm Ritter (1776–1810) had observed in 1803, on a lead plate voltmeter, the principle according to which this kind of secondary cell works.

Lead-acid batteries are still the most common.

Alkaline storage batteries (1914)

Around **1914** the ingenious American inventor **Thomas Edison** (1847–1931) developed the first alkaline storage battery, so called because the electrolyte is not acid but basic: nickel–iron and nickel–cadmium storage cells.

The iron–zinc battery was developed in 1941 by the Frenchman H. André.

The energy storage efficiency of this kind of battery is two to three times superior to that of the lead battery.

Dynamo (1871)

On **17 July 1871**, the French Academy of Sciences gave an enthusiastic welcome to the Belgian **Zénobe Gramme** (1826–1901). He had been living in France since 1856 and had invented the dynamo, (short for dynamo-electric generator), which was a generator of continuous and completely reversible electric current.

With his invention, which marked the beginning of electrical technology, Gramme combined many of the discoveries and inventions made since the invention of the magneto. Although an early version of the dynamo had been constructed by the Italian Antonio Pacinotti in 1860, it had not advanced beyond the experimental stage. Gramme's dynamo was developed by the German, Friedrich von Hefner-Alteneck, in 1873.

Alternator (1878)

In **1878** the French company **Gramme**, founded by the Belgian Zénobe Gramme and the Frenchman Hippolyte Fontaine, industrially manufactured the first alternators. A German company founded by Werner von Siemens (1816–92) began producing them at the same time.

STORAGE BATTERIES IN SPACE

The French battery company, Saft, has developed new nickel–hydrogen storage batteries, chosen by CNES (the National Centre for Space Studies) and the ESA (European Space Agency). They will provide a new source of energy and will be used to feed the communications satellites and orbital stations such as *Colombus*, *Spacebus* etc.

The nickel–hydrogen storage battery is the result of many years of research in the fields of electrochemistry and metallurgy, and will enable a considerable saving of more than 50 per cent of mass energy compared with the nickel–cadmium battery used so far.

CELLS

Name	Date	Inventor
Voltaic cell	1800	The first electric cell was invented by the Italian Alessandro Volta.
Thermo electric cell	1821	Also called thermoelectric couple, invented by the German physicist Thomas Seebeck (1770–1831).
Daniell cell	1836	Named after the British chemist John Frederic Daniell, inventor of the first impolarisable cell.
Bunsen cell	1843	Developing the work of the British physicist, William Robert Grove, the German chemist, Robert Wilhelm Bunsen (1811–99) invented a more efficient version of the Daniell cell.
Bichromate cell	1870	Developed by the German physicist Henrich Ruhmkorff, it initially bore the name of the original French inventor, Grenet.
Leclanché cell	1877	Developed by the French engineer Georges Leclanché, and improved much later by the French physicist Charles Féry. Still in use today.
Dry cell	1887	Invented by the English physicist, Hellesen.
Fuel cell (Bacon cell)	1936	Invented by the American, Bacon, in 1936, it was not produced commercially until 1960 in the USA. A cell fuelled by the hydrogen/oxygen reaction is used in space missiles.

An alternator is a device that transforms mechanical energy into electrical energy; it produces an alternating current.

Transformer (1882)

The principle of the transformer, which enables the parameters of electrical current to be modified (i.e. voltage, strength), was demonstrated by the English scientist, **Michael Faraday** (1791–1867), who built a transformer in his laboratory.

The modern transformer was invented simultaneously by the French chemist and physicist **Lucien Gaulard**, the American **William Stanley** and the Englishman **John Dixon Gibbs** between **1882** and 1885.

Transmission of electricity (1882)

In 1882 the French engineer **Marcel Deprez** (1842–1918) carried out the first transmission of electricity through high voltage cables. On **25 September 1882** he transmitted a continuous electric current from Miesbach to Munich along a telegraph wire. On 6 February 1883 an amazed crowd, gathered at Porte de la Chapelle in Paris, saw an electric motor start up at the same time as an apparently unconnected dynamo situated next to it. In fact the motor was connected by an electricity line which ran to Le Bourget and back to Paris.

Magneto-hydrodynamic generator (1959)

In **1959** the **Avro Research Laboratories** in Massachusetts (USA) built the first magnetohyrodynamic generator in a purely experimental form. The generator was capable of directly transforming calorific energy into electrical energy. It was a practical demonstration of theoretical work carried out ten years earlier by the Swedish physicist Hannes Alfvén, Nobel prize winner in 1970.

Nickel–iron battery (1983)

The improvement of storage batteries was a constant preoccupation of car manufacturers. One of the lines of research was directed towards new galvanic couples, nickel–iron in particular, which had been considered by the American **Thomas Edison** towards the end of the 19th century.

Since the appearance of the first electric Nissan Micra at the end of **1983**, many prototypes have been produced by Japanese car manufacturers who are convinced of the superiority of this type of battery in terms of durability and power. But they are not as yet being mass produced. In Europe similar efforts are being made notably by Mercedes and Peugeot, the latter in collaboration with Saft, a French company specialising in different types of battery, which provided the battery for the prototype of the electric Peugeot 205.

Petroleum

Origins (ancient times)

It seems that hydrocarbons were first used in the Middle East. According to the Bible, Noah used pitch to caulk his ark and Moses' mother used it to coat her son's cradle before she set it afloat upon the Nile.

Pitch was also used by the Egyptians for preserving their mummies and by the Chinese for heating houses, cooking and making bricks as well as for lighting. In Europe, the earliest references to the use of oil are in connection with the greasing of chariot wheels and, in particular, medicine. Oil and bitumen were used in ointments for the treatment of lumbago, blows and swellings.

Distillation

The origins of distillation processes, which were often discovered by chance in the very early stages of the history of petroleum, are tied in with its first uses.

In 1556 the German mineralogist Georg Bauer, also known as Georgius Agricola, explained how to distill caulking tar from petroleum in his work *De Re Metallica*. In 1650 the distillation of crude oil brought to boiling point enabled varnishes, waxes, grease and lamp oil to be obtained. The first English edition of *De Re Metallica* was published in 1912 by Herbert Clark Hoover, who became President of the United States in 1929.

Petrol (1855)

In **1855** the American chemist **Benjamin**

Silliman carried out a series of experiments on distillation based on the studies of the composition of oil made by European chemists. As a result, he obtained a number of products: tar, lubricating oil, naphtha, paint solvents and petrol, which was used as a stain remover and considered of minor importance.

Oil lamp (1840)

Outside China, oil was not used as a source of lighting until relatively recently when a colourless liquid was extracted from crude oil which, when lit, burnt with a flame strong enough to provide light without giving off too many unpleasant fumes. This was known as burning oil. In the **1840s** in Scotland **James Young** discovered that an excellent lighting oil could be obtained from shale rocks of central Scotland.

Paraffin (kerosene) (1859)

In 1859, an American, George Henry Bissell, distilled crude oil and marketed part of the distillation under the name kerosene.

Bissell's contractor was Edwin L. Drake, known as Colonel Drake, who will be remembered for carrying out a drilling operation which went down in history. At Titusville on **27 August 1859**, when a depth of 23m *75.5ft* was reached, the oil shot skywards and started what was to be known as the black gold rush.

Continuous distillation (1873)

As the quantities of crude oil to be treated increased, it became necessary to develop processes that would enable continuous distillation to take place, i.e. in the same place and on the same site. Such processes were developed by the Russian **A. A. Tavrisov** in Baku, USSR in **1873** and by the American Samuel Van Sycle at Titusville, USA, in 1877.

In 1880–81, the Nobel brothers developed a new method of continuous distillation by placing the stills in a cascade.

The distillation column was finally perfected in 1926 by an American company, Power Speciality Company.

Cracking process (1891)

The cracking process is a refining process which enables more fuel to be obtained from a given amount of petroleum, and a certain number of heavy residues from the distillation process to be converted into lightweight products such as petrol and gas oil.

Thermal cracking, invented by the Russian **Vladimir Chukhov** in **1891**, was used industrially after the First World War.

In 1913 W. M. Burton, head of manufacturing at Standard Oil of Indiana patented a cracking process involving high temperatures and high pressure. In 1915 Jesse Dubbs improved the process, and Shell became the first licensee.

Catalytic cracking, from which the products obtained are far superior to those obtained from thermal cracking, was invented by the Frenchman, Eugène Houdry in 1930.

Catalytic reformation (1949)

Catalytic reformation, which replaced thermal reformation conceived in the 1930s, was invented by **V. Haensel** in **1949** for the Universal Oil Products Company (SA). This operation allowed the composition of petrol to be modified, specifically to increase the octane levels.

The first off-shore oil rig was patented in 1869.

Seismography (1914–18)

Seismography is a method of prospecting for oil that is carried out on the surface. It was developed by the German **D. L. Mintrop** during the **First World War**. He used it to locate Allied artillery emplacements. After the war, Mintrop used the method to study underground geological formations, and set up the first seismic exploration company in the United States.

The seismographic method consists of artificially provoking slight disturbances at ground level. The resulting waves traverse underground rock layers and some are reflected back. Waves returning to the surface are analysed to determine the make-up of various rock strata.

Seismography has been greatly advanced by the introduction of data processing. The Geophysical Analysis Group (GAG) and the Massachusetts Institute of Technology (MIT) carried out the first experiments between 1953 and 1957.

Drilling for oil (2nd century BC)

Drilling can be traced back to the Chinese who, from the **2nd century BC**, exploited petroleum by a drilling process using bamboo tubes and bronze pipes.

Drilling techniques developed with the expansion of industrialised civilisation. One of the first patents registered was for the rotary drilling method perfected in 1844 by the Englishman Robert Beart, which consisted of a boring bit with toothed rollers rotating around a central rod. The method was improved by the Frenchman Rodolphe Leschot in 1863 and then in 1887 by the patent of Chapman which describes the modern rotary method. Turbo-drilling was introduced by the Russians in 1922, and electro-drilling in 1949 by the American company Electrodrill Corporation.

Since 1986 the French company Elf Aquitaine has hoped to be able to exploit the oil in the Paris region by using 'detector vans' and horizontal drilling methods. The first exploratory drilling took place in August 1988.

Off-shore drilling for oil (1869)

On **4 May 1869 Thomas F. Rowland** filed the first patent for a fixed platform. The first working off-shore oil well was built off the California coast in 1897.

On 21 August 1928 the Italian-born American Louis Giliasso filed a patent for a submersible barge which was constructed in 1933.

The first drilling ship, the *Submarex*, was constructed by the American company Cuss in 1953. The first drilling ship capable of dynamic positioning was the *Cuss 1* by Global Marine which was deployed in 1961.

MICROWAVE DRILLING

Danish researchers at the Lundtofsletten Institute of Chemistry, north of Copenhagen, have developed a new measuring instrument to increase the profitability of drilling for oil. The instrument, which is based upon the microwave principle, can measure with unprecedented accuracy what is known as the boiling point, i.e. the point at which the oil–gas mixture reaches a level which prevents the oil from rising to the surface. An injection of gas has the effect of artificially delaying the process and enabling more oil to be extracted. The instrument could increase the profitability of many oil deposits throughout the world.

The first self-elevating mobile platform (conceived in 1869 by Samuel Lewis) was the Delong No. 1, built in 1954. As the Blue Water 1 constructed in 1962, it was the first semi-submersible platform.

Poseidon and Skuld (1992)

Two projects are currently under consideration which deal with the automatic exploitation of oil deposits lying below the sea bed, with a view to reducing as far as possible costly human presence on drilling platforms. The Poseidon project, which involves a completely submerged drilling station linked to the mainland, has been developed by the Institut Français du Pétrole (the French Petroleum Institute), the French company Total and the Norwegian company Statoil. The first Poseidon station should be operational in 1992.

The Skuld project, named after the Scandinavian goddess of the Future and of Necessity, and developed by Elf Aquitaine, Total and Statoil, is less ambitious in terms of drilling potential, approx. 100m *328ft* compared with Poseidon's 1000m *3282ft*, and should be operational by 1990. Another Skuld project, known as Super Skuld and with greater drilling potential, is also under consideration.

Well logging (1927)

The first attempt at electrical core sampling, the method by which the nature of the substrata are determined by taking electric measurements, was carried out on **5 September 1927** during a test drilling by a team from the company **Schlumberger Ltd.** under the direction of Henri Doll.

Previously, rock samples had been taken by mechanical coring, and the cylindrical shape (core) of the piece of earth or rock taken for the test drilling gave the process its name.

Electrical core drilling was so successful that, as a result, Schlumberger Ltd is today a powerful multinational company.

Subterranean storage (1916)

It was the German company **Deutsche Erdöl** that, in **1916**, patented the first process for storing hydrocarbons in a gaseous form underground.

Oil tanker (1886)

The first tanker designed specifically for the purpose was a German ship, the *Glückauf*, launched in **1886**. In 1861 barrels of oil extracted at Titusville, Pennsylvania, were transported in a sailing ship, the *Elizabeth Witts*.

Methane tanker (1959)

The first methane cargo ship, the *Methane Pioneer*, sailed out of the Gulf of Mexico for England on **31 January 1959**. Constructed for the North Thames Gas Board, it contained 2200 tonnes of liquid methane.

The firemen of the oil wells (1915)

On 17 April 1861 the Henry Rouse oil well, to the east of Oil Creek, exploded and caught fire killing 19 people and wounding ten. It took three days to put to put the fire out, by covering the well with earth.

It was around **1915** that **K. T. Kinley** had the idea of exploding a charge of dynamite to blow out the fire, a method which was adopted on an increasingly spectacular scale by his son, Myron Kinley, and later by Paul 'Red' Adair.

Engines and ignition devices

Induction coil (1841)

The induction coil was invented in **1841** by the French physicists, **Antoine Masson** and **Louis Breguet**. It is a device which makes use of the phenomenon of electromagnetic induction to produce an high voltage alternating current. It is fitted to all ignition systems in all internal combustion engines.

Glass capped spark plugs (1988)

The new 'Colourtune' spark plug, manufactured by the British company **Gunson Ltd**, has a transparent glass cap which enables better regulation of the ignition as the colour of the spark can be checked. This can be done by replacing just one of the spark plugs with a 'Colourtune' plug. A yellow spark means that the mixture is too rich and that the engine is wasting energy, whereas a blue spark, obtained by regulating the carburettor, indicates a correct mixture.

Carburettor (1893)

In **1893** the German engineer **Wilhelm Maybach** invented the modern injection carburettor. It is an indispensable part of the engine as it is within the carburettor that the fuel mixture is prepared from air and petrol vapour before being taken in by the cylinder.

The injection carburettor was used for the first time in an engine with two parallel cylinders, known as the Phoenix engine manufactured by Maybach and Daimler. It was one of the first operational petrol engines intended for use in cars, and was extremely successful.

The carburettor that existed before the development of the injection carburettor was known as the surface carburettor.

Throttle valve (1893)

The carburettor was improved the year it was invented (**1893**), by **Karl Benz**. Benz installed a throttle valve to regulate the amount of air and gas supplied to the engine. This allowed the speed and power of the engine to be adjusted.

THE TWO-STROKE ENGINE OF TOMORROW . . .

Since the models presented by certain companies e.g. DKW, Saab, Wartburg, Suzuki, just after the Second World War, the two-stroke engine has been virtually excluded from the car manufacturing industry and reserved for lightweight machinery such as mopeds, motor saws, outboard motors and lawn mowers.

But in 1973 Ralph Sarich, an engineer of Yugoslavian origin who had settled in Australia, received a prize for his invention of an ingeniously designed thermal engine, with a central piston which described a sort of orbit. With a substantial grant from the Australian government, he went on to set up a company, the Orbital Engine Company, which carried out research into the updating and adaptation of the two-stroke car engine.

The new engine, compared with its four-stroke counterpart, is half the size, three times less heavy and less expensive as the result of discarding two hundred parts of the timing gear.

Agreements have been signed with Ford and General Motors who are carrying out road tests and who will certainly not waste any time in marketing its advantages. In particular it is much less polluting compared with the old two-stroke engines and uses between 20 and 30 per cent less petrol.

PUT A MICROCHIP IN YOUR MOTOR

This 'micro motor' has a diameter of 70 micrometres *.003in* and is only a few micrometres thick, and yet it works because it is a proper motor.

It was developed by the American, Richard Muller, at the University of California, Berkeley, by applying the same techniques used in electronics to create integrated circuits. The uses envisaged relate to the field of medicine, such as the development of instruments to be used in microsurgery, mechanisms that could move through the arteries and scrape off fatty deposits, or even 'intelligent' tablets which would release, at a given moment, through microscopic valves, the exact dosage of a medication.

In industry, these motors could be used to circulate coolants through electronic circuits which, because they are becoming increasingly compact and increasingly powerful, often present problems of overheating in large computers. There should also be great opportunities for their development in the field of space research, where there is always a demand for maximum performance and smallest possible bulk and weight.

Separate combustion chamber (1890)

The first engines to have separate combustion chambers were built in **1890** by the British engineer **Herbert Ackroyd-Stuart**. This system was later refined on certain hot bulb engines.

The combustion chamber plays a very important role in all diesel engines as it ensures that an adequate fuel mixture is achieved.

Antechamber (1909)

The German engineer **Prosper L'Orange**, from the **Benz** company, invented the antechamber in **1909** and completed its development in 1919.

In a parallel development at the Swedish company Svenska Maskinverken, engineer Harry Laisner worked on a similar project from 1913.

Rotochamber (1919)

In **1919** the English engineer **Harry Ralph Ricardo** constructed his famous T-shaped combustion chamber with a cylinder head, the part of the cylinder where the gas is compressed.

At the time, many manufacturers were trying to use air turbulence to produce an improved fuel mixture. In 1904, a basic model had been developed by Stricland and in 1920, Taylor gave it a hemispherical shape.

Air reserve chamber (1926)

The air reserve chamber, another type of separate combustion chamber, was designed in **1926** by the German engineer **Franz Lang**.

Water cooled engine (1823)

In **1823** an Englishman **Samuel Brown** invented a water cooling system. In 1825 Brown founded a company that built several engines of this kind, one of which was fitted on a car and another on a boat.

Radiator (1897)

The radiator was invented by the German engineer **Wilhelm Maybach** in **1897**. After numerous attempts, Maybach perfected a honeycomb radiator. The first examples were built by the German company Daimler Motoren Gesellschaft, where Maybach worked.

Turbo engines

Turbocompressor (1905)

While building a gas turbine, a young Swiss engineer, **A. Büchi**, envisaged replacing the combustion chamber with an internal combustion engine. This would eliminate the inconvenience caused by high pressure and high temperatures.

His research led to supercharged diesel engines. In **1905** he patented the first turbocompressor, and in 1908 Buchi went to work for Sulzer Frères in Switzerland. This company brought out the first turbocharged engine in 1911. It was a four-stroke single cylinder diesel engine with pneumatic fuel injection.

ENGINES

Name	Date	Inventor
Piston engine	**1673**	The Dutch scholar Christiaan Huygens demonstrated the first piston engine before Colbert and the French Academy of Sciences.
Steam engine	**1712**	In 1712, the Englishman Thomas Newcomen built the first practical steam engine.
Closed circuit, hot air engine	**1816**	Invented by Robert Stirling of Great Britain and manufactured industrially in 1844. Used today in submarines and spacecraft.
Electric engine	**1822**	Barlow's wheel, 1822, named after its inventor, can be considered as the first electric engine. In 1838 the German physicist Moritz Hermann von Jacobi equipped a paddle steamer with an electric engine (an AC engine). In 1873 the reversibility of the dynamo (its functioning as a motor) was demonstrated at Vienne by the French engineer, Hippolyte Fontaine.

The first Harley-Davidson motorcycles had 3hp, then 4hp, single-cylinder engines.

Turbocharged diesel lorries (1953)

The Swedish company **Volvo** was the first to equip lorries with turbocharged diesel engines in **1953**.

Petrol turbo (1967)

One of the most important events in the history of the motor car in recent years is the use of the turbocompressor in private cars.

From **1967 SAAB** was the first to use turbocharging on its standard engine, producing greater power from a smaller amount of energy because of the system of supercharging applied to this type of engine.

In 1974 BMW followed suit and subsequently introduced the turbo-diesel in its 5-24 in 1983.

Turbocompressor for aircraft engines (1917)

In **1917** a French engineer **Auguste Rateau** built the first turbocharged aircraft engine. It produced 50hp and weighed 23kg *51lb* and turned at 30 000rpm.

Rocket engines

Origins

Used initially in warfare and then for fireworks, the rocket engine was re-introduced into Europe by the English general, Sir William Congreve, around **1800**. Until the 20th century its only propellant was gunpowder. The propellant or propergol is the substance which generates the energy required to propel the rocket. Rocket engines propel by reaction the device to which they are attached, by ejecting hot gases produced by a propellant carried on board the missile. The only rocket engine used today is the internal combustion rocket engine which uses chemical fuels and oxidisers: a substance which combines with another substance to produce combustion. Electric and atomic rocket engines are still at the project stage.

Solid propellant rocket engines

The first rockets were simple cylinders made from cardboard or wood and closed at one end. The principle remains the same, although from 1800 onwards the cylinder was made from metal.

During the First World War, the American Robert Hutchings Goddard (1822–1945) improved rocket efficiency by 65 per cent by increasing the speed of the escaping gases. There were subsequent improvements to efficiency. The solid propellant engines are less complicated than those using liquid propellant but are much less flexible.

Cooling of rocket engines (1933)

In **1933** the Austrian engineer **Eugen Sänger** invented a new cooling system for liquid fuel rocket engines, known as regenerative cooling. It became the most widely adopted system.

Sänger constructed several rocket engines which were used to back up the take-off of the first aeroplanes equipped with turbojet

Open circuit, hot air engine	**1851**	John Ericsson, USA. The engine was used in a liner.
Two-stroke engine without preliminary compression	**1860**	First internal combustion engine to operate successfully. Invented by the Belgian born French inventor, Etienne Lenoir.
Atmospheric engine	**1867**	Invented by the Germans, Nikolaus Otto and Eugene Langen. The most widely used internal combustion engine until the invention of the four-stroke engine in 1876.
Brayton engine	**1873**	Developed by the American, Brayton. The first engine to operate successfully with preliminary compression of the fuel mixture.
Four-stroke internal combustion engine	**1876**	In 1876 the German, Nikolaus Otto (1832–91), patented the first gas-operated four-stroke internal combustion engine. The patent constituted one of the great dates in the history of the motor car.
Double piston engine	**1878**	Invented by the German, Ferdinand Kindermann for the Hannverscher Maschinenbau company, and almost at the same time by Linford of Great Britain in 1879.
Two-stroke internal combustion engine	**1879**	Invented by the English engineer, Dugald Clerck to bypass the patent by Otto (see above). It was less efficient and more polluting but had a much higher fuel rating. It is still used for mopeds and racing bikes.
Four-stroke petrol engine	**1883**	In 1883 the Frenchman, Edouard Delamarre-Debouteville and Léon Malandin built the first petrol-fed four-stroke engine. It was experimental and was not produced commercially. On 29 January 1886, the German engineer, Carl Benz, patented the first truly efficient four-stroke petrol engine.
Synchronous motor	**1885**	These polyphase motors were a practical application of the discovery of the revolving magnetic field by the Italian physicist Galileo Ferraris. At the same time, Nikola Tesla, an American of Croatian origin, perfected the polyphase alternator.
Revolving cylinder engine	**1887**	Invented by the Frenchman, Millet. The most famous was the Gnome engine, built in 1908 by the French engineers Louis and Laurent Seguin. The amalgamation of the Gnome and Rhône companies produced a nine-cylinder engine which was the most widely used aeronautical engine of the First World War.

The radial engine, invented by Fernand Forest in 1888, has been widely used in aviation.

High-power gas engine	1888	Invented by the Frenchmen Léon Malandin and Edouard Delamarre-Debouteville, it developed 100hp. In 1900, the 1000hp barrier was crossed. These motors were mainly designed for industry.
Radial engine	1888	Also invented by the Frenchman, Fernand Forest, this 12-cylinder, four-bank parallel radial engine (i.e. three cylinders per bank) was widely used in aviation.
Four-cylinder linear engine	1889	Invented by Fernand Forest in France and Wilhelm Maybach in Germany, this engine with its four-stroke cycle is currently the most widely used engine in the European car manufacturing industry.
V engine	1889	Invented by the German engineers, Gottlieb Daimler and Wilhelm Maybach. In 1889 the engine was installed in a car designed by Maybach which is today considered as the first modern car.
Engine with pre-combustion chamber	1890	Invented by the English engineer, Herbert Ackroyd-Stuart. Forerunner of the surface-ignition engine, also known as the hot bulb engine, which was perfected in 1902 by the Swedish engineer, Rundölf.
Opposed cylinder engine	1895	The flat-twin, the first horizontally opposed cylinder engine, developed by the French industrialist Albert de Dion and manufacturer Georges Bouton. In 1896 the German Carl Benz constructed a similar engine. A famous example of this type of engine is the Volkswagen engine of just before the Second World War with its four horizontally opposed cylinders.

engines, but he soon decided to work on the development of a jet engine which would effect a take-off unaided.

Monopropellant rocket engine (1935)

In **1935** the German scientist **Helmuth Walter**, constructed the first monopropellant rocket engine. This is a liquid fuel rocket engine which uses only one propellant and for this reason is referred to as monopropellant.

In February 1937, a Walter rocket supplying 100kg *220lb* of thrust was attached to a German Heinkel Kadett plane as a take-off engine. This was the first rocket-assisted take-off in the history of aviation.

The V2 rocket engine (1942)

Rocket-engine research in Hitler's Germany was carried out under the guidance of Hermann Oberth. Later it was directed by Oberth's student Wernher von Braun (1912–1977). Work was first done at the military base of Kummersdorf, then transferred to the Peenemunde base in 1937.

The infamous V2, which used liquid fuel, made its first flight on **13 June 1942**. Designed by **Wernher von Braun**, it was equipped with a rocket engine built by W. Thiel. (Thiel was killed in 1943 during a bombing raid on the Peenemunde base.)

The engine on the V2s that terrorised London in 1944 and 1945 was the last in a series of rocket engines designed initially to power the V1. The rocket attained a velocity of 760m *2493ft* per second. Escape velocity for leaving the earth's gravitational field (for example, in order to put a satellite into orbit) is 11m *7 miles* per second, or 15 times faster than the V2.

After the war, the V2 engine served as a model for both the Americans and the Russians, and von Braun emigrated to the United States to contribute to its rocket programme.

Jet engines

As well as rocket engines, there is another type of engine, the jet engine, which uses air as an oxidiser and which is more correctly referred to as an air breathing jet engine.

Ramjet engine (1913)

In **1913** the French engineer **Lorin** patented the first ramjet engine but it was never produced commercially. After the First World

JET REACHES TEN TIMES THE TEMPERATURE OF THE SUN

For a few seconds, a record temperature of one hundred million degrees Celsius i.e. ten times the temperature of the sun, was reached by a research team working at Joint European Torus (JET) at Culham in Great Britain, in October 1988.

Diesel engine	1893/97	In 1893, the German engineer, Rudolf Diesel (1858–1913) constructed the first prototype of the engine named after him. This four-stroke engine was mass produced from 1897 onwards. The two-stroke diesel engine, perfected by the German engineer Hugo Güldner, is currently the most powerful internal combustion piston engine. In 1986, eight Japanese car manufacturers set up the Clean Diesel Laboratory, an institute for research into the diesel engine, which aims to develop, within six years, a new and less polluting type of diesel engine.
Rotary engine	1956	Perfected by the German engineer, Felix Wankel and constructed by NSU. It is used by the Japanese company Mazda.
Ceramic engine	1995?	One of the formulae for the future. The low thermic conductibility of this material enables considerable savings in energy to be made. Many experiments are being carried out in Japan by Nissan and Isuzu, in France by Peugeot and in Germany by Porsche. In 1985, Nissan began marketing a ceramic turbo-compressor

War, the project was continued by another French engineer, René Leduc, who constructed the first prototype in 1936. It was not until 1949 that a machine propelled by a ramjet engine made its maiden flight. In spite of some impressive performances the ramjet engine was replaced by the turbojet engine for various technical reasons such as efficiency, ease of adjustment etc. But as a result of recent research by seven engineers from ONERA, the French National Office for Space Study and Research, and Aérospatiale, the ramjet engine has been revived and may well propel the space-craft of the future.

Turbojet engine (1930)

It was the Englishman **Frank Whittle** who first attempted to construct a turbojet engine between 1928 and 1930.

The turbojet engine is a jet engine: the exhaust gases, produced by reaction, create the thrust of the engine and consequently propel the machine containing the engine.

The first British plane equipped with a turbojet engine was ready to fly on 15 May 1941. The turbojet engine of the Gloster Whittle E28 used a turbocompressor and produced a thrust of 375kg *827lb*. In Germany, Hans Pabst von Ohain started similar research in 1936 and a Heinkel fighter plane made its maiden flight in 1939.

Concorde is equipped with Olympus turbojet engines, constructed by the English company, Bristol Siddeley.

Turbofan engine (1940)

In **1940** the American company **Metropolitan Vickers** developed a turbofan engine. This

GAS AND STEAM TURBINES

Type	Date	Inventor	Origins	Characteristics
Gas turbine	1791	John Barber	GB	Remained at the project stage
Steam turbine	1884	Sir Charles Algernon Parsons	GB	Steam, reaction turbine. First turbine built industrially.
Steam action turbine	1889	Carl Gustaf de Laval	Sweden	Low power, single-stage turbine.
Steam turbine with pressure stages	1879–1900	Auguste Rateau	France	Enables slower rotational speeds and higher effciency.
Internal combustion gas turbine	1903	Armengaud	France	First internal combustion gas turbine to be built. Preceded by patents of Peer (USA, 1890) and Hordenfeldt and Christophe (France, 1894).
Combustion/gas turbine	1904/5	Stolze	Germany	Same turbine invented in France by Armengaud and Lemale.
Internal combustion gas turbine with preliminary compression	1909/10	Hans Holzwarth	Germany	The first gas turbine to be used in industry.
Closed cycle gas turbine	1940	Prof Ackeret Dr Keller	Switzerland	Developed by the Swiss company Escher-Wyss under the name of the Escher-Wyss aerodynamic turbine.

engine is intermediate between a turboprop and a turbojet. It has two advantages over these two other engine types: in-flight fuel consumption is 20 per cent less, and there is less noise from the exhaust.

The turbofan engine has been widely used. The American firm Pratt & Whitney has built turbojets of this type for the Boeing 707, Boeing 720, Boeing 727, the Caravelle, and the Douglas DC-8. General Electric and Rolls-Royce have produced turbofan engines as well.

Turbojet with afterburner (1945)

In **1945**, the British company **Rolls-Royce** constructed the first turbojet with an afterburner. Afterburning allows greater thrust to be obtained without substantial increase in engine weight. This increased thrust is particularly useful in take-off.

The biggest drawback of afterburners is the very high fuel consumption. That is why they are only used very briefly, at take-off, and to obtain peak speeds with fighter aircraft.

Pulse-jet engine (1940)

In **1940** the principle of the pulse-jet was discovered by the German **Paul Schmidt**. He was trying to develop a ramjet that could start up under its own power. The pulse-jet was immediately put to use on the German flying bomb, the V1.

A JET PROPELLER: THE PROPFAN

After ten years' research carried out under the auspices of NASA, the 'propfan' project was disclosed in 1985. It combines a contra-rotating propellor and turbine which produces a very considerable saving on fuel.

The first demonstration flight of the turbojet engine with a jet propeller, took place on 20 August 1986 in a Boeing 727. However, despite very satisfactory results, Boeing, the world's leading aircraft constructor, seemed to want to shelve the project. It was McDonnell Douglas who continued the trials with the MD-80, resuming flights on 18 May 1987 in the United States, and taking it to the Farnborough Airshow in England in the autumn of 1988.

Today, more than 160 flights have been carried out with machines equipped with propfan, of which 140 were on the MD-80.

The aim is to produce standard aircraft. Currently being studied are the MD-91, which will carry 114 passengers, is powered by two General Electric GE36 engines and has a fuel consumption estimated at 40 per cent less than that of a conventional aircraft, and the MD-92.

Lockheed tests an eight-blade propfan for NASA.

IGNITION SYSTEMS

System	Date	Inventor	Origins	Characteristics
Flame transfer	**1836**	William Barnett	GB	Enables the pre-compressed fuel mixture to be ignited at regular intervals. Replaced in 1900 by electric ignition.
Ignition tube	**1855**	Alfred Drake	USA	Although used widely between 1880 and 1905, it lacked precision at moment of ignition. Not usable for vehicles with variable speeds.
Magneto-electric generator	**1880**	Giesenberg	Germany	Low tension magneto. Developed in France in 1882 by Fernand Forest.
Electric ignition by battery and spark coil	**1883** **1883**	Etienne Lenoir Karl Benz	France Germany	The most widely used system in internal combustion engines.
High voltage magneto	**1902**	Gottlieb Honold	Germany	The first high voltage magnetos were constructed by Robert Bosch, specialists in the manufacture of electrical equipment for engines. After 1925 the system was replaced by ignition by storage battery and spark coil.
Starter motor	**1931**	Maurice Gondard	France	The first starter motor for cars by the creator of the Solex carburettor.

Nuclear energy

Atomic pile (1942)

The first atomic pile was constructed, below the football stands at the University of Chicago, under the direction of Italian-born physicist **Enrico Fermi** in December **1942**.

The energy produced by atomic fission is given off in the form of heat. This heat is then recovered and transformed, for example, into electrical energy. During fission neutrons are released as well as heat, and these in turn induce more fission. However, in an atomic pile this reaction can be controlled.

Mox (1970)

This is a fuel mixture of uranium and plutonium which was first used in the 1970s. Mox (Mixed Oxide) comes from the reprocessing of natural and enriched uranium based fuels burnt in nuclear power stations.

Mox provides a solution in terms of a recycling process for nuclear waste. It still contains 0.8 per cent of uranium 235 (more than uranium in its natural state), and is all the more interesting because after spending three years in a nuclear reactor, the waste has been converted into a fuel and can be reprocessed and re-used. Experiments have been carried out in particular in the nuclear power stations of West Germany.

Isotopic separation (1922)

In **1922** the English physicist **William Francis Aston** carried out the first isotopic separation in a laboratory. In order to achieve this, he had used a mass spectrograph which sorted the atoms by using a magnetic field. Isotopic separation involves isolating a particular isotope from a given substance. In fact, a substance never naturally presents itself in isolation, but is always accompanied, usually in very small proportions, by substances with the same atomic number but containing different numbers of neutrons.

Nuclear moderator (1942)

In **1942 Enrico Fermi**, Italian physicist and Nobel prize-winner in 1938, used graphite as a moderator in the first 'atomic pile', or nuclear reactor, in Chicago. The probability of fission occurring depends on the energy of the neutron which collides with the fissile nucleus. For the main fissile nuclei, uranium 235 and plutonium 239, the probability increases when the extremely high kinetic energy of the neutron decreases. In order to moderate this energy, substances known as moderators are used, which are composed of lightweight nuclei which, on impact with the neutrons, dissipate their initial energy and slow them down, without capturing them too often.

The only moderators which can be used are carbon, in the form of industrial graphite; hydrogen in water or certain hydrocarbons; deuterium in heavy water and pure or oxydised beryllium.

Nuclear power stations (1951)

The first nuclear power station to produce electricity was the **EBR-1** in America, which created 300kW, opened in **1951**. The first civilian nuclear power station was opened in June 1954 at Obninsk in the USSR when a 5000kW (5MW) production reactor was started up. Prior to this, the United States had developed a nuclear powered engine for military purposes to be used in a submarine and which had an equivalent capacity. But as it was a prototype, and a military project, this first atomic power station did not receive international approval. Britain's first electricity-producing power station was Calder Hall (Unit 1), Cumbria, opened in 1956.

Like all machines, nuclear power stations have a limited life span. It is estimated that after 20 to 40 years' service, the installations of the reactor are worn out and become dangerous. Already 135 installations throughout the world have been shut down.

At the end of 1987, the most productive nuclear power stations in the world were three West German stations which each had an annual production of about 10 billion kW/h.

The smallest nuclear power station in the world has a diameter of 2m *6.5ft* and is 2m *6.5ft* high. It weighs 6 tonnes and is capable of producing 20kW/h of electricity for 20 years. It was developed in the United States, in the laboratories of Los Alamos in New Mexico, and is intended to be used as a generator in areas where access is difficult.

AN AIRCRAFT TO PREVENT NUCLEAR DISASTERS

As a result of the Chernobyl disaster in April 1986, the Netherlands has equipped itself with a special plane with a view to ensuring national safety. It is capable of measuring radioactivity in the air and can be mobilised immediately in the event of a major nuclear accident in one of Europe's nuclear power stations.

With authorisation, the plane can fly over the countries of Eastern Europe, while special agreement allows it to fly over Western European countries, immediately and without authorisation, to the scene of a potential nuclear disaster.

Nuclear reactor (1951)

The various types of reactor and the different materials which could be used were developed between 1940 and 1945, but the industrial development, which began in 1951 in the United States, required detailed preliminary study.

The largest reactor in the world is the 1450MW Ignalina station in Lithuania, USSR.

It is interesting to note that in Europe in 1985, electricity supplied by nuclear powered generators stood at 31 per cent. Today the production of nuclear electricity

Calder Hall was opened in 1956 by HM the Queen and marked the beginning of the commercial generation of Magnox nuclear power stations.

The European tokamak, JET, will be replaced in 1992 by the NET.

THERMONUCLEAR REACTION . . . IN A TEST TUBE

A discovery has been made which, if it proves to be true, will be crucial to the future of energy.

In March 1989, two scientists, Martin Fleischmann from Southampton University in England, and Stan Pons from Utah University in the United States, claimed to have obtained the reaction of nuclear fusion in an ordinary test tube. Such a system would be capable of replacing experimental reactors such as JET. It is not hard to imagine the scepticism of scientific circles. All the 'big names' tried to repeat the experiment: Massachusetts Institute of Technology (MIT), European Centre for Nuclear Research (CERN), Atomic Energy Commission (CEA) and so on, so far without success.

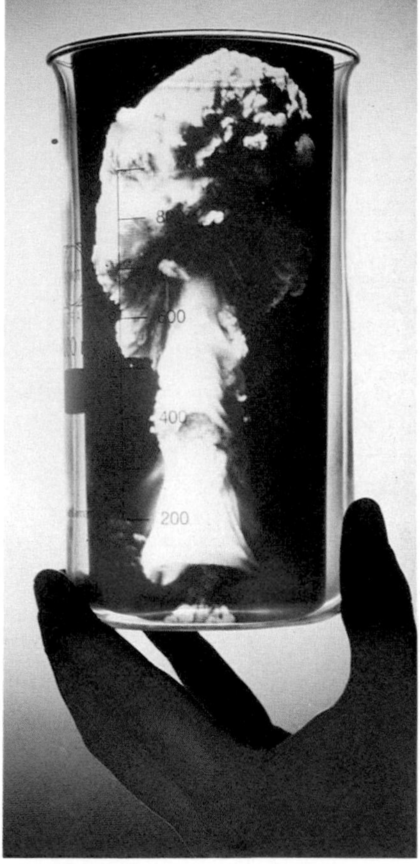

Experiments carried out by Martin Fleischmann and Stan Pons on thermonuclear fusion remain highly controversial.

has reached 71 per cent in France, 60 per cent in Belgium and 31 per cent in Germany and just 17 per cent in Britain; within the European Community, the production of nuclear powered electricity has risen to 456 billion kW/h.

Breeder reactor (1959)

The first breeder reactor was commissioned at **Dounreay** in Scotland in **1959**. A breeder reactor is a reactor which produces more nuclear fuel than it consumes, so that the depletion of the world's stock of fissile uranium (i.e. seven per cent of natural uranium), does not present a threat.

In 1955 France started to consider the development of such a system and, in 1975, the Phénix (Phoenix) breeder reactor (250MW) came into service at Marcoule on the river Rhône. The reactor was named after the bird which, according to Greek mythology, was able to rise repeatedly from its ashes. The first breeder reactor came into service in the United States in 1963, in the USSR in 1968, and in West Germany in 1977.

Thermonuclear reaction

Thermonuclear fusion

The great hope for the energy of the future is thermonuclear fusion of hydrogen, with a view to reproducing on earth the reactions which take place in space and which result in the fusion of hydrogen atoms.

Research is directed mainly towards the possibility of recuperating energy produced in this way, and building electricity generating stations around the fusion reactors.

The Bethe Cycle (1938)

In **1938** the German-born American physicist **Hans Albrecht Bethe** discovered the nuclear transformation cycle, named after him, which explains the energy of the sun and the stars.

Lawson's criterion (1957)

In **1957** the British physicist **John Lawson** stated the conditions which must be fulfilled by a plasma of deuterium or tritium before the phenomenon of nuclear fusion can take place, and be maintained without an external supply of energy.

The Lawson criterion defines the relation that must be reached between the density of the plasma and the length of time of its confinement, the temperature having to reach 100 million K before the reaction can take place.

Today, plasma physicists consider that the construction of fusion reactors has a good chance of success.

Magnetic confinement (c.1949)

Research on thermonuclear reaction controlled by magnetic confinement began independently in the United States, the USSR and Western Europe towards the end of the **1940s**. In the early stages the research was carried out secretly, and some definitive work on plasma physics was done at the Kuratchov Institute in Moscow. For the first time, in 1968, physicists at the Institute brought a plasma to a temperature of 12 million K in a machine called a tokamak.

Tokamaks (1963)

The first tokamak (an abbreviation of the Russian name *Toroidal Kamera Magnetic*, a machine for magnetic confinement which enables plasmas to be studied while in a state of fusion) was invented by the Russian physicist **Lev Andreevitch Artsimovitch** and first used in **1963**.

In the 1970s tokamaks were constructed in most laboratories. In 1978, the Princeton tokamak in the USA reached a temperature of 70 million K, but was still a long way from the Lawson criterion.

In 1983, the Alcato C tokamak at the

Massachusetts Institute of Technology (MIT), USA, exceeded the temperature threshold of the Lawson criterion but only by 15 million K. Today, huge tokamaks like the JET (Joint European Torus), installed at Culham in Great Britain in 1983, are being built in European laboratories. The aim is to exceed both the threshold of the Lawson criterion and a temperature of 100 million K in order to demonstrate the feasibility of controlled nuclear fusion.

NET (1992)

The programme of the Joint European Torus (JET) is to be continued until **1992** when the Next European Torus (**NET**) will take over. By the beginning of the next century, NET should have developed a European fusion reactor.

ITER (2015)

The European Community, the United States, Japan and the USSR, which brought its Tokamak-15 into service in 1988, are going to co-operate between now and the end of 1990 in the development of a pilot study for an International Thermonuclear Experimental Reactor (**ITER**). Their aim is to demonstrate the 'technological feasibility' of thermonuclear reaction which should enable a non-polluting and virtually inexhaustible source of energy to be developed commercially by around **2015–25**.

The Americans have launched a major programme of wind-driven generators in response to shrinking oil reserves.

New sources of energy

Origins

The history of the human race has been characterised by the search for sources of energy. In the 1970s the oil crisis revived this preoccupation and research was directed towards new sources of energy or new ways of developing old forms of energy. By the end of 1988 an entire programme was proposed within the framework of the European Community to promote renewable sources of energy such as solar, biomass etc.

Windmill (10th century)

The earliest recorded windmills are those used in Iran during the second half of the 7th century, but the idea did not reach Europe until the **10th century**. A windmill at Weedley, near Hull, Humberside, is probably the earliest British example, dating from 1185.

Windmill is the generic term for any mechanism used to harness the kinetic energy of the wind in order to operate a machine, in particular a millstone.

Wind engines (1876)

A modern version of the traditional windmill, the wind engine or aerogenerator, converts the kinetic energy of the wind into mechanical energy and, more precisely in the case of aerogenerators, into electrical energy. Wind engines were first mentioned during the second half of the 19th century. Since then, far from becoming obsolete, the use of wind power energy has continued to develop, particularly in countries which have very windy regions. For example, at the end of 1986, Holland began to construct the largest fleet of wind engines in Europe: 25 windmills, 73m *240ft* high, which can produce enough energy to supply the needs of a town with a population of 10 000.

In the United States, more than 300 000 aerogenerators provide 300 million kWh every year. This is the country which in the 1970s built the most powerful wind engine: it produces more than 2MW.

Hydrogen (1766)

Hydrogen, the most plentiful substance in the universe, could provide the basis of energy in the 21st century.

Discovered in **1766** by the English physicist and chemist **Henry Cavendish**, hydrogen reacts violently when it comes into contact with certain substances, for example chlorine and oxygen. This property makes it a much more energetic fuel than petroleum.

Hydrogen does not exist in a free state on earth and exists in only very small quantities in the atmosphere. However, it is found in many substances, the most common of which is water.

The future of hydrogen as a fuel depends on finding the solutions to certain questions including how to produce large quantities as economically as possible, and transporting and storing it safely.

At present the main problem is that of producing large quantities of hydrogen from water and solar energy.

Biomass

Biomass is the oldest source of energy used by man. With the exception of solar energy, until the 18th century it was our main source of warmth.

The present forms of its development are extremely varied, the most developed being biogas and biofuels.

Ethanol (antiquity)

This is the technical name for ethyl alcohol often known simply as alcohol. The method for obtaining it dates back to antiquity, but its use in industry to produce energy has developed noticeably since 1973.

Ethanol is obtained by fermentation and then refined by distillation. The raw materials are often sugars or starches of vegetable origin.

Methane (1776)

This gas, discovered in **1776** by the Italian **Count Alessandro Volta** (1745–1827), is produced during the decomposition of organic matter by fermentation. The first use of biomethane dates back to 1857 when a methane plant was built in a leper colony near Bombay in India.

Methane is particularly used in industrial and urban heating. It is currently being tested to decide its suitability as a car fuel.

Methanol (1661)

Methanol or methyl alcohol was discovered in **1661** by the Anglo-Irish philosopher and chemist, the **Honourable Robert Boyle**, among the distillation products of wood. In 1812 Taylor established a connection with alcohol, but it was the two French chemists, Dumas and Péligot, who determined its composition in 1835. Like ethanol, methanol is a liquid fuel.

Experiments are currently under way to examine the possibility of using wood alcohol to replace lead in petrol.

Geothermal energy (1818)

Geothermal energy is an inexhaustible source of energy produced by the earth's heat. The oldest geothermal installation is in Larderello in Tuscany where in **1818** the Frenchman **F. de Larderel**, after whom the village is named, exploited the jets of natural steam, *soffioni*, given off from the ground. Today the *soffioni* of Larderello feed a group of power stations which produces 250 MW.

There is a distinction between low energy geothermics which is sufficient to heat houses etc, and high energy geothermics which produces mechanical energy that can then be converted into electric energy.

Synfuel of New Zealand was the first company to transform methanol directly into petrol.

In terms of high energy geothermics, more than 2500 MW are currently being produced throughout the world, particularly in North and South America i.e. the United States, Chile and Mexico, and in New Zealand. Japan is only just beginning to exploit its vast resources.

Heat pump (1927)

A heat pump is a device that recovers heat from any free source such as air, ground water etc. It converts mechanical energy into thermal energy. It transfers the heat accumulated through use of a conventional system of radiators and heat pipes. The only energy expended is that needed to run the compressor for transferring the heat. More energy is recovered than is consumed, which makes this a particularly attractive process.

Credit for having thought of the process goes to Lord Kelvin. In 1852 he had already set down the principles of a thermodynamic machine that produced heat as well as cold. Nevertheless it was 75 years before the first heat pump was built by **T. G. N. Haldane** in 1927. He used it to heat his office in London, and his house in Scotland.

Solar energy

Origins (1878)

As early as **1878** a Frenchman **Augustin Mouchot** presented a small solar generator at the Paris Exposition. The solar plant furnished enough energy to operate a steam engine. However, this pioneering work was not followed up until similar research was undertaken after 1945. Later the energy crisis in the 1970s prompted the governments of principal industrial nations actively to encourage the work.

Photovoltaic cell

Origins (1839)

In **1839** the French physicist, **Antoine Becquerel** (grandfather of Henri Becquerel who discovered radioactivity), constructed the first photovoltaic cells. However, they were never developed during his lifetime.

Development (1954)

It was the development of space research during the early 1960s that brought about the manufacture of photoelectric cells. In **1954** American scientists at the Bell Laboratories, **G. L. Pearson**, **C. S. Fuller** and **D. M. Chaplin**, developed a solar battery which consisted of tiny silicon cells whose transformation capacity, although tripled within a few years (6 per cent in 1955 and 18 per cent in 1978), gave a relatively low rate of efficiency for the high costs of production. However, in the last ten years or so, work carried out on lowering production costs has enabled the system to be used to supplement weak spots in the world's electricity network such as at isolated telecommunication relay stations, in lighting and pumping water in arid and sunny places, etc.

Amorton technology (1987)

The Japanese company **Sanyo** has conceived a photovoltaic cell using the amorphous i.e. non-crystalline properties of silicon, and have thereby developed a complete technology, known as Amorton technology, which enables solar energy to be used to an extremely high level of efficiency.

Solar furnace (3rd century BC)

The idea of the solar furnace can be traced back to the Greek scholar Archimedes (278–212BC). While defending Syracuse, he armed the soldiers with concave shields which, by concentrating the sun's rays, made the Roman troops under the consul Marcellus believe that they were facing 'soldiers of fire.'

The French chemist, Antoine Laurent Lavoisier (1743–95), father of modern chemistry, was the true inventor of the solar furnace. He concentrated the sun's rays in order to achieve the combustion of a diamond in an atmosphere containing oxygen without the use of fuel.

It was not until 1946 that an experimental plant was constructed at Meudon in France where the first large scale solar furnace enabled a temperature of 3000°C *5432°F* to be reached.

Solar collector (17th century)

Modern solar collectors use the greenhouse effect discovered in the 17th and 18th centuries: the sun's energy passes through glass warming the air, and is trapped there.

It was during the 17th century that greenhouses first appeared in France where the *Jardin du Roi* (the King's Garden), now known as the *Jardin de Plantes* (the Plant Garden),

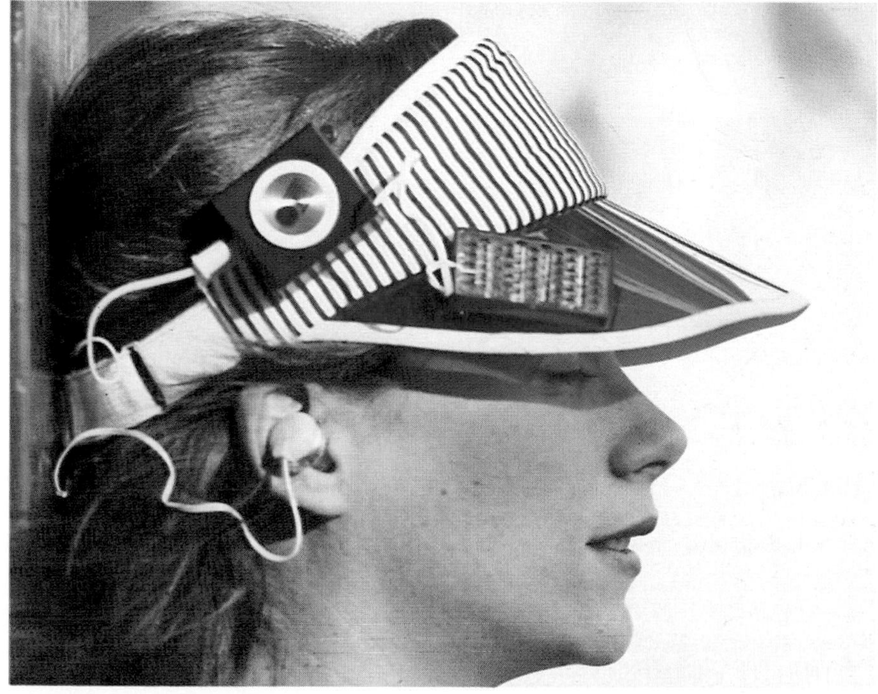

This radio is solar powered.

SUN POWER THE THERMOCHEMICAL CELL

This new cell with its extremely high accumulative capacity was developed by scientists at the Max Planck Institute for Coal Research in West Germany and launched in 1988 by the company Bomin-Solar. It is capable of transforming between 70 and 80 per cent of the solar energy collected into usable forms of energy, electricity in particular, at a cost which is substantially lower than that of existing systems such as the photovoltaic system.

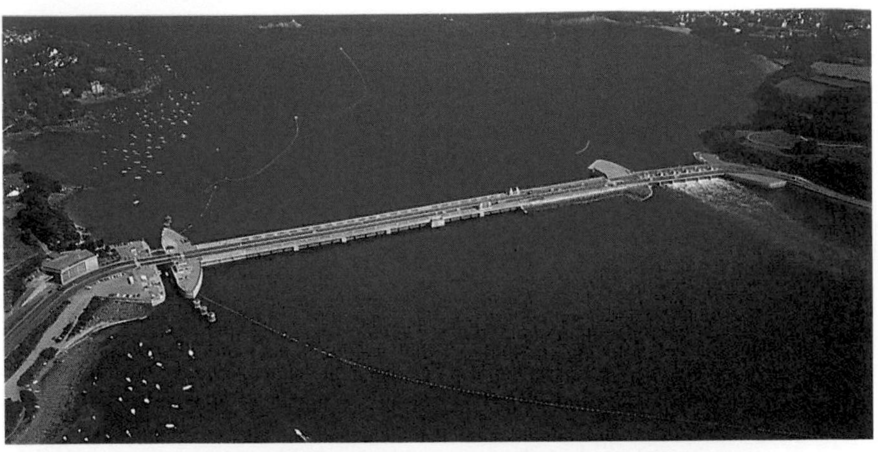

The tidal power station on the river Rance in Brittany, northern France, was the first of its kind.

was created in Paris in 1653. In England, they were constructed on a large scale from the end of the 18th century.

During the 1920s, a large number of solar collectors were constructed in California and then abandoned in favour of petroleum. The energy crisis of 1973 revived interest in them as a source of energy.

Solar thermal power plant (1960)

The first solar thermal power plant was constructed in Ashkabad, capital of the Soviet Republic of Turkmenistan, in **1960**.

The most powerful plant is Solar 1 at Barstow in California (USA) which produces 10MW and began functioning in 1982.

The oceans as a source of energy

Wave energy (1875)

The idea of using wave energy seems to have been considered scientifically for the first time in **1875** by the Australian **R. S. Deverell**.

In 1889 a very basic form of installation known as Ocean Grove was operating off the coast of New York. In 1980, the Japanese Foundation for Naval Construction started to build a prototype power station, which they called Kaiyo, off the coast of the island of Iriomote.

A power station using wave energy (1985)

In February **1985** the Norwegians started to build, to the west of the port of Bergen, the first power station in the world to use wave energy to produce electricity. It will produce enough electricity to supply 70 average-sized homes in Norway i.e. 1.5 million kW/h. The Norwegian government is also investing in another project which consists of using wave energy to force large quantities of water into a funnel so that it can be used by a standard hydraulic power station.

Tidal power station (1966)

In 1955 the French engineer Albert Caquot laid the foundations of a project for a tidal power station in the bay of Granville, and incorporating the bay of Mont-Saint-Michel. The idea was not a new one. A scientific commission had already met in 1919 to discuss the basic operational principles of an electricity power station using tidal energy. In 1943, the technical committee considering the sea as a source of energy recommended that the Mont-Saint-Michel scheme should be abandoned. The project was considered far too ambitious and a decision was taken to begin with a smaller project: the tidal power station on the estuary of the river Rance, opened on **26 November 1966**, which has an output of 240MW.

At the moment there are only three tidal power stations in the world: in France, the USSR and Canada where the Bay of Fundy in Nova Scotia has record tides of 20m *65.6ft*. But new projects are being considered, in particular the project for the biggest estuary dam in the world producing tidal energy, to be constructed on the river Mersey, near Liverpool, UK. If the project proves viable, by 1995 the dam could be providing 0.5 per cent of the electricity used by England and Wales. Another British project, on the river Severn, could provide 5 per cent of the electricity required by Great Britain.

Industry

Architecture and construction

Origins

It was in the Neolithic period (between the 9th and 4th centuries BC) that architecture really began. Curiously, this art sprang to life in a part of the world lacking those classic building materials, wood and stone. In the Near East the clayey earth was used to build the first walls, either in the form of puddled clay or unfired bricks, around the 7th century BC.

Two series of inventions were to play a decisive role in the development of architecture: iron tools, whose use became generalised from the 10th century BC; machines such as the pulley (with the block and tackle) and the winch, of which even the most modern machines are only sophisticated examples.

The lever dates back to prehistoric times. The pyramids could not have been constructed without the lever, the mathematical principles of which were described by Archimedes in the 3rd century BC.

The first architect (2800 BC)

The first architect whose name we know is the Egyptian **Imhotep**. It is to him that we owe the world's first large scale monument constructed entirely in stone: the Step Pyramid at Saqqâra. It was built for the Pharaoh Zoser around **2800 BC**.

Winches, pulleys and cranes

Medieval builders used winches and capstans which were probably very similar to those used by the Romans. The latter introduced many mechanical systems to their building sites such as hydraulic and wind powered cranes and hoists.

The origin of the crane is still quite mysterious, except for a three-pulley crane system which is attributed to Archimedes (287–212 BC).

The construction of the Tower of Babel as imagined in 1470.

An ancient device – the crane – used in the construction of the ultra-modern Arche de la Défense in Paris.

The pulley, together with the crank and the winch, is first mentioned in *Mechanica* a work of the Aristotelian school (4th century BC). The invention of the pulley and of the tackle block is attributed to Archytas of Tarentum (430–360 BC) who was an army general, philosopher, mathematician, creator of automata and an innovator in the field of mechanics.

The level (1573)

The water level is described for the first time in the works of a Polish geometrician, **Strumienski**, and would therefore date from **1573**.

The spirit level is the invention of the Frenchman Melchisedech Thévenot (1620–92), an unconventional character, great traveller and keen researcher in the fields of science and technology. It was in his correspondence with the Dutch astronomer Christiaan Huygens between 1661 and 1662 that the first reference to his spirit level can be found. But Thévenot would only publish a description of his instrument in 1681.

Aqueduct (antiquity)

The aqueduct was born out of necessity in the Middle East and the Mediterranean basin. Cultivated land and entire towns were irrigated by artificial channels that crossed bridges and ran under ground. In 703 BC the Assyrian king Sennacherib had a 50km *31 mile* long aqueduct built to supply Nineveh. In Egypt, the canals leading from the Nile were often so wide that boats could navigate them.

The arched aqueduct, such as the Pont-du-Gard at Nîmes (AD 23), was developed mainly in Rome and throughout the Roman empire.

The longest aqueduct today is the Water Project in California, which was built in 1974 and is 1329km *825 miles* long.

Bridge (Neolithic Age)

In the late Stone Age tree trunks and piles of stone were used to construct simple bridges. The idea of building stone arches began with the Sumerians around 3500 BC. The Romans were masters of constructing arches notably thanks to their discovery of natural cement.

In China at the beginning of the 7th century the engineer Li-Chun built the first arched bridge in stone at Zhaoxian, in the province of Hebei.

Suspension bridge (6th century)

The Chinese were building suspension bridges in the 6th century, although they

A BIRTHDAY TUNNEL

Isambard Kingdom Brunel, the son of a Frenchman who fled to England during the French revolution, built the London to Bristol railway tunnel in 1825. This 2880m *9448ft* long tunnel is perfectly straight and oriented in such a way that on Brunel's birthday, 9 April, the rays of the rising sun shine through from the east to the west. Besides this peculiarity, the Box tunnel is still regarded as one of the best examples of its type.

existed in the form of draw-bridges and foot-bridges made from creepers before that.

If one excepts a chain foot-bridge built over the Tees in 1741, the first metal suspension bridge was constructed on the Merrimac river in Massachusetts in 1809. Currently the largest bridges are the Verrazano bridge at the entrance to the port of New York, which spans 1420m *4658ft* (1964), and the Humber bridge (1981) whose central bay of 1542m *5059ft* is the longest in the world. But numerous projects are either in progress or being planned: a 13.1km *8.1 miles* long bridge (completed in 1988) links the Shikoku and Honshu islands in south-east Japan. The bridge is in fact 11 bridges that 'jump' from islet to islet. A 3300m *10826ft* long suspension bridge between Sicily and the Italian mainland should be completed by 1995. A 4000m *13123ft* long bridge is due to be completed between Le Havre and Honfleur in France by 1992.

Primitive suspension bridges have been used since the Stone Age.

First metal bridge (1773)

The first metal bridge was built between **1773** and 1779 by the British iron-founder **Abraham Darby**. The Iron Bridge at Coalbrookdale was made of cast-iron and its central arch spanned 30m *98ft.*

Tunnel (1826)

The first railway tunnel was built in England in **1826** on the Liverpool to Manchester railway.

The Mont-Cenis tunnel, the first tunnel through the Alps, was begun in 1857 under the supervision of Germain Sommeiller. During the digging he invented the first pneumatic pick, which allowed the boring of the tunnel to be greatly speeded up.

One of the most famous tunnels in Europe, and one of the most important technical achievements of the 19th century, is the St-Gotthard tunnel which links Germany to Italy via Switzerland. Its construction began in 1872, lasted ten years and was carried out under very difficult conditions.

A new transalpine railway tunnel is under construction: the work, financed by Switzerland, should begin in 1994 and, according to which plan is adopted, will last 14 to 17 years.

The world's longest tunnel (1988)

This railway tunnel, which was officially opened in March 1988, links the main Honshu island to Hokaido via the Tsugaru strait in northern Japan at 240m *787ft* below sea level.

The 53.9km *33 mile* long tunnel was bored through volcanic rock – a technological achievement that has required 25 years to complete, cost an astronomical sum of money and has claimed several lives.

But now, instead of a four-hour crossing which was often made dangerous by storms and typhoons, today's passengers travel at 110km/h *68mph* and complete their journey in just two hours.

Sky-scraper (1885)

The sky-scraper was invented in **1885** in Chicago by the engineer and architect **William Le Baron Jenney**. The upward expansion of the town had begun in the 1850s, principally as a means to escape from the

The elegantly designed bridge at Coalbrookdale was built from over 380 tonnes of cast-iron.

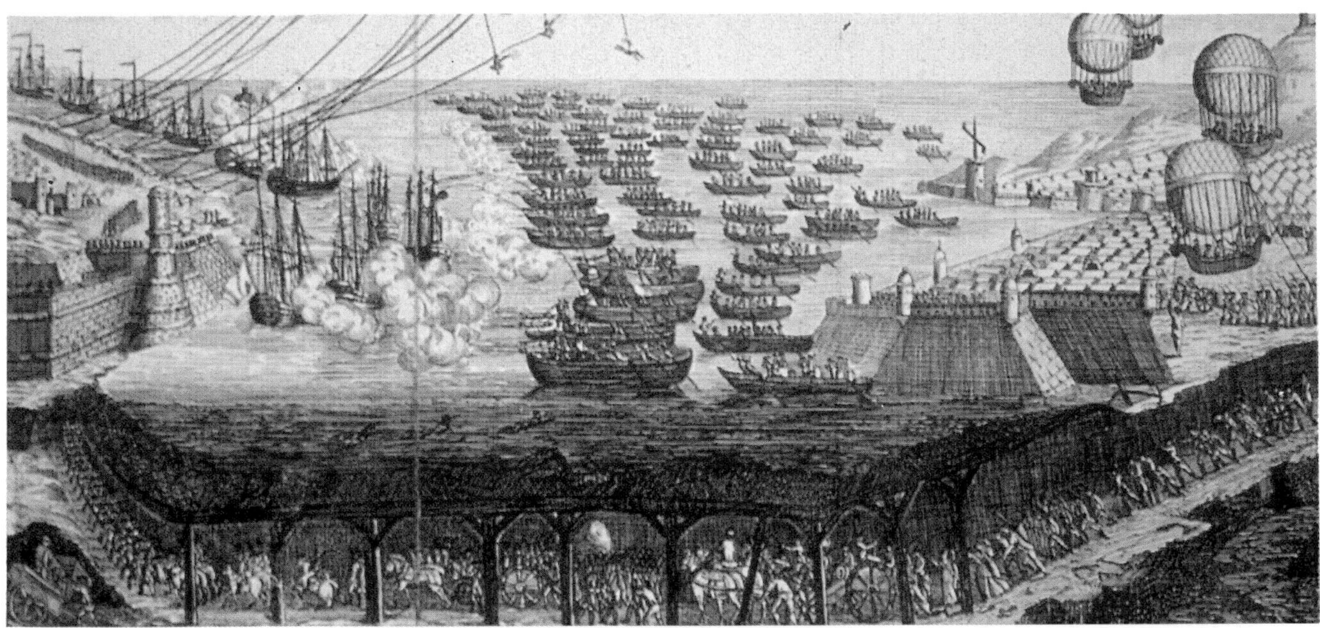
Some methods of getting from one side of the Channel to the other as projected in the 19th century.

CHANNEL TUNNEL (1993)

The first proposal for a tunnel under the English Channel was made in 1751 by a French engineer, Nicholas Desmaret. Since then, a further 25 possible schemes have been suggested including one made to Napoleon in 1802 that involved a cobble-stoned passage to be used by horse-drawn carriages. In 1860 there was a proposal to construct a 6m *20ft* high 32km *20 mile* long four-track jetty. In 1882 and 1974 some construction work began but both projects were abandoned for strategic, political and financial reasons.

The Anglo–French Channel Tunnel Treaty was ratified by the British and French governments in 1987 and tunnel boring began in November of that year by Eurotunnel, who contracted the design and construction to the Transmanche Link group.

The Tunnel is planned to open in 1993 carrying passenger, freight and vehicle rail traffic. There will actually be three tunnels: two parallel railway tunnels will run either side of a central service tunnel. The single track railway tunnels – one for each direction of travel – will run 50km *31 miles* underground, of which 37km *23 miles* will be under the Channel at an average depth of 40m *130ft*.

If all goes to plan the constructors say 30 million passengers and 15 million tonnes of freight will be carried annually. However, all is not going smoothly on the financial front: Eurotunnel revised its estimated cost of construction from £5.23 thousand million to £7 thousand million two years after work began.

Despite the sophisticated technology involved, the Channel Tunnel venture has had its fair share of problems.

water which transformed the roads into quagmires. But it was perhaps also the fear of fire (Chicago had been devastated by one of the greatest fires in history in 1871) and the necessity to build quickly and high which persuaded architects to replace timbers with a steel backbone which was light, strong and solid. It was thus that William Le Baron Jenney devised the internal structural skeleton which bore the weight of the entire building; the external wall had nothing to support.

The first sky-scraper, the *Home Insurance*, had only ten floors (two were subsequently added). One of the most famous is the *Empire State Building* in New York City (architects Shreve, Lamb and Harmon), which was opened on 1 May 1931. For a long time it was the highest in the world at 381m *1250ft*, with 102 floors. Its summit had been planned as an embarkation point for airships.

Tallest sky-scraper

The tallest is the Sears Tower in Chicago which has 110 floors and is 443m *1453ft* high – 475.10m *1558ft* if one includes the antenna.

Materials

Iron (2000 BC)

The so-called Iron Age marks the period when iron replaced the use of bronze approximately **2000 BC**. The discovery of iron was, however, made much earlier. A recent translation of a text from the Fayoum region shows that the Egyptians were capable of extracting iron ore some 3500 years BC, but the technique used was rather rudimentary.

The use of iron was developed in Asia Minor (Mesopotamia). It was introduced into Greece towards 1200 BC, where it permitted the growth of architecture: buildings could be made with blocks of stone joined by metal bolts. Iron also allowed beams of great length to be constructed because metal girders spread weight evenly. The Propylaea of the

British engineers such as Joseph Whitworth, Joseph Bramah and Henry Maudslay were at the forefront of developments in machine tools. Here is an adzing and planing machine.

MACHINE TOOLS

Type	Date	Inventor	Country	Characteristics
Wood-turning	1900 BC	—	Egypt	This is the first machine tool that we know of. The drilling machine is probably contemporary.
Planing machine	1751	N. Focq	France	Invented for planing iron, it was superseded by a tool made by the French locksmith Caillon in 1805.
Planing machine	1835	Joseph Whitworth	UK	Improved the previous models.
Slide lathe	c.1751	J. de Vaucanson	France	Tool used for the manufacture of loom parts.
Reaming machine	1775	John Wilkinson	UK	Polishes and adjusts the diameter of a cylinder while it is being drilled.
Hydraulic press	1796	Joseph Bramah	UK	Provides great pressure which can be applied progressively.
Precision lathe	1797	Henry Maudslay	UK	This type of lathe is one of the oldest and most significant of machine tools.
Circular saw	1799	A. C. Albert	France	The patent for this tool was however not granted to this inventor.
Circular saw	1816	A. Brunet and J. J. Cochot	France	Circular saw with tempered steel teeth.
Milling machine	1818	Eli Whitney	USA	Also invented a cotton seeder in 1792.
Filing machine	1826	James Nasmyth	UK	Replaced chisel work.
Steam power hammer	1839	James Nasmyth	UK	It was used to forge a drive shaft for the steamship *Great Britain*.
Grinding machine	c.1840	inventor not known	UK	Towards 1842, the French Malbec built a grinding tool by using quartz and a rubber compound.
Steam roller	1859	Louis Lemoine	France	A very important invention used in road building.
Hammer-drill	1861	G. Somelier	France	Invented during the construction of the Mont-Cenis tunnel.

Crystal Palace took just nine months to erect from prefabricated cast-iron columns and girders and sheets of glass.

Acropolis in Athens are a magnificent example of this.

But the use of iron in architecture is mainly a 19th century creation. Some important examples of the style include the Coal Exchange in London (1847), the domed Reading Room of the British Museum and the Crystal Palace.

Eiffel Tower (1889)

The Eiffel Tower was built by the engineer **Gustave Eiffel** with the help of Maurice Koelchin, Emile Nouguier and the architect Stephen Sauvestre who was responsible for its decoration. It is a prime example of metal architecture and was inaugurated on 31 March 1889.

Originally 312.27m *1025ft* high (now 320.75m *1052ft* including transmitters at the top), the tower's steelwork weighed 7340 tonnes, bringing its total weight to 9700 tonnes (inclusive of all installations). Over the years this load was increased to 11440 tonnes but between 1981 and 1983 a load reduction programme bought the total weight down to 10000 tonnes and eventually, with the replacement of the third lift, to 9000 tonnes. Originally built to last 20 years, the tower was due to be dismantled in 1910 but its scientific role saved it from destruction: the tower held laboratories for the study of astronomy, biology, meteorology and atmospheric pollution, plus radio and television transmitters.

In 1964 it was classified as a historic monument.

Steel (antiquity)

Since antiquity, certain blacksmiths – notably among the Hittites – have produced tools and weapons from iron mixed with small quantities of carbon. In 18th century England, ironfounders led by Abraham Darby produced small amounts of steel. But it was the discovery of a direct conversion process by Sir Henry Bessemer (1813–98) that paved the way towards industrial production of steel.

Stainless steel (c.1912)

Stainless steel was developed between 1903 and **1912** thanks to the simultaneous efforts of Harry Brearly in Britain, F. M. Buckett in America, and Benno Strauss and Edward Maurer in W. Germany.

The ribs of the transept roof are raised into place. In 1854 Paxton's Palace was dismantled and reassembled in Sydenham, south London, where it remained until it was destroyed by a fire in 1936.

Prefabricated panels (1888)

On **12 December 1888**, Frenchman **Georges Espitallier** patented the first prefabricated panels: standard size units made of varnished compressed cardboard and slag-wool. This wool is derived from blast-furnace slag and was developed by the Welshman Edward Parry.

The idea of prefabrication was not however completely new: the Crystal Palace built in 1851 for the Great Exhibition, was a gigantic prefabricated conservatory which was designed by British engineer and architect Sir Joseph Paxton (1801–65), a pioneer of iron architecture. It took six months to assemble the parts which came from a number of different factories.

Composite materials

Composite materials are made from two or more elements which have complementary properties: often one material withstands traction and the other compression. These compounds have been used for centuries: the first one was probably daub, a mixture of hay and mud. Reinforced concrete was the first modern composite material.

The first very high performance composites were produced in 1964 but have only had industrial applications since 1970. In 1972, aramide fibre was created by Du Pont. It is lighter and tougher than carbon fibre but may soon be replaced by an even stronger polythene fibre developed by the Dutch manufacturer DSM (1984). Metallurgists have also discovered the benefits of reinforcing metals with carbon fibres.

Besides these spectacular 'super' materials which are used only in small quantities (a few thousand tonnes per year), numerous low cost composite materials are also being developed. The most recent of these are FITs which appeared in 1984. These are thermoplastics reinforced with long fibres. The resulting material is very strong and easy to manipulate.

Carbon-carbon and ceramic-ceramic

There are two main groups of so-called thermostructural materials: carbon-carbons which are composed of carbon material covered by carbon fibres and of which Carbone Industrie, a subsidiary of the company SEP (Société Européenne de Propulsion) is the second largest manufacturer in the world. The second group is made of composite materials based on a ceramic matrix (ceramic-ceramic and ceramic-carbon). These were developed by SEP in 1981 and their licence was sold to the American company Du Pont in 1987.

Unique plants have been built to manufacture these materials which are used in fields that range from space research to medicine. Some types of biocompatible carbon compounds are excellent as bony artificial materials for use in surgery.

Concrete

Mortars (200 BC)

Lime has been known to man since antiquity, but lime mortars – used to bind stones together – appeared only around **200 BC** when the **Romans** began adding volcanic ash to lime to produce a more resistant concrete. It was called pozzolana.

Hydraulic cement (1756)

Medieval builders seem to have lost the Roman recipe for making mortars. Their own were of poor quality until the 12th century, when some improvements were made.

Cement was only re-invented in the middle of the 18th century when, in **1756**, the English engineer **John Smeaton** rediscovered its principle: the presence of clay in limestone. And 30 years later, the furnace masters Parker and Wyatts perfected a new cement by burning nodules of clayey limestone.

Portland cement (1824)

In **1824** a mason from Leeds, **Joseph Aspdin**, patented Portland cement, so called because it was made by burning a mixture of clay and chalk the colour of Portland stone.

Reinforced concrete (1892)

The non-combustibility of concrete was used by the French engineer **François Hennebique** as a means of fire prevention; this had also been a major factor in the development of iron architecture. He developed a concrete slab and later a monolithic structure similar to timber. These inventions, patented in **1892**, led to a widespread use of reinforced concrete and opened new perspectives in architecture.

Glass

Venetian glass (10th to 15th century)

There were glass-makers in Venice as early as the end of the 10th century, and in 1271 the profession gave itself a statute. The glass-makers were obliged to work on the island of Murano because of the pollution caused by their craft. This isolation helped to protect the secret of the glass-blowers' craft; indeed, the craftsmen were not allowed to leave the country. The invention of Venetian glass is attributed to a glass-maker called Beroverio in 1463.

Crystal-glass (17th century)

Industrial production of crystal-glass was developed by glass-makers in England in the 17th century. **George Ravenscroft** took out a patent in **1674**. Crystal had previously been produced by Venetian craftsmen and by the Bohemians (16th century) whose crystal was brighter than the Venetians'.

Wired glass (1893)

In **1893** the Frenchman **Léon Appert** opened the way to the fabrication of wired plate glass. Around 1910 two engineers, completely unknown to each other, simultaneously developed plate glass: Emile Fourcault in Belgium, and Irving Colburn in America. In 1952 Alistair Pilkington developed a manufacturing method that was much cheaper than the preceding ones: a layer of molten glass was poured onto a layer of molten tin to produce a perfectly smooth and shiny glass.

Laminated glass (1909)

In 1903 the French chemist **Edward Benedictus** began research on laminated glass and patented his invention in **1909**. The glass was commercialised in 1920 and was first used for car windscreens. Some laminated glass is even resistant to shots from automatic weapons.

Pyrex (1884)

In Germany in **1884 Carl Zeiss** invented a glass that contained boracic acid and silica and which was extremely resistant to heat. In 1913 J. T. Littleton of the American company Corning Glass had the idea of using the glass for crockery. C. Sullivan and W. C. Taylor carried out research which led to the development of Pyrex in 1915.

Float glass (1958)

This was invented in **1958** by **Sir Alistair Pilkington**. The production process is still a secret even though patents have been ceded to other manufacturers. The glass has many applications: car windscreens and glass walls for example. The latter are made from Kappafloat and Planar, 'energetic' glasses which let solar heat in and prevent internal energy from escaping. Extra thin float glass is being developed for use in optics, photography and aerospace.

It is thanks to float glass that glassmaking, which for a long time was a traditional activity, has become a heavy industry. Float glass is produced in a giant oven capable of producing between 500 and 600 tonnes of glass a day. There are 110 such ovens in the world, 33 of which are in Europe.

The Jacob Javits Center was constructed in Manhattan, New York, in 1986 by Ieoh Ming Pei, the same architect who built the Pyramid at the Louvre in Paris.

Clocks and watches

Escapement (725)

It seems that the first escapement mechanism was invented by a Chinese **I. Hing** in **725**. The escapement is one of the most important parts of a timepiece: it controls the transfer of energy from the motor to the hands, and it provides the oscillator with energy which compensates for that lost through friction. The most commonly used kind of escapement nowadays consists of toothed wheels and an anchor, invented by the Frenchman L. Perron in 1798.

Spiral spring (1675)

In **1675** the Dutchman **Christiaan Huygens** (1629–95) invented the balance wheel and spiral spring oscillator. The introduction of the spiral spring into watch design had an effect analogous to that of the pendulum into clocks – another of Huygens' innovations, dating from 1657.

Chronometer (1736)

A watchmaker and astronomer **George Graham** (1673–1751) first used the term chronometer as applied to a small portable pendulum. In 1736 John Harrison (1693–1776), another English watchmaker, made the first naval chronometer, in wood, and perfected it in 1761.

Winder (1755)

In **1755** the French author **Beaumarchais** invented a watch that could be wound without a key for Madame de Pompadour. She could turn a ring mounted on the face with her fingernail. It was not until 1842 however that the Swiss A. Phillippe succeeded in producing a mechanism which allowed the watch to be wound and the hands to be repositioned too. The self-winding watch was invented by a French watch-maker A. L. Perrelet in 1775. The first self-winding wrist watches were patented by H. Cutte and J. Harwood in 1924.

Electric clock (1840)

The electric clock was perfected in **1840** by the Scotsman **Alexander Bain**. At the same time, the English physicist Sir Charles Wheatstone (1802–75) invented electric distribution of time from a so-called mother clock.

Alarm clock (1847)

The first modern alarm clock was invented by the French clock-maker **Antoine Redier** (1817–92) in **1847**. The Braun Voice Control was the first alarm clock to stop ringing at the sound of a voice. It appeared in 1985.

Quartz clock (1920)

Work on the use of quartz as a resonator in clocks began in **1920**, but it was not until 1929 that the American clockmaker **Warren Alvin Marrison** perfected the first clock with a resonator of this type. Quartz watches were commercialised for the first time in 1969 by Seiko. In 1988 there was a major change in the way quartz watches work: the battery was replaced by a tiny dynamo which creates energy to compensate for that consumed. Two companies are behind this development: Seiko, who have been working on the project since 1973, and the French company Jean d'Eve.

Waterproof watch (1926)

In **1926 Hans Wilsdorf** and his Rolex team developed the first completely waterproof watch case. In 1927 Wilsdorf gave an Oyster watch to Mercedes Gleitz, a typist from London, who swam the Channel with it strapped to her wrist.

Atomic clock (1948)

The principle of the atomic clock was laid down by the American chemist **Williard F. Libby** (1908–80), winner of a Nobel Prize for Chemistry in 1960. An atomic clock uses the energy changes within atoms to produce extremely regular waves of electromagnetic radiation.

The LASSO atomic clock

LASSO is an acronym for Laser Synchronisation from Stationary Orbit. It is a project which will allow the synchronisation of atomic clocks across the globe thanks to a device installed on board the Meteosat P2 satellite launched in June 1988 by Ariane. Due to the shape of the earth, clocks situated on different continents vary in distance from the centre of the planet and therefore do not rigorously indicate the time in the same way.

The system, which is managed by the European Space Agency, should allow the Bureau International de Poids et Measures at Sèvres in France, to determine a new 'international atomic mean time'.

Metallurgy

Etruscan furnace (7th century BC)

The Etruscans have left us eloquent testimony of their metallurgical techniques. Layers of iron ore and of charcoal was piled over a hollow to a height of about 2m *6ft*. The whole thing was then covered with clayey mud and a flue was made at the top of the dome. Holes were pierced lower down the dome to ensure a good draught. When the smelting was complete, the furnace was demolished and the metal retrieved from the process was then worked by a blacksmith.

Blast furnace (13th century)

The antecedents of blast furnaces date from the **13th century**. In the 14th century, waterwheels were used to work bellows. The inventor of the modern blast furnace was Abraham Darby (1711–63), an iron-master from Coalbrookdale. Pig-iron (an alloy of iron and carbon at more than 2.5 per cent) can be obtained directly from iron ore in a modern blast furnace. The second step forward was made by a British engineer born in Germany, Karl Wilhelm von Siemens (1823–83) who, in 1857, had the idea of using the heat from the gases inside the furnace to heat the air sent in by the bellows. The open-hearth process was patented in 1858.

Coke (1735)

The metallurgist **Abraham Darby** was the first to use coke as a fuel in blast furnaces in **1735**. Coke is obtained by heating coal in a confined space at between 900° and 1150°C.

Bessemer converter (1856)

Henry Bessemer (1813–98) perfected a

A sample of molten iron is taken at a blast furnace.

METALLURGY: A STRANGE INVENTION

It is highly probable that metallurgy was invented purely by chance. The most likely explanation would appear to be that when Prehistoric man discovered fire, which according to the most recent discoveries made in Africa he did more than a million years ago, he also discovered traces of the most easily smelted metals in the ashes of his hearth. It was in such a way that extractive or chemical metallurgy was 'invented'. Copper, lead and pewter were among the first metals to be extracted from their ores, some 5000 years ago. Bronze, an alloy of copper and pewter, was known to the Egyptians in 3700 BC. The Chinese are thought to have been using iron in at least 2000 BC, after previously using cast iron for a very long period.

In order to produce iron, cast iron must undergo a chemical process to eliminate the carbon which makes it brittle. It was not until 1784 that it became possible to produce large quantities of iron industrially by means of the process known as puddling, invented by the English manufacturer, Henry Cort.

In 1856, the English engineer and inventor, Sir Henry Bessemer (1813–98) invented the refractory-lined furnace which enabled good quality steel to be produced.

Another step forward was made in 1878 when electrometallurgy was discovered simultaneously by a Frenchman, Paul-Louis Héroult and an American chemist, Charles Martin Hall. This in turn enabled aluminium, magnesium, sodium and calcium metallurgy to be developed.

converter in **1856** that allowed him to remove the carbon from pig-iron and to produce steel of a relatively good quality. This discovery, perfected at Bessemer's own cutlery factory in Sheffield, was the basis of the rise of the steel industry.

Welding (4000 BC)

Heterogeneous welding, which enables two pieces of different types of metal to be joined together, dates back to about 3500 BC. Autogeneous welding is more recent, dating from about 1500 BC. Until the end of the 19th century, the only method of welding iron and steel was by forging. In 1877, the American engineer and inventor Elihu Thomson invented resistance welding. In 1807, the Englishman, Sir Humphry Davy (1778–1829), invented the electric or carbon arc. In 1885, Bernados invented the carbon arc torch which enabled a filler metal to be used in welding. In 1890, the Russian, Nikolai Gavrilovich Slavianov, developed the process of arc welding with a consumable electrode and in 1904 Kjeliberg invented the coated electrode. But it was not until 1920 that arc welding became widely used.

Today electron beam welding is being used much more extensively. This was a process developed in 1954 by M. Stohr, an engineer at the Saclay Centre for Nuclear Research. Since 1970, laser welding has also become more common.

A heavy electric current passes between the rod and the metal. Heat is generated, and the metal welds.

Transforming lead into gold

It is tempting to link the work of alchemists with that of metallurgists. However, whereas metallurgists have tried to obtain purer and purer metals which are fully adapted to the uses planned for them, alchemists have always worked on the transmutation of metals: particularly of lead into gold.

We now know that transmutation is a phenomenon which occurs naturally: uranium changes into lead by radioactivity over thousands of years, after passing through various stages. Gold can be obtained from lead, at the expense of a great deal of energy, in a particle beam accelerator.

In some ways, one could argue that the work of alchemists was closer to that of nuclear physicists than metallurgists.

Textiles and new materials

Breathing fabric

Thanks to processes patented in Japan synthetic fibres can now be produced that measure less than ten micrometers in diameter: that is to say, approximately three times finer than the fibres used in conventional textiles. The advantages of microfibre fabrics are that their threads are woven together so tightly that they prevent molecules of water from seeping in, but are loose enough to let vapour through. In other words, these fabrics are totally waterproof – which is not the case for most coated or treated fabrics – and they allow perspiration to evaporate.

Carbon fibres (1880)

The first carbon fibre was obtained by calcining a bamboo stalk. This was done by the American **Thomas Edison** who used the fibre as a filament for his glow-lamp in **1880**.

The fibre currently used for the manufacture of composite material was invented in Japan by A. Shindo in 1961 and was later perfected, also in Japan, in 1969. This fibre is obtained by burning very pure polyacrylonitrile fibres in a vacuum. The resulting fibre is 15 times tougher than the best steel, weight for weight. Its principal applications are in sport and aeronautical engineering.

Fibreglass

Contrary to popular belief, fibreglass is already an old invention. In 1836, a Frenchman, Ignace Dubus-Bonnel, deposited a patent for the 'weaving of glass, pure or mixed with silk, wool, cotton or linen, and made pliable by steam'. He also included a sample with his patent request, woven on a Jacquard loom. He could thus produce imitation gold or silver brocades by combining silk with a weft of glass fibres which had been coloured with metal particles. Dubas-Bonnel's fabrics won him prizes at the 1839 Paris Exhibition, and the inventor produced the draperies which decorated the hearse used for the reburial of Napoleon's ashes at the Invalides in 1840.

Despite its success on that occasion, the new fibre was subsequently forgotten, probably because of the high costs of production, and would only reappear around 1950.

BIOTEXTILES: THE FABRICS OF THE FUTURE

Grafted fabrics have arrived. New types of fibres and fabrics can be made by 'grafting' molecules onto conventional (natural or synthetic) materials. The results can range from fluoride activated cottons which are permanently waterproof; fabrics in elastomer encased fibres which remain crease-free; synthetic textiles which are self-sterilising with anti-bacterial drugs; fabrics that filter highly selectively.

Furthermore, a number of traditional operations such as drying and fabric preparation can be achieved by the use of these techniques. One of the methods adopted by the Institut textile of France uses particle radiation: electrons penetrate synthetic material and ionise polymeric molecules which separate and give rise to free radicals which are then bound to the desired molecules.

Currently, most composite materials are made lighter and stronger by the addition of fibreglass. Its uses vary from car bumpers and bonnets, to rocket engine parts, to the masts of surfboards. Some 300 000 objects are today manufactured with fibreglass.

Materials that remember (1960s)

Made from special alloys these amazing materials which 'remember' (originally metals but now also plastics) allow the creation of variable shape objects. Thanks to their so-called memory, these materials resume the shape they were first given whenever the temperature at which they were moulded is recreated. This phenomenon had been observed for the first time in 1932 by American researchers, but had remained a scientific curiosity for many years with no obvious industrial application.

In 1960, **William Buehler**, an engineer at the Naval Ordnance Laboratory of White Oak (USA), developed nitinol, a nickel/titanium alloy, capable of changing shape according to the temperature. Since then, these alloys have been developed in the fields of space and areonautics. In 1969, Raychem developed the first nickel/titanium circuit joints for the Gruman F14 fighter plane. The shape-memory alloys are now enjoying a spectacular breakthrough in many fields: a patent is deposited every two days. The Japanese firm Walcoal has, since 1985, sold a flat bra which, once worn, adapts to the body's shape thanks to an underwire in the memorising alloy, and the heat of the body.

Memory plastics

It would seem that plastic with a memory, an invention being developed by the Japanese group Nippon Zeon, will become the new super-material of the future. It has the same properties as the metal alloys but, thanks to a great molecular mass and to its particular chemical links, it is as elastic as rubber. So, for example, in the future dented car bumpers

SPINNING

Type	Date	Characteristics
Spindle	antiquity	Probably the first tool used to spin wool and linen.
Spinning wheel	1500	Seems to have originated in the Middle East. It was introduced into Europe towards the beginning of the 16th century.
Flying shuttle	1733	The Lancashire weaver John Kay developed this spinning machine which was twice as efficient as previous models. Up to then, it had taken two people to weave a wide cloth as the shuttle had to be thrown from one side of the frame to the other. Kay's device returned the shuttle automatically. As a result, the quality of the fabric was improved and a single weaver could work on a broad cloth. This saved on labour costs and doubled productivity.
Spinning jenny	1768	Invented by the Lancashire mechanic Thomas Higgs who named it after one of his daughters, and perfected by James Hargreaves.
Water-frame	1769	Higgs also invented the water-frame in 1769 which used hydraulic energy. The Preston man Richard Arkwright appropriated this invention and became rich at Higgs' expense. The machine made continuous spinning possible and is the basis for modern mechanical spinning.
Mule jenny	1779	This was invented by Samuel Crompton to solve the problems associated with spinning cotton. The machine combined features from Arkwright and Hargreaves' machines: it spun cotton into yarn and then wound the yarn onto spindles.
Ring throstle	1828	The ring throstle was invented by J. Thorpe and perfected in 1833 by W. Mason. It is still the most commonly used machine for spinning wool, combed cotton and synthetic fibres.

The spinning jenny wove yarn on several spindles simultaneously, thereby greatly increasing productivity.

The water cutting technique was used for the first time in 1971. Flow System's machine can cut cakes – and sheets of steel. The water jet travels at almost three times the speed of sound and is very precise.

KNITTING MACHINES

Type	Date	Inventor
Stocking frame	1589	William Lee invented the first loom for making stockings. After a number of attempts, he managed to open a mill in Nottingham with the help of his associate, Aston. This showed that his machine could produce materials as fine as silk and ten times faster than by hand.
Rip-stop knitting	1775	This invention is attributed to the British man Crane and allowed the manufacture of ladder-proof knitwear. The first patent covering this type of knitting was not filed until 16 years later by Dawson, another Briton.
Circular loom	1798	Developed by the Frenchman Decroix for the manufacture of seamless stockings. Despite its origins, the loom was first used in England in 1806 where it was introduced by Sir Marc Isambard Brunel (1769–1849).
Power loom	1832	A mill using a rectilinear loom (producing four garments at a time) powered by hydraulic energy was opened in the US by E. Egberts and Timothy Bally in 1832. However, the use of non-human energy had previously been tried in England in 1818. In 1838, a hosiery factory using steam powered looms was opened by the Briton John Burton in Germantown (USA).
Rectilinear shuttle loom	1857	It was successively perfected by the Briton Luke Burton in 1857, by Arthur Paget in 1861 and by William Cotton in 1863 (patent for a loom which produced between two and 12 stockings at a time).
Home knitting machines	1860	They appeared during the 1860s with the machines produced by the Britons John Aiken and Herrick around 1865.

STEEL CUTTING BY WATER JET

This revolutionary technique invented and patented by the American Norman Franz in 1968 has rapidly established itself across the USA. Since the 1970s, water jet cutting has been adopted by car manufacturers (at General Motors, a hundred machines cut parts by this method) and was quickly taken up by other industries such as those working with steel, fabrics, chipboard and even minerals. Despite its advantages this process has been slow to establish itself in Europe, although Sweden and West Germany have used it for a number of years.

The process consists of cutting material with a jet of pure water sometimes mixed with polymers or abrasive substances under a pressure of 2000–4000 bars. Programmed by an optical reader and by a digital console, it allows a variety of substances to be cut with great precision.

Thanks to the possibility of combining several jets, water cutting is neater and more economical than using lasers.

will reshape themselves when heated to the temperature of manufacture.

Biometal (1986)

The United States have pioneered the work on biometals, particularly at the Naval Research Laboratory, since 1979. However, while working on shape-memory alloys in **1986**, **Dai Homma**, vice president of the Toki Corporation of Japan, discovered a true biometal. He found that thin wires (a few tenths of a millimetre in diameter) of a metal with a peculiar crystaline structure contract under direct heat (Joule effect). This type of material could be used to activate robots, thereby making them lighter and greatly simplifying their design.

Anti-noise noise (1988)

An original technology was developed by the acoustics laboratory of the **CNRS** of Marseilles in France in **1988**. Based on the principle that a noise has to be heard to be treated, the laboratory has developed a noise-reducing helmet which works by active sound absorption.

Noise passing through the helmet's plastic shell is fed to an electronic device which then treats it and generates a signal which reduces the noise strength. This combination of noise/noise-destructor brings noise levels down by 20 to 40 decibels according to the frequencies.

A similar device has also been developed in the USA by G. B. B. Chaplin who invented this technological paradox. His patents are expoited by Noise Cancellation Technologies. The electronic 'silencer', still at a prototype stage, should find numerous applications in industry for the attenuation (or even the suppression) of engine and machine noise. It should be introduced onto the car market in 1992.

Laser

Nova (1986)

The most powerful laser in the world, the Nova, was put into service in **1986**. The **Lawrence Livermore National Laboratory**, directed by the University of California for the American Department of Energy co-operated with some 200 private companies and dedicated six years of research to the laser's construction. This $176 million laser will be able to unleash, by nuclear fusion, almost as much energy as that found at the centre of the sun. At the moment it produces a hundred thousand million kW in a thousand-millionth of a second.

The fastest laser on earth (1987)

A team of American researchers at **GTE** has developed a laser capable of commutating 22 thousand million times per second. What use is it going to be? A minute semi-conductor at the heart of this laser is theoretically capable of transmitting 200 to 400 video signals simultaneously via an optical fibre. Optical fibres are mainly used to carry telephone communications, but the faster lasers allow optical fibres to be used for the transmission of images.

Gas discharge lasers (1959)

It was **R. Gordon Gould** who in **1959** invented the gas discharge laser which is used in compact disc players and bar code readers. However, the patent he applied for in 1959 was only granted by a federal judge in December 1985. Note that the Russian physicists Nikolai G. Bassov and Alexander M. Prokhorov developed the first gas discharge and semi-conductor lasers as early as 1960.

Medical laser (1960)

The first ruby medical laser was developed by **Theodore H. Maiman** in **1960**. It was used by Doctor Freeman, as early as 1964, for the treatment of lesions on the retina.

In 1967, researchers at the American Bell Laboratories, D. R. Henriott, E. I. Gordon, D. A. S. Hale and W. Gromnos developed the Light Knife which cuts and cauterises wounds at the same time.

Other medical lasers have followed: since 1970, the carbon dioxide laser, which was discovered in 1964, has been used for endoscopic surgery, the destruction of tumours and the removal of tattoos. In 1978, the argon laser was introduced. The Nd-Yag laser, tested on animals in 1975, has been used on humans since 1976. Then there is the *Soft laser*, whose principle was discovered by the Italian doctor Tarrantini who worked on it from 1979 in collaboration with the French doctor Bernard Sillam. This laser is used for sports injuries and rheumatology. The laser's name was invented by Doctor Jarricot.

More recently, pulsed Yag lasers have revolutionised eye surgery. Developed at the beginning of 1980 by Professor Danielle Aaron of the Rothschild Foundation, this laser means surgery can take place without having to open up the eyeball. Since the early 1980s a photo-chemical therapy by laser can treat certain cancers.

The American Theodore H. Maiman displays a ruby laser which he first built in 1960.

Laserfilm (1985)

The manufacture of a new film, sensitive to laser beams and used for the reproduction of video discs, began on **19 November 1985**. It was developed by engineers of the American company **McDonnell Douglas** after years of research.

Laser weaponry (1985)

Shortly after the construction of the first laser in 1960, the American government commissioned research in the field of laser weaponry from the Hughes Aircraft Company. Missile guidance applications were the first to be developed, and this technology is now completely mastered. Research then turned to weapons which could destroy missiles, aircraft and satellites. The first so-called 'killer' laser for the **Strategic Defense Initiative** programme was tested (officially and successfully) on **6 September 1985**. The test, which simulated the normal conditions of operation for Russian missiles, was carried out against a Titan 1 rocket.

Anti-laser protection (1988)

The Americans are currently working on a new laser-resistant material known as the Laser Shield, which was created by the California-based aerospatial company Harlamor-Schadek.

Holograms

Holographic credit card (1984)

In **1984 Visa** launched an international credit card in the USA which bore a hologram. Besides their aesthetic appeal, holograms reduce the risks of counterfeiting.

Holograms (1986)

Thanks to a combination of holography and computing a team from the Massachusetts Institute of Technology (MIT), directed by **Stephen Benton**, has managed to represent a car in three dimensions. This technique could reduce the design time of a car from five years to 18 months.

Holographic banknote (1988)

On the occasion of the Australian bicentenary, the **Commonwealth Reserve Bank** and the **Organisation de Recherche Scientifique et Industrielle** co-operated to launch a new $10 bill by using a new type of hologram which makes it virtually counterfeit-proof. The banknote is made of special plastic sheets and holds a holographic image which changes colour according to the angle of view. The new banknote also has raised marks which makes it easy to identify by touch.

Holographic stamp

The first holographic stamp was launched in January **1988** by the Austrian post office. 3 040 000 stamps were issued.

Holography became practicable after the development of laser technology in the 1960s.

EVERYDAY LIFE

In the kitchen

Plate (ancient times)

The plate was known to peoples of the ancient world, especially the Romans. But it disappeared during the Middle Ages and was replaced by bowls and wooden trenchers.

Plates reappeared in 1530, in silver, at the banquet celebrating the marriage of King Francis I, of France (1494–1547) to Eleanor of Hapsburg.

Carton (1951)

It was **Ruben Rausing**, a Swedish industrialist specialising in packs for dry foodstuffs (flour, sugar, etc.) who revolutionised the packaging of liquids and drinks in **1951**. Combining the most highly developed paper, aluminium and plastic technologies, he created the tetrahedral carton, a totally new form of packaging in form, manufacture and cost price.

In 1961 Ruben Rausing and his Tetra Pak company made their decisive expansion with the first aseptically filled cartons of UHT-treated long-life milk. Following dairy products, a whole range of goods including fruit juices, soups, cream and wine, were to be packaged in 'brick' form.

This has brought about a real revolution in our daily life, with Tetra Pak now present in 98 countries and selling more than 40 thousand million packages every year.

Beer (4000–3000 BC)

Man has been drinking beer for thousands of years: in India (c.3200 BC), in China (c.3000 BC) and above all in the Middle East where the Sumerians reserved 40 per cent of their cereal crop for brewing. In Egypt, beer was considered as a national drink. It was, however, very different from the beverage we know today. Less liquid than our beer, it resembled a kind of drinkable 'bread' but had nonetheless a high alcohol content (13 to 15 degrees proof) and was generally made from barley. In the Middle Ages monks introduced hops into the recipe and from the 12th century professional brewers appeared.

Alcohol-free beer (19th century)

Alcohol-free beer was created in France at the end of the **19th century** (between 1873 and 1900).

Tinned food (1795)

In **1795** the Frenchman **Nicolas Appert** (1749–1841) invented a brilliant method for preserving food. *Appertisation* consisted of sterilising foodstuffs in hermetically sealed containers away from circulating air. This process was not yet that of tinning food and involved jars covered with five layers of cork.

Appert won a competition organised by the French government, with a prize of 12000 francs. He was thus able to perfect and industrialise his method.

The Admiralty tested his product in 1804. Samples were sent to Brest and kept for three months before being tasted. The results were convincing: in his report the maritime prefect said that 'the beans and peas, prepared with or without meat, have retained the freshness and pleasant flavour of fresh vegetables'.

Tin cans (1812)

In 1810, Pierre Durand patented a metalled vessel for preserving food. The patent was bought for £1000 by the Englishmen **Bryan Donkin** and **John Hall**, who combined Durand's preserving process with Appert's. Tin cans were first made in **1812** in a preserving factory built at Blue Anchor Lane, Bermondsey, London.

Tin cans opened by a key (1866)

The American **J. Osterhoudt**, of New York, invented on **2 October 1866**, a tin can with a key fixed onto the top. The key is simply loosened and then turned to open the can.

The tin can has been reinvented in the form of a transparent plastic container with a metal top.

Cocoa powder (1828)

Cocoa powder was made for the first time by the Dutchman **Conraad Johannes van Houten** in Amsterdam in **1828**. He invented a way of obtaining a soluble cocoa powder which could be used to make a hot drink. Up to that time, the only way to make it was by melting pieces of chocolate.

Ovaltine (1904)

In 1865 **Doctor George Wander** set up a laboratory in Berne, Switzerland, to manufacture concentrated barley malt. In **1904** Wander produced a healthy drink from the malt added to milk with cocoa, eggs and vitamins. He called it Ovomaltine. In 1909 the British factory was set up in King's Langley and the company applied to register the Ovomaltine trademark. A clerk wrongly transcribed it as Ovaltine – and the name stuck.

Kit Kat (1937)

Britain's favourite chocolate bar (it has been the bestseller for several years) celebrated its jubilee year in 1987. It began life in 1935 as Chocolate Crisp, becoming Kit Kat two years later. Production had to stop during the war because of milk shortages and when the bar was relaunched in 1945 the familiar wrapper changed to a blue one to show that the chocolate was made without milk.

Coffee (15th century)

According to legend the stimulating qualities of coffee were discovered by a goat herder in the Yemen. He noticed that his herd wouldn't sleep at night after eating the red fruit of the coffee bush. We know for a fact that coffee was drunk in Aden in 1420. It was adopted in Syria and Turkey and by 1615 coffee had reached Venice. Coffee houses became popular in the 18th century; perhaps the most famous is Edward Lloyd's in London where underwriters and merchants carried out much of their insurance business.

Instant coffee (1937)

The first attempts at the production and marketing of instant coffee took place in America in 1867, but with no great success.

It was the Swiss company **Nestlé** who gained a lion's share of the market by creating Nescafé in **1937**. Nowadays there are 100 different varieties of Nescafé in the world.

Espresso (1946)

Espresso coffee has been around since the late 19th century but it only became popular in Europe with the invention of the **Gaggia** coffee maker in **1946** in Italy.

Coffee grinder (1687)

The invention of the coffee grinder in **1687** contributed to the diffusion of this brand new drink. The electric coffee grinder was invented in 1937 by the Kitchen Aid division of the Hobart Manufacturing Co. of America. The first model sold for $12.75.

Champagne (c.1690)

It is impossible to attribute the invention of Champagne (which may well have been

Ovaltine's posters have always emphasised the nutritional qualities of the drink.

Kit Kat is Britain's favourite sweet.

produced by accident) to one person. However, around **1690 Dom Pierre Pérignon** (1638–1715), a monk from an abbey in Champagne, invented a process that produced the effervescence. Dom Pérignon also dared to go against all the rules of wine-making and blend several wines together. He also had the idea of keeping the wine in tightly corked bottles which ensure that the carbon dioxide that is produced during fermentation adds sparkle to the wine.

Chocolate (1819)

A 23-year-old Swiss, **François-Louis Cailler**, made the first bars of chocolate at Vevey in **1819**. Small scale production of chocolate had begun in France and Italy after the Spaniards returned from South America with the recipe. At that time it was a drink prepared from roasted, crushed cocoa beans.

In 1879 another Swiss, Rodolphe Lindt, built a chocolate factory in Berne. In those days blocks of chocolate were hard and had to be crunched; they also left a gritty sensation in the mouth along with a bitter aftertaste. Even when heated up, the chocolate remained thick and heavy. Because of this Lindt invented a machine that kneaded the chocolate for a long time; he then had the idea of adding cocoa butter to it. The chocolate we know today was born, and Lindt patented his invention in 1880.

Coca-Cola (1886)

Coca-Cola was invented in **1886** by **John Pemberton**, a 50-year-old chemist from Atlanta, Georgia. He decided to develop a syrop that would be original and thirst-quenching. Working relentlessly in the back room of his 'drug store', he produced a mixture containing cola nut extract, sugar, a little caffeine, cocoa leaves with the cocaine removed and vegetable extracts. (The syrop's exact composition is still a closely guarded secret.) A few months later, an assistant mistakenly served a customer Coca-Cola mixed with soda water: that proved to be the little touch that made the drink a success.

To market his new drink, Pemberton formed a partnership with Frank Robertson whose elegant handwriting was used for the Coca-Cola trademark.

In May 1985 'New' Coke was introduced and the old formula was retired. Coca-Cola drinkers were outraged and the original recipe was revived.

Coca-Cola bottle (1913)

With the immense success of Coca-Cola, there followed many imitations. In July **1913** the company's managers realised that the only way to put an end to this was to give Coca-Cola a bottle that would be absolutely original. They entrusted the task to the glass-maker from Indiana, C. S. Root, who charged one of his assistants, Edward, to research the drink. Edward found an illustration of a cola nut which he copied. The design could not be used as it was as the bottle would not stand up. The base of the nut was therefore cut off. The technical director of the factory, Mr Samuelson, reproduced the truncated nut in glass and decorated it with vertical lines. The trademark on the bottle made the fortune of C. S. Root.

Coca-Cola first sold its famous bottles in packs of six in 1923.

Breakfast cereals

Corn flakes were popularised in 1898 by the American **Will Keith Kellogg**. Before that, the American Henry D. Perky of Denver, Colorado (USA), was the first, in 1893, to have the idea of making ready prepared breakfast cereals. Having met a man who nursed his stomach pains by eating boiled wheat soaked in milk every morning, Perky came up with a product made from wheat, which he called Shredded Wheat.

Will Keith Kellogg was employed by his elder brother, a doctor at the Battle Creek Sanitarium in Michigan. Working at night in the hospital kitchen, Will boiled wheat in an effort to help the doctor search for a digestible substitute for bread. One day in 1894, after a batch of boiled wheat accidently was left to stand, the brothers tried again. Unknowingly, they had tempered the wheat by letting it stand. The compressed wheat was flaked off rollers with blades devised by Will. Thus was the modern-day breakfast cereal born. Granose Flakes were launched in 1894, which became Corn Flakes in 1898.

Croissant (1683)

The crescent roll was invented in **1683**, in Vienna, by the Pole **Kulyeziski**. The city had been under siege by an immense Turkish army led by Kara Mustafa. The famished

This cake was baked some 4000 years ago to accompany an Egyptian into the underworld. Its amazing degree of conservation is due to a happy accident. It was cooked inside two moulds which fitted perfectly together and within which a vacuum was created as the cake cooled.

Viennese were finally saved by Charles de Lorraine and the King of Poland, John III Sobieski. Kulyeziski, having taken a decisive part in the final victory, was given the stocks of coffee abandoned by the routed Turkish army and was authorised to open a cafe in Vienna. This he did, and to accompany his coffee, he had a baker make small milk bread rolls in the shape of crescents to commemorate the victory over the Turks. They were immediately successful.

Carbonated mineral water (1741)

Fizzy mineral water was invented in Whitehaven, Cumbria, by **William Brownrigg** in **1741**. He had the idea of adding carbonic acid (which produces bubbles) to ordinary spring water and then bottling it. The bubbles would appear when the bottle was opened.

Choc-ice (1922)

The first choc-ice was invented by an American from Iowa, **C. K. Nelson**. He called it an Eskimo-pie and patented it on **24 January 1922**.

Goose liver (antiquity)

Although goose liver is a delicacy we most often associate with France *(foie gras)*, the ancient Greeks enjoyed eating the livers of geese more than 2000 years ago.

In 52 BC, the Roman consul Metellus Pius Scipio had geese held in dark pens and force-fed with figs to obtain what he felt was a goose liver of perfect quality.

Microwave oven (1945)

On **8 October 1945** high frequency radio waves made their official entry into the kitchen when an American, **Percy Le Baron Spencer**, applied for a patent for what was to be the microwave oven. It simply involved applying the principle of radar to cooking food. Spencer, who was a physics engineer at Raytheon, one of the world leaders in radar equipment, noticed one day that the energy given off by the tubes used for radar produced heat.

This electromagnetic energy gave him the idea of putting a handful of maize in a paper bag and placing the bag within the field of the tube. The maize immediately burst, transforming itself into popcorn. He melted chocolate in the same way. Raytheon developed a cooking programme for microwaves and patented the first cooking apparatus of this type, the Radarange. This machine had a power of 1600 watts. It was heavy, awkward and expensive and was originally intended for use in hospitals and military canteens. In 1967 the Amana company, a subsidiary of Raytheon, put the first household microwave oven on the market.

Refrigerator (1913)

The **Domelre**, manufactured in Chicago in **1913**, was the first functional household refrigerator. In 1918 the American **Nathaniel Wales** designed a device that was widely marketed under the name of Kelvinator. The Frigidaire trademark appeared one year later in 1919.

The Swedes Carl Munters and Balzar von Platen succeeded in constructing a silent and functional refrigerator. They filed their first patent in 1920, and developed a condenser device in 1929. Mass production began in 1931 with Electrolux, in Stockholm.

In 1926 the American company General Electric manufactured a hermetically sealed unit and, in 1939, it introduced the first dual-temperature refrigerator. This allowed frozen foods to be kept in one compartment.

Refrigerated meat transport (1877)

In **1877** the Frenchman **Charles Tellier** (1828–1913), who had designed a method of meat preservation using dry cold, won an international competition for transporting meat between the Old World and the New.

The *Frigorifique*, a three-masted ship equipped with a steam engine, took 105 days to cover the 12000km *7456 miles* between Rouen and Buenos Aires. The carcasses of ten beef cows, 12 sheep and two calves were kept refrigerated in dry air and travelled very well. At about this time also the first meat transport by refrigerated wagon took place, which was developed by Gustavus F. Swift (1839–1903). Until then animals had to be transported on the hoof.

Deep-freezing (1924)

Industrial deep-freezing was launched by the American **Clarence Birdseye** in **1924**. He applied preservation techniques used by the Labrador Inuit, which he had observed in 1912 and 1915. In 1924 he set up the Freezing Company which used the methods of the time to deep-freeze up to 500 tonnes of fruit and vegetables a year. In 1929 he realised that the deep-freezing process would have to be speeded up and invented a freezing machine with two plates, which chilled the product on both sides. Lastly, in 1935, he invented his multiple plate freezer, which is still used today.

Dehydration (1945)

The dehydration process, designed by the American **Howard** in **1945** and patented in 1949, put the finishing touch to freezing techniques. It made it possible to reduce the mass to be transported by about 50 per cent.

Pre-cooked frozen meals (1945)

In **1945** the American **Maxson** was the first to offer pre-cooked frozen meals to airline passengers.

These were followed by TV dinners, which started to develop in a spectacular way in 1954. In the United States in 1960, 215 million dishes were prepared.

Fork (6000 to 3000 BC)

Distant precursors of the modern fork have been unearthed at the diggings made at the site of Çatal-Hüyük in Turkey. It then seems to vanish, and the first indication of its 'reinvention' is its mention in certain inventories drawn up in the 14th century. It was probably brought to the West by the Italians. The 1307 inventory of Edward I mentions seven forks, including one in gold. In 1380, the inventory of Charles V of France mentions 12 forks, some of them 'decorated with precious stones'.

It was not until the first half of the 17th century that the fork reached English dining tables.

Toaster (1909)

The first toasters were marketed by the **General Electric** company of Schenectady, New York, in **1909**. The machine quite simply consisted of bare wires wound around mica strips.

The first prototype of the toaster that grills on both sides and ejects the toast was developed by Charles Strite of Stillwater, Minnesota, and patented in 1919.

Camembert (1790)

It is not known whether **Marie Harel** got the recipe from her mother, Marie-Catherine Fontaine, or from a non-juror priest from the Brie district whom she sheltered at the start of

REFRIGERATION

Name	Date	Characteristics
Water vapour refrigeration machine	1755	In this year the Scotsman William Cullen first obtained a little ice from water vapour in a vacuumed bell jar. In 1777 Gerald Nairne added a little sulphuric acid, which accelerated the process. In 1866 Edmond Carré took the process out of the experimental stage by developing a machine which was immediately successful, particularly for chilling carafes.
Compressed ether refrigeration machine	1805	A prototype was presented in Philadelphia (USA) by Oliver Evans. Its chief innovation lay in the introduction of a closed-cycle system. This process was patented in 1834 by the American Jacob Perkins. The first industrial machine was designed by a Scottish immigrant to Australia, James Harrison (patent granted in 1855).
Air refrigerating machine	1844	The principle of reducing air pressure was known as early as the 18th century. The American John Gorrie applied it in his machine in 1844. He was a doctor in Florida and invented this machine to relieve his patients. He obtained a British patent in 1850 but, strangely, his invention caused a scandal in the United States, where some people accused him of competing with God by using his machine to make ice at any time of year. Nevertheless after a great struggle he obtained his patent in 1851.
Absorption refrigerating machine	1859	In 1859 the Frenchman Ferdinand Carré, Edmond's brother (see above), patented a machine in which the fluid, having generated coldness, is absorbed by another substance rather than being drawn up by a compressor. This was the first absorption refrigerating machine. The absorption refrigerator was perfected in 1944 by the Swiss company Sibir.
Ammonia compressor refrigerating machine	1872	It was an American of Scottish origin, David Boyle, who obtained the first patent for a compressor using ammonia. But it was the German Karl von Linde who made it successful with two machines, developed in 1876 and 1877, which were soon on the market.
SO_2 compressor refrigerating machine	1874	In 1874 Raoul Pictet, a Swiss professor of physics in Geneva, used liquid sulphur dioxide (SO_2) in a system of refrigeration by compression. In 1876 Pictet's machine was used for London's first artificial skating rink.

Early domestic refrigerators were cooled by blocks of ice kept in a separate slate-lined wooden cabinet.

the Revolution. What we do know is that her cheese was very successful at the Camembert markets in the Department of Orne in France. Her daughter and son-in-law, who were excellent business people, took over from her, establishing the cheese's reputation and giving it its name.

McDonald Hamburgers (1948)

In 1940 two Americans, the brothers **Maurice** and **Richard McDonald**, set up a hamburger stand next to their cinema near Pasadena in California. In **1948** they had the idea of making it self-service, stressing the quality of the hamburgers served there.

By 1952 they were known throughout southern California, where they had established subsidiaries. Ray Kroc, a dealer in restaurant equipment, offered to sell franchises for them in the rest of the country. In 1962, when there were 200 establishments, Kroc bought the McDonalds' share of the business for $2.7 million. Today there are 10300 McDonalds restaurants in 50 countries, including the USSR.

Ketchup (1876)

The ketchup we know today was invented by the American **Henry Heinz** in **1876**, but its origins go back to ancient times. The Chinese were probably the first to prepare a sauce called ketchup or *ke-tsiap*, a sort of brine marinade for fish or shellfish.

It was introduced into Europe at the end of the 17th century by the British, who had come across it in Malaysia, where it bore the name *ketchap*, and adapted it to the ingredients available in Britain. It was then taken to the United States by long distance navigators from Maine.

Condensed milk (1858)

Although the Frenchman Nicolas Appert had the idea of condensing milk in 1827, the process was not applied industrially for another 30 years. It was not until **1858** that the American **Gail Borden** set up the first factory producing sweetened condensed milk in the United States.

In 1884 work by Meyenberg made it possible also to produce evaporated milk that contained no added sugar.

Mustard (4th century)

In the footnotes of history we find the story of how the Gauls introduced the Romans to the mustard seed when they occupied Rome in the 4th century. When these seeds were ground up with vinegar or wine must, the result was *mustum ardeo*, from which our name mustard comes. It was not until the Middle Ages that mustard found its way on to every table.

In the 13th century the town of Dijon first became famous for the quality of its mustards using verjuice, an extract of unripe grapes, to make this most sought-after condiment.

As centuries passed and tastes changed, mustard was flavoured in different ways: with vanilla in the 17th century, then with orange blossom or violet water, then, in the 19th century, with various spices or herbs like tarragon, chervil, chives, lemon or red fruit.

A taste for spicy food developed in the British centuries ago, and piquant sauces like Lea & Perrins are still popular.

Non-stick pan (1954)

The non-stick pan was invented by the Frenchman **Marc Grégoire** purely by chance. Mr Grégoire, a research engineer, was trying to perfect his fishing rods in **1954** when he discovered the processes which make it possible to encrust metal with Teflon. His patents were applied to kitchen utensils and with these he founded the Tefal company in 1956, which went on to produce its famous frying pan. Tefal is still the uncontested leader in non-stick utensils, having sold 25 million frying pans, casserole dishes, etc.

The electric frying pan dates back to 1911, when it was launched by Westinghouse.

Tonic water (1840)

In **1840** the British company **Schweppes**, founded by the German Jakob Schweppe, developed Indian tonic water. Schweppe had begun making soda water from distilled water

The benefits of mineral water have been promoted for many years, but it was only in 1986 that the first bar selling 1200 different brands was opened.

charged with carbon dioxide in 1792.

Tonic water is a soda containing sugar and quinine. The idea of putting quinine in soda originated in the Indian Army as part of the fight against malaria.

Sugar (antiquity)

The extraction of cane sugar goes back to antiquity, and the plant was probably first cultivated in India: the Greeks and Romans referred to sugar as 'Indian salt' and 'honey of India'. The Christians probably brought it to the West during the Crusades. It has been proved that during the 12th century there were mills in Sicily that made 'honey canes'. Centuries ago the Chinese knew not only how to extract cane sugar but also how to refine it, an operation that was developed much later in the West.

Sugar beet (1812)

During the Continental System, Napoleon considered it of the utmost importance that France should be able to manufacture sugar, by extracting it from beet. On **2 January 1812** he was overjoyed to hear that the manufacturer **Benjamin Delessert** (1773–1847) had succeeded in doing this at his factory in Passy.

Napoleon went to Passy that very day and decided to establish imperial factories and to devote 600000 arpents, or a similar number of acres, to sugar beet cultivation.

Delessert's predecessors were the German chemist Margraff, who started his work in 1747, and a German of French origins, Achard, who began in 1799. The juice was extracted by rupturing the cell walls of the root. But it was impure, people did not like the sugar and productivity was low. In 1864 the Frenchman Robert, a sugar manufacturer in Moravia, obtained the juice by diffusion, which did not require rupturing the cells.

Aspartam (1965)

Aspartam, the synthetic sweetener, has 200 times the sweetening power of sugar. It is a product of biotechnology and was discovered by scientists in the American **Searle Laboratory** in **1965**. Although it has considerable sweetening power, it does not add calories or cause dental caries, and has an excellent taste in foods and drinks.

In the United States its use has been permitted in foods since 1982 and it is also used in West Germany, Great Britain and Switzerland.

Saccharine (1879)

Saccharine was discovered by **Constantin Fahlberg**, an American who was working under the direction of Professor Ira Remsen at Johns Hopkins University in Baltimore. He published the results of his work on **27 February 1879**. Recent studies seem to prove that saccharin can be hazardous to health but only if consumed in extremely large quantities.

Tea

Two different legends recount the discovery of tea. Shang Yeng was the Emperor of China around 2737 BC. As a health measure, he ordered his subjects to drink nothing but boiled water. One day, leaves from a nearby tree fell into his own simmering water and the Emperor was delighted by this new drink.

A horse for tea

The second legend is set circa AD 520. According to Japanese tradition, an Indian prince, Bodhidharma, who had become an ascetic, went to China to teach Zen Buddhism. To keep himself awake during long hours of meditation, he cut off his eyelids and threw them away. At the place where they fell there soon grew a bush. When the master's disciples came to meditate with him, they picked the leaves of this tree and made an infusion from them to keep themselves awake. It was a tea plant.

During the Ming dynasty in China, tea was used as money. A good horse would cost 68kg *150lb* of tea.

Vacuum flask (1906)

In **1906** the Scottish physicist and chemist **Sir James Dewar** (1842–1923) invented a thermal isolation device which made it possible to keep gases liquified. The vacuum flask is the everyday version of Dewar's device.

Corkscrew (17th century)

Towards the end of the **17th century**, the use of watertight corks made corkscrews indispensable. We do not know who invented the first ones but in 1795 an Englishman called Samuel Hershaw developed the screw and nut corkscrew.

Tupperware (1945)

The little airtight plastic boxes were the brainwave of the American **Earl W. Tupper**. He was a former chemist at Du Pont and invented Tupperware in **1945**. The particular feature of real Tupperware is that they are not sold in the traditional way but through 'parties', and are guaranteed for ten years.

Fatless meat (1988)

This was invented by the Australian butcher **Dallas Chapman**. His process, the result of 15 years of research, makes it possible to reduce fat content by 96 per cent, cholesterol by 30 per cent and calories by 85 per cent.

The meat obtained in this way looks like sausage meat. It is fairly tasteless and can be flavoured chicken, pork or beef, on request.

Whisky (1494)

The first mention of a spirit made from malted barley dates from **1494**. The monk **John Cor** distilled it for his abbey.

Following the civil and religious wars which ravaged Scotland for centuries, the production of malted barley spirit, in other words whisky, remained secret until 1823, when a certain John Smith established an official distillery in the valley of Glenlivet.

Irish whiskey (5th century)

In the **5th century**, Catholicism was very strong in Ireland. Monks were travelling Europe and the Near East in all directions spreading the good word. Among them was Saint Patrick, who brought the first still and the art of distillation back from Egypt. From then on, as well as the Gospel, the Irish monks taught the art of distilling spirits. Using barley and pure water, they developed a spirit called *uisge beatha* ('water of life' in Gaelic), from which their cousins the Scots were the first to benefit.

The art of distillation was brought to Ireland by Saint Patrick.

An excellent product . . .

In 1170 the Norman soldiers of Henry II, King of England and Duke of Normandy, discovered *uisge beatha*. They found the product excellent to drink but the name difficult to pronounce, so they changed it to whiskey. Historical documents dating from 1276 make Old Bushmills in Northern Ireland the oldest whiskey distillery in the world. It is still in operation today.

Bourbon (1789)

This alcohol, the American cousin of Scotch and Irish whiskies, was invented in **1789** by **Pastor Eliza Craig** in Bourbon County in northern Kentucky. The refinement of this whiskey, distilled from corn mash and malted barley, was carried out by Doctor James Crowe in Franklin County, Kentucky.

Yogurt (1542)

Originally from Asia Minor (Persia and Turkey), yogurt first appeared, according to legend, during biblical times. It was revealed to Abraham by an angel and was supposedly responsible for the patriarch's long life. Not until the First World War did the recipe, rediscovered in Bulgaria, reappear in France.

The French are the world's top consumers of yogurt, getting through 12.7kg *28lb* a year each!

Wheatless bread (1989)

A new bread without wheat, using manioc, rice, millet, maise or sorghum, was presented in January **1989** by the experts of the **United Nations Food and Agriculture Organisation**. This is a technological innovation which could enable many developing countries to produce their own cereals and bread.

Chewing-gum (1869)

In 1848, an American, J. Curtis, marketed spruce resin for the first time. Spruce is a native tree of Maine, USA.

In 1860, spruce resin was abandoned in favour of chicle, a gumlike substance obtained from the sapodilla, a tree which grows in Yucatan. Chicle proved a better medium for flavours such as mint and aniseed. It appears that it was first marketed by an American, T. Adams.

On **28 December 1869 William F. Semple** of Ohio was granted a patent for chewing-gum, made from a mixture of rubber and other substances. It met with amazing success and was produced commercially from 1900.

Artificial flavouring (1874)

In **1874** two Germans, **Doctor Wilhelm Haarman** and **Professor Ferdinand Tiemann**, synthesised vanillan, the principal component in vanilla husk. Two years later Karl Reimer conceived a chemical compound which fully reproduced the flavour of vanilla.

Margarine (1869)

Margarine was invented in **1869** by the Frenchman **Hippolyte Mège-Mouriès**, following a contest launched by Napoleon III to come up with a replacement for butter. An artificial butter that would be economical and conservable and would not go rancid represented an uncontestable advantage for the nutrition of the army and navy.

Mège-Mouriès' method consisted in processing animal fat (essentially tallow) from which he obtained a paste whose colour and consistency were close to butter and which did not have a disagreeable odour. He christened his product margarine because of its pearly colour (pearl is *margaron* in Greek). Later, thanks to improvements in Mège-Mouriès' process, margarine was made from vegetable fat.

Food processor (1947)

This kitchen appliance, destined to equip millions of kitchens throughout the world, has its roots in a **1947** design by the Englishman **Kenneth Wood**. Marketed as the Kenwood Chef, it was composed of a powerful and sturdy engine block to which a large number of accessories could be fitted: mixer, citrus squeezer, mincer, slicer and shredder, pasta and ravioli maker, food mill, can opener, etc. Its multiple uses allowed it single-handedly to replace a large number of small appliances.

Sandwich (1762)

John Montagu, fourth **Earl of Sandwich** (1718–92), is said to have invented the sandwich in 1762. A devoted gambler, the earl one day refused to leave his gambling table for lunch. His cook prepared a small snack for him consisting of a slice of meat between two slices of buttered bread. The sandwich rapidly became very popular throughout the British Isles but did not spread to continental Europe until the following century.

Freeze-dried food (1946)

In 1946 to 1947, the American **E. W. Flosdorff** demonstrated that the process of freeze-drying, which was already known and used, could be applied under proper conditions to products such as coffee, orange juice, or meat.

Freeze-drying achieves dehydration through refrigeration: the water content solidifies faster than the other elements in the product and is eliminated in the form of ice.

Freeze-drying was invented by the Frenchmen Arsène d'Arsonval and F. Bordas in Paris, in 1906, and rediscovered by an American, Shackwell (1851–1940), in St Louis, Missouri, in 1909. The process was first applied medically.

It was not until 1955 that freeze-drying entered the food industry, where it was applied to Texas shrimps and Maryland crabs.

In the home

Aerosol (1926)

In **1926** the Norwegian **Erik Rotheim** invented the aerosol. He discovered that a product could be projected in a fine spray by introducing a gas or liquid into the container to create internal pressure.

On 22 August 1939 Julian S. Kahn from New York invented the disposable spraycan which could contain an aerosol. But this idea did not have its first commercial application until 1941. In that year two Americans, L. D. Goodhue and W. N. Sullivan, manufactured an insecticide in aerosol form.

Aerosols are now out of favour because of popular concern about the damage being done to the ozone layer by CFCs.

Air conditioning (1911)

The American **Willis Carrier** invented air conditioning in **1911**. In 1902 he had studied the regulation of air humidity at a Brooklyn

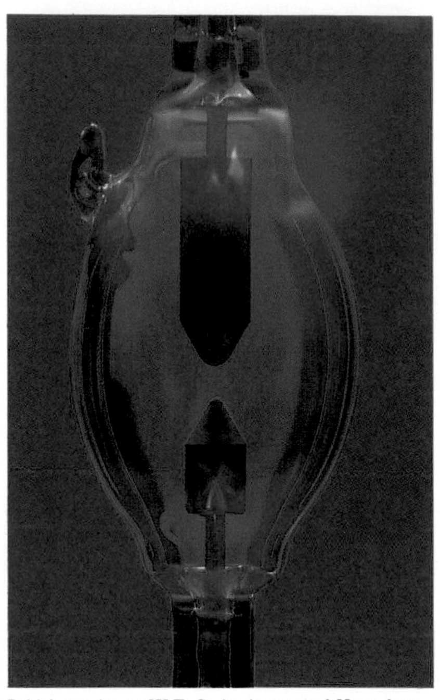

British engineer W.E. Staite improved Humphry Davy's arc light by advancing the electrodes mechanically as they burned away.

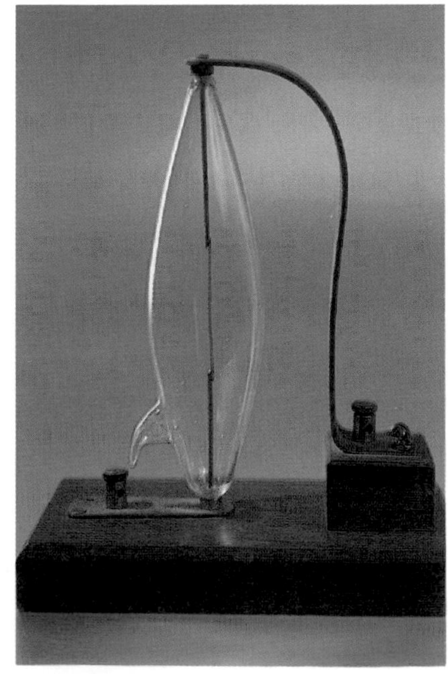

Incandescent lamps were developed almost simultaneously by Joseph Swan of Newcastle-upon-Tyne and Thomas Edison in the States. They later joined forces to produce light bulbs in Britain.

printing press. By 1904, this led him to devise an air-conditioning system whereby the air is 'washed' by sprays of water. This system is still in use today. Continuing his research, in 1911 Carrier devised an air humidity graph, that allowed him to make a rational estimate of air-conditioning requirements.

Artificial ventilation (1555)

Air conditioning can obviously be dated back to artificial ventilation and the person classically cited in relation to this is the mineralogist **Agricola**, who, in **1555**, described procedures for mine shaft ventilation.

We know that Leonardo da Vinci was interested in this problem too. But we must also remember the method of cooling by water evaporation in old Andalucian houses, where air entered after passing through vegetation and a fountain on the patio.

The Grand Palais *in Paris was lit by the first neon tubes in 1910.*

LIGHTING

Type	Date	Inventor	Country	Characteristics
Oil lamp	1804	Aimé Argand	France	Argand was the true inventor of these lamps which take their French name from the Swiss pharmacist Antoine Quinquet. Argand made the first one in 1804 in England. Quinquet modified the chimney, which he made of crystal.
Safety lamp	1816	Sir Humphry Davy	UK	Invented for miners. In 1811 Davy also discovered the principle of the electric arc.
Arc lamp	1847	W. E. Staite	UK	After Davy's first model of an arc lamp, which remained at the experimental stage, Staite's lamp was the first really effective lamp. It was perfected in 1848 by another Englishman, W. Petrie.
Yablochkov's candle	1876	Pavel Niolayevich Yablochkov	Russia	An improved arc light. In 1872, Yablochkov was asked to light the Moscow–Kursk railway, on which the Tsar was to travel, so that any potential terrorists would be unable to take advantage of the dark. The insulating metal melted as it was used. Victoria Embankment in London became the first street in Britain to be permanently lit by electricity in 1878.
Incandescent lamp	1878–9	Thomas A. Edison Sir Joseph Swan	USA UK	After a series of court cases for patent violations, the two inventors joined forces in 1883. The incandescent lamp was based on Joule's Law.
Tungsten filament	1910	William D. Coolidge	USA	Patented 30 December 1913.
Neon tubes	1909	Georges Claude	France	The first tubes were used to light the Grand Palais on 3 December 1910. In 1912 Jacques Fonsèque used neon for the first advertising sign on the Boulevard Montmartre *(Palace Coiffeur).*
Fluorescent tubes	1872–1923	Various scientists		The first fluorescent lighting was installed in towns in 1933.
Litec bulb	1965	Donald Hollister	USA	Instead of a filament it has an electromagnet which draws the current through a built-in electronic device.
SL lamp	1980	Philips	Netherlands	Based on the same principle as the fluorescent tube. It can produce the same amount of light as an incandescent lamp using 25 per cent of the energy and lasts five times as long.

Early apparatus (1919)

In **1919** the first air conditioned cinema opened in Chicago at the same time as Abraham and Straus, a large air conditioned store in Brooklyn, New York.

The use of independent home air conditioning systems was first foreseen in the United States in 1926 by H. H. Schutz and J. Q. Sherman. In 1931 they applied for a patent for an air conditioner to supplement windows.

Matches (1831)

The first primitive matches were developed at the end of the 17th century. They were simply small sticks of wood dipped in melted sulphur and were developed by the Irishman Robert Boyle (1627–91). They needed to touch something burning in order to catch light. The first real matches to light without contact with fire were simultaneously (and independently) invented in **1831** by the Frenchman **Charles Sauria** and the Austrian **Stephen von Roemer**. The first factories were set up in Vienna in 1833.

The substance used (white phosphorus) was highly toxic for the workers and the product was dangerously inflammable, which led chemists to continue their research.

Red phosphorus

The discovery of red phosphorus in 1847 enabled the Swede Lundström to create so-called safety matches, which need to be struck against a special material. These new matches first gained recognition at the Universal Exhibition in Paris in 1855.

Today the 42 thousand million matches we use annually have a tip that contains no phosphorus; red phosphorus is contained in the special striking strip.

Lift

Origins (1743)

The first known lift was built at Versailles in **1743** during the reign of Louis XV and was for use by the King himself. It was installed on the outside of the building, in a little courtyard, and enabled the monarch to go from his apartments (on the first floor) to those of his mistress, Mme de Châteauroux (on the second). Using a system of counterweights it was possible to move the lift without too much effort.

Mechanical lift (1829)

The first lift of this type was built in London, in the Coliseum in Regent's Park, in **1829**. It could take about ten passengers. The public were invited to go up into a replica of the dome of Saint Paul's cathedral to admire a panorama depicting London. It was thus more of an attraction than a real means of locomotion.

Otis lift (1857)

The first lift for public use was inaugurated on **23 March 1857** in New York. It was built by the American **Elisha Graves Otis** for R. V. Haughtwout & Co, a five-storey store on Broadway. E. G. Otis had already presented the first safety lift with a brake in 1852 in New York.

Hydraulic lift

The Frenchman **Léon Edoux** (1827–1910) installed two lifting devices using hydraulic pistons 21m *69ft* high at the Paris Exhibition in 1867. He called them *ascenseurs*.

The lifts using hydraulic pistons, which became widespread in the United States after 1879, went 20 times faster than Otis' 1857 lift. Their development was held up by the difficulty of digging very deep foundations. Nevertheless in 1889 Edoux managed to build a lift that travelled 160m *525ft* up the Eiffel Tower.

Electric lift (1887)

The first electric lift was built by the German firm **Siemens & Halske**, for the Mannheim Industrial Exhibition in **1887**.

It reached a height of 22m *72ft 2.4in* in 11 seconds. In one month it carried 8000 passengers to the top of an observation tower overlooking the exhibition.

The first electric lift to rise over 200m *656ft 4.8in* was built in New York in 1908.

Today the fastest lift (600m *968ft*/min) is in the Sunshine Building in Tokyo.

Vacuum cleaner (1901)

Hubert Cecil Booth designed and patented the first vacuum cleaner in London in **1901**. His Vacuum Cleaning Company provided a cleaning service with uniformed employees.

A much lighter machine was developed by an American inventor, Murray Spengler. He sold the rights to his vacuum to William B. Hoover who launched the Hoover Model 'O'.

Carpet-sweeper (1876)

The history of the mechanical sweeper starts at the end of the 17th century. In 1699 the Englishman Edmund Hemming invented a broom of this type to sweep the streets. Other Englishmen, James Hume in 1811 and Lucius Bigelow in 1858, can also be considered as creating forerunners of the carpet sweeper.

Allergic to dust

The Bissel was invented by the American **Melville R. Bissel**, who patented it on **19 September 1876**. He was the owner of a porcelain shop in Grand Rapids (Michigan). Mr Bissel suffered from an allergy to the dust produced by the straw he used to pack his pots. To cure this, he designed a broom with a cylindrical brush which pushed the dust into a container. The local success of this device led Mr and Mrs Bissel to set up the Bissel Carpet Sweeper Company to market their product. Their name has now been passed on to posterity, since it is often used to refer to the sweeper.

Dry cleaning (1855)

The first dry cleaner's was founded in **1855** in Paris by the Frenchman **J.-B. Jolly**. He discovered the principle of dry cleaning accidentally when he tipped a bottle of turpentine over a dress. He noticed that the dress was not stained, but on the contrary, cleaned.

The vacuum cleaner demonstrated here in 1910 was all the rage with wealthy housewives.

CENTRAL HEATING

Type	Origins	History
Hot air	**antiquity**	Used in China from ancient times. It was also widely used by the Romans in thermal baths (hypocaust) and private houses. Hot air heating was mentioned by Seneca (1st century AD). It reappeared in the 19th century.
Steam heater	**18th century**	The first model was installed in a factory by the Scotsman James Watt.
Hot water heater	**1777**	This system was already known to the Romans. In 1899 the radiator made of assembled elements first appeared in the United States.

A German from Leipzig, Ludwig Anthelin, made a further step forward in 1897. He discovered the use of carbon tetrachloride, which is much less flammable. Unfortunately this product attacks the respiratory tracts, so it was replaced in 1918 by trichlorethylene.

Flushing system (1590)

As early as **1590**, an English courtier **Sir John Harington** had invented a practical water-flushing system for cleaning toilets, but his invention did not end the reign of chamber pots.

It was not until 1775 that the British inventor Alexander Cummings patented a flushing system. In 1778, another Englishman, 30-year-old Joseph Bramah, invented the ball-valve-and-U-bend method still used today. But it was not until the end of the 19th century, with the advent of running water and modern plumbing, that the flushing system found its way into most homes.

Pocket ashtray (1989)

This was invented by a German, **Rainer Jakob**, not only to prevent smokers from scattering their ash everywhere, but to enable them to extinguish their cigarettes automatically. It is a cigarette case which incorporates an ashtray section. Once the case is closed, there can be no risk of fire or burns. It was presented at the **1989** inventors' exhibition in Geneva.

Detergent (1916)

The first synthetic detergent, Nikala, was invented in Germany in **1916**. It enabled water to penetrate fibres, but it could not itself remove dirt. The invention of modern detergents, that indispensable part of our daily life, goes back 50 or 60 years in some cases. The oldest, Lux flakes, was developed by the British manufacturer Unilever in 1921. Also from Unilever are Vim (1923), Persil (1932), Omo (1952), Skip (1959), Lux washing-up liquid (1959), Sun (1965) and Cajoline (1972). The famous detergent Ariel was created in 1968 by Proctor & Gamble (USA), who in 1982 created Vizir, the first machine-washing liquid.

Bleach (1789)

In **1789** the chemist **C. L. Berthollet** discovered the bleaching properties of Javel water. In the 19th century the pharmacist Antoine Labarraque (1777–1850) discovered its disinfectant properties and introduced it into hospitals.

Escalator (1892)

The escalator was born in 1892 from the combined efforts of the Americans **Jesse W. Reno** and **George H. Wheeler**. It was first given the name escalator by Charles D. Seeberger in 1899. The project for a mechanical stairway first gave rise to a moving slope, which proved rather dangerous. Reno perfected this slope by replacing it with rotating steps. He patented his invention on **15 March 1892**. Later he designed ribbed platforms that passed through the teeth of a comb fixed to the ridge of each step. The first public escalator was presented for the first time at the 1900 Universal Exhibition in Paris, before being installed in the United States in Gimbel's Department Store in Philadelphia. From 1922 onwards Reno and Wheeler's escalator was installed in numerous shops and public offices. The first spiral escalator was put into operation in 1985 in a Japanese shop by the Mitsubishi Electric Corporation.

Fire extinguisher (1816)

In **1816 Captain George Manby** from Norfolk developed a fire extinguisher which worked using compressed air. The first fire extinguisher using a chemical base was invented in 1868 by the Frenchman François Carlier. His extinguisher contained bicarbonate of soda and water. A bottle filled with sulphuric acid was attached inside, near the cap. To use the extinguisher, one had to break the bottle with a needle, thus freeing the sulphuric acid. A chemical reaction then produced carbonic acid, which forced the water out and helped put out the fire.

In 1905 the Russian Alexander Laurent hit upon the idea of mixing a solution of aluminum sulphate and bicarbonate of soda with a stabilising agent. The bubbles so formed contained carbonic acid. They floated on oil, petrol, or paint, and prevented contact with the air, thus with oxygen.

Ironing

Origins (4th century)

The origins of the iron are very old. In the 4th century the Chinese were using a kind of receptacle with a brass shaft containing embers. In the West the ancestor of the iron was the 'smoother' made of wood, glass or marble. It was used cold until about the 15th century, since the use of starching gum meant it could not be heated.

The first mention of the iron does not

This escalator in a San Francisco department store cost £2 million to design and build. Its spiral serves the shop's ten levels in a manner that looks complicated but which is quite simple.

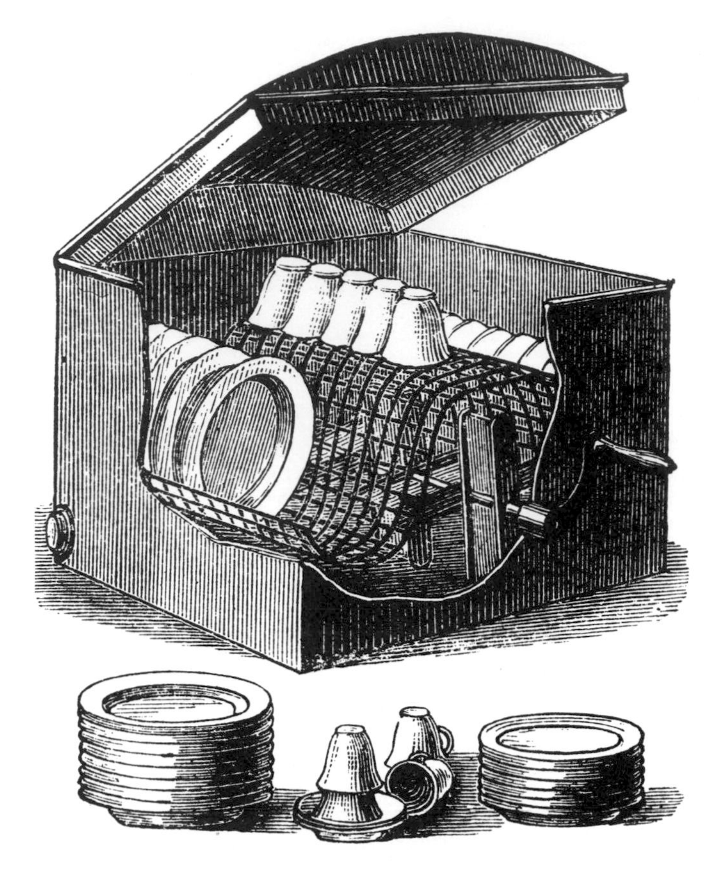

The dishwasher, like many household appliances, was conceived in the United States where there was a shortage of domestic staff.

appear until the 17th century. After this there were irons that were thrown on the fire, hollow irons filled with embers and lastly the classic laundry irons heated on stoves (19th century). Then other means of heating them were found, such as hot water, gas, or alcohol, until the first electric iron was perfected in 1882.

Electric iron (1882)

The electric iron was invented and patented by New Yorker **Henry W. Seeley** on **6 July 1882**. However, it could not be used at the time as homes were not then connected to an electricity supply.

Dishwasher (c.1850)

The first model of a mechanical dishwasher was developed between **1850** and **1865** in the United States.

The electrically powered dishwasher first appeared in 1912.

In 1932 an appropriate detergent, Calgon, was discovered. This facilitated the dishwasher's development.

The automatic dishwasher first appeared in 1940, also in the United States. It was not exported to Europe until about 1960.

Sewing machine (1830)

In **1830** the Frenchman **Barthélémy Thimonnier** invented the first sewing machine to work in a regular and useful way. It already included all the elements of today's machines.

The following year Thimonnier set up a company in Paris. But he met with violent opposition from workers. Thimonnier had to return to his native town of Amplepuis and to his trade as a tailor.

In 1834, at the same time as the Frenchman was carrying out his research, an American, Walter Hunt, designed a machine with two threads and a shuttle. Twelve years later Elias Howe replaced the hook on this machine by a needle with a hole in it.

In 1845 Thimonnier and his new partner Magnin made a second application for a patent for a machine which could do 200 stitches a minute and then, on 5 August 1848, another application for a machine made of metal, the *couso-brodeur*, which could do chain stitch. A little earlier Thimonnier had applied for a British patent, which he almost immediately gave over to a Manchester company. In 1849 the Englishmen Morey and Johnson bought an American patent. The American machine had a hooked needle, like Thimonnier's. In 1851 the Great Exhibition was held in London. By an unbelievable stroke of bad luck, the *couso-brodeur* arrived in London two days after the judges had examined the exhibits.

The Singer (1851)

In **1851** the American **Isaac Singer** was the first to build and market a sewing machine for domestic use in Boston, Massachusetts. His needle was taken from Howe's machine. Singer became far more famous than Thimonnier, who died a ruined man in 1857, after finally having the consolation of presenting his machine at the Universal Exhibition in Paris.

Washing machine (19th century)

Replacing the steam boiler, which itself replaced the washtub of the Middle Ages, the washing machine appeared during the **19th century**.

Composed of a wooden bin that was filled with soapy hot water, the first mechanical washing machines used heavy blades to stir the washing. This principle of tossing clothes inside a rotating cylinder still governs the operation of modern machines. One of them operated in 1830 in an English laundry. Around 1840, in France, an industrial model with double sheathing, four compartments, and a draining plug was designed; it was driven by means of a crank.

Electric washing machine (1901)

The first electric washing machine was invented and developed by the American **Alva J. Fisher** in **1901**.

It was not until the start of the Second World War that electric vertical-tub machines with built-in turbo-washers or with a vertical axle fitted with blades began to be mass-produced in the United States. Horizontal drum machines appeared in 1960.

Electric heater (1892)

The first electric heater was patented by the Englishmen **R. E. Bell Crompton** and **H. J. Dowsing** in **1892**. They attached a wire to a cast iron plate and protected the whole with a layer of enamel.

In 1906 the American Albert Marsh from Lake County (Illinois) invented an alloy of nickel and chrome which could be heated red hot without melting. This resistant alloy proved to be an ideal element for the construction of electrical heating apparatus.

Then in 1912 the Englishman C. R. Belling from Enfield in Middlesex perfected a refracting clay around which a nickel-chrome alloy wire could be wound. The same year he made the first Standard electric heaters.

Pocket clothes dryer (1987)

This clothes dryer was invented in **1987** by the American **Lucy Binger**. It consists of a hair dryer and a plastic bag. The tip of the dryer is slipped into the specially designed opening of the plastic bag, which contains the clothes to be dried. The other end of the bag has holes in it to allow air to pass through. This economical and portable clothes drier, ideal for travellers, is only suitable for small articles such as socks, stockings, etc.

Lock (antiquity)

The oldest locks seem to have first appeared in ancient Egypt. These massive locks made of hard wood worked using combinations of cylindrical pins of different lengths, which fitted into grooves cut in immobile components that acted as keys.

Safety locks

Safety locks date from the 18th century (the Englishman Robert Barron's throat lock in 1778, the pump lock by another Englishman Joseph Bramah in 1784). They were perfected in the 19th century, notably by Alexander Fichet (unhookable lock patented in 1829 and 1836), Charles-Louis Sterlin and Eugène Bricard (two bolted lock, 1829) and Linus Yale (pin lock, called the Yale lock, patented 6 May 1851, which was inspired by the ancient Egyptian locks).

The skeleton key was presented at the Chicago Exhibition in 1894 by Alfred and Jules Bricard.

Another American invention that we wouldn't like to live without: the electric washing machine was developed in the early years of this century.

Anti-theft electronic dog (1986)

The British Bulldog has all the advantages of a guard dog without the disadvantages. The device runs off the mains and is equipped with a built-in microphone. As soon as it detects a human presence it makes barking sounds identical to those of a real guard dog. It was invented by the Dutchman **A. van de Haar** in **1986**.

Automatic door (1987)

This is a Chilean invention, a mechanical device (a steering wheel) which enables the driver of any vehicle to open a door at the back of the vehicle without getting out or opening the window. Its advantages are comfort (for example when it is raining) and security. It was perfected and patented in **1987** by **Raul Espinosa Marty**, from Santiago de Chile.

Anti-asphyxia alarm (1988)

In **1988** two American doctors, **Kurt Shuler** and **Gerhard Schrauser**, who were horrified by the increasing number of accidents caused by carbon monoxide, developed a series of alarms using warning lights which come on when the level of CO in the air becomes dangerous. These are the world's first battery-operated alarms which can work anywhere.

In the bathroom

Bath (antiquity)

In antiquity, the Greeks and Romans used baths made of marble or of silver. In the Middle Ages people bathed in simple wooden tubs.

This one-piece bath at the Ben Youssef mosque in Marrakech dates from the 10th century. Mass-produced cast-iron bathtubs didn't appear in Britain until the 1880s.

Ambroise Paré's bath

The ancestor of the individual bath with heated water is the steamroom designed by the French surgeon Ambroise Paré (c.1509–90) for hydrotherapeutic purposes in the second half of the 16th century. But it was another 200 years before this comfort became a normal part of life. In the 18th century metal baths (already in use in the preceding century) became more widespread. At the end of the 18th century a special varnish, developed by a Parisian craftsman called Clément, made it possible to varnish sheet metal and thus to manufacture bathtubs at an accessible price, which led to their quickly replacing the traditional wooden or marble tubs which had been used until then. It seems the first hotel to provide a bathroom for every room was the Mount Vernon Hotel, Cape May, New Jersey (USA) in 1853.

Toilet paper (1857)

Toilet paper was invented in America by **Joseph Cayetty** in **1857**, but for a long time it remained a luxury item.

More than 70 years separates these two car-showers. The first dates from 1915 and the second is a more recent design by the Australian, Keith McLaren.

Bathroom scales (1910)

Scales for domestic use were invented in **1910** by a German company, **Jas Ravenol**. They were marketed under the name of Jaraso. The first American scale of this kind is attributed to J. M. Weber (patented in 1916). The first scale with a digital display was introduced by the Hanson Scale Co. in 1964. In 1984, H. S. Ong, a native of Malaysia, designed a talking scale.

Toothbrush (15th century)

The toothbrush appears for the first time in a Chinese painting from the end of the **15th century**. It appeared in Europe in the 17th century.

The first nylon toothbrush was Doctor West's Miracle Toothbrush, manufactured by Du Point in the United States in 1938.

The first patent for an electric toothbrush goes back to 1908, but they did not become popular until the 1930s.

Nivea cream (1911)

In **1911** a Hamburg chemist, **Paul Beiersdorf** (who also invented the sticking plaster in 1882), invented a skin cream. It was marketed by the company he set up in 1882. The cream, which was white as snow, was called Nivea.

Eau de Cologne (1709)

The Italian **Farina** family are credited with creating eau de Cologne in Germany in **1709**.

Perfume

Perfume originated in the East where refined manners and a wealth of vegetation were both to be found. Initially wood and scented resins were burnt on altars, and then perfumes were put into dishes to increase the flavour of food or, failing that, to act as an aphrodisiac.

The perfume industry, which has been well developed in France since the 17th century as a result of the impetus given by Colbert, was transformed by chemical discoveries and, at the end of the 19th century, saw the introduction of synthetic products. Today, these form between 50 per cent and 80 per cent of the component of perfumes, extending the range of natural scents and ensuring an improved stability.

Dry perfume (1984)

The first dry perfume was invented in **1984** by **Franka Berger** of France who, from 1985, commercialised it under her own name in America and Japan where it was immediately successful.

In 1989, Franka Berger invented the first mousse perfume for the body.

Perfume-lighter (1988)

Following the success of their famous biros, lighters and razors, on **22 February 1988**, **Bic** brought out four perfumes presented in the form of a mini-spray. The container, designed by Joël Desgrippes, is not much bigger than a lighter and can be used 300 times.

Kleenex (1924)

The first disposable paper hankies were produced in **1924** by the **Kimberley-Clark Co.** of Neenah, Wisconsin (USA), under the name of Celluwipes. The product was later renamed Kleenex-'Kerchiefs and then shortened to Kleenex.

Razor (12th century)

Obviously no one person can claim to have invented the razor. Man has always shaved, with seashells, shark's teeth, and later, bronze blades.

The razor, properly speaking, dates from the **12th century**. The steel razor was created at Sheffield, England in the 18th century.

Safety razor (1895)

In **1895**, the American **King Camp Gillette** patented the safety or mechanical razor, whose distinguishing feature was its double-edged replaceable blades. He marketed his product through the company he founded in Boston, in 1901. The Gillette Safety Razor Company has undergone slow but steady expansion since then.

Gillette introduced the twin-blade GII in 1971 and the first swivel head razor in 1975.

More than a thousand million people use Gillette razors worldwide.

Electric shaver (1928)

The first electric shaver was developed by the American **Colonel Jacob Schick**. The first model was patented and marketed in **1928**.

Lubricating razor (1986)

Launched simultaneously in 1986 by Schick (Schick Pivot Plus) and Gillette (Contour Plus),

An early eau de Cologne label.

this razor is designed with a blade containing a lubricating substance which gives a smoother shave without scraping the skin.

Lipstick (17th century)

At the beginning of the 17th century, women coloured their lips with a fairly harsh, slightly scented pomade coloured with the juice of black grapes and alkanet (dyers bugloss).

Later, cerates, a kind of salve, were produced which had a base of wax and oil. The same principle is used in the manufacture of Rosat, a product used to prevent lips becoming sore and cracked.

In the 20th century, chemists came to the rescue of the cosmetologists. They succeeded in manufacturing sticks of rouge which were easily moulded and did not have any adverse effects on the lips or mouth.

Soap (2500 BC)

For a very long time man had been aware of the cleansing properties of oily and fatty compounds, associated with vegetable ash. A form of soft soap, probably of a very indifferent quality, existed in Mesopotamia in 2500 BC. It reached Rome and Gaul in the 4th century, *sapo* being a word of Gaulish origin. From the 13th century onwards, it was known how to make a solid soap using such fatty raw materials as soot or oil and alkalis such as plant ash or, better still, ash from seaweed which contains sodium carbonate. Until the end of the 18th century, there was very little development in soap-making techniques, and it remained a rare and expensive commodity.

Marseille soap came into being in 1791 with the invention of a new process for the manufacture of sodium carbonate by Nicholas Leblanc.

Facial care for men (1975)

The first facial care for men was introduced in **1975** by **Yves Saint Laurent**.

Clothes

Nylon stockings (1938)

The stocking market was revolutionised by the appearance of the first nylon stockings in **1938**. Nylon was developed by a team of researchers from the American company Du Pont led by **Doctor Wallace Carothers** (1896–1937), who committed suicide a year before stockings appeared. The first nylon stockings reached Europe in 1945.

Bikini (1946)

On the **3 June 1946**, the Frenchman **Louis Réard** presented an extremely daring two-piece in his swimwear collection. He called it the bikini because he considered it as explosive in its own way as the American atomic bomb which had been exploded four days earlier on the Pacific island of Bikini. The novelty was such that none of his professional models would present it and Réard had to appeal for help to a dancer from the Casino in Paris, Micheline Bernardini. His creation was patented and, duly protected, the word bikini soon entered the dictionary.

Press-stud (1886)

The press-stud was invented on **29 May 1886** by an industrialist from Grenoble in France, **Pierre-Albert Raymond**. The patent was requested in the name of the company Raymond & Guttin, the latter being the co-inventor. The metal press-stud was initially used as a fastener in the local glove industry.

Many imitations were produced worldwide, but with each attempt the company won its case. It extended its products to the international level, and is still in existence today.

Screw-on button (1988)

The first button without thread, that doesn't have to be sewn on but which is screwed into place, was developed by the Swedish sculptor, **Gudmar Olovson**. Presented at the Paris Trade Fair in **1988**, this ingenious device can be used on any kind of material. You merely screw the button into place, preferably near a button hole, and then open out the end of the screw and the button is firmly fixed. To remove it, you simply reverse the process!

Shoe (4000 BC)

The museum in Romans, France, has a pair of Egyptian shoes made of papyrus, dating back to 4000 BC. But the shoe was undoubtedly invented much earlier, although the distinction between the right and left foot only dates from the middle of the 11th century. Until then, respectable footwear had to be made to measure.

The pump, which was originally a basic type of slipper for wearing indoors, appeared in the 16th century. Its popularity increased during the 18th century when it became extremely fashionable. From 1900, it played an important part in the emancipation of women by replacing the ankle boot and becoming from then on a comfortable town shoe.

The Dorelot (1985)

Françoise Douez, a French psychologist, has invented a new type of clothing for newly born babies in an attempt to avoid excessive handling and the often traumatic experience of having successive layers of clothing passed over their head. The *Dorelot*, as it is known in France, wraps round the baby and is held in place by Velcro-lined tabs. The French patent was registered in 1985 and since then it has been patented in 47 countries.

Pin (1817)

In **1817**, the American **Seth Hunt** informed the Patent Office of his invention of an automatic machine for manufacturing one-piece pins, with body, head, and point. His machine began operation in 1824, when Samuel Wright filed a patent in England. England had been the home of the industrial pin since John Tilsby had founded the first large pin works in Gloucester in 1625.

Hunt's machine was improved in 1838 by the Englishmen Henry Shuttle Worth and Daniel Foote Taylor, from Birmingham. Their pin was less dangerous.

Safety pin (1849)

The invention of the modern safety pin is attributed to the American **Walter Hunt**, who developed it in **1849**. It appears that fibulas and brooches, used in ancient Crete to attach draped clothes, were made according to a related principle.

ITEMS AND ACCESSORIES FROM ANTIQUITY

Objects	Origin	History
Needle	**prehistoric**	Used in Egypt, Greece and Rome. In 1370, the German city of Nuremberg was an important centre for the manufacture of polished steel needles.
Stockings	**Ancient Rome**	In the 16th century, the strips of cloth used until then were replaced by knitted hose. The stocking trade revolutionised their manufacture at the end of the same century.
Button	**3000 BC**	An ancient invention found in the Indus Valley (3000 BC) and in Scotland (2000 BC). Materials became more varied in the Middle Ages.
Cravat	**Ancient Rome**	Of military origin (the Roman *focale*). Became more widely worn during the 17th century. The introduction of the fashion is attributed to either the Swedes c.1600 or the Croatian Army c.1668.
Fan	**antiquity**	Lotus leaves in Egypt, peacock feathers in Rome. The folding fan appeared towards the end of the Middle Ages.
Mirror	**Ancient Egypt**	Many mirrors have been discovered in the tombs of the pharoahs. The invention of the crystalline mirror is attributed to the Venetians during the 18th century.
Comb	**prehistoric**	Used as early as 8000 BC in Scandinavia. Used in Ancient Egypt. The manufacture of tortoise shell combs began in the United States in 1780.

Zipper (1890)

Around **1890**, the American **Whitecomb Judson** devised a quick zipper system based on interlocking small teeth. The idea was ingenious but its practical application not simple. Judson filed his patent in 1893 and entered into partnership with a lawyer, Walter, to found a company.

In 1905 machines to manufacture zippers were in operation, but their products were far from perfect. It was not until 1912 that Judson's invention provided full satisfaction to its users: this was due to the improvements made by the Swede Gideon Sundback.

The American zipper was marketed by the Goodrich Company, which used it on snow boots.

Macintosh (1823)

It was the Scotsman **Charles Macintosh** (1760–1843) who, in **1823**, was the first to succeed in producing a waterproof cloth that could be used to manufacture clothing. His cotton fabric, imbued with a mixture of rubber and turpentine, maintained its full flexibility. The Macintosh became *the* waterproofed overcoat worn by men during the 19th century.

Jeans (1850)

Jeans were created around **1850** by **Oscar Levi Strauss** for pioneers in the West of America. These hard-wearing trousers were originally cut out of a blue cloth which served for tenting. This cloth had been imported from Nîmes, France's traditional production centre, hence the term denim. The famous copper studs first appeared on the pockets in 1860.

The earliest mention of the word *jean* dates from 1567 and appears to be a corrupt form of the word genoese, from Genoa. A twill cotton fabric was also manufactured in Genoa, where sailors wore pants made of this material. Denim is a heavy cotton cloth having an ecru woof and an indigo warp.

It was a Scot, Charles Macintosh, who invented the material that bears his name, and another Scot, Alexander Sykes, who developed raincoats for sheep.

Miniskirt (1965)

The miniskirt was created by the dress designer **Mary Quant** in her store, Bazaar, on King's Road, London in the spring of **1965**.

Almost simultaneously, the French fashion designer Courrèges was creating a line that was very architectural, quite short and futuristic. Wearing miniskirts, opaque tights and small helmets, fashionable women resembled astronauts. But this was a haute couture collection, whereas Mary Quant's skirts were enthusiastically taken up by a generation of young women.

Handkerchief (2nd century BC)

The handkerchief is thought to have appeared in Rome during the 2nd century BC, but it did not become part of everyday life. It did not reappear until the 15th century in Italy where the distinction was made between the handkerchief used for the nose, which was slipped into the pocket, and the handkerchief used for the face, which was held in the hand. It reached France during the 16th century and was immediately adopted by the Court and actors, becoming as indispensable to tragedy as the fan can be to comedy. It became widely used during the 17th century.

Umbrella (2nd century BC)

It would seem that the umbrella was a Chinese invention. In fact it already appears on prints as early as the **2nd century BC**. From China it spread to Persia and from there it was

This is what a strip of Velcro looks like in close-up. The hooks were inspired by a burdock seed head.

taken to England by Sir Jonas Hongway, an 18th century globe-trotter. But in spite of the British climate, it was not an immediate success. The term *parapluie* (literally, against the rain) had already been coined in Paris in 1622. The folding umbrella was invented by a Frenchman, Jean Marius, in 1705.

In December 1970 Maurice Goldstein of France registered the idea of an inflatable umbrella. In 1978, Claude James, an Englishman living in France, patented a similar idea.

Rucksack (1936)

On **22 January 1936**, the patent was issued for the first modern rucksack, an improved version of the Tyrolean rucksack developed by the company, **Lafuma**. The main innovation was that the metal grid no longer protruded but was incorporated into the rucksack. This was the forerunner of all modern rucksacks.

Bra (antiquity)

There is proof that the bra existed in Ancient Rome. It is depicted on ceramics where the female gymnasts wear the *strophium*, a sort of scarf wound over the breasts in order to provide support.

In 1805, a band of elastic material was worn below high-waisted muslin dresses to keep the bust firm. It crossed at the front and was fastened at the back of the neck. In 1889 Herminie Cadolle invented the bra as we know it, but it did not become widely used until the 1920s. According to current statistics, the average English woman buys 3 bras per year compared with 1.3 and 5 bought by her French and American counterparts.

Velcro (1948)

Velcro is a Swiss invention whose discovery dates from **1948**. Returning from a day's hunting, the engineer **Georges de Mestral** often noticed that burdock seed heads clung to his clothing. Under the microscope he discovered that each of these heads was surrounded by minute hooks allowing them to catch onto fabrics. It then occurred to him to fix similar hooks on fabric strips which would cling together and serve as fasteners.

Eight years were needed to develop the basic product: two nylon strips, one of which contained thousands of small hooks, and the other even smaller loops. When the two strips were pressed together, they formed a quick and practical fastener. The invention was named Velcro (from the French *velours* velvet and *crochet* hook). It was patented worldwide in 1957.

At the hairdresser's

Babyliss (1959)

In **1959**, the first electric curling tongs, known as Babyliss, were invented by two Frenchmen, **René Lelièvre** and **Roger Lemoine**.

Hairstyle video (1988)

You can now see yourself on a screen with the hairstyle of your choice, without losing a single lock of hair. This has been made possible by the use of a video and a graphics sheet linked to a micro-computer, designed for use

The Video-Look now makes it possible to foresee the disasters that might befall at the hairdresser's.

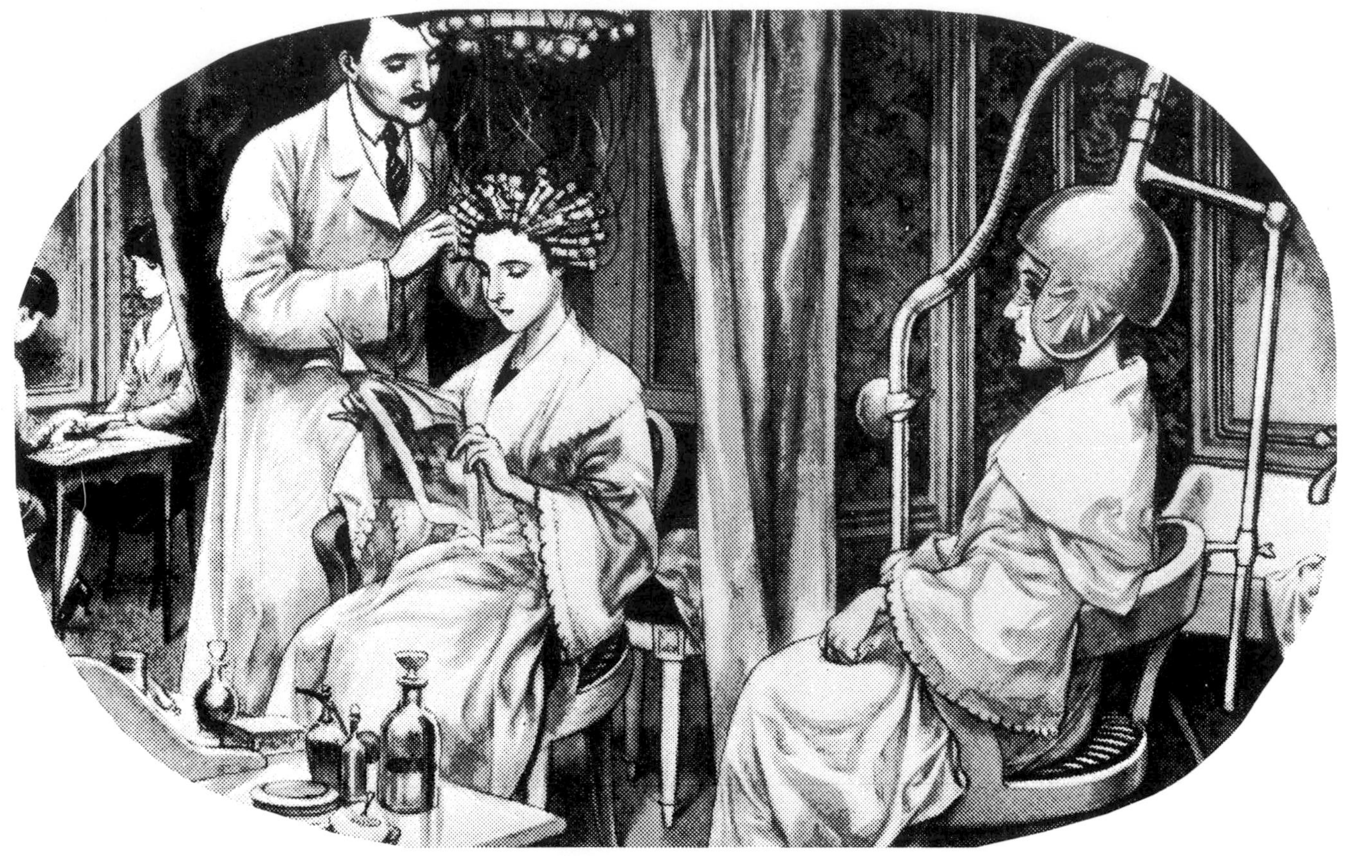

In 1929, having a hairdo was quite an operation.

by professional hairstylists and beauticians and for beauty counselling. The Video-Look was invented in **1988** by **Alain Saulnier** of France.

Hair colour (1909)

The first conclusive tests were carried out in **1909** by the French chemist, **Eugène Schueller**. He founded a company which in 1910 became L'Oréal. In 1927, the hairdressing industry was revolutionised by the invention of Imédia, a dye which was manufactured from organic colouring agents and which offered a wide range of natural shades.

Hair lacquer (1960)

The year **1960** saw the emergence of new concepts in the shape of hairstyles. It was the era of backcombed hair. The L'Oréal laboratories developed an original formula lacquer with a synthetic polymer base and then a revolutionary system of micro-diffusion by aerosol. Elnett lacquer actually created an invisible net which held the hairstyle in place but which disappeared when it was brushed out.

The set (c.1870)

The modern set was invented by **Lenthéric** of France around **1870**. The waves were no longer obtained by heating the hair with curling tongs, but by drying them with warm air. During the 1920s, the Parisian hairstylist, Rambaud, realised the advantages of combining the techniques of the permanent wave and the set. After cutting and perming the hair, Rambaud, rolled it and secured it with curling pins. It was then dried under a hot air drier.

The result was surprising. The hot air permanent applied on its own produced tight little curls which did have the advantage of not dropping out, but which were not particularly attractive. Rambraud had replaced it with a soft and loosely waving hairstyle.

Hot air perm (1906)

On **8 October 1906**, **Nestle**, a German hairstylist living in London, demonstrated a new method of curling hair. He then went to live in the United States where his new invention, Permanent Waving, was extremely successful.

Hair washer (1988)

Japanese manufacturers have just discovered that women have ideas that can prove useful. A 24-year-old Japanese woman, **Fumiko Akutsu**, a factory worker for Mitsubishi, was responsible for the invention of a personal hair-washing machine for which she won first prize in a competition for new ideas run by Mitsubishi in **1988**.

It should be pointed out that many Japanese homes do not have bathrooms, showers or running hot water. This tiny machine has been extremely successful: 20 000 were sold in three months.

Hair dryer (1920)

The two earliest models appeared in Racine, Wisconsin (USA), in **1920**. They were the Race, made by the **Racine Universal Motor Co.**, and the Cyclone, made by **Hamilton Beach**. They were manual models.

In the winter of 1951, Sears, Roebuck & Co. marketed the Ann Barton, the first helmeted model for home use.

Shampoo (1877)

The term *shampoo* originated in England in **1877**. It was derived from *champo*, a word in Hindi which from *champna* means to massage or to knead.

Originally brewed by hairdressers, shampoos were made by boiling soft soap in soda water. But it was not until after the Second World War that shampoos came into general popular use.

Leisure

Metal detector (1931)

Those who have scoured the beaches for treasure buried beneath the sand are quite familiar with these small instruments. They were invented in **1931** by the American, **Gerhard Fisher**, and were originally intended for industrial and geological use. The chance discovery of some coins dating back to the American Civil War, opened up a whole new dimension. Since then, they have been continually improved. A new range was patented in 1985 by the Fischer Research Laboratory. They are light, easy to handle and capable of

It was an American clergyman, Clement C. Moore, who developed the modern Father Christmas around 1822. The Laplanders now organise trips by Concorde to visit Santa at home.

detecting any kind of precious metal on land and even in shallow water. The latest model, the 1280 X Aquanaut was designed for underwater treasure hunting and is operated by a battery which will last for up to 40 hours.

Sun-tan parasol (1984)

In **1984** the Englishman **John Sear** developed a sunshade which makes it possible to sunbathe without suntan oil, without sunglasses and with complete protection from the sun. It is made from Solmax, a filter screen invented in 1982 by another Englishman, J. A. Cuthbert. Tests have proved that this filter, in violet coloured plastic, will, as long as it retains its colour, exclude 90 per cent of the sun's harmful rays while letting through 75 per cent of the UVA rays which tan the skin. In this way it eliminates the risks of sunburn and sore and inflamed eyes.

Cigarette (16th century)

The cigarette was invented by beggars in Sevilla, Spain, at the end of the **16th century**. They had the idea of rolling the tobacco salvaged from cigar butts inside a small cylinder of paper. Cigarette use spread to Portugal, Italy, England, and France.

The first industrial cigarette factory opened in Havana in 1853.

DIY

A pouring lip for paint (1987)

The pouring lip was invented in **1987** by the American company **Spill Bill**, and makes life a lot easier for painters and diy enthusiasts. It is made of flexible plastic and fits onto the rim of the paint tin once the lid has been removed. The paint can be poured easily and cleanly into another receptacle and the brush can be rested on it without running down the side of the tin. Once the paint is dry, it is easily removed from the lip.

High temperature paint (1988)

Al-Coat is a special high temperature paint developed in **1988** by the Swiss company, **Lorito**. It resists extremely high temperatures and strong sunlight without bubbling and provides protection against ageing and cracking. It can be applied with a paint gun or a brush to boilers, stoves, radiators, pipes and car exhausts. It dries within 20 minutes.

Electric hand drill (1917)

After developing presses for postage stamps, machines for attaching boot buttons and machines for printing bank notes, two young American inventors, **S. Duncan Black** and **Alonso G. Decker** revolutionised popular diy in **1917** by manufacturing the first rotary hand drill. Their portable drill weighed 11kg *24lb 4oz*. After the Second World War, much lighter and greatly improved hand drills enjoyed a rapid and widespread popularity, to the point of becoming a basic and indispensable item for all diy enthusiasts.

Gas operated drill (1986)

This new wireless hand drill was invented in **1986** by the Japanese company **Tanaka Kogyo**. It has a two-stroke engine, a fuel tank and is completely safe to use, even in damp weather. The absence of electrical wiring makes it safe and easy to use.

Plane (c.1200 BC)

The plane appeared during the Iron Age, around 1200 BC. The earliest depictions we have date back to the Gallo–Roman era, in countries occupied or influenced by the Romans.

A variation of the plane, the trying-plane, is one of the earliest tools. The best preserved are those which were carefully worked, sculpted and polished, between the 16th and the 18th century. Many trying-planes are real museum pieces.

Saw (antiquity)

The saw would appear to have been invented by the Egyptians during the period corresponding to the Bronze Age, around 3500 BC to 1200 BC. Roman craftsmen produced a wide variety of saws, including the frame saw, which until the 14th century formed the basis for this type of equipment. The metal saw appeared in the 15th century.

Screwdriver (17th century)

This is a tool which has been continually developed since it was first used at the end of the 17th century. The screw has become more important since the middle of the 19th century with the development of machine tools. For a long time, producing screws was a laborious task since the grooves had to be filed by hand.

The Englishman John Whitworth was responsible for the standardisation of the thread in 1841.

Dovetail screwdriver (1978)

The Dovetail Screwdriver was invented in **1978** by the American, **Earl Benitz**. It is a screwdriver which can do everything at once: it can hold the head of the screw while starting to bore the hole and finally insert the screw firmly and accurately. Its steel shank is unlike that of a standard screwdriver. Instead of being flat it is dovetailed and fits into the slot in the screw. From here on, the job can be completed single-handed. It can be used in even the tightest of corners.

Brace (15th century)

The forerunners of the brace, i.e. all tools used for piercing, existed in prehistoric times. The Egyptians used an instrument derived from the hand-drill. The auger, which only dates back to the middle of the Bronze Age, found its final form during the Iron Age.

The screw auger, from which the brace was developed, seems to have been used in Scandinavia from the 11th to 13th century, mainly on the structures of river going vessels. The brace, in the form that we know it, appeared in the first quarter of the 15th century. The first known depiction of it is on the painting of the *Annunciation* by Robert Campin, Master of Flémalle.

Nail (18th century)

The first machines to make nails date from the **18th century** (patents of the American **Ezekial Reed** in 1786 and the Englishman **Thomas Clifford** in 1790). However, the artisan's nail goes even further back in time, since the oldest known nails were found in Mesopotamia and date from approximately 3500 BC.

The hammer (prehistory)

This is one of the oldest tools. A stone may have been the first hammer. The first decisive improvement was the addition of a handle. The stone axe hammer of the Neolithic Age already had a handle, as does the current hammer. However, stone was a fragile material. It was not until the Bronze Age that the metallic hammer made its appearance and was used to hit all objects.

Commerce

Coining (7th century BC)

Coining, i.e. the affixing of an official mark to a gold ingot, was invented in the middle of the 7th century BC in Asia Minor during the reign of Ardys, King of Lydia.

The coins were small ingots made of a mixture of gold and silver known as *electrum*, found in its natural state in the river which flowed through Sardis, the capital of Lydia. The name of the river, the Pactolus, later became synonymous with abundance and wealth.

The first coins to be made from this mixture, were shaped like flattened pellets. On one side they bore a triangular or square hallmark and on the obverse side, an Assyrian-style lion whose nose was decorated with a sort of shining globe.

Traveller's cheque (1891)

The traveller's cheque was invented in 1891 by **American Express**. The first cheque was counter-signed on **5 August 1891** and the system spread rapidly throughout Europe, contributing to the development of tourism and international exchanges.

Since 1985, traveller's cheques have existed in ECUs, the European Currency Unit.

Disappearing cheques (1988)

Some American inventors (who carefully conceal their identities) developed cheques that self-destruct a short time after being cashed. They have cost the banks $70 000!

Watermark (1282)

The translucent mark printed on a piece of paper, known as the watermark, appeared for the first time in Fabriano, Italy, in **1282**. The Italian word, *filigrana* or filigree, used originally to refer to a piece of delicate work in gold or silver. For centuries, paper makers used this method, known as the clear watermark to mark their products.

Dark watermark

Towards the end of the 18th century, the appearance of the watermark changed due to the appearance of a dark line, the dark watermark. Combined with the clear watermark, it provided increased protection.

Shaded watermark

Finally, by combining the raised or clear watermark with the sunk or dark watermark an Englishman, William Henry Smith, invented the shaded watermark in 1848. This transformed the art of the watermark by making it possible to produce more complicated compositions with an almost unlimited fineness of detail and richness of half-tones.

Credit card (1950)

The first company to organise payment by credit card was founded by an American **Ralph Schneider** in **1950**. The first two hundred members of the Diners Club were able to dine on credit in 27 New York restaurants. The Bank of America was the first to introduce a bank credit card, the Bankamericard, in 1958.

Biometry

It is currently (theoretically) possible to check the identity of a person without any risk of error by using the techniques of biometry, based on the biological parameters of the human being i.e. fingerprinting, retinal printing, voiceprinting etc.

This type of technology is of interest in the areas of defence, the nuclear industry, banks, large companies etc. Nearly a thousand organisations are currently interested. The first operational biometric system was the Eye-Dentification 7.7.

Christmas shoppers bustle outside Bourne & Hollingsworth in Oxford Street in 1919.

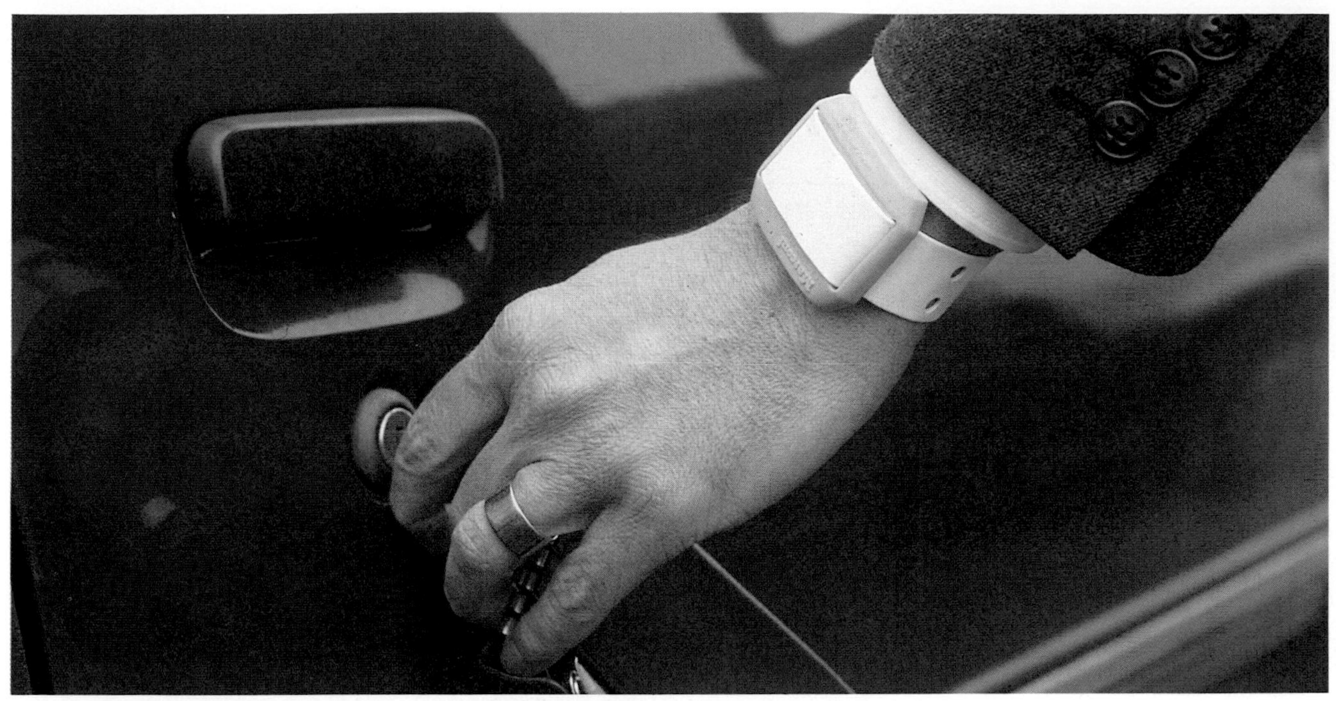

The electronic security bracelet is worn by more than 10000 Americans.

'Eye prints' (1984)

After fingerprinting comes 'eye printing', launched in September **1984** by the American company **Eye Dentify**. It is a rapid system of identification by a retinal image which sweeps the back of the eye and enables a more accurate identification than by the standard fingerprinting method. To date, a little over 400 EyeDentification System 7.5s have been installed, 280 of which are in the United States. This makes the company the world leader in biometric systems.

IDX 70 (1987)

The main competitor of the EyeDentification system is the IDX 70 produced by the American company Identix. This is a system for the identification of fingerprints, recorded on a memory card. The invention is based on a process known as global vision developed by researchers at the University of Berkeley and by the work of Rockwell for the FBI in 1962.

Identification by fingerprinting was invented by the English physiologist and cousin of Charles Darwin, Sir Francis Galton (1822–1911), in 1892.

Geometry of the hand (1987)

A device based on the evaluation of the geometry of the hand, known as the Palm Recognition System, has been developed by **Mitsubishi** and is used only in Japan.

Sign-on (1987)

The Sign-on system is based on the recognition of the dynamics of the signature, in other words the movement made during signing. It was brought onto the market by the English company **Alan Leibert Associates** in **1987**. Another, much earlier, British system was also based on the (non-dynamic) recognition of the signature. Verisign was developed in 1973 and marketed in 1982 by the Transaction Security Division of Analytical Instruments Ltd.

Department stores (1824)

The first department store was the *Belle Jardinière*, founded in Paris in **1824** by **Pierre Parissot**. His publicity and sales methods revolutionised commercial practice. He insisted that his merchandise was sold at 'a fixed price and for cash' whereas trade had always been based on two main principles: the negotiation of the price between the seller and the buyer and buying on credit. Harper's Building, designed by the architect John B. Corlies, opened in New York in 1854, followed in 1858 by the Crystal Palace Bazaar, the first department store to be opened in London.

Anti-theft label (1988)

Metostop was invented by the Swedish company **Esselte** and has been responsible for a 70 per cent reduction in the thefts carried out in department stores. It looks very ordinary and should, in theory, be able to protect all types of merchandise: books, records, cassettes, as well as alcohol and perfumes, products which until now have proved difficult to defend.

Tele-shopping (1985)

The first television sales channels were introduced in the United States in **1985**. Viewers are invited to telephone their orders for items shown on their screens during various shows and games.

Car hire (1918)

Car hire was introduced in **1918** by a second hand car dealer from Chicago. He hired out 12 vehicles. His company was bought in 1923 by the president of the yellow taxi cab company of Chicago, John D. Hertz who renamed the unsteady business the Hertz Self-Drive System.

Scales (c.3500 BC)

In about 3500 BC, the Egyptians weighed wheat and gold using scales which consisted of two pans suspended from a beam. The scales with unequal arms, were invented in Italy, probably in Campania, c.300–200 BC.

Bill of exchange (4th century BC)

In the **4th century BC**, the Greeks invented the bill of exchange. Isocrates' (436–388 BC) *Discourse on Banking* bears witness; he refers to this means of payment allowing one to travel without taking along large sums of money.

Travellers handed over a sum of money to their local banker. In return, the banker gave them a letter. Upon presentation of this letter to a banker at the traveller's destination, the traveller would be given the money required.

The bill of exchange was the forerunner of the bank note.

Banknotes (1658)

Originally, bankers at medieval fairs delivered registered receipts to their depositors. Then, in about 1587 in Venice, it became possible to transfer these receipts through the practice of endorsement. It was this endorsement that helped spread the use of paper money.

The first bank to issue banknotes was the **Riksbank** of Stockholm in **1658**. The Bank of England was founded in 1694.

Safe (1844)

In **1844**, the Frenchman **Alexandre Fichet**, a locksmith born in 1799, invented the first modern safe, replacing strongboxes with secret compartments which were used until then but which were not resistant to fire or theft.

In 1829, he had already applied for a patent for a burglarproof lock of his own invention.

THE INVENTIONS WE CANNOT DO WITHOUT

There are a number of inventions which have become such a part of our daily lives that we cannot imagine what life would be like without them. Some of the most recent include:

Swatch watch

The ultimate in interchangeable and disposable watches which can be co-ordinated with a shirt or a tie. The Swatch watch has revolutionised our attitude to watches which used to be serious, even austere, by making them a 'trendy' fashion accessory. The Swatch was invented in Switzerland by Ernst Thomke and two engineers, Jacques Müller and Elmar Mock.

'Post-it' notes

This invention has invaded our daily lives, in the office, at home, everywhere. It was invented purely by chance in 1970. Doctor Spencer Sylver of the American company 3M was involved in research on a completely different product when he discovered an adhesive which sticks without sticking. He sent samples of his discovery to other laboratories in the 3M group, but no use could be found for this surprising product.

It was not until ten years later that Arthur Fry, another research worker in the 3M group, found a use for what was to become the Post-it, again purely by chance. He was a member of a choir and was trying to find a way of marking the pages of his music book without damaging the paper. And this was why, in 1980, he put a thin layer of this famous 'unknown' adhesive onto the page markers of his score . . . and it worked! The little pastel coloured pieces of paper which stick, unstick and can be re-stuck at will, came into being. The name Post-it was invented in 1981.

Pre-prepared salads

This is indeed an invention! It came into our lives in the 1980s with pre-prepared vegetables, cabbages, grated carrots, radishes etc. They are known as the fourth range, a term which is straight out of science fiction, the two previous ones being tinned and frozen food, and the first being the traditional forms of food.

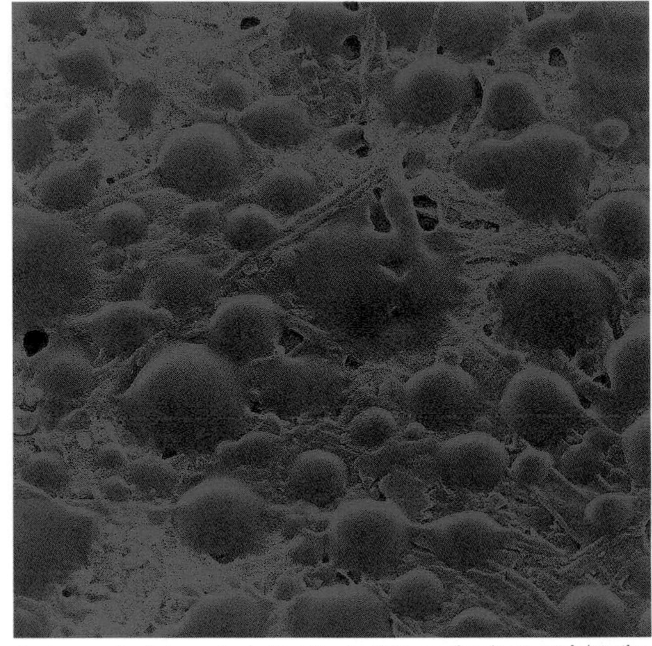

A microscopic photograph of a Post-it note. Spheres of resin are sunk into the paper; each time the paper is pressed, bubbles burst and glue is released.

Frozen foods

Frozen foods were commercialised in 1924 by the American Clarence Birdseye. Initially basic or not very elaborate products were sold, such as frozen vegetables and fish in breadcrumbs. Pre-cooked frozen meals were introduced by Findus, one of Birdseye's competitors.

The disposable razor

The disposable razor was invented in 1975 by the French company, Bic, the famous biro manufacturers. The company was directed by Baron Bich who was of the opinion that half a razor blade was sufficient for a shave and that with the saving made on the other half, it was possible to manufacture a handle.

Scotch tape (1925)

An American, Dick Drew, invented adhesive tape in 1925. Then a young assistant at the 3M laboratory at St Paul, Minnesota, Drew asked car manufacturers to test the first samples of waterproof abrasive paper. At that time, car-body builders had to paint cars in two tones. Paint was applied by spray gun, the difficult part was to separate the colours clearly and distinctly. Glued together newspapers were used, but it often happened that when the bands were removed, the fresh paint also came off. Drew studied the problem and, with the encouragement of management, sought a solution: adhesive masking tape. Five years later cellulose adhesive tape appeared.

But why was it originally called Scotch? Because, so as to make it easier to put on and peel off, 3M delivered to the car-body builders a tape, only the edges of which were self-adhesive. Suspecting that this was done to save adhesive tape, the workers called the bands Scotch tape.

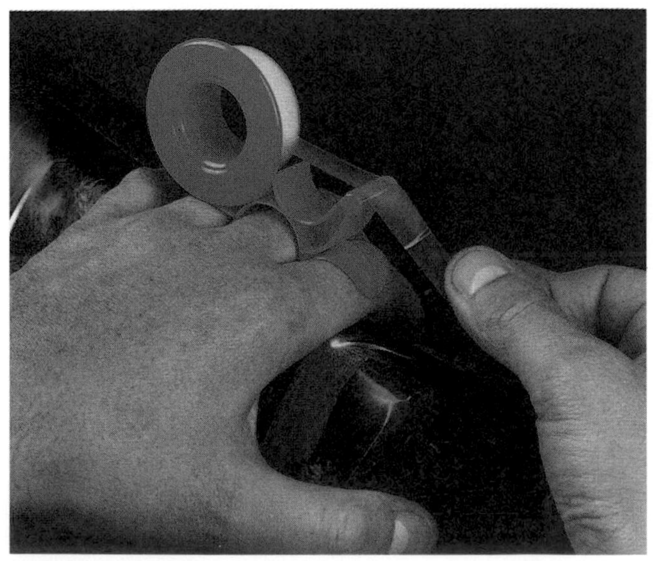

This ingenious sellotape-ring leaves both hands free. It was invented by a Swiss inventor, Roland Beuchat, and was presented in Geneva in 1989. It is often the simplest ideas which are the best.

Stock exchange (1450)

The stock exchange, that is, a place where financial transactions occur, was created about 1450. Until then, merchants and bankers got together at fairs. In the middle of the 15th century, a family of Bruges bankers, the Van de Bursens, opened its house to these transactions. Over the entrance portal was a frontispiece depicting three engraved purses. Antwerp in turn opened an exchange in 1487. It soon became the largest in Europe. Sir Thomas Gresham (c.1519–79) founded the Royal Exchange in 1568, which became known as the Stock Exchange in 1773.

Supermarkets (1879)

The first shops of this type were born in the United States in **1879**. They were introduced into Britain in 1909. In those days, they had a plain and rather drab appearance, not at all like the flashy style that we know today.

Cash register (1879)

The cash register was invented by the American **James J. Ritty** on **4 November 1879**. He owned a saloon in Dayton, Ohio, and the constant quarrels with his customers exasperated him.

During a boat trip to Europe, he noticed a machine that registered the number of times a propeller turned. This machine triggered the idea for his cash register, which served both as a printing–adding machine and a till.

Supermarket trolleys (1937)

In Oklahoma City (USA), the owner of the Humpty Dumpty Store invented the first supermarket trolley on **4 June 1937**. He had remarked that his customers had trouble lugging all their purchases through the different departments. He converted folding chairs into carts: the feet were mounted on wheels, a basket replaced the seat, and the back served to push the vehicle.

In the office

The 1964, the Soviet Patents Office received its first (Russian) photocopier for ordinary paper. In October 1988, an agreement was signed between Rank Xerox Ltd and Vnesghtorgizdat, the Russian publicity agency, for the opening, in the same year, of the first public photocopying service in the Soviet Union.

Carbon (1806)

Carbon paper, used to obtain several copies from one document, was invented by the Englishman **R. Wedgewood**, who patented it on **7 October 1806**. The process he described employed a thin sheet of paper saturated with ink and dried between sheets of blotting paper.

Photocopy (1903)

Photocopy refers to the process of rapid reproduction of a document by the instantaneous development of a photo negative. It was invented by the American **G. C. Beidler** in **1903**. Beidler, an office clerk, noticed the constant need for copies of documents, so he developed a machine for replacing laborious manual or typed copies and patented it in 1906. The first photocopy machine was marketed by the American company Rectigraph in 1907, but it was not until the 1960s that the photocopy became commonplace.

Xerography (1938)

On **22 October 1938**, the American, **Chester Carlson** (1906–68) produced the first xerographic image (from the Greek *xeros* meaning dry and *graphein* to write). He called the new process Xerography and patented it after several improvements had been made. Between 1939 and 1944, 20 companies refused his patents. In 1944, the Battelle Memorial Institute, a non-profit-making organisation based in Columbus, Ohio (USA), signed an agreement with Carlson and began to develop xerography.

In 1947, Battelle signed an agreement with a small photographic business, Haloid, which later became Xerox.

In 1959, the first photocopier, the Xerox 914, was brought onto the market.

For a long time Xerox has had no competition in the photocopying market, making $15 thousand million in 1987, but in the last few years its virtual monopoly has been threatened by the Japanese. Toshiba, for example, has machines which can produce a hundred copies per minute.

Colour photocopier (1973)

The Japanese company **Canon** developed the first colour photocopier which was brought out in Japan in **1973**.

Laser colour copier (1986)

The laser colour copier, presented in **1986**, was also invented by **Canon**. It offers a good quality quadrichrome reproduction i.e. in four colours, on ordinary paper. The image can be reproduced in several ways by enlargement, reduction, changing the dimensions.

Pocket photocopier (1986)

This tiny, extra-light photocopier, the KX Z40X, is an ideal size, barely 16cm *6.2in* long by 7cm *2.7in* wide. It was developed by **Panasonic** in **1986**. It can reproduce any kind of document and operates for up to 20 minutes. It has no wires, no batteries and can be recharged from a storage battery.

Photocopying currently accounts for $20 thousand million on the world market.

Paper clip (1900)

The paper clip, a metallic clip that allows one to attach sheets of paper together, was invented in **1900** by a Norwegian, **Johann Waaler**, who patented his invention in Germany.

GAMES, TOYS, SPORTS

Games & Toys

Billiards (14th century)

The first billiard game took place in the **14th century**. It was played on the ground, a little like croquet. Henri de Vigne conceived and built the first billiard table for Louis XI. Several variations of the game exist: American billiards, or pool, which currently numbers 50 million aficionados throughout the world and which overshadowed traditional billiards and snooker for some time.

Tarot (1457)

No one knows who created tarot or when and where it originated. In **1457**, St Anthony, in his *Treatise on Theology*, makes the first known reference to tarot. In any case, tarot is considered the precursor of modern card games. No one knows either what the word means or what language it comes from. The Italians transmitted it under the name *tarocco*. Some say the origin of the word is Egyptian (*tar* meaning way and *ro* meaning royal); others claim it has its origins in Hebrew (a derivation of Torah, the Law). The letters mixed up give *rota*, which means wheel in Latin: hence the Wheel of Fortune.

Poker (1000 BC)

Contrary to what one might think, poker was not created in Chicago at the time of Prohibition. The principle of the game first appeared 3000 years ago in another game called *As*, which was played in Persia. The role of pairs, threes, full houses and fours was already established, as was that of one of poker's essential elements: bluff.

The game was introduced into Europe by the Crusaders and developed in Spain under the name of *primero*. French versions of *primero*, called *bouillotte* ('kettle') and *brelan* ('threes'), were imported into Louisiana by French settlers. These games travelled up the Mississippi as the West was won and a new variety began to take over from the former types. Poker (so called because players 'poke' their opponents to make them flare up like a fire) was born.

Whist (1743)

This card game of British origin, the ancestor of bridge, was described for the first time in **1743** by **Edmund Hoyle**, in a treatise he wrote about it.

Bridge (1850)

Bridge grew out of whist and first appeared around **1850** in Istanbul. The game consists of a battle between two camps, hence the name 'Bridge', since each player is partnered by the player sitting opposite. Bridge has been and continues to be the object of many scientific studies. The two pioneers of the game were the American Ely Culbertson and the Frenchman Albarran, who started a revolution in the game in 1925, notably by inventing the points system by which players evaluate the strength of their hands before the bidding starts.

Dominoes (c.2450 BC)

In the National Museum of Baghdad (Iraq) there are objects made of bone dating from **c.2450 BC** that were found in Ur in Chaldea and that archaeologists think are similar to our dominoes. However it was not until the 18th century that this game appeared in Europe, reaching Britain via France around 1795. The term domino comes from a similarity with the black garment of the same name worn by priests in winter over their white surplice. Dominoes were ebony on one side and ivory on the other.

Ranchers in the American Midwest have invented a new game; Cow Chip Bingo. Cows are let loose in a corral divided into numbered squares. If the place where the cow stops corresponds to a number you chose, you've won.

GAMES AND TOYS OF ANTIQUITY

Game or toy	History
Draughts	Its complex rules were fixed in 1723.
Swing	The invention of the swing is attributed to the Roman god Bacchus. Swings long retained symbolic and religious significance.
Ball	Probably of Greek origin. It was adopted by the Romans, who took it to Gaul.
Marbles	As old as humanity. The game was played in Greece (using acorns, olives, etc.) and in Rome.
Hoop	Probably of Greek origin. Highly recommended by Hippocrates as a game that encourages sweating.
Kite	Invention attributed to the Chinese general Han Sin (2nd century BC).
Dice	Plato attributes the invention of dice to the Egyptian god Thoth.
Ur game	Considered the world's oldest game, discovered in Ur in Mesopotamia. It is probably the ancestor of backgammon.
Puppets	Existed in Egypt and China, where they featured in religious celebrations.
'Heads or tails'	Ancient Greece. The Romans introduced the coin.
Lead soldiers	The Romans had military toys and model soldiers.
Spinning top	The principle has been known for thousands of years.

Chess (6th century)

The earliest mention of the game of chess is made by the Persian **Karmanak** (590–628). The game seems to have originated in northern India, around the year **500**, but the rules we use today were established in Europe around 1550. The 16 pieces, king, queen, bishop, knight, rook or castle and pawns, are arranged on a chequered board with 64 squares and manoeuvred against the opponent.

The expression 'checkmate' comes from a phonetic deformation of the Arabic phrase *al shâh mat*, which means 'the king is dead'.

Computer chess

In 1950 Claude Shannon set out the basics for programming games of strategy. In 1959 a computer programmed by the American Arthur Samuel played draughts. The first chess-playing microcomputer, the Chess Challenger 3, was built by **Fidelity Electronics** in the United States. Human supremacy is gradually being worn away, but we cannot yet predict when a computer will be a chess grand master.

Pinball (1930s)

The distant ancestor of pinball is the game of billiards described by Charles Dickens in 1836. The great economic crisis of 1929 and Prohibition favoured its development. In 1931 pinball went into mass production and soon acquired the features it still has today: in 1932 the tilt was invented by Harry Williams, and the Ballyhoo was launched by a young Chicago businessman, Richard T. Moloney; in 1933 the machines were electrified, while the Rockelite, created by the American company Bally, was the first to be brightened by luminous scores, and bumpers were invented in 1936.

Electric pinball machine (1938)

It was **Samuel Gensberg**, a Pole who emigrated to the United States at 18, who invented the first electric pinball machine around **1938**, called the Beamlight. The first modern pinball machines appeared in 1947, with Chicago Coin's Bermuda and H. Mabs' Humpy-Dumpty. The first electronic game was made in 1976.

Frisbee (1948)

The Frisbee was invented by students at Yale University (Connecticut) in 1947, who played with aluminium flan cases. These came from a Bridgeport baker, Joseph Frisbie, who was a regular supplier to the University. In **1948** a young American just out of the army, Fred **Morrisson**, applied for a patent for a similar disc in plastic. Later he granted the licence to the Californians Wham-O (inventors of the hula-hoop, see below), who, having heard about the origins of the game, called it Frisbee.

Go (1000 BC)

Tradition has it that the game of go was invented 3000 years ago by the Chinese Emperor Yao, to encourage his children to think logically. Go was introduced to Japan via Korea towards the mid 8th century, and it was in Japan that it had its greatest success. The object of the game is to capture one's opponent's territory.

Go was at its peak of popularity in the 17th century and a former Buddhist monk called Hon Inbosansa, a go champion, founded an official go academy in Edo (Tokyo).

Othello (1974)

Goro Hasegawa, a Japanese, invented Othello in **1974**. This game of strategy, which pits two players against each other, consists not in eliminating an opponent's men by removing them from the board (as in chess or draughts), but in turning them over to take possession of them. Othello, which also exists in an electronic form, is the most widely played game in Japan after the game of go.

Hula-hoop (1958)

The plastic hula-hoop was invented in **1958** by **Richard P. Knerr** and **Arthur K. 'Spud' Melvin**, who owned the Wham-O Manufacturing Company of San Gabriel, California. In six months, the inventors sold 20 million hula-hoops in the United States at $1.98 each, a turnover of about $40 million.

Snakes and ladders (17th century)

The oldest version known today is a carved wooden board of Venetian origin, dating from **1640**.

Computer games (1980s)

These are the great revolution of the **1980s** in the field of games.

All the games can be played by two people, with the computer acting as referee, or they can be played by one person against the game program. The computer is a formidable electronic adversary. All the great classic games of strategy figure high on the list among the programs on offer, for example chess, draughts, backgammon, go and Othello.

Video games (1972)

The first video game was invented in **1972** by an American engineer who was then aged 28:

A MILLION IN ONE DAY

You know about Dragon Quest 3, which sold a million in a single day in Japan, and Zelda, which sold a million in one month in the United States. But do you know who is hiding behind these games? It is a Japanese firm from Kyoto called Nintendo. They used to specialise in card games. After launching electronic mini-games in 1983, they nearly went down with them. But Nintendo's directors played the creativity card to the full. They have four in-house teams of designers who compete with independent inventors. The latter were behind half the games launched by Nintendo. In 1988, Nintendo sold seven million consoles in the United States: a figure to set businessmen dreaming. And it is to them that Nintendo is offering a new service: a console that can be connected to the Stock Exchange index.

Nolan Bushnell. It was a very basic game, a kind of tennis on screen. But Bushnell persevered and his game soon became widely played in games arcades. He set up his own company, Atari.

In 1976 Warner bought Atari for $28 million. In the same year Atari launched cassettes for use on domestic equipment. Video games had come into the home. The first two big successes were Space Invaders, presented on 16 June 1978 by the Japanese company Taito Corp., and Pac Man, which came out of a collaboration between two companies, Namco (Japan) and Midway (USA).

Games on videodisc (1982)

Computer games with some elements (such as the backdrops) stored on videodisc first appeared in **1982**.

Until now only large Hollywood film companies (Paramount, Universal, Columbia, Warner) have been able to invest the $2 million necessary for the creation of each game.

Electronic games that interact with television (1987)

In the United States in **1987** two new television programmes were launched simultaneously with two new electronic toys: Captain Power (a television series produced by **Landmark Entertainment Group** with toys by **Mattel**) and Techforce, created by Nolan Bushnell's company **Axlon**.

The television programmes contain coded sound signals which make the toys, such as robots, trucks or aeroplanes, move. The toys can also shoot and, for example, bring down an aircraft on screen.

Flight simulator for PCs (1988)

With the Flight Simulator 3.0 from **Microsoft**, any owner of an IBM PC, PS or a compatible machine can experience the sensations of a jet pilot landing at an international airport or talking to a control tower over the Pacific.

3-D screen (1986)

In **1986** the Japanese firm **SEGA** invented the Master System console. Its selling point is a system of glasses containing liquid crystals which create an amazing 3-D effect when the wearer looks at the screen. The first available 3-D game is Missile Defence.

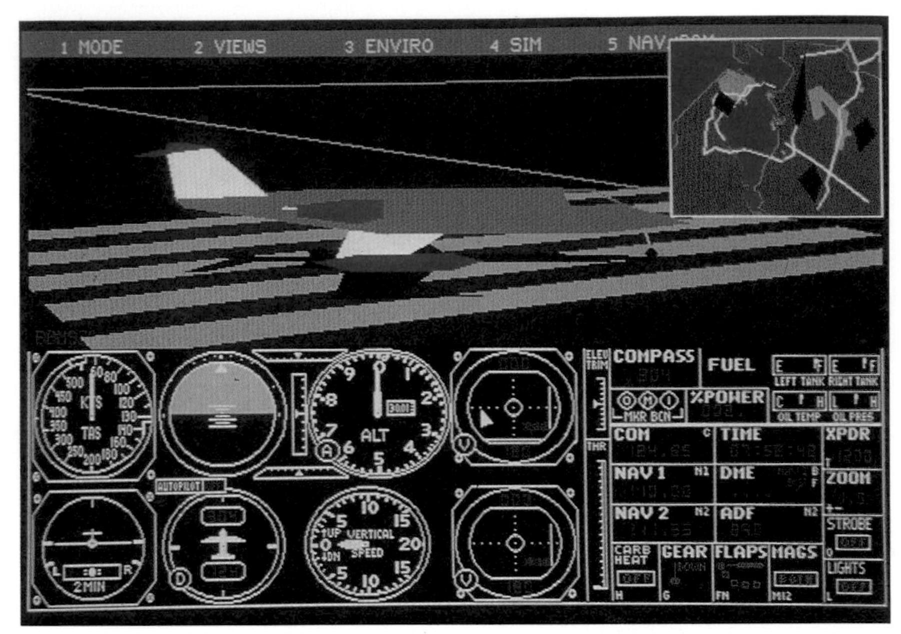

Owners of IBM and IBM compatible machines can now experience the sensations of piloting a jet plane thanks to Microsoft's Flight Simulator program.

Wargames (1780)

These are simulations of historical or entirely fictitious military conflicts. The contemporary form of the wargame was invented by **Helvig**, the Duke of Brunswick's master of pages, in **1780**. In 1837 von Moltke, the general in charge of the Prussian army, made the playing of wargames part of military training. After 1870 all nations followed this example. Then wargames were popularised in the United States in 1953 by Charles Roberts. The mass production of the game Tactics enabled him to set up the Avalon Hill company, which today dominates the wargames market.

Dungeons and Dragons (1973)

Dungeons and Dragons is the first modern role-play game and is the fruit of the imagination of an American travelling salesman, **Gary Gigax**.

Gigax was bored with wargames and so, with his friend Dave Arneson, designed a simulation game where the players would not be obliged to move pieces on a board. Each participant takes on a character. The universe Gigax created was inspired notably by the famous novel, *The Lord of the Rings* by J. R. R. Tolkien.

Having perfected the rules of his game, Gigax tried in vain to sell it to the big games companies of the United States. In the face of so many rejections, he became a part-time shoe mender before producing and selling his game himself in **1973**. Today TSR Hobbies, of which he is Managing Director, is the most prosperous of games producers.

Mystery weekends (1986)

Crazier and crazier: the Canadian company **Blyth and Co**. finances murders in large hotels. About 20 guests are invited to play detectives. Mingling with them incognito are actors whom they must unmask. Americans, Canadians and Britons fight to be included in these weekends (great thrills guaranteed for about £400). And some take it all so seriously that it can lead to tragedy: in the United States one player pushed another off a train . . .

Video Cluedo (1988)

The first great detective parlour game with video backup, Video Cluedo, was previewed at the 7th International Festival of detective films. It also won the Grand Prize for toys in **1988**. There are victims and murderers and each player takes on a character. Players have to solve the mystery and guess the secret identity of each of their opponents. The

YOUNG MILLIONAIRE INVENTORS

The field of games and toys is one of the few where young inventors can rapidly make an impressive fortune. This was the case with Frank Hornby (Meccano, 1900), Ole Kirk Christiansen (Lego, 1955) and Richard P. Knerr and Arthur K. Melvin (Hula-hoop, 1958) and it is still going on today. More recent successes are Nolan Bushnell and Atari (sold for $28 million in 1976), Gary Gigax and Dungeons and Dragons, Yuri Horii and Dragon Quest (1988), Erno Rubik and his cube (more than 100 million sold), the brothers Chris and John Haney and Scott Abbott with Trivial Pursuit (more than 50 million sold) and Rob Angel, whose Pictionary sold more than 13 million in 1987/8. Whatever their nationality, these inventors all have one thing in common: they found it very difficult to find someone who would have confidence in them and had to launch their games themselves.

At mystery weekends, actors mingle with guests trying to discover whether or not the butler did it . . . Role-playing holidays like these are immensely popular.

game was invented by the American company **Kenner**.

Crosswords (1913)

The crossword game was invented in 1913 by an American journalist born in Liverpool: **Arthur Wynne**.

Wynne worked in the games department of the *New York World* and was always looking for new puzzles. He remembered a game from Victorian times which he had played with his grandfather called the Magic Square. By reconstructing the square, including black squares and adding a list of 32 definitions, he invented the crossword, the first of which appeared on **21 December 1913** in the *New York World*'s weekly supplement. His definitions were descriptive and very simple.

Pictionary (1986)

Many young inventors have emerged in the games field over recent years, have founded their own companies and made a fortune.

One of these is the Canadian **Rob Angel**, who at the age of 29 created Pictionary, based on a game from his childhood. One player draws a word and the others have to guess what it is. By the end of 1988 his game had sold more than 13 million units and is today distributed in 26 countries.

Trivial Pursuit (1981)

Invented in **1981** by three young Canadians, **Chris** and **John Haney** and **Scott Abbott**, this game requires knowledge of the subjects geography, history, art and literature, sport, science and entertainment. From two to 36 players may take part at a time, but the game is usually played by six people. The game was a huge success in Canada and now 50 million of them have been sold throughout the world.

Scrabble (1948)

In **1948, James Brunot**, from Newton, Connecticut, patented the game of Scrabble. The source of his patent was a game called criss-cross, in which players had to make up crosswords on a piece of cardboard using wooden letters. Criss-cross was invented in 1931 by Alfred M. Butts.

Jigsaw puzzle (c.1760)

The jigsaw puzzle, a picture glued on to some kind of backing and cut into irregular pieces which must then be reassembled, came into being simultaneously in France and in Britain around **1760**. It was originally an educational toy. In France in 1762 a man called **Dumas** began selling cut up maps which had then to be put back together. In Britain **John Spilbury** stuck a map of England on to a thin layer of mahogany. He cut the jigsaw along the borders between counties which were then sold separately. Spilbury died at the age of 29, without having made a success of his idea. After this, jigsaws became historical. In 1787 the Englishman William Darton produced a puzzle with portraits of all the English kings, from William the Conqueror to George III. The player had to know the order of succession off by heart to be able to do the jigsaw.

In 1789 Wallis simplified this game and produced a history of England in colour, which provided the model for all later puzzles that required more observation and patience.

Lego (1955)

The Lego bricks we know today first came on to the market in **1955**. They were designed after the Second World War by the Dane **Ole Kirk Christiansen**, a former carpenter who had retrained in toy manufacturing. He formed the name Lego from the Danish *leg godt*, which means 'to play well'.

The Lego World Cup took place for the first time in Billund, the town where its inventor was born, in August 1988. Some 175 000 children from 14 different countries took part in the qualifying rounds. The Italians and the Japanese proved the most brilliant.

Adventure book-game (1984)

The first interactive books arose out of the same literary tradition as role-play games and were created by **Steve Jackson** and **Ian Livingstone** from the British Games Workshop.

Meccano (1900)

Meccano was invented by the Englishman **Frank Hornby** in Liverpool in **1900**. Hornby

After years of rejection, Trivial Pursuit became an enormous success.

This model of the port at Copenhagen was constructed from three million Lego bricks.

started dreaming up wonderful inventions when he was a little boy. He made several unsuccessful attempts, then got married, settled to work for a meat importing company and brought up his two sons. It was for them that he invented the toy Mechanics Made Easy which, in 1907, changed its name to Meccano. Hornby wanted a toy which would encourage children to build things rather than destroy them. Frank Hornby set to work and thought of the different elements needed to construct a crane. In the first stage he thought of making metal strips to similar formats with regularly spaced holes so that they could be assembled. Then he made nuts and bolts which would fit the holes and designed small pulleys.

Hornby was also the creator of the clockwork train set.

Monopoly (1933)

Monopoly was invented in **1933** by the American **Charles Darrow** (1889–1967), who was unemployed during the great depression which followed the stockmarket crash of 1929. The game was soon very successful.

However, Monopoly was inspired by a game invented in 1904 by a young woman called Lizzie J. Magie. When the Parker company (Kenner-Parker Brothers) bought the rights to Monopoly from Darrow in 1935, they gave Mrs Magie $500.

Monopoly has been translated into 19 languages and has earned more than £700 million. At the end of 1988 the USSR finally permitted Monopoly to be sold within its borders. Glasnost in action.

Model cars (19th century)

The first small model cars appeared at the same time as their life-size namesakes at the end of the last century. The first cars on a 1/45 scale were launched by the British company **Matchbox**.

Teddy bear (19th century)

The Americans, Germans and Russians all say they were the first to invent the teddy bear.

Toys representing bears have long been made in Russia. Tsar Nicholas II gave a wooden bear to President Loubet of France at

The Marquess of Bath owns a fine collection of teddy bears.

the time of the Franco–Russian treaty of 1892.

According to other sources, while Theodore 'Teddy' Roosevelt was President of the United States (1901–9) his son had a little bear that he loved. When the bear died, the boy was so unhappy that a handyman had the idea of making a toy that looked like the bear for him. The President's son was called Teddy, hence the name teddy bear.

Third version: in 1902 the German Richard Steiff designed the first teddy bear, with boot-button eyes and grey mohair fur. An American trader bought three thousand of them from him. These soft toys appeared dressed as pages at the wedding of Theodore Roosevelt's daughter. And we are back with the Teddy bear.

Dolls (antiquity)

Dolls have been around almost as long as children. They were made from wood, terracotta, and more rarely, wax and ivory. The ancient Greeks and Romans had dolls with movable limbs. The torsos of Greek dolls were often made of burnt clay to which the limbs were attached by cords. The Romans were fond of rag dolls.

In 19th century Europe, Saxony produced most of the torsos for dolls which were made from papier-mâché. Nuremberg and London, among other cities, specialised in making porcelain dolls. These would be dressed in Paris before being sent as far afield as China.

Talking dolls (1820s)

The first talking dolls were created in the **1820s** by the inventor of the metronome: **Johann Maelzel**. In around 1887 Thomas A. Edison (1847–1931), the inventor of the record player, adapted a phonograph with round discs to go inside a doll.

Barbie Doll (1958)

Barbie was created in **1958** by **Mattel**. It was the first doll to have an adult's body and a whole wardrobe of miniature clothes. Ken, Barbie's boyfriend, appeared in 1961. In 1963 Barbie's friend Midge came on to the market followed by a younger sister, Skipper, Allan etc. Barbie has continued to follow fashion and is as popular as ever; in 1989 a Barbie doll was bought somewhere in the world every ten seconds.

Barbie's English competitor is Sindy who also has a boyfriend, called Paul, a wardrobe of clothes, and a fan club.

Playmobile (1974)

In 1970 **Hans Beck**, head designer at **Geobra-Brandstätter** in Dietehofen near Nuremberg (West Germany), began developing an entirely new world of toys. Playmobile was presented in 1974 and at once achieved fame and fortune, to such an extent that Geobra became the leading toy manufacturer in the Federal Republic.

Playmobile consists of little people and their entire environment all made of ABS, a particularly tough plastic. The range of models just keeps on growing.

In 1979 the Hungarian academic Erno Rubik made a fortune with the cube that had us all baffled. In 1988 he produced an equally puzzling clock.

The Koosh ball was invented by American Mark A. Button. It doesn't roll or bounce, but it is great fun to throw and juggle.

Rubik's clock (1988)

In 1979 a Hungarian academic, **Erno Rubik**, invented a fiendish cube, of which more than 100 million have been sold throughout the world. This made him one of the few Hungarian multi-millionaires, with three cars including a Mercedes. With his royalties, Rubik set up a small design company in Hungary and funded a foundation for inventors.

In **1988** he launched the mind-blowing clock. The trick is to use the knobs to set all the hands of 18 clocks to midnight when each knob controls several clocks at once.

Sport

Athletics

In ancient Greece races and throwing competitions had pride of place in the Olympic Games. Modern athletics really began to develop in the mid 19th century in Great Britain. The first championships were organised in 1866 by the Amateur Athletic Club.

Marathon (1896)

In the first of the modern Olympic Games, held in Athens in **1896**, the marathon was won by a Greek shepherd, **Spiridon Louis**.

In so doing he updated the feat of Phidippides, who ran from Marathon to Athens to announce Miltiades' victory over the Persians under Darius the Great, on 13 September 490 BC. Legend has it that the exhausted messenger died on arrival, after running 40km *24 miles* without stopping. In 1924 the distance he had run was fixed at 42.195km *26 miles 385 yd*.

Gymnastics (antiquity)

Bodybuilding was practised by the Greeks and Romans.

In the 19th century, the Swedes led the way with a gymnastics technique invented by P. H. Ling (1776–1839), which was the first attempt to approach physical education scientifically.

The end of the men's 800m in 1984 at the 23rd Olympic Games in Los Angeles. In 1988 139 nations were represented.

Fosbury flop (1968)

By his own account, the American **Richard Fosbury** created his famous highjump technique almost involuntarily. When he could not manage to execute a western roll correctly, he started from a scissors jump and gradually developed the backwards jump which earned him an Olympic gold medal in Mexico in **1968**. Today he is a civil engineer in Idaho and designs mountain bike tracks.

SHOEMAKER TO CHAMPIONS

In 1920 the Dassler brothers Adolf, known as Adi, and Rudolph, who were shoemakers in a little town near Nuremberg, created their first sports shoe. After this Adi designed running shoes with spikes, straps under football boots and crampons that screw in. His slogan was 'innovate and communicate'. He was quick to go to the Olympics to offer his shoes to the champions: Arthur Jonath in 1932 and Jesse Owens in 1936. The rest is history: today Adidas shoes are worn by the greatest athletes, and a few others.

Ball games

Cricket (1727)

Games using sticks and balls are ancient and the oldest sport of this type in Britain was known as knur and spell. This was of Norse origin and was played before the 11th century.

A primitive form of cricket was being

Yorkshire bowler Wilfred Rhodes in action in 1930. He remains cricket's most prolific wicket-taker.

played during the reign of Elizabeth I, but the rules were not codified until **1727**. In 1730 the first match was played in London at the Artillery Ground in Finsbury. By 1760 the prime exponents of the game were the members of the Hambledon Cricket Club who played on Broad Halfpenny Down in Hampshire. In 1787 Thomas Lord founded the Marylebone Cricket Club (MCC) which moved to its present site in 1814.

Hockey (1883)

Derived from medieval games played with sticks, the rules of hockey were codified in England in **1883**.

Baseball (1750)

According to legend, baseball was invented in **1839** by **Abner Doubleday** in New York (USA). It is doubtful that this is so.

The game actually has its origins in England where, around **1750**, children were playing rounders, from which baseball evolved.

In 1845, one of the founders of the first baseball club, the Knickerbocker Club of New York, defined the rules of the game. His name was Alexander Cartwright, and he invented the diamond-shaped playing field. Up until then the game had been played on square or pentagonal fields.

Basketball (1891)

Basketball is one of the few sports that did not come about as a result of a slow evolution. In **1891**, **James Naismith**, a professor at the International YMCA College in Springfield, Massachusetts (USA) decided to develop a sport that could be played indoors, at night or in winter. He nailed two baskets to the opposite walls of a gym and set down the rules. The first game took place on **20 January 1892**. The teams were made up of seven players each and the game was played over three 20-minute periods.

Skittles (antiquity)

We find the earliest mention of skittles in the *Odyssey*. At that time they were made of stone and square at the top.

In medieval Germany, the skittle symbolised the devil who had to be knocked down. This is probably why Luther recommended playing skittles after church, giving each skittle the name of a sin. Since then a great number of variations on the game have been created in every country and region.

Tenpin bowling (1874)

Tenpin bowling is the modern form of skittles, whose rules were codified in **1874** in the United States. There is one important difference: the skittles are righted automatically.

Association football (1848)

The first attempt to set out the rules of football was made in **1848** at Cambridge University. It was at Oxford University that the name 'soccer' is said to have originated. In 1863 the Football Association was formed.

American football (1880)

American football appeared at the end of the last century and is today played by tens of millions in the United States. It was probably around 1870 at Harvard University that a type of football crossed with rugby was first played. The rules of American football were fixed in **1880** by **Walter Camp**.

A golfer and his caddy at the end of the 18th century.

Rugby (1823)

Rugby was invented in November **1823** by **William Webb Ellis**, a pupil at Rugby School (Warwickshire, England).

The story goes that during a fast-moving football match, he picked up the ball and ran towards the opponents' goalposts clutching it to his chest. The new game using hands soon expanded, but the rules remained vague. In 1871 the Rugby Union was founded, which established the rules for the 15-player game.

In 1893 the Yorkshire Union proposed that payments to players should be allowed but were outvoted. However, just two years later, some northern clubs broke away to form Rugby League in which payments were permitted.

Golf (15th century)

Golf was invented in Scotland and it gained such popularity there that King James II banned the sport in March 1457 as he felt that his subjects were wasting too much time playing the game. However, golf remained the national sport and Mary Queen of Scots was the first woman to play the game.

The first golf balls were made of leather stuffed with feathers. Balls made from gutta percha appeared in the middle of the 19th century.

Computer golf course (1979)

In **1979**, to make it possible to play golf in the city, the American companies **Optonics** and **Texas Instruments** developed a computer simulation of a golf course. The course unfolds on screen, depending on the ball's trajectory. Part. T. Golf recreates all the conditions encountered on a real course.

Golf ball 'retriever' (1987)

Every player has had the bitter experience of losing a ball in the water. Thanks to the Sonar Retriever, invented by the Americans **D. J. Allison** and **R. J. Spielman**, they will now be able to get them back.

Volleyball (1895)

Volleyball was invented in **1895** by the American **William G. Morgan** and first came to Europe during the First World War with the arrival of the American troops.

Cycling (1868)

Cycle races quickly followed the invention of the bike. The first race was held on **30 May 1868** in the Saint Cloud park in Paris over a distance of 1800m, *1 mile 209yd*. The race was won by an Englishman, James Moore, who also won the first town-to-town race between Paris and Rouen in 10h 40mins.

Tour de France (1903)

The Tour de France, invented by Henri Desgranges, took place for the first time on **1 July 1903**. There were six stages and 20 out of the initial 60 runners completed the course. The first race was won by Maurice Garin. Henri Desgranges was also the inventor of the *maillot jaune*, the yellow shirt, introduced in 1919.

Mountain bike, or all-terrain bike (1973)

In **1973** fans of sport, space and adventure in Marin County, California (USA), customised their road cycles so that they could take them on the nearby mountain slopes. Many of them became very skilful at this and soon the Canyon Gang was to be seen haring down the steepest slopes.

Gradually small-scale manufacturers began to build these machines, which were soon called mountain bikes.

High speed tricycles (1986)

On **11 May 1986** a human-powered machine (an enclosed cycle) broke the 29m *95ft* a second barrier, thereby winning the $18 000 prize set up in 1984 by the company Du Pont. The winner was the Gold Rush, ridden by Fred Markham. The Gold Rush is an improvement on the Easy Racer, a widely sold bicycle, and was designed by **Gardner Martin**. It weighs 14kg *30lb 14oz*. It has an aluminium frame, kevlar streamlining, mylar brake discs, a lexan windscreen and spandex filled wheels. The former record was held by the Allegro, invented in 1984 by the American Don Witte.

Cycletouring (1896)

Cycletouring came into being in Italy. The first official route was from Rome to Naples and it was covered by nine cyclists in **1896**. The first Audax certificates were created by Vito Pardo in 1898: they were awarded to cyclists capable of covering 200km *124 miles* between sunrise and sunset.

Swimming, sailing and watersports

Swimming

In all civilisations the origins of swimming are lost in the mists of time, but the first organised competitions started as recently as the 19th century: the first took place in London in 1837.

The Australians held a contest in 1846 (Sydney Championship) and soon afterwards organised the first world championships.

In 1875 the British swimmer Captain Matthew Webb was the first to cross the English Channel, in 21h 45mins.

Crawl (1902)

The indigenous peoples of the South Pacific swam the crawl. It was adapted by two Australians, **Syd** and **Charles Cavill**, who introduced it to Europe in **1902** and to the United States the following year.

Butterfly (1926)

This variant of breaststroke, which is very tiring but very fast, was developed by a German swimmer, **Eric Rademacher**, in **1926**. The Americans used it in breaststroke races. It was banned before becoming an official stroke in its own right.

Water polo (1859)

Water polo is a combination of swimming and a ball game played in teams (11 players per side, seven of whom are in the water). It first appeared in Britain in **1859**.

Rowing (antiquity)

Rowing's origins date back to antiquity. Virgil described a rowing race in the *Aeneid*. The Olympic sport of rowing is practised in more than 50 countries. One of the oldest races is the one that is held every year since 1829 between coxed eights from the Universities of Oxford and Cambridge.

Canoe-Kayak (1865)

The canoe is descended from the hollowed-out tree trunk, a primitive form of boat used by the Native Americans of Canada, while the kayak was first used by the Eskimos. It differs from the canoe in that it has a watertight 'skirt' which is tightly attached round the waist, and in the use of two paddles. The British Royal Canoe Club was founded in 1865.

Rafting (c.1950)

Rafting is an American invention. At the end of the Second World War, the American Army Surplus bought small inflatable dinghies designed for disembarcation. Though flexible and easy to handle, they were also long, heavy and motorised and were used to carry groups of tourists through the gorges of the Grand Canyon. Some fans did not think this activity much of a sport; however, it developed when the inflatable dinghies were shortened and strengthened and the motor

Rafting is a sport that guarantees adventure and excitement.

replaced by oars, so that these boats, now called rafts, could take the most impressive rapids.

'Tubbing' is a typically American sport descended from rafting. The dinghies are replaced by enormous truck inner-tubes, used as individual rafts.

Yachting (17th century)

Yachting was established in Britain around **1664** by **King Charles II**, who got a taste for navigation while he was in exile in Holland.

Yachting, the sport of kings, then spread to the other European courts. Louis XIV of France had mini-ships built to sail on the great lake at Versailles, and ordered a Venetian gondola to be carried over the Alps for his use.

Yachting for pleasure also attracted the upper-middle classes of Cork City in Ireland. The Water Club was set up there in 1720 and its descendant, the Royal Cork Yacht Club, is the oldest yacht club in the world.

Centreboard (15th century)

At the end of the **15th century** the **Dutch** began to equip their boats with lateral centreboards, two streamlined wooden surfaces which could be lifted for tacking. A few hundred years earlier, the Incas were equipping their rafts with *guaras*, or small centreboards which were stuck between two balsa trunks.

In 1774 in Boston, Massachusetts (then a British colony), a British officer, Lieutenant Schank, who had heard about the Inca system, built a dinghy with a vertical centreboard which ran right along the keel. In 1811 the American brothers Joshua, Henry and Jacob Swain applied for a patent for a pivoting centreboard without ballast which was immediately very successful in the United States.

It was an Englishman, Uffa Fox, who popularised the sliding centreboard, which was capable of remarkable speeds. In 1928 his Avenger won 52 out of 57 regattas. Since then, the hull shape of centreboard boats has changed little. A sailing boat like the 505, designed in 1953, is still in the forefront of progress.

Fin keel (1840)

The complexity of centreboards and the space taken up by their shaft led boat builders to place fixed centreboards under their keels. These fixed centreboards made of strong sheet metal to ballast the boat, first appeared in Bermuda around **1840**.

At the end of the 1870s a type of boat called the Houari with a non-ballasted keel was developed in Marseilles. The bulb-shaped ballast at the end of the keel was invented by the Englishman E. Bentall on his 15.4m *50ft 9in* boat *Experiment*, which was launched in 1880.

But the first to make fin keels really effective was Nathanael Herreshof with his 1891 boat *Dilemma*. This sailing boat brought about a real revolution and was the forerunner of the modern monohulls.

Catamaran

The catamaran appeared several thousand years ago off the Coromandel Coast in India. The catamaran was basically a raft made with three tree trunks of different lengths. Its name comes from the Tamil word *kattumaram*: tied timber. The term is now applied only to boats that have two similar hulls. These derive more immediately from the huge canoes, each hewn from a single trunk and temporarily joined together, used by Polynesians from Tahiti and Hawaii to cross the ocean.

In 1662, the Englishman William Petty, having learned of the Pacific style of sailing, built two catamarans and raced them successfully against single hulled yachts.

The first American catamaran was the *Double Trouble* built by John C. Stevens in 1820. Unfortunately, its name turned out to be prophetic.

The first person to come up with a seaworthy and manageable sporting catamaran was Nathanael G. Herreshof, who worked from 1876 to 1881 on the project.

The first modern high sea catamaran was launched from Hawaii in 1947 by Brown, Kumalae and Choy. It was called *Manu Kai* and measured 12m *39ft* in length, and was the first vessel to exceed 20 knots.

Hobie Cat (1968)

In **1968**, after many years of development, the first Hobie Cat appeared bearing the name of its inventor, the American **Hobart L. Alter**. The Hobie 14 (14ft) is an ultralight craft, very strong and easy to handle. Its two highly streamlined asymmetrical hulls are joined by two metal poles and a trampoline. It is ideal for pleasure trips and for racing, as its light structure (fibreglass hulls, metal mast) allows it to reach high speeds. Some 200 000 of these wonderful boats have already been sold.

These ships left Portsmouth on 14 May 1987 to commemorate the departure of the first convict boat to Australia two hundred years before.

CROSSINGS, REGATTAS AND RACES

Type	Date	Characteristics
Solo crossing	**1601**	The earliest mentioned example (which possibly did not happen) dates from 1601, when the Dutch surgeon Henry de Woogt asked for a permit to make a crossing from Vlessingen to London. In 1869 Empson Edward Middleton sailed solo around Britain on board a 7m *23ft* long yawl. The first person to sail solo across the Atlantic was Alfred Johnson in 1876. The first solo trip around the world was made by the American captain Joshua Slocum, from 1895–98.
America's Cup	**1851**	This is the world's oldest sporting trophy (in any sport). It was during a regatta around the Isle of Wight on 22 August 1851 that the schooner *America* won the silver Cup. In 1983, after 132 years of American domination, the Cup was won by the Australians. In February 1987 the Cup was returned to the United States through the efforts of Dennis Conner on *Stars and Stripes*.
Solo transatlantic race	**1891**	The first solo transatlantic race left Boston on 17 June 1891. Some 70 years later, in 1960, the first solo transatlantic race against the prevailing wind took place, from Plymouth (UK) to Newport (USA).
Atlantic record	**1905**	This competition was organised on the initiative of the Kaiser and was won by the American Charlie Barr and the schooner *Atlantic* (56-man crew). In 1917, when the winner donated his gold cup to the American war effort, it was discovered that the cup was not real gold! In 1988 the competition was won by Serge Madec, on board the 23m *75ft* catamaran *Jet Service*, in seven days, 6h 30mins (New York to Lizard Point, UK).
Blue Ribband	**1937**	A symbolic trophy given to the fastest Atlantic crossing by passenger liner between Bishop Rocks (UK) and the Ambrose lighthouse (USA), a distance of 2938 miles. It has been held by the *Normandie*, the *Queen Mary* (1938) and then the *United States* (1952), for a crossing of 35.59 knots in three days, 10h 40mins. This record was beaten in 1986 by the British *Virgin Atlantic Challenger II*, owned by Richard Branson, in three days 8h. In June 1990 the *Hoverspeed Great Britain* completed the journey in three days, 7h 55mins.

Clayton Jacobson from California invented the jet ski in 1974. The first races were held in Europe in 1988.

Trimaran (1786)

The first boat with three hulls, derived from the Polynesian *latakoïs*, was built by the Scot **Patrick Miller** in **1786**. It was essentially a steam-propelled trimaran with a paddle wheel between each hull.

In 1868, the first inflatable boat to cross the Atlantic, the American sailing ship *Non-Pareil*, had three hulls. This crossing was to prove the efficacy of 'tri-hulls' as rescue boats.

Around 1943, a Russian immigrant to the United States, Victor Tchechett, decided to improve ships with three hulls; he coined the word trimaran. It wasn't until the end of the 1960s, however, with the designs of Derek Kelsall, and especially those of the American Dick Newick, that trimarans were really refined.

Proa (1933)

It was **Nathanael Herreshof** who, around **1933**, had the idea of applying the techniques of craft with balancers used in the Pacific to pleasure craft. These sailing boats, which originated in Malaysia, are laterally asymmetrical.

While Herreshof was satisfied with making models, in 1970 Englishman Rod Macalpine-Downie designed a sailing boat with a balancing pole, the Crossbow, which reached a speed of 31.09 knots in 1975.

Marconi rigging (1921)

Around 1820 the 'leg of mutton' sail appeared in the Bermudas. Instead of being trapezoidal like the sails of the time, the sail was triangular. The English imported the design to Europe around 1880, and it became known as 'Bermuda rigging'. At the same time, it was determined that lengthening the sail influenced the boat's speed. In 1895, a small American yacht owned by William P. Stephens, the *Ethelwynn*, had a single-sail Bermuda-rigging mast, but the mast was shrouded to give it greater rigidity. This streamlined, shrouded rigging was called 'Marconi rigging' by analogy with the design of radio antennas. The first great yacht equipped with Marconi rigging was the 34.2-metre-long *Nyria*, designed by **Charles E. Nicholson** in **1921**.

Fan sails (1989)

An Italian engineer, **Giovanni Valzani**, has designed a sail with a revolutionary shape. It has four lateral masts stemming (like a fan) from a fixed central mast. This system allows great flexibility in the arrangement of sails. As it can swivel right round on its axis, it also makes it possible to catch the wind without having to tack.

Sail folding system (1988)

A young American, **Martinus Van Breems**, has created a system which facilitates the folding of the mainsail when it is lowered. His *Dutchman* was immediately successful and equips the yachts produced by Hunter Marine and Beneteau.

Waterproof ship's log (1989)

The idea is catching on: a ship's log that is resistant to bad weather and particularly water. But it took long years of research at **Sogetic** to develop a waterproof synthetic paper that could be written on in biro or pencil without problems. However, felt tips and fountain pens should not be used, as the paper does not absorb ink any more than it does water.

Ultra-light aircraft allow the pilot a virtually unhampered view of the countryside.

Shooting

Archery

Although the bow and arrow was invented in prehistoric times, the first sporting archery competitions took place at the end of the Roman Empire. In the Middle Ages the importance of the bow and arrow as a weapon gave rise to contests of skill, but archery did not really become a sport until the beginning of the 19th century. It became an Olympic sport in 1900.

Shooting (1814)

The origins of shooting as a sport go back to the institution of live pigeon shooting in Britain in **1814**. Clay-pigeon shooting was invented, again in Britain, in 1880.

Airborne sports

Hang-gliding (1948)

In **1948** the American **Francis Melvin Rogallo** designed a supple, flexible wing made of woven wire, covered with a silicon-based coating. His invention was taken up by various organisations, notably NASA, before being abandoned.

By 1964 increasingly functional delta-shaped wings had appeared. That year Bill Moyes, an Australian engineer who had worked for NASA, designed a 4.5m² *5.38sq yd* delta wing. On 4 July 1969 Moyes' partner and compatriot Bill Bennett set off on water-skis pulled by a dinghy, then released himself from the dinghy and flew over the statue of Liberty.

The autonomous take-off hang-glider which did not require another vehicle to pull it took over from the towed version as a result of the work of inventor Dave Kilbourne. Today hang-gliding has become an international sport.

Ultra-lights (1975)

In 1970 the popularity of hang-gliding began to grow. But people who lived in flat areas had to travel to be able to practise this sport. This is why hang-gliding fans thought of motorising the wings. In **1975** the first viable prototype ultra-lights appeared more or less simultaneously in Australia, France, and the United States. The Americans Mauro, MacCormack, Rotec and Chlurzaczik sold launching machines.

And advances continue: performance is improved and models become safer, but often more complex.

Parachute (1946)

The first parachute worthy of the name was patented in 1802, but parachuting as a sport did not appear until after 1945. World championships have been held since 1951. The sport is largely dominated by the Soviet Union, which holds almost all the world records: altitude 25 808m *84 702ft* P. Dolgev, 7 June 1960; greatest number of jumps 10 000 by A. Ossipov, etc.

Combat sports

Wrestling (antiquity)

Weaponless combative sports are among the oldest known to mankind. Wrestling was popular in ancient Egypt and Mesopotamia and was practised in India around 1500 BC.

Grand Champion Chiyonofuji (The Wolf) performs the dohyo-iri *or ring-entering ceremony.*

Indeed, it is mentioned in the epic Sanskrit poem, *Mahabharata*.

Wrestling was a common pastime in ancient Greece and images of the sport were very popular. It was probably one of the first Olympic sports, there are many references to wrestling in Europe in the Middle Ages.

Sumo (200 BC)

Rice farmers were probably practising sumo around **200 BC** and the sport's first grand patron was Emperor Suinin in 23 BC. Sumo continued under imperial patronage until 1185, but under the shoguns, public matches were banned. Sporting sumo was allowed again after 1600 and in 1684 Ikazuchi Gondaiya, a masterless *rikishi* (wrestler), proposed rules and techniques to control the sport. It was around this time that the hard clay ring or *dohyo* was introduced. In 1889 the Japanese Sumo Association was formed.

Aikido (1925)

In 1925, having been a major figure in martial arts, **Morihei Ueshiba** (1883–1969) invented a new fighting technique: aikido. The aikidoka gives fighting a spiritual significance beyond its utilitarian aspects, channelling the attacker's aggression and proving its uselessness. Ueshiba defined the aim of aikido thus: to destroy the attack, not the adversary.

Boxing (3000 BC)

Boxing is an ancient sport; a decorative fresco in Iraq dating from about 3000 BC shows boxers with their fists wrapped in pieces of leather. In ancient Greece, the gloves were adorned with pieces of iron and contests often ended in death.

In modern times, the first great champion was James Figg (born in 1696), who opened the first boxing school on Tottenham Court Road in London in 1719. Another Englishman, Jack Broughton (1704–89), established the rules for fighting barefisted. He set the dimensions of the ring and prohibited blows below the belt.

The ninth Marquess of Queensbury and boxer Arthur Chambers drew up the rules that are used today: the wearing of gloves, the break after three-minute rounds, the ten-second count for a floored boxer etc. The first fight to take place under these rules happened in Cincinnati, Ohio (USA) on 19 August 1885.

Judo (1882)

Judo (which means 'gentle art' in Japanese) was invented in **1882** by **Jigoro Kano**. Born in 1860, Kano first devoted himself to jujitsu, but he wasn't very strong. He decided to compensate for this handicap by developing his body and spirit. Kano perfected a method of attack and defence that brought about victory by using suppleness rather than strength. Judo was immensely popular in Japan and then spread throughout the world. There are now more than 15 million judoists worldwide, including 500000 black belts.

Karate Do (1916)

Karate is based on the use of the human body's natural weapons.

It is said to have been first practised by Bodhidharma, a Buddhist monk who went to China from India in the 6th century. The combat technique was forgotten and not revived until the 16th century in Okinawa, Japan.

But it was on this island that the master **Gichin Funakoshi** (1869–1957) was born. He devoted his life to the development of his art and the first historic demonstration of the technique took place at Kyoto in **1916**. At the time this method of combat was known by the name 'Okinawa Te'. Funakoshi gave it a more meaningful name: Karate, from *Kara* empty and *te* hands. He added the suffix *do* which means way.

The first world championship took place in Paris in 1970.

Surfing and gliding

Ice yachting (1790)

In **1790** the American **Oliver Booth** put a sail and ice skates on a packing case. He travelled on the frozen surface of the Hudson in winter, at Poughkeepsie, where the first ice yachting club was set up. The largest ice yacht ever was 21m *68ft 11in* long and carried 100m² *120sq yd* of sail; it was built in 1870 for John E. Roosevelt of New York. In 1938, on the ice of Lake Winnebago (Wisconsin, USA), John D. Buckstaff reached the fastest speed of any wind-propelled machine: 230km/h *143mph*.

Sand yachting (1910)

The sand yacht is both a sport and a means of transport. It was used in ancient times by the Egyptians, the Romans and the Chinese. In the 19th century workers on the Kansas Pacific Railway used sand yachts to supervise the railway line.

In **1910** the Belgians **Frank** and **Ben Dumont** launched the sport of sand yachting. Their

Wind power in action: these wheeled boats are propelled across the desert by the wind in their sails.

Skateboarding began in the 1950s when American surfers used the boards to practise on land. Now the sport has its own rules and its own champions.

machines were faster than the cars of the day. In 1925 the form of the contemporary sand yacht – two wheels at the back and one in front which does the steering – was set. In the Californian deserts one often sees large sand yachts doing speeds of more than 120km/h *74mph*.

Sail train (1988)

In November 1988 a 42-year-old engineer, **Christian Nau**, took his sail train across the highest railway in the world, the Altiplano in the Andes (altitude 4881m *16019ft*), going through 66 tunnels and over 59 bridges.

His train consists of three connected wagons with solid steel wheels and bird-like wings.

He is now preparing to attempt to break the world speed record for sail travel on rails, on the high speed train line of northern France.

Surfing (1778)

Surfing probably first appeared in Hawaii. The earliest description of it was made in **1778** by the English navigator and explorer Captain James Cook (1728–79). However, surfing did not reach California until around 1900, and did not become popular until the beginning of the 1950s, under the influence of films and music (particularly the Beach Boys in the 1960s). Today surfing is a real institution in the United States, as important as baseball, American football, boxing and basketball.

Funboard (1979)

The funboard was created by **Arnaud de Rosnay**, **Robby Naish**, **Matt Schweitzer** (son of the inventor of windsurfing) and **Mike Waltze**, a group of windsurfing fans who were hungry for ever stronger sensations. They customised surfboards, making them lighter and smaller, and set off to tackle the enormous waves of the Hawaiian islands.

Morey-Boogie (1971)

The inventor of Morey-Boogie is the Californian **Tom Morey**. In **1971** he created a strange, wide, almost rectangular surfboard, which was also softer, to help prevent accidents without affecting performance. You use this board lying on your stomach and wearing flippers (you can try to stand up if you like!). It is an excellent way to approach waves even if you are not a seasoned surfer and the Morey-Boogie requires no less athleticism. For the last few years it has been an extraordinary craze in the United States, where competitions sometimes have more spectators than the big surf meetings. The most exciting feat to execute is 'el rollo' a complete turn inside the tunnel of a wave.

Speed sail (1977)

Invented in **1977** by the French windsurfer **Arnaud de Rosnay**, the speed sail is a kind of land-borne windsurfboard on roller skates. It makes for a spectacular competitive sport, with the best reaching speeds up to 140km/h *86mph*. Since it was invented, the sport has become important outside France; the 1987 World Cup was held at Daytona Beach in Florida (USA).

Barefoot waterskiing backwards is strictly for the experts.

Nordic skiers train on the road in summer.

Waterskiing (1922)

In **1922 Ralph Samuelson** invented waterskis. He loved snow skiing and wanted to have the same sensations in the summer. However, he never tried to develop or market the sport he had created.

In 1925 Fred Walker of Huntington, New York, who had seen Samuelson's skis, created and patented Akwa-skees.

Wind surfing (1958)

Wind surfing was invented by the Englishman **Peter Chilvers** in **1958**. But this Rolls-Royce mechanic was happy to sail around by himself and never exploited his invention commercially.

In 1964 American Newman Darby had the same idea and mounted a sail on a surf board. Four years later two Californians, Jim Drake and Hoyle Schweitzer, who had never heard of their forerunners, also put a sail on a board so as to be able to sail in calm weather. But they added a keel, articulating joint and wishbone boom, thus giving the sailboard its definitive form.

Originally made of wood, the sailboard has become lighter and lighter thanks to the introduction of new materials that allow speeds of more than 50km/h *30mph* to be achieved.

Roller skating (1759)

In **1759**, a Belgian manufacturer of musical instruments, **Joseph Merlin**, invented roller skates. Invited to a ball at Carlisle House in London, he had the idea of making a gliding entrance while playing the violin at the same time. Unfortunately, he hadn't thought about the problem of brakes and he crashed into the mirror at the end of the grand entrance, smashing it and his violin to pieces and seriously wounding himself.

In the sporting world, the ice skater J. Garcin created roller skating at the beginning of the 19th century, at first simply for summer training. In 1863, an American, James L. Plimpton, of New York, patented the first roller skates with four wheels.

Winter sports

Ice skating (1850)

Ice skating has existed for centuries in northern Europe. At first, blades made of bone were fitted into wooden soles; probably at the beginning of the 17th century these blades became metallic. The world's first skating club was formed in Edinburgh in 1642.

In the United States, in **1850**, the first true skates appeared, with iron blades. Immediately after the European tour by the American Jackson Haines, the American-style skates were adopted world wide. The first world figure skating championships took place in 1896 in St Petersburg, now Leningrad, USSR.

Ice hockey (1855)

Some see ice hockey as the descendant of a game played by the Dutch on frozen canals in the 17th century. But the real birthplace of the sport was **Canada**, where several towns claim to have invented it. The modern rules were fixed at Kingston, Ontario, in **1855**.

Bobsleigh (1890)

The history of the bobsleigh begins in December **1890** when **Wilson Smith** had the idea of joining two toboggans together with slats in St Moritz. That same winter, the local blacksmith used the design of this prototype to build the first bobsleigh fitted with four iron runners.

Skiing

The origins of skiing are Nordic; it was a mode of transport for thousands of years. The great Vasa Race commemorates the feat of one of the kings of Sweden, who escaped from Denmark and reached the Dalecarlia forest on skis with his partisans in 1520.

In 1880 the Norwegian Sondre Nordheim had the idea of making the front of his skis curved.

In 1888 Fridtjof Nansen crossed Greenland from east to west in 39 days and published an account of his expedition. Reading this gave the Austrian Zderski the idea of shortening skis and giving them metal attachments to hold shoes. The first downhill race was held on 6 January 1911 at Arlberg in Austria and organised by a former officer in the Indian Army, Lord Roberts of Kandahar (1832–1914). It was another Briton, Arnold Lunn, who created the slalom in Murren in Switzerland.

Monoski (1973)

The monoski was invented by a surfer. American **Mike Doyle** was the uncontested king of water surfing in the 1950s. In 1972 he took up snow skiing. His idea was to find, on snow, the same sliding sensations that are found on water. He built the first single ski in **1973** in only a few hours.

This prototype was transparent and as wide as nearly three normal skis. A pair of standard bindings keep both feet parallel. The single, renamed mono, provides extraordinary lift on powdery snow.

Doyle inaugurated his monoski in 1973 on slopes in Jackson Hole, Wyoming. For the occasion he wore a Hawaiian shirt!

Ski jump (1808)

The first known ski-jumper was **Olaf Rye**, who is said to have jumped 9.5m *31ft 2in* in **1808**. The first official record was held by Haugen in 1914, with 46.30m *151ft 11in*.

Ski flying (1860)

This sport first appeared in California, in the Sierra Mountains. In **1860** six skiers wearing skis 3m *9ft 9in* long hurled themselves forward much to the delight of those laying bets. At the end of the 19th century, the Norwegians were achieving speeds of over 100km/h *62mph*. In April 1988 the French skier Michael Prüfer reached the astounding speed of 233.741km/h *145.25mph*.

Steve Zaleznik practises ski flying in the Sierra Nevada Mountains of the Yosemite National Park.

Artistic skiing (1950)

The origins of artistic skiing go back to the first dangerous jumps said to have been performed in 1807. In 1920 a German, Fritz Rauel, took figures from skating and adapted them to skiing. But it was in **1950** that a Swede, **Stein Eriksen**, fixed the rules of this sport. The first world cup was held in 1980.

Ski boots (1893)

The first ski boots appeared in **1893**. They were made of reindeer skin with the fur outside and were directly inspired by the Eskimo shoes brought back by the Norwegian arctic explorer, Fridtjof Nansen (1861–1930) from his Greenland crossing in 1888/9.

Boots with hooks (1962)

These were developed by the Frenchman **Martin**, who sold his patent to the Swiss company Henke. The French company Le Trappeur bought the patent from them and launched the first leather Martin ski boots with five hooks.

Plastic-covered boots (1968)

The principle of the plastic shell was developed in **1968** by the American company **Lange**.

Conform'Able System (1976)

In **1976** in Grenoble the Frenchman **Loïc David** invented the Conform'Able system, which consists of making soles moulded directly to the foot to fit all sports shoes (ski boots, tennis shoes, cycling shoes, etc.). The Americans, Austrians and Japanese have become great fans.

Heated ski boots (1986)

The first were developed by the Swiss company **Raichle** and marketed under the name RX-Hot.

Ski lift

George Encil left Austria in 1930 at the age of 20, saying he was going to make his fortune in Canada. The country was hard and he spent some difficult years there. But he was working at the foot of the Calgary mountains and would watch the skiers climbing the slopes on foot before skiing down them. This gave him the idea of designing a mechanical lift. Soon Encil Skilifts were being built all over the world.

Safety bindings (1948)

In **1948**, after suffering two broken legs in the same year, a French engineer and ski fanatic, **Jean Beyl**, designed a safety binding and set up the Look company.

In 1966 the French firm Salomon launched the first safety heelpiece which releases the heel on falling. The next year Salomon marketed the first elastic double pivoted toepiece. Today this company, headed by Georges Salomon, is the world leader in bindings, making more than two million of them per year.

Derby Flex (1985)

In **1985** an Italian engineer, **Ambrosio Bettosini**, developed a little pad made of foam rubber and aluminium to fit between the ski and the boot. The Derby Flex absorbs vibrations and spreads them out, thus making better performances possible in giant slaloms and downhills on frozen snow. The International Ski Federation first banned, then permitted them. Since then manufacturers have been trying to copy this pad and include it in their skis. And the inventor can do nothing, since he did not patent his Derby Flex.

Unforgettable mittens (1988)

The Swiss **William Wegmuller** observed that many skiers forgot or lost their mittens. He therefore invented mittens which are attached to the ski poles. But the skier still has to hang on to those.

Recco system (1974)

This is a system for finding avalanche victims using radar. It was developed in **1974** by **Magnus Granhed** from the Polytechnic Institute in Stockholm. It consists of a strip of reflective material worn by the skier and a portable search unit. The reflective strips do not wear out and need no maintenance. The Recco system can be used from a helicopter and makes it possible to explore a hectare *2.5 acres* in five to seven minutes. Most big ski resorts are now equipped with it.

Artificial snow (1935)

In **1935** the first artificial ski slope, with a ski jump, was set up in Boston, Massachusetts. Then it was the turn of Madison Square Gardens in New York in 1936 and, three years later, Los Angeles.

Snow maker (1976)

The snow maker came into being by chance, more than 30 years ago in the United States. The story goes that an irrigation company was in the habit of using water sprays in very cold weather to prevent freezing. And one day, the spray turned by chance into snow. . . . The most important patent in this domain is that of **Armand Marius**, on which today's snow makers are based.

Snowmax snow inducer (1988)

Snowmax is a protein produced by a subsidiary of **Kodak**. It considerably improves the efficiency of snow makers by raising the temperature at which the water crystallises into snow to approximately 5°C. It was very useful during the Calgary Winter Olympics in 1988.

Snowfall simulator (1986)

In June **1986** the Japanese company **Sugar Test Instrumental Co.** perfected a snowfall simulator which can reconstitute a 30cm *12in* thick snow cover in 24 hours. It makes it possible to have snow of different qualities which are very similar to those of natural snow.

Motor sports

Drag racing (1930)

Drag racing first appeared in the United States in **1930**, before spreading to Britain and then to the other European countries. Drag races are competitions of pure acceleration, carried out over a distance of a quarter of a mile (402m), between two single-seaters. These cars are extremely light in relation to the power of their engines, which may easily be as much as 1000hp, enabling them to accelerate from 0 to 350km/h *217mph* in seven seconds.

Jet motorbike (1985)

In **1985 Douglas J. Malewicki**, a prolific Californian inventor, created the most powerful motorbike ever built. It is powered by a T58-GE-8E turbine helicopter engine from General Electric. Ridden by Bob Cornell, this bike reached 331km/h *206mph* over a quarter of a mile in North Carolina on 28 September 1985.

Formula 1 (1950)

The first World Championship Grand Prix took place at **Silverstone** on Saturday **13 May 1950**. It was won by an Italian, Giuseppe Farina in an Alfa Romeo Tipo 158. Formula 1 cars are single-seaters with up to 3000cc or boosted 1500cc engines. Grand Prix races cover a minimum distance of 300km *186 miles* and a maximum of 320km *199 miles* within a time of two hours. Petrol tank capacity is limited to 220l *48 gal*.

Since 1988 turbos have been banned.

Central engine (1955)

In **1955** the British builders **Cooper** took their Racer 500 as a basis to develop a sports car with a central Coventry Climax engine, the Cooper Type 60, which was adapted to Formula 1 and in which the driver Jack Brabham won the title of World Champion racing driver in 1959 and 1960. In 1961 Ferrari, Lotus and BRM adopted the central engine.

Cooper's invention revolutionised the manufacture of all competition cars (from Formula 1 to simple go-carts), which now all have central engines.

Direct fuel injection (1960)

In **1960** the German **Kugelfischer** perfected direct fuel injection using a mechanical pump. This system was adopted on all competition cars a few years later, but did not appear in mass-produced models until 1975. The firms Lucas (UK) and Bosch (West Germany) did much to spread its use by developing electronic control.

Superkart (1985)

Superkart was created in **1985**. Go-karts now have bodies that improve their aerodynamics, driven by 250cc engines with gear boxes, and race on car racing tracks at more than 240km/h *149mph*.

Stock cars (1969)

Stock cars were developed in Europe after a special track was opened in the United States in **1969**. This is the 4.3km *2 miles 1183yd* Talladega ring, in which speeds of 320km/h *200mph* are regularly reached with bursts of up to 350km/h *217mph*. There are stock car races every week in the United States, where the cars used are mass-produced models which have been customised at great expense in order to beat speed records. In Europe the cars are stripped to the maximum, the aim being chiefly to eliminate the other competitors by pushing them off the track.

All terrain vehicle (ATV) (1967)

John Plessinger, a student at the University of Michigan, created in **1967** a motorised tricycle that was designed to cope with rough ground. The first models didn't have any suspension but the centre of gravity was low which lent the vehicle greater stability.

Plessinger sold his patent in 1969 to Sperry Rand, who launched the Tri Cart. However, it was the Japanese company Honda who began marketing ATVs in 1973 and made them popular.

Rally driving (1907)

At the start of 1907 the French daily newspaper *Le Matin* gave out a formidable challenge: 'Are there any drivers prepared to go from Peking to Paris?'. The race left Peking on **10 June 1907** and the victorious *Italia*, driven by Prince Borghese, reached Paris a few months later on 11 August, having overcome unbelievable difficulties. The next year there was an even crazier race: New York to Paris via Alaska and Russia. It was won by the American car *Thomas Flyer*. In 1909 the Transcontinental (New York–Seattle) was won by a Ford.

Paris–Dakar (1979)

In **1979**, taking up (and perfecting) an idea from the creator of the Ivory Coast–Côte d'Azur race Jean-Claude Bertrand, the rally driver **Thierry Sabine** launched and organised the Paris–Dakar long distance race. In the first year there were 70 competitors.

Despite the accidental death of Thierry Sabine during the 1986 race, the Paris–Dakar has continued and remains the top race of its category.

Tennis

Lawn tennis (1874)

Tennis was invented in 1873 by **Major Walter C. Wingfield**, who patented the new game in **1874**. He introduced a certain number of rules borrowed from an Indian game, in particular the practice of playing on grass. Wingfield called his game *Sphairistikè*, a Greek word meaning 'ball game'. The word tennis comes from the French. In the earlier *jeu de paume* ('palm game'), the server cried *Tenez!* ('Here!') to warn the other player that play was about to start.

In 1875 the All-England Croquet Club decided to set aside one of its lawns for tennis. The game proved so popular that the club changed its name to the All-England Croquet and Tennis Club a year later. In 1877 the first Tennis Championship was held at Wimbledon.

Tennis racket

The tennis racket comes from that used in the old French game of *jeu de paume* ('palm game'). It changed little until the 1960s.

Gut strings (1875)

In **1875** an English tennis racket manufacturer went to see the French manufacturer **Pierre Babolat** in Lyon and asked him if it was possible to make violin strings long enough (21 feet per string) to string rackets. The Frenchman saw a new market opportunity. After a few months of research the first gut strings appeared in Lyon.

Steel tennis racket (1960)

The former French champion **René Lacoste** (born in 1904) invented the steel tennis racket in **1960** and continued to perfect it with the help of his son François until 1980.

More than six million steel rackets have been manufactured throughout the world.

Half-court tennis was invented in Australia and is an example of how sports and games evolve.

Stringset string tensor (1988)

Perfected by the Swede **Bengt Petersson**, this is a set of accessories which can easily be fitted on to the strings of a racket and have a triple effect. They prolong the tension of synthetic strings, hold the strings in place and reduce vibrations.

Half-court tennis (1970s)

Half-court tennis was invented in **Australia** in the **1970s**. It is played in a smaller space (a third the size of a tennis court) with rackets of about 50cm *20in*, which have very short handles and are less tightly strung than normal tennis rackets. The balls used are the same size, but softer. The half-court method was promoted in Australia by players like Tony Roche and Allan Stone. From Australia, half-court spread to the United States and then Europe.

Badminton (1860s)

Badminton is descended from the ancient racket game of battledore and shuttlecock. Our version of the game is named after the Duke of Beaufort's country seat at **Badminton** in Gloucestershire where it was first played in the **1860s**. It is said that one rainy day a cord was stretched across the hall for an indoor game of battledore, and modern badminton was created.

In 1893 the Badminton Association was formed in England, and the rules codified.

Squash (1886)

Opinions differ as to the date that squash first appeared. Some say 1815, others 1830. According to the latter, squash was invented by two English gentlemen who were imprisoned for debt, and thought up the game so that they could get some exercise in prison.

Squash is a sport derived from tennis, and was not recognised in its own right until **1886**. It differs from tennis in that it is played in an enclosed space, there is no net and both players face the same way.

Table tennis (19th century)

Derived from a medieval game, like lawn tennis and badminton, table tennis developed in Britain in the second half of the **19th century**.

The first known mention of table tennis is to be found in the catalogue of a sports equipment manufacturer, F. H. Ayres, dating from 1884. The oldest patent that has been traced dates from 1891. At about the same time (1890) the Englishman James Gibb brought the idea of celluloid balls from America. Another Englishman, M. Goode, invented rubber bats with small raised points on the surface (1924). As for the name Ping-Pong, it comes from the sound made by the ball and was patented around 1891 by John Jacques of Croydon, England.

Strangely, the earliest national championships took place not in England but in Hungary, in 1897.

Theme parks

Disneyland (1955)

This was the first modern theme park and is still the most popular. Mickey Mouse, one of its heroes, celebrated his 60th birthday in 1989 and hasn't a wrinkle to show for it. The first Disneyland, near Los Angeles, was opened on **17 July 1955**. It was followed by Orlando Park in Florida, and a third was opened in Tokyo. Between them, the three parks draw 40 million visitors a year!

Eurodisneyland

This is by far the most ambitious of the current theme park projects. More than £1000 million have been invested in this park which covers an area of 1945 hectares *4806 acres* and hopes to attract ten million visitors a year from 1992 onwards. It is no small investment for the Paris region which, with the help of Mickey Mouse, the pirate river and several spectacular scenic railways, is hoping to confirm its position as the European tourist capital.

Equestrianism

Horse racing (1400 BC)

Beginning about 1400 BC, the ancient Hittites of Anatolia gave themselves up to frenzied horse racing. Horse racing became an Olympic sport in 648 BC. As for the English, they were already importing Arabian horses under the Roman occupation, and their first race took place in AD 210 at Netherby in Cumbria. English horse racing in its modern form began in 1600 in Newmarket. Each horse carried a uniform weight of 63.5kg *140lb*.

English thoroughbreds (1680)

The English thoroughbred stock was created by a group of horsebreeders whose aim was to produce faster horses. They crossed selected English mares with Arabian stallions: the Byerley Turk born in 1680, the Darley Arabian born in 1702 and the Godolphin Arabian born in 1734. All English thoroughbreds are descended from these three stallions.

Polo (1869)

Polo, the name is Tibetan in origin, was imported to England from Bengal in **1869**. This sport combines horseriding and play with a mallet and ball.

Mountaineering

Crampons (1561)

The first record we have of crampons being used relates to an Italian shepherd called **Grataroli** in **1561**.

The first four-point crampons appeared in the 18th century. In 1908 the Austrian Eckenstein developed a ten-point version. Subsequently, Grivel, a blacksmith at Courmayeur who had worked with Eckenstein added two horizontal points to the front, so making it possible for a climber to ascend a slope of more than 45 degrees.

In the years 1969 and 1970 the Austrian manufacturer Stubai sold crampons that had four spikes at the front. In 1971 Simond took out a patent for crampons that had all the spikes aligned so that they grip everywhere and provide greater security.

Free climbing (1970)

The Belgian Claudio Barbier was the first to take up rock climbing without artificial aids at the end of the **1960s**. He tackled the Steyr cliff over the Meuse river. In France Jean-Claude Droyer, a Parisian, launched this sport, whose most famous exponent is **Patrick Edlinger**. The climbers try to climb cliffs, rocks or rock faces already equipped with pitons. They thus give themselves something to hold on to, but do not use any equipment to climb.

Vibram sole (1938)

The Italian alpinist **Vitale Bramini** invented a moulded rubber sole in **1938** that was to revolutionise mountaineering; it was called Vibram.

Before this sole was produced, climbers used studded leather soles that made climbing on rock very difficult, if not impossible. They therefore carried various overshoes made from crepe, rope or felt, but these were ineffective on snow or ice. Vibram was universally adopted and greatly increased climbers' safety and comfort.

A GREAT CHAMPION, A GREAT INVENTOR

Why did René Lacoste sign the shirt he created in 1925 with a crocodile? He says that the American Press gave him that nickname following a bet: the captain of the French team for the Davis Cup had promised Lacoste a crocodile-skin suitcase if he won an important match. The nickname stuck.

But Lacoste was not satisfied with being simply part of the 'musketeers' team that won the Cup for France. He had a long and brilliant career in the car and aviation industries. Moreover it was thanks to techniques he learned from aeronautics that he developed the first metal racket in 1960.

Umbrella light (1987)

Now you can walk through the rain and fog without danger, thanks to an umbrella with a light invented in **1987** by Californian **Lawrence A. Lansing**. Not only does the umbrella allow pedestrians to see better in the dark, it also makes them more visible to motorists, and thus less likely to be knocked down when crossing the road or walking along a country lane. Particularly recommended for children!

Jacket that changes colour (1987)

The days of reversibles are over: now there's no need to turn your jacket inside out. In **1987** the Italian **Massimo Osti** created a garment made with crystals which change as the temperature drops. A white jacket turns blue, a pink one khaki, and a yellow one green, at the first hint of winter.

Aquarium ties (1988)

Gary Fisher (and with a name like that it's hardly surprising!) is an original and inventive stylist. His speciality is plastic ties containing live goldfish! A former architectural draughtsman, Gary Fisher was unemployed when he had the idea of creating highly individual ties in **1988**. He began with goldfish, then extended his range. Today he has original models on offer, called 'professional ties', which contain the typical objects that go with a particular job. Thus if you are a carpenter, your transparent tie might, for example, contain a screwdriver, nails and a crocodile clip (in miniature of course), and if you are an accountant you could have a tie featuring a few pocket calculators. The possibilities are endless.

Underwater museum (1989)

All divers, whether or not they use breathing apparatus, should make sure to visit the underwater museum that was set up in **1989** by the former manager of a boat maintenance company. This museum is located at a depth of 12m *39ft* off the coast at Antibes. Visitors

This fish, which comes from the Nile, discharges an electrical current. It was displayed at an exhibition in France in 1989 and the charge was used to power the two clocks at the top of its tank.

Frenchman Stéphane Seckler created the car-plane to overcome the problems of getting from airport to home.

enter without a ticket and there are no security guards at this exhibition of seven statues by sculptor Vincent Duglas.

The boat that follows the shape of the waves (1850)

In the 19th century industry, science and imagination often came together to produce bizarre creations. The famous ironmaster Henry Bessemer, inventor of the converter, also invented a first-class lounge for transatlantic liners which was designed to spare rich passengers the torments of seasickness. It was mounted on a universal joint and remained horizontal at all times, even in the worst storms. But this was not the best invention of its type: the Connector, also built in Britain in **1850**, was a ship with three jointed sections that followed the movement of the waves completely. It seems, however, to have remained at the prototype stage.

Car umbrella-holder (1989)

Where do you put a wet umbrella when you rush to your car for shelter in a downpour? In

Thanks to its three articulated sections, the Connector *could, in theory, follow the movements of the waves.*

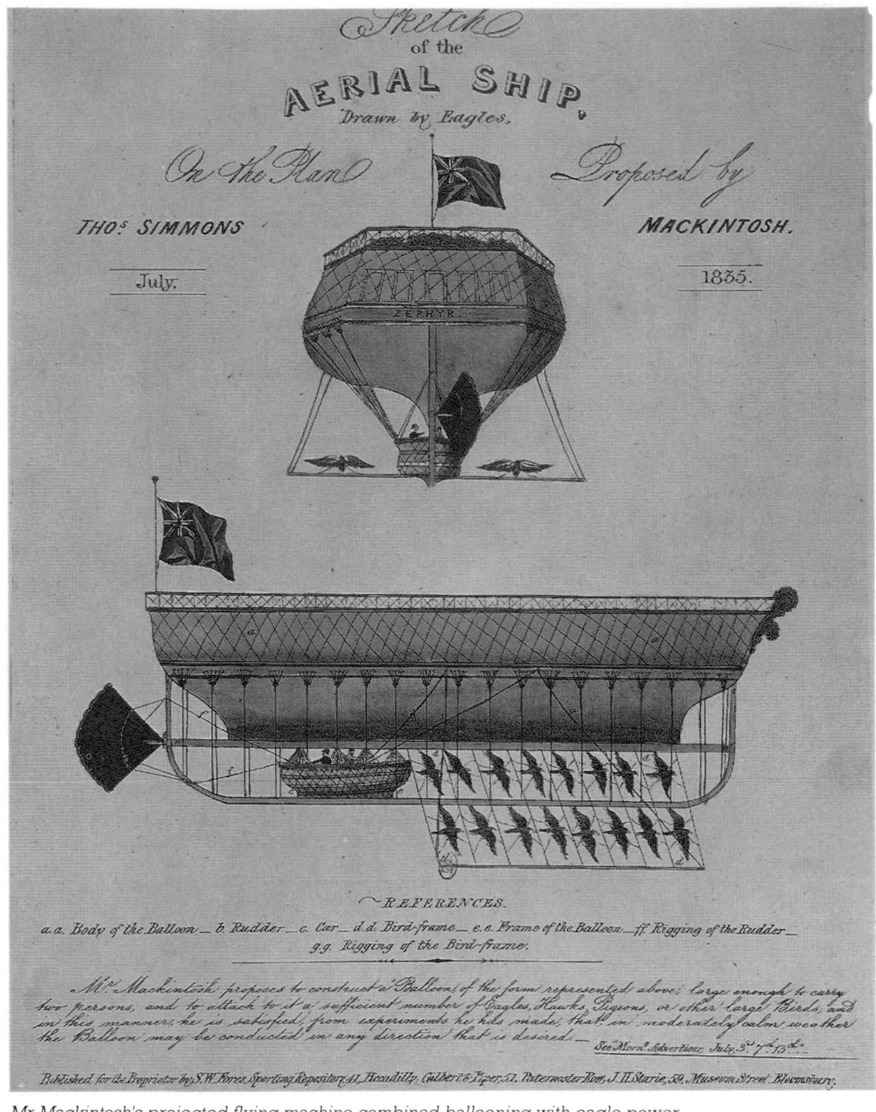

Mr Mackintosh's projected flying machine combined ballooning with eagle power.

1989 the Italian **Vincenzo Borriello** designed a device that fits diagonally into the car door, enabling umbrellas to be tidied away and preventing wet seats and carpets. This accessory can easily be fitted to mass-produced cars.

Car that walks like a crab (1988)

The Crab Motor is the brainchild of Taiwanese inventor **Ching Tang Huang**. Taipei, the capital of Taiwan, obviously has traffic problems as bad as those of any other city in the world, because here is another invention born out of the need to park somehow, somewhere. The Crab Motor can move from left to right, or turn round on the spot whenever the road is as narrow as the car. An additional advantage is that there is no longer any need to use a jack to change the tyres: they can be raised individually.

Video Cat (1987)

It is now possible to have the pleasure of owning a cat or dog without the grind of having to open tins, change the litter tray or go for walks at set times, thanks to a Video Cat (or Video Dog).

In **1987** the American company **Creative Programming Inc.** of New York brought out video cassettes on which you will find the animal of your choice. It lives with you and plays, mews or barks at your convenience. So, you can find out whether you are made for each other before committing yourself to a proper adoption.

The idea must have found favour, because Creative Programming Inc. are now offering a Video Baby, with the following slogan: 'Have the wonderful experience of motherhood without the disadvantages'.

A biscuit for koala bears (1988)

The koala bear's basic food, fresh eucalyptus, is hard to find in winter. At the Taronga zoo in Australia **Professor Ian Hume** decided to find a solution to this problem. In **1988**, after years of research, he developed a substitute food: a

Koalas are now happy thanks to Professor Hume's eucalyptus biscuits.

In search of well-being, Doctor Juan Abascal's patients enjoy a period of mental flight.

special eucalyptus biscuit which now makes up three-quarters of the food consumed by koalas in captivity.

The smell of success (1989)

They say that money has no smell, but it seems that smells can influence the financial affairs of human beings. Such is the view of the Japanese firm **Matsushita** who, in **1989**, began selling cassettes that give off scents such as mint, jasmine, or lemon for up to 15 minutes. These are intended to have a beneficial effect on business people and their employees. For example, before going out to make a deal it is a good idea to inhale a stimulant such as eucalyptus, followed by a tranquiliser like lavender on returning to the office.

Computer Cushion (1988)

Working at a computer can often be very tiring. With a stooped back, a stiff neck and arms bent over the keyboard, the operator sometimes ends up looking like a praying mantis according to the American inventors of the Computer Cushion, the **KB-Pillow Company**. In **1988** they invented a cushion on which the keyboard rests, and which can be placed in the operator's lap. The cushion allows operators to straighten their backs and bend their arms in the way that suits them best.

Journeys to the heart of the brain (1988)

Since **1988** a former museum security guard, **Randy Adamadama**, has been offering 45 minutes of total happiness for $10 at the Paradise Shopping Center at Costa Madera, California. Wearing headphones and special glasses that project soothing colours, patients are led towards their 'inner self' by means of a machine that regulates the vibrations of the brain. The results are said to be physical and mental relaxation and a sense of well-being. Another similar method is that of the Mental Flight, offered by Doctor Juan Abascal on board his Synchro-Energizer in Florida.

Parasol for books (1988)

The American **Gregory Ainsworth** loves reading in the sun, but doesn't enjoy the glare and resulting tired eyes. In **1988** he invented a little parasol made of plasticised paper in the shape of a fan, which keeps the pages of any size of book in the shade.

Sexual regenerator (1987)

The Japanese inventor **Hiroshi Majima** has applied the ancient oriental principle of the alternation of heat and cold for his invention. The S-Charger is an electrical device, designed exclusively for men, which renews their sexual vigour. It consists of a base with two poles – one topped with silver and cold, the other golden and hot. The whole thing is connected to a transformer which controls the current of electricity. One minute of treatment is enough: the penis, placed first on one pole then the other, is restored to its former potency.

Dumb-bells that quench the thirst

It is no longer impossible to run and drink at the same time thanks to the water-filled

A French company created this rather unusual egg roll after two years of research that cost £500,000. The yolk and the white are cooked separately and then reassembled.

A simple way of avoiding dehydration in runners.

These odd-looking riders are in fact radio-controlled models.

dumb-bells invented by Canadian **Daniel Paquette**. His idea consists of little water bottles in the shape of dumb-bells that have a stopper and a handle that goes round the hand. Runners and walkers can now exercise their arm muscles with no risk of dehydration.

Robot jockey (1987)

Super-jock was invented by American **Dave Kim**, manager of a ranch in the Midwest of America. The jockey is made up of a radio-controlled robot with a motorised spring mechanism which activates a whip. It has two speakers that relay the instructions from the radio-controller. The device is covered with a plastic shell painted to look like the silhouette of a jockey.

Record-breaking inventor

He is the Japanese inventor Yoshiro Nakamats, who has 2,360 patents to his name. That is more than Thomas Edison – inventor of the phonograph and the fluorescent light – who took out only 1,200.

Among Doctor Nakamats' creations there is a little of everything, ranging from the serious – such as the floppy disc – to the totally unexpected, such as talking paper (Nakapaper) and Cerebrex, the Brain Chair. Cerebrex is a robot that tests Nakamats' mental and physical faculties and which stimulates his creativity. In fact, Cerebrex is an intelligent robot who, unlike other machines of his kind, is able to create a 'friendly' relationship between his master and himself, thanks to a computer.

Cerebrex won the Grand Prix at the 10th Inventions Fair at New York in 1986.

Cards for shy people (1987)

Help is at hand for those shy people who would like to express their feelings (in few words!) in a restaurant, a lift, between trains, on the underground, etc. These cards, which are about the size of a credit card, will convey your message for you. They were invented in **1987** by American **Diana Amsterdam** and come in sets of ten. Each one tells a different story: 'You are beautiful' or 'You are irresistible' for example. All the would-be suitor has to do is add his or her name and telephone number.

In 1929 a Hungarian inventor imagined a bizarre machine that could prevent someone from falling asleep.

The Japanese inventor Yoshiro Nakamats developed the thinking chair or Cerebrex. It is supposed to exercise one's intellectual abilities and stimulate creativity.

Parcels wrapped in this paper should be opened with care: the dollar-bills are real and can be cut out and used.

Yoga for cats (1986)

A Japanese, **Shigenori Masuda**, has discovered that yoga is an excellent exercise for domestic animals, particularly for cats. He has even developed certain positions for cats and dogs who are overweight and have a tendency to laziness.

The director of a health clinic for cats and dogs in Japan, Shigenori Masuda, teaches them yoga.

Language and literature

Artificial languages

Shorthand (antiquity)

The invention of shorthand goes back to ancient times.

The Greek Xenophon (c.430–355 BC) could record his conversations with Socrates thanks to his semiology (writing by symbols). In Rome Marcus Tullius Tiro developed one of the first systems for abridged writing in order to take down Cicero's speeches.

Use of shorthand disappeared subsequently only to re-emerge in the 17th century. The initiative came from England, where, in 1602, J. Willis wrote a treatise on shorthand.

Various methods have since been introduced. Of these, the most famous are those by Isaac Pitman (1837) and John Robert Gregg (1888).

Argot (17th century)

The first mention of the word argot dates back to the **17th century**. However, different slang languages – secret languages used by thieves and bandits to protect their own from the curiosity and indiscretion of third parties such as informers and policemen – had been in existence for a long time. One of the earliest mentions of slang language can be found in the *Jeu de Saint Nicolas* by Jean Bodel written around 1200. Moreover there is evidence of these secret underworld languages throughout Europe from the 17th century, such as cant in England and *Rotwelsch* in Germany.

Esperanto (1887)

Esperanto was invented by a Pole, **Lazarus Ludwig Zamenhof** (1859–1917), in **1887** as an international language in the spirit of peace and communication between nations. It must be said that Esperanto has not enjoyed any real success, but it may soon experience unexpected glory as the pivot language for translation machines. That is according to BSO, a Dutch company that intends to launch a translation system at the beginning of the 1990s for messages sent in English and received in French by passing first through Esperanto.

Esperanto replaced another international language, called Volapük, which was invented in 1880 by Johan Schleyer (1831–1912), a priest living near Constance (Germany).

Alyoundi (1987)

Alyoundi, a new international language, was invented by a Russian engineer called **Alexander Kolegov**. According to its inventor, this language is suitable for international communications systems and data-based translation into national languages, and can take into account all grammatical categories. Thus in the future it could come to be used as a reserve language for the purposes of storing human data in computers with unlimited memory.

Talking to monkeys (1988)

Professor David Premack of the University of Pennsylvania has developed an experimental language enabling him to teach chimpanzees to 'talk' and communicate with him. Recognition of this extraordinary success, the result of studies carried out since 1965, came when he won the Fyssen Foundation award in Paris in April **1988**.

Punctuation (2nd century BC)

The invention of punctuation is due to **Aristophanes of Byzantium** (257–180 BC). The Greek grammarian directed the famous Library in Alexandria and developed a system comprising three signs corresponding to our full stop, semi-colon and colon. It wasn't until the 16th century, however, with the invention of printing, that its usage was really respected. However, the rules of punctuation remained highly unreliable until the middle of the 19th century.

Writing materials

Papyrus (c.3000 BC)

The **Egyptians** are usually credited with the invention of papyrus around **3000 BC**. The stem of a reed cultivated in the Nile Valley constituted the raw material.

Parchment (2nd century BC)

It was, no doubt, the commercial and cultural rivalry between the Pharaoh Ptolemy and the King of Pergamum in Asia Minor, which, between **197 and 156 BC**, was the origin of the invention of parchment. The Pharaoh, taking umbrage at the growing reputation of Pergamum as a cultural centre, must have stopped providing it with papyrus, thus obliging the scribes of Pergamum to invent a new material. Parchment is made from the skin of sheep, goat or calf (vellum).

Paper (2nd century BC)

The earliest example of paper was discovered 25 years ago in a tomb excavated by archaeologists in the Xian region of China. Analysis has shown that this paper had been made from hemp fibres mixed with a small quantity of linen, made during the period of the Han dynasty in the **2nd century BC**. However, it was not until the reign of Emperor Hedi (AD 88–106) that paper suitable for writing purposes began to be used. The inventor of this was Cai Lun, one of the Emperor's eunuchs who developed an inexpensive method using bark from trees, linen scraps, old rags and disused fishing nets.

The techniques rapidly improved and paper reached neighbouring countries first (Vietnam, Korea and Japan), and then Arab countries in the 8th century.

The Silos missal, from near Burgos in Spain, is the oldest manuscript on European paper. It dates back to the beginning of the 11th century.

Very long paper (1799)

On **18 January 1799 Nicholas Robert**, employed in Paris at the bookshop and printers François Didot, obtained a patent 'for the manufacture of an extraordinary paper, measuring between 12 and 15m *39ft 4in* to *49ft 3in* long, without the need for any workman and by purely mechanical means'. This invention was improved by an Englishman, Bryan Donkin, in 1803. It enabled the paper mill to free itself from the traditional method of manufacturing paper, that is to say, sheet by sheet in a tank.

The main producer of paper is the United States, with a production of 64 million tonnes a year.

Digital paper (1989)

Digital paper, sometimes called optical paper, is intended for use in data processing. In terms of composition, it bears no relation to traditional paper. It is composed of four layers of materials placed one on top of the other. One of the materials is polyester which is fine yet strong.

The first sheet of digital paper was manufactured in **1989** by the British company **Image Data**, a subsidiary of ICI.

Pencil (16th century)

In **1564** the discovery of graphite in Cumbria led to the invention of lead pencils. In the 18th century the Cumbrian graphite mines had become a royal monopoly the exploitation of which was subject to many rules since graphite was also used in cannon foundries. Each workman was searched upon leaving the works and theft was a capital offence punishable by hanging.

The interruption of relations between France and Britain in 1792 led the French engineer Jacques Nicolas Conté to invent graphite and clay pencils covered in cedar wood. Demand crossed international borders and his pencils were soon to reach all parts of the world.

Brazil is the main manufacturer of pencils in the world, producing 4.5 thousand million per year.

Ink (2500 BC)

It was the Chinese who invented ink in **2500 BC**. It was made with smoke, glue and aromatic substances.

Evidence has been found in Egyptian hypogea of papyrus covered in black or red ink applied with a reed and even a quill pen.

Clean ink (1985)

It was for the purpose of daily newspapers that the American **Rodger L. Gamblin** patented, in **1985**, an ink which does not run and which does not soil readers' fingers.

Assuero Cortani began work in a shop selling typewriters at 12 years old. Now, more than 60 years later, he is one of the world's leading collectors of the machines.

TYPEWRITERS

Inventor	Date	Country	Characteristics
William Petty	**1660**	GB	Manual system with two pens. This is the forerunner of the typewriter.
Henry Hill	**1714**	GB	Queen Anne granted a patent for this machine which was never built.
Pelegrino Turri	**1808**	Italy	Earliest example of typewriting, although the machine has not survived.
Xavier Progin	**1833**	France	First modern circular machine with bars bearing the characters.
Alexander Bain	**1841**	GB	The familiar inked ribbon was devised.
Malling Hansen	**1870**	Denmark	First typewriter marketed.
Christopher Latham Sholes	**c.1870**	USA	The 'literary piano', prototype of a machine sold to Remington which was marketed as from 1874 under the name Remington Model 1.
C. C. Blickensderfer	**1889**	USA	First portable typewriter in a case: the Blick.
Dr T. Cahill	**1901**	USA	First electric machine. The company went bankrupt after having built 40 examples at a cost of $3925!

R. G. Thomson	1933	USA	Develops the Electromatic, launched by IBM in 1933.
IBM	1965	USA	Launches the first electronic typewriter with a memory (magnetic tape), the 72BM.
Olivetti and Casio	1978	Italy Japan	They market the first electronic typewriters with live memories, fitted with daisy wheels, not golf balls.
Matsushita	1984	Japan	First typewriter without keys. The Panaword keyboard is a sensitive sheet. The user writes his message by hand, on a screen.

Anti-fraud felt-tip pen (1985)

The French company **Reynolds** has, since **1985**, sold a felt-tip (and ball-point) anti-fraud pen which contains unerasable security ink. This pen was designed by Reynolds at the request of the banks to make the falsification of cheques more difficult.

Propelling pencil (1915)

The first automatic propelling pencil, called the Ever-Sharp Pencil, was invented in **1915** by the Japanese **Rokuji Hayakawa**, founder in 1912 of a company to which the propelling pencil gave its name. The Sharp Corporation has since broadened its activities, particularly in electronics.

Rubber (18th century)

Made from a rubber base that erases pen or pencil marks, the rubber is believed to have been invented in the mid-**18th century** by a Portuguese physicist named **Magalhaens**, or Magellan (1722–90), who perfected numerous instruments for use in physics and astronomy. A rubber was mentioned for the first time in 1770 by the British chemist J. Priestley. Today, plastic and synthetic rubber are commonly used to manufacture erasers.

Fountain pen (1884)

Nobody knows exactly how far back the revolutionary idea of adding an ink reservoir to a quill pen goes. In her *Memoirs*, Catherine the Great of Russia noted that she used an 'endless quill' in 1748. Was she referring to one of the first pens?

Between 1880 and 1900, fountain pen inventions proliferated; more than 400 patents were registered.

The inventor of the first proper fountain pen was the American **Lewis E. Waterman**, an insurance broker, who had had enough of almost losing contracts due to malfunctioning pens. He threw himself into the task of solving the problem of ink flow and, on **12 February 1884**, obtained the first patent for what was to become the Waterman Regular.

The ink cartridge was invented by M. Perrand, director of Jif-Waterman, in 1927 and patented in 1935.

Ballpoint pen (1938)

The ballpoint pen was invented in **1938** by **Laszlo Biro** (1899–1985), a Hungarian journalist. During a visit to the print shop of the magazine for which he wrote, Biro was impressed by the advantages of quick-drying ink. He proceeded to make a prototype of a pen based on the same principle.

To escape the Nazi threat, he settled in Argentina in 1940 and there developed his invention. He patented it on 10 June 1943, and his pens were sold in Buenos Aires starting in 1945 and adopted by the RAF in 1944 to resolve the problems pilots faced, flying at high altitudes.

Bic (1953)

In **1953** a French baron, **Bich**, developed an industrial process for manufacturing ballpoint pens that dramatically lowered the cost of production. The Bic was born, and each year we buy three thousand million of them.

Erasable ballpoint pen (1979)

It was the American firm Gillette who, in **1979**, launched the first erasable ballpoint pen, the Eraser Mate. One result of this product was that banks in the United States cautioned their depositors against using the pen to write cheques.

Felt-tipped pen (1963)

It is the Japanese firm **Pentel** to whom we owe the invention of the felt pen with an acrylic tip. It was invented and marketed in **1963**.

Pentel also invented the first felt-tipped ballpoint pen in 1973, the Ball Pentel, and, in 1981, launched the first ceramic nib; the Ceramicron.

Printing

Origins (868)

Printing was already a widespread practice in China under the Tang dynasty (618–907): books on magic, scholastic manuals, etc were produced. The discovery in the Dunhuang caves of a copy of the Diamond Sutra printed in **868** gives us the name of the printer, Wang Zhe.

Chinese printing prospered under the Sung dynasty (960–1279), and in the year 1000 an important Buddhist Sutra was published.

In 1041 the Chinese Bi Sheng made mobile characters out of fired clay.

The casting of metal characters was developed mostly in Korea around 1392.

Rotary press (1845)

The first rotary press was patented by the American **Richard Hoe**, in **1845** in the United States. It was first put into operation the following year.

Gutenberg press (c.1447)

Around **1447**, the German printer **Johanne Glensfleisch** called **Gutenberg** (c.1398–1468) developed, along with his associates, the technique of moveable characters. In addition, he perfected the material necessary for the quality and conservation of characters: an alloy of lead, antimony and tin.

Around 1455, in Mainz, Gutenberg printed the *Biblia sacra latina*, known as the 42-line-per-page Bible. It was the first Latin edition of the Bible in moveable characters. Gutenberg's business partner, Johann Fust, took him to court for repayment of a loan advanced earlier, gained possession of the moveable type characters, and took all the Bible's profits. However, Gutenberg was able to start up a new printing business by 1465. In 1477 William Caxton (c.1422–91) set up the first printing press in England.

Improvement of the printing press

Around 1800, Lord Stanhope got rid of wood completely in presses and used metal instead. He multiplied newspaper productivity by ten, reaching 3000 sheets per day.

The first printing press run by a steam machine was developed by the German F. Koenig in 1812. Later, Koenig, in association with A. Bauer, constructed the first cylinder press, which was followed by many others. In London notably, a single machine printed overnight the 4000 copies of the London edition of *The Times*.

Xylography

This is one of the oldest and simplest methods of printing an illustration using a block of wood. Combined with typography, it led to the production of beautiful illustrated books at the end of the Middle Ages (14th to 15th centuries).

Intaglio (1450)

This process uses engraved or etched surfaces to hold the ink. The first method was copper-plate engraving, developed in **1450**.

Engraving with burin on a copper plate was followed at the beginning of the 16th century with the use of aqua fortis (the old name for nitric acid). Aqua fortis produces depressed lines in the plate at points where the protective varnish has been scraped away.

Lithography (1796)

In **1796**, the German typographer **Alois**

Senefelder (1771–1834) invented lithography, a method of printing by transfer. He realised that a drawing done with a soft-lead pencil on limestone (lithography comes from the Greek word for stone, *lithos*) is water-resistant. However, the unmarked stone absorbs water. If a coat of greasy ink is spread over the surface of the stone, it doesn't stick to the wet spots, only to the greased areas. The stone has therefore only to be placed in a press and it reproduces the initial drawings.

Senefelder rapidly improved his method. Instead of water, he used a solution of gum-arabic and nitric acid, which is completely impervious to printer's ink.

Photogravure (1822)

We owe the invention of photogravure to **Nicéphore Niepce** around **1882**. We can say that, chronologically, Niepce invented photogravure before photography. In fact, his invention first gave rise to a printing form (engraved copper plate) capable of producing existing images by means of a press. Replacing man-driven tools by light, photogravure contributed to the growth of typography and gave rise to heliogravure and offset printing.

Heliogravure (1875)

The heliogravure printing process was invented by the Austrian **Karl Klietsch** in **1875**.

An industrial intaglio process, heliogravure became widely used in the printing of magazines and catalogues.

Offset (1904)

Invented by the American lithographer **W. Rubel** in **1904**, offset came out of the lithographic process he perfected. The word offset covers the same transfer technique (direct apposition) as lithography. Offset is not done on stone but on a sheet of zinc.

Publishing

Book publishing

Veritable publishing houses already existed in ancient Greek and Roman times. Athens and Rome boasted printed works of which several hundred copies were published.

It was, of course, only with the invention of the industrial print shop that publishing really took off, and the bookshop came soon after. The latter was born in London with the bookstore of Wynkyn de Worde, successor to Caxton, and publisher of the first book in English to be produced in England in 1495.

But it was not until the late 16th century that bookshops began to specialise in selling books from one field or another, and it was only then that publishers charged them with the task of distributing their products.

Cookery book (AD 62)

The cookery book dates back to the famous treatise *De re coquinaria* published by the Roman gastronome **Apicius** in **AD 62**. It describes the feasts of the Emperor Claudius I (10 BC–AD 54) and his successive wives, Messalina and Agrippina.

Library (1700 BC)

The first libraries appeared in Chaldea in **1700 BC**. The first books were baked clay tablets. In 540 BC Pisistratus endowed Athens with the first public library.

Dictionary (600 BC)

The oldest dictionary to have been found dates back to **600 BC**. It comes from

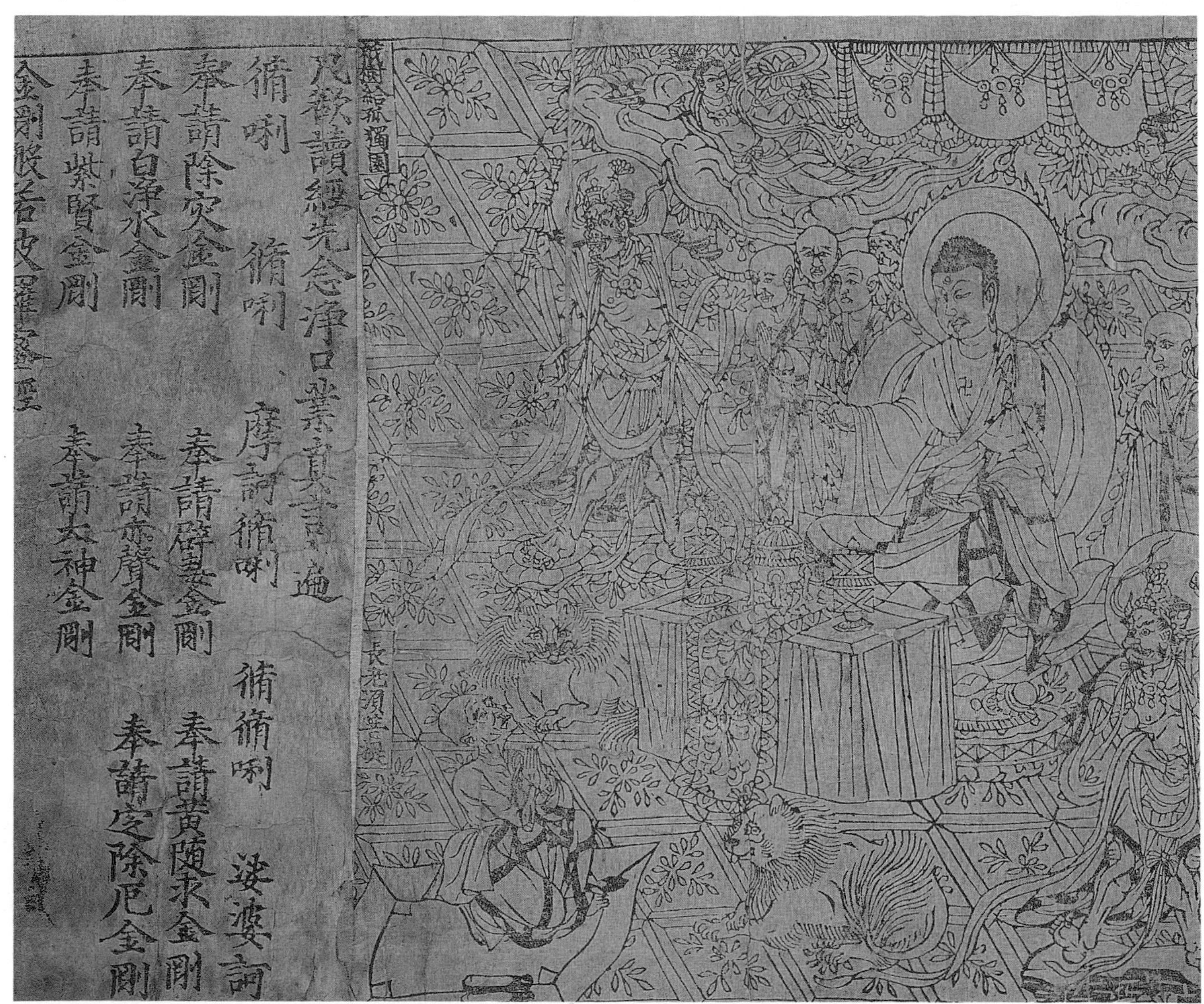
凡欲讀經先念淨口業真言一遍
脩唎 脩唎 摩訶脩唎 脩脩唎 娑婆訶
奉請除灾金剛 奉請辟毒金剛 奉請黃隨求金剛
奉請白淨水金剛 奉請赤聲金剛 奉請定除厄金剛
奉請紫賢金剛 奉請大神金剛
金剛般若波羅蜜經

The Diamond Sutra is the oldest known printed book. This Buddhist prayer was produced in AD 868.

William Caxton presents the Duchess of Burgundy with a copy of The Recuyel of the Historyes of Troye. *It was printed in Cologne between 1473 and 1474.*

A THOUSAND ARCHITECTS FOR ALEXANDRIA

The library in Alexandria contained more than 500 000 volumes when it was destroyed in 47 BC by a fire started by Julius Caesar's legionaries.

The former President of the University of Alexandria, Lufti Duwidar, has launched a project to build a new library 'joining together all the knowledge of humanity in all the languages of the world'. The project is estimated to cost £100 million and is to be financed by donations from all over the world.

Some 1200 architects have entered the international architects' competition. The award-winner will have one year to prepare his/her detailed sketches of the building, and construction is due to begin in the early 1990s. The opening is planned for June 1995.

WHO INVENTED THE PAPERBACK?

Penguin produced the first mass-market books in 1935 with the introduction of the paperback. Priced at just 6d (2½p) they were within the price range of large numbers of people who might not before have considered buying books.

Mesopotamia and is written in Akkadian, the language of the Assyrians and the Babylonians. In China, the Hou Chin dictionary did not appear until 150 BC.

In 1480 the English printer, **William Caxton**, published the first bilingual English–French dictionary for tourists. It had 36 pages.

The first polyglot dictionary, based on Latin and Italian, then extended to German, English and French, was the famous *Dictionary of the Latin Language* (1509). It was the work of the Italian scholar, Ambrogio Calepino.

In 1755 the English lexicographer Doctor Johnson (1709–84) produced his *Dictionary*.

Book (2000 BC)

Although the book appeared in China in **2000 BC**, it was only between the 2nd and 4th centuries that it appeared in the West in the form we know. During this period, it went from being the volume (a scroll of papyrus or parchment) which was not very manageable, to the more portable codex (a volume of manuscripts where sheets are inserted and folded together).

Encyclopaedia on videodisc (1986)

Since **1986** the Japanese company **Pioneer** has been marketing a videodisc encyclopaedia, the Knowledge Disc, in the United States. It is a videodisc read by laser that contains the text of 20 volumes of the US Academic Encyclopedia.

Newspaper (1605)

The first gazette to come out regularly appeared in Antwerp in **1605** under the title *Nieuwe Tijdinghen*. Its creator was the printer **Abraham Verhoeven**.

It was widely imitated in Europe and, from **1609**, two weekly gazettes appeared in Germany, one of which, *die Relation aller*

TYPESETTING MACHINES

Type	Date	Inventor	Country	Characteristics
Machine with moveable characters	1041	Bi Sheng	China	First moveable characters made of clay.
Machine with metal characters	1403		Korea	Some 30 years later, the introduction of a phonetic alphabet in Korea was to make the characters easier to manipulate.
First European machine	c.1447	Gutenberg	Germany	He invents a mould for the casting of metal characters representing the letters of the alphabet. Printed by moveable type.
Linotype	1886	Ottmar Mergenthaler	USA	A machine which makes printing letters by casting characters line by line. This machine was built to an order placed by the editor of the *New York Herald Tribune*.
Monotype	1887	Tolbert Lanston	USA	Casts and typesets individual characters from hot metal. It enables the typesetting of 9000 characters per hour.
Photocomposition	1953	Louis Moyroud and René Higonnet	USA	The first phototypesetters are put into operation; invented by two Frenchmen living in the USA.

A 16th-century Persian prince is seen here teaching his children to read.

fürnemmenund gedenkwürden Historien, published in Strasbourg, mentions in its 37th edition the invention of Galileo's telescope.

The first newspaper in English was the *Corrant of Italy, Germany etc.* printed in Amsterdam by George Veseler in 1620.

The first daily newspaper was brought out by the German Thimotheus Ritzch from Leipzig. It appeared for the first time in 1650 under the title *Einkommende Zeitungen*. Subsequently called the *Liepzig Journal* (*Leipziger Zeitung*), it continued to appear until 1918!

Visual arts

Painting and drawing

Origins (40 000 BC)

It is tempting to assume that the first 'canvas' was the human body. Archeologically, the most ancient examples of this art are to be found on cave walls dating from the Upper Palaeolithic era (40 000–12 000 BC). Handprints (sometimes outlined with the ancestor of the airbrush, a hollow bone from which red ochre powder was blown), and markings made by fingers dipped in dye later evolved into images of animals and people.

Canvas (2000 BC)

Canvas (linen, cotton or hemp) is an ancient material. A fragment of linen was found in Egypt that dates from between 2000 and 1788 BC.

A cloth or canvas base offered a cheaper and lighter base than wood, particularly for large paintings. However, its use did not spread until the late 15th century and early 16th century. In the Middle Ages, it was used only for painted processional banners.

1990: World literacy year

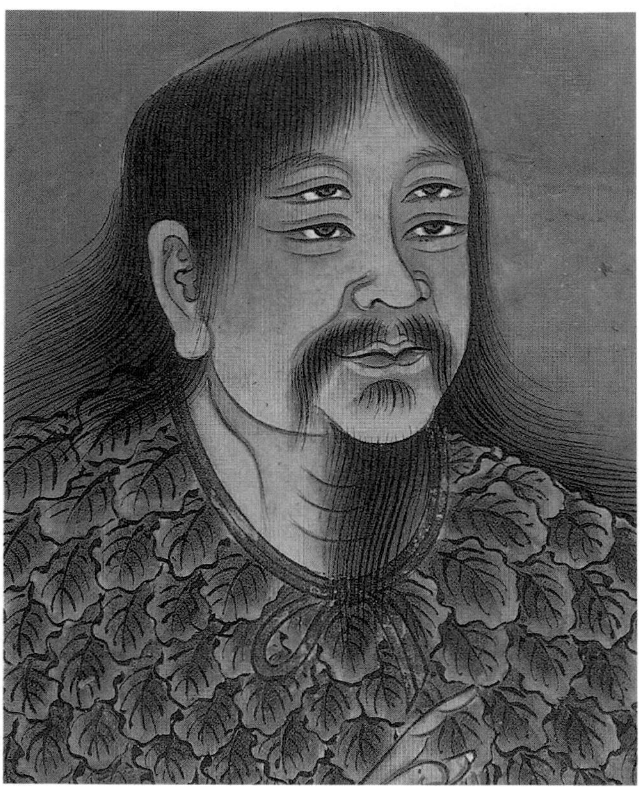

T'sang Kie, the hero with two pairs of eyes, is said to have created Chinese characters after studying traces left by birds' feet in the sand.

900 million people are illiterate

There are about 900 million around the world who can neither read nor write. Almost 98 per cent of these illiterate people live in developing countries: 666 million in Asia, 162 million in Africa, 44 million in Latin America and the Caribbean. As far as industrialised countries are concerned, there are more than 50 million adults who have difficulty in reading, writing and doing simple arithmetic.

To take part fully in society it is essential to know how to express oneself and to be able to communicate in a way other than by speech. Further, to be able to understand what others write without having to go through an intermediary is the first stage to being free.

Writing gives thoughts and ideas a structure as well as giving us a way of recording them. The invention of writing seemed so extraordinary that more than one civilisation claimed that it had been bestowed on them as a gift from their gods.

Historically, the three most ancient forms of writing (around 3000 BC) are as follows: in the Middle East, cuneiform; in Egypt, hieroglyphics; and in the Far East, Chinese ideograms.

Cuneiform

Originally semi-pictographic, cuneiform is the most ancient writing system documented. It was developed by the Sumerians. The term cuneiform refers to the extremely angular aspect of the signs. A young teacher from Göttingen in Germany, Georg Friedrich Grotenfend (1775–1853) was the first person, in 1802, to present a translation of cuneiform. Soon afterwards Henry Creswicke Rawlinson (1810–95), a British major working for the East India Company, also solved the enigma.

Hieroglyphics

The appearance of hieroglyphics, the most ancient and characteristic form of Egyptian writing, coincided with the unification of Egypt: around 3000 BC. Their use was maintained until the 3rd century AD.

Hieroglyphics are picture symbols; hieratic writing is a cursive form of hieroglyphics that was used by the priests until AD 500. Demotic writing also came from the same source and gave seven of its letters to the Coptic alphabet which was used in Egypt after its conquest by the Arabs in 641.

In 1822 the French Egyptologist J. F. Champollion (1790–1832) deciphered the hieroglyphic and demotic characters inscribed on the Rosetta stone, a basalt slab dating from the reign of Ptolemy V (196 BC).

In 1984 Michael Hainsworth developed software for processing texts written in hieroglyphics.

Ideograms

Chinese writing is said to have been invented by the country's emperors, some 3000 BC. The most ancient documents discovered date from between 1400 and 1200 BC. Korea, Japan and

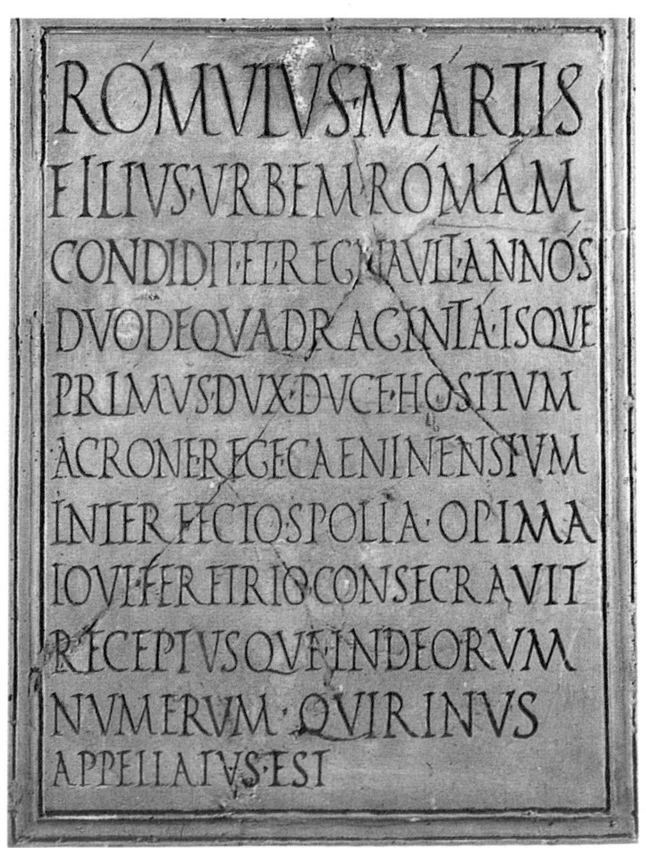

The Latin alphabet dates from around the 7th century BC.

Vietnam later adopted aspects of the language. Chinese is the only ancient language still in use, and it often forms the only link between the one thousand million Chinese who speak many different dialects.

Greek alphabet

The Greeks borrowed their alphabet from the Phoenicians and adapted it to their language. However, the fundamental innovation of the Greek alphabet was the introduction and rigorous notation of vowels. It is the first two letters in the Greek alphabet, *alpha* and *beta*, that are the root of the word alphabet.

The alphabet

These abstract signs representing language and thought date back to the second half of the second millennium. Discoveries made in Râs Shamrah, in the Middle East, confirm the theory that it was the Phoenicians who invented the alphabet during this period.

Comprising 22 signs, the Phoenician alphabet is at the root of all Western alphabets.

Latin alphabet

The earliest evidence of Latin writing we have dates back to the 7th and 6th centuries BC. The Latin alphabet derived originally from the Greek and was passed to us by the Etruscans. The Roman Empire was to impose this alphabet on the whole of the Western world. A century before Christ, the Latin alphabet comprised 21 letters to which y and z were added from the Ionian alphabet. In the Middle Ages j, u, and w were introduced, to give us the alphabet we use today.

Arabic alphabet

The first proven inscription in Arabic dates back to AD 512.

This language, which had semitic origins and is the sister of Hebrew, is also related to Phoenician writing. The alphabet comprises 28 letters, all consonants, three of which serve as long vowels as well as diphthongs.

Cyrillic alphabet

It was a theologian and missionary of Greek origin Saint Cyril (c. AD 827–869), nicknamed the philosopher, who in 862 invented the Cyrillic alphabet for the purpose of translating the Bible for Slavic peoples to whom he wished to preach the Gospel.

This Glagolitic alphabet (from *glagolu*, meaning word in old Church Slavic) was modified towards the beginning of the 10th century through the introduction of 24 Greek letters. The present Russian alphabet derives from it.

The religious schism which occurred at the beginning of the 9th century gradually split the Slavic world into two alphabetic areas: Russians, Ukrainians, Bulgars and Serbs adopted the Cyrillic alphabet along with Greek Orthodoxy; the Poles, Czechs, Slovaks, Slovenes and Croatians adopted the Latin alphabet and Roman Catholicism.

Sign language

The first sign language alphabet was compiled in 1620 by J. P. Bonnet. He was a private tutor at the Spanish Court which, at that time, held many deaf people.

Braille

In 1829 Louis Braille (1809–52), a professor at the French Institute for the Blind, published a writing system based on raised dots. Braille had been accidentally blinded at the age of three by his father's tools.

Raised letters for the blind had been proposed by another Frenchman, Valentin Hany, some years earlier, but Braille's system of six dots was easier to read with the fingertips.

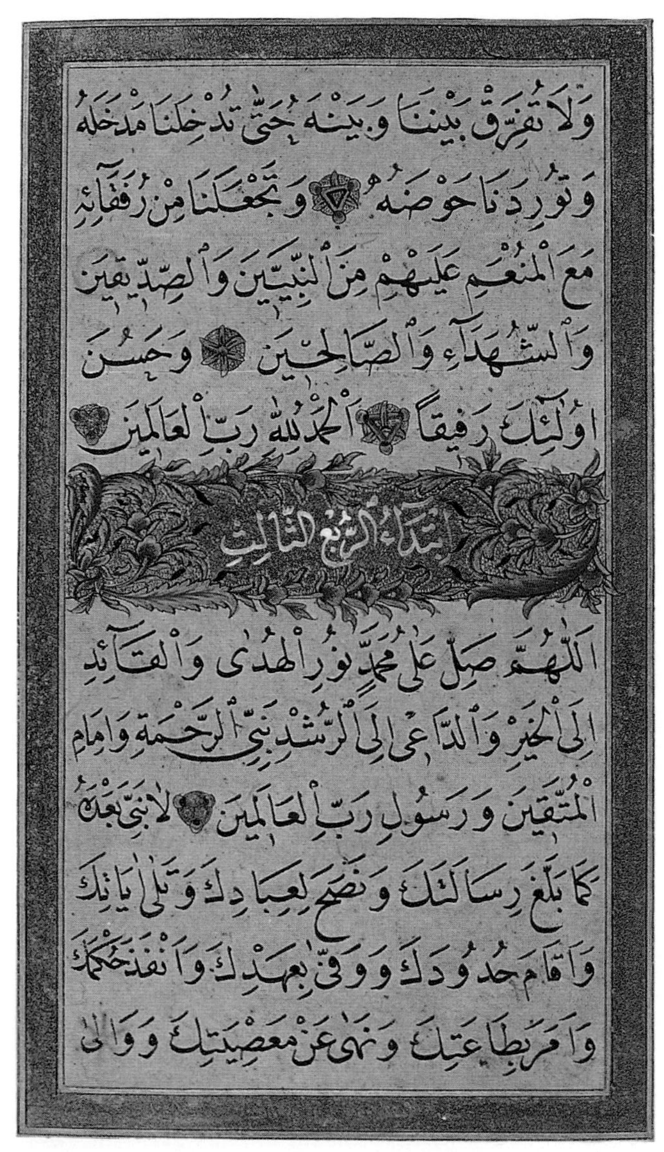

This page is from a Koran that dates from the 19th century.

Hieroglyphics were first used around 3000 BC.

Audiovisual

Radio

Coherer (1888)

The Frenchman **Edouard Branly** (1844–1940), doctor of sciences and of medicine, and Professor of Physics at the Catholic Institute in Paris, detected radio waves in **1888**. After two years of research and improvements, he presented his radio detector to the Academy of Sciences on 24 November 1890. This device converted radio waves into usable electric current. Edouard Branly established that the radio waves could be detected tens of metres away and even through walls. Wireless telegraphy was about to be born.

The Englishman Sir Oliver Joseph Lodge (1851–1940) perfected and named the apparatus the coherer and it is under this name that it has been passed down.

Syntony of circuits (1894)

In **1894** the Englishman **Sir Oliver Lodge**, a professor at the University of Birmingham, introduced a new idea: that of tuning.

It seems obvious today that the receiver should be tuned to the wavelength of the transmitter from which one wishes to tap a signal. By applying the work of Lord Kelvin, Lodge, a forerunner in this field, established this system of tuning which for many years has been called Syntony. Another pioneer of syntony in the United States was the Croatian Nicola Tesla (1856–1943).

Aerial (1895)

In **1895** the Russian **Aleksandr Stepanovick Popov** invented the aerial.

As assistant professor at the School of Torpedos in Kronstadt, he had used the methods of Branly and Lodge for the purposes of detecting distant storms. He noticed that sensitivity was increased when he used a long vertical wire for the reception of waves produced by lightning. A lightning conductor was therefore the first aerial.

It was to Aleksandr Popov that we owe the first radio electric link in morse transmitted at a fair distance of 250m *273yd*. The first words transmitted were 'Heinrich Hertz' on 24 March 1896.

Although Popov was the inventor of the aerial, in 1891 Branly had shown that by equipping his apparatus with long metal rods their range would be improved.

Wireless telegraphy (1895)

It was the Italian **Guglielmo Marconi** (1874–1937) who was able to bring together previous research and who invented wireless telegraphy and the radio.

At the age of 21, in **1895**, he succeeded in making the first wireless link over a distance of 2400m *1 mile 866yd*.

On 28 March 1899 he carried out the first telegraphic transmission between Dover and Wimereux – a distance of 50km *31 miles* – as well as, on 12 December 1901, with the co-operation of the Englishman Sir John Fleming (1849–1945), the first radiotelegraphic link across the Atlantic between Cornwall and Newfoundland, 3400km *2113 miles* apart.

First portable radio (1922)

It was in **1922** that **J. McWilliams Stone** of Chicago (USA) invented the Operadio, the first portable radio receiver. It cost $180 and weighed nearly 10kg *22lb*.

Radio control (1974)

In **1974** the German company **Blaupunkt** experimented with the ARI radio control system (the car radio tunes in automatically to the station transmitting information on road traffic). Since then, ARI has been adopted in several European countries such as Switzerland, Spain and West Germany. In the future it is expected to give rise to a whole range of car radios transmitting on the Hertzian wavelength on FM. The first of its type, developed by Grundig, was presented in 1988.

Stereophony (1881)

The first transmission in stereo took place following the initiative of Clément Ader, the pioneer of aviation, at the time of the first exhibition on electricity, held in Paris in **1881**. For the event, Ader had invented a telephonic 'stereoscopic' system which, each evening, enabled an enthusiastic public to follow performances at the Opera, some 3km *2 miles* away.

However, stereophonic sound, such as we know it today, emerged in the 1930s when numerous systems were developed. After several experiments carried out during the decades which followed, the definitive method was adopted in 1964.

The Edison effect (1883)

In 1883, **Thomas Alva Edison** (1847–1931) invented the first incandescent light bulb.

This bulb reveals the Edison effect: a metal heated until red-hot emits an electron cloud. Radio tubes would make use of this effect and would permit the transmission of sounds.

Diode (1904)

Marconi's collaborator, **Sir John Ambrose Fleming** (1849–1945), created the diode in 1904. He placed a plate in front of a heated wire (the filament) in a vacuum tube. The diode was the first radio tube, but it did not much advance the wireless sets of the time.

Radio tube (triode/Audion) (1906)

Lee DeForest (1873–1961), an American, invented the first triode in **1906** and named it the Audion.

With the Audion, the radio was equipped with an extremely sensitive apparatus. Due to the auxiliary electrode, it became possible to gauge transmission power with precision and thus to transmit voice vibrations, music, and other sounds in all their subtle nuances.

Lee DeForest's invention is the basis not only for radio but also for television, radar, and the first computers.

Crystal set (1910)

In **1910**, the work of two American researchers, **Dunwoody** and **Pickard**, on crystals led to the invention of the crystal set, which was the first radio. Galena is a lead sulphur crystal that, combined with some simple elements, permitted thousands of amateurs to build their own wireless sets and to receive the first radio broadcasting transmissions.

Superheterodyne receiver (1917)

This was invented in **1917** by the American **Edwin H. Armstrong**.

At the time, it was very difficult to achieve

FROM THE INVENTION OF LIGHT TO THE INVENTION OF SOUND

It was the British physicist James Clark Maxwell (1831–79) who demonstrated the existence of electromagnetic waves. This discovery was at the root of the invention of radiotelegraphy, some 25 years later.

Maxwell demonstrated that light was the result of electromagnetic vibrations of a certain wavelength. His theory, put forward in 1865, allowed scientists to predict the propagation, reflection and diffraction of light. Moreover it showed how electromagnetic waves other than light waves could be propagated.

In 1888, at the age of 31, the German Heinrich Rudolf Hertz (1857–94), Professor of Physics at the Polytechnic of Karlsruhe, detected, created and used electromagnetic waves. After becoming familiar with Maxwell's theories, Heinrich Hertz manufactured a rudimentary transmitter called the resonator. With the help of this resonator Hertz verified all of Maxwell's theories. For the first time, waves, subsequently called hertzian waves, were produced and subsequently detected at a distance.

This invention was in turn to enable Guglielmo Marconi to invent the wireless or radiotelegraphy.

THE SMALLEST RADIO IN THE WORLD

It measures 1.5 × 1 × 0.2cm ⅝th × ⅜th × 1⁄16th in, that is to say, about a fifth of the size of current radios. It was built thanks to new technology developed by a Japanese telecommunications company, Nippon Telegraph and Telephone Corp., who introduced it in March 1989. This same technology could also serve to create a miniature mobile telephone.

proper receiver adjustment as receivers were equipped with a great many buttons, and tuning in to a different frequency involved complicated manipulations. The superheterodyne, or variable frequency, receiver allowed the listener to search for stations with a single button. It considerably simplified receiver adjustments and also facilitated their industrial manufacture.

Presently, 99 per cent of radio and television receivers, radar connections via satellite, etc. employ the principle of the superheterodyne.

Car radio (1922)

The invention of the car radio can be attributed to the American **George Frost**, who, in **1922**, at the age of 18, installed a radio in a Model T Ford. The first industrially produced car radio was the Phildo Transitone, manufactured by the Philadelphia Storage Battery Company in 1927.

Frequency modulation (FM) (1933)

The American **Edwin H. Armstrong** began studying the principle of frequency modulation in 1925. This consists in modifying the frequency (or wavelength) of a transmission to adapt it to the rhythm of sound variations.

He studied the principal transmission and reception circuits in 1933 (his first patent was taken out on **24 January 1933**) and he demonstrated that noise can be decreased by increasing the frequency band (contrary to what takes place in amplitude modulation). In 1938 General Electric installed the first transmitter of this type in Schenectady, New York, and this was followed by the Yankee Network. FM was also adopted in 1939 for police patrol radios in the United States.

Transistor (1948)

A new era in radio technology began in **1948** when three AT&T scientists, **John Bardeen**, **Walter Brattain** and **William Shockley**, demonstrated the results of the work that would win them the Nobel Prize for Physics in 1956. They had invented the transistor, which replaced the vacuum tube, revolutionised the field of electronics and founded whole new industries.

Transistors detect, amplify, rectify and switch currents. They are tiny (millions can be mounted on a fingernail-size microelectronic chip), relatively cheap, and use very little power. These properties have made possible modern computers, space flight, hearing aids, electric guitars and many other common electronic devices.

The first transistor radio appeared in August 1955: the Sony TR-55.

Television and video

Nipkow scanning disc (1884)

It was a German student, **Paul Nipkow**, who invented the cutting up of images into lines. His 'electric telescope', patented in **1884**, was a pierced disc which turned in front of the object to be analysed and detailed all its points, line by line. This method, called the Nipkow disc, was the basis of television. It made mechanical television possible until 1935.

Mechanical television (1923)

The Scottish engineer **John Logie Baird** (1888–1946) was one of the pioneers of television. In **1923** he applied for a patent for the use of the Nipkow disc within a mechanical television system. The first experiment was successfully carried out three years later, in 1926. The image obtained comprised only eight lines. He later invented the 240-line mechanically scanned system of television.

Mechanical television used a Nipkow disc (or one of its derivatives) for capturing images. For reception another disc was used which was synchronous with the first, linked to a neon bulb.

In 1928 he developed a television system in colour, although it was still mechanical. In 1930 he established an experimental mechanical television network with the help of the BBC.

From 1930 onwards John Logie Baird marketed his Televisor, the first mass market transmitter. However, his efforts proved unsuccessful.

Cathode-ray tube (1897)

It was in **1897** that the German physicist **Karl Ferdinand Braun** (1850–1919) invented the cathode-ray tube. It was a kind of vacuum tube in which a fluorescent screen is bombarded by a stream of high-energy electrons. This cathodic oscillator won him the Nobel Prize for Physics in 1909.

Electronic television (1926)

Virtually unknown in Europe, the Japanese **Kenjiro Takayanagi** almost certainly invented electronic television before Zworykin. In **1926** Takayanagi succeeded in transmitting and picking up the image of a Japanese character. The image comprised 40 lines and 14 frames per second. Only the camera used was mechanical.

At the beginning of 1928 Takayanagi managed to transmit images of hands and faces.

In 1934 during a trip to the United States he met Vladimir Zworykin.

In 1936 Takayanagi succeeded in perfecting an improved television system (441 lines) which was entirely electronic.

Having been virtually unknown for many years, Takayanagi's works were one of the revelations of the Tsukuba exhibition in 1985 in which a modern replica of his laboratory, destroyed in the war, had been reconstructed.

The beginning of television

It was in London in **1932** that the **BBC** undertook to transmit the first regular television programmes. These were based upon the mechanical method of John Logie Baird perfected in 1932.

The first real television station was built in Berlin, Germany, in 1935, in anticipation of the Olympic Games. In 1936, in the United States, the Federal Communication Commission opted for a system of 441 lines. In the same year, NBC carried out experiments with the help of an iconoscope from the Empire State Building in New York.

The first live journalistic reporting was undertaken in Britain by the BBC in 1937 at the coronation of King George VI.

Flat screen (1985)

The first model to be truly operational was designed at the beginning of **1985** by **Matsushita**. The following three types of technology are in competition with each other: liquid crystals (used for pocket television sets), plasma screens (already used for certain portable computers) and electroluminescent screens.

In 1988 an agreement was made between the CEA (Commissariat à l'Energie Atomique) and the JRDC (Research Development Corporation of Japan), a Japanese organisation to help specifically with innovation, for the purpose of developing and marketing a flat colour screen on a grand scale. The technology employed was fixing matrices using liquid crystals, developed both in France by the LETI division of the CEA and in Japan by Professor Tatsuo Uchida of the University of Tohoku; the other methods are those developed by Stanley Electric Co. Ltd

TELEVISION FOR BLIND PEOPLE

A Boston television station, WBGH, has developed a system thanks to which blind people can receive, through a special stereo television channel, a description of the action and the sets narrated during pauses in the dialogue. This programme was extended at the beginning of 1989 to nine other local television channels. WGBH was the first, 15 years ago, to introduce subtitles for the hard of hearing to the United States.

21st CENTURY TELEVISION: 'HIGH DEFINITION'

In 1968 the NHK (Nippon Broadcasting Corporation), that is to say, Japanese television, began research into 'high definition television' and in 1974 were joined by Sony in this work. Sony engineers worked with those of the NHK in the development of a new system called Hi-Vision, which would give television a quality of image comparable to that at the cinema: images of 1125 scanning lines on 60 hertz. However, this would need a change in production, cameras and the whole stock of current television sets. As far as Europeans were concerned this expense was excessive. They have launched a counter-attack under the Eureka programme: a high definition system with images of 1250 lines that is compatible with existing networks (Pal, Secam, NTSC) and in particular with the standard D2-Mac Packs (1984). So the 600 million or so television sets in service in the world won't have to be replaced. This system has been available since 1988.

Doubtless it will take another ten years before high definition television systems, be they Japanese Hi-Vision or the European and American (ACTIV, Advanced Compatible TV), become fully operational. The first live high definition re-transmission organised by the national Japanese channel NHK took place at the time of the opening ceremony of the Seoul Olympics on 17 September 1988.

concerning the use of colour filters. Another type of television set is one with squared corners, the surface of which is almost flat. These appeared almost simultaneously at Thomson, Toshiba and Grundig in 1985.

Double scanning television set (1988)

The 625 horizontal lines which form a television image change 50 times every second producing a slight blinking effect and visual fatigue for the spectator. By increasing the scanning frequency to 100Hz (doubling the 50Hz scanning) the **Sony** (KV-FX 29B) and **Grundig** (Monolith 70-100 HDQ) television sets launched in **1988** produce a more stable image which is more restful for the eyes.

This improvement has been made possible by the digitilisation of the video signal received and by its temporary memorisation. This same digitalisation makes it possible to take excellent stills of the image.

Video Walkman (1988)

Like its audio predecessor, this is also a **Sony** invention. Marketed in Japan since August **1988**, it comprises a case weighing 1.1kg *2lb 7oz* containing a television receiver with liquid crystals and an 8mm format video recorder.

Ampex solution (1953)

It was the Californian corporation **Ampex**, who, in **1953**, resolved the problem of tape consumption by adopting a system which made it possible to maintain an acceptable speed. This method is still in use today on professional or mass media video recorders.

The team of Ampex researchers was led by Charles P. Ginsberg and Charles E. Anderson. They were joined by a 19-year-old student called Ray Dolby, who was soon to become a household name.

On 2 March 1955 Charles E. Anderson carried out a very convincing demonstration of a method for recording sound by modulating the frequency. After further improvements, this video recorder was finally launched on the market in April 1956 under the name Ampex VR 1000.

Video cassette recorders (1970)

In addition to professional video recorders, manufacturers also designed models intended for the mass market. It was for this purpose that video cassette recorders were developed (as opposed to tapes).

Towards the end of the 1960s **Matsushita**, **JVC** and **Sony** together developed the standard U-Matic. The first models were launched on the market in **1970**. Subsequently the standard U-Matic gained in prestige to the point where, today, it is considered to be the standard professional recorder.

In October 1970 Philips launched its VCR, an apparatus aimed at the mass market.

Mass market video formats

There are two main formats of mass market video recorders and they are incompatible with each other. They are as follows:

Betamax (1975)

Invented by the Japanese company **Sony**, the Betamax was launched in **1975**. Today Betamax has 10 per cent of the world market. In January 1985 Sony launched a new version of the Betamax format, called High Band, with a better quality image.

VHS (1975)

The **VHS** format (Video Home System) was launched by JVC in October **1975** and marketed as from 1976. The VHS now holds a dominant position in the world market, with more than 80 per cent of sales.

A variation of the VHS model, the VHS-C was launched by JVC in 1982. It was then intended for the portable video and uses reduced size video cassettes which can be re-read on a traditional VHS recorder thanks to an adaptor.

In July 1985 JVC launched the VHS HQ (High Quality). In March 1987 JVC brought out a Super VHS in Japan with an increase of horizontal lines on the image from 240 to 400.

In 1988 a technological agreement was made between ten Japanese companies (such as Sony and Matushita) for the purposes of developing a Super-8mm capable of competing with JVC's Super VHS.

Digital video recorder (1985)

In May **1985**, at the Symposium of Montreux, **Sony** introduced the first video recorder with a digital recording facility. This model was solely for professional use.

Since then digital video recorders have become commonplace. The effects which can be produced are very varied. For instance, it is possible to superimpose a reduced secondary image on to a corner of the screen. It is also possible to choose the speed of the tape with faster and slower options, etc.

Window video recorders (1987)

In **1987** about ten firms such as Hitachi, NEC, Toshiba, JVC, Zenith and RCA introduced VHS recorders with a microprocessor that can produce the following special effects: it can stop at a particular image; allows simultaneous viewing of recorded film and television (one of the two images selected appears on a corner of the screen); lets the viewer watch one channel and have eight other programmes in little windows on the screen.

Pencil for video recorders (1987)

This is a **Panasonic** invention (Japanese Matsushita group). It is a system of monthly advance programming using a light pen and infra-red transmission. The pen is used to scan the bar codes which hold all the information about television programmes and the video recorder is ready to record on the specified date at the exact moment required.

Video recorder with a screen (1990)

This is a **Panasonic** invention, called Maclord AV Gear. It is a portable video recorder weighing 1500g *3lb 5oz* measuring 24 × 15 × 10cm *9.45 × 5.91 × 3.94in*. Designed for use with S-VHS-C cassettes, it has a liquid crystal screen measuring 7.5cm *2.95in*. The maximum recording time for the apparatus is an hour. It should reach Europe sometime in **1990**.

POSTAGE STAMP VIDEO SCREEN

Less than 3cm *1in* across, this is a video screen the size of a postage stamp. It is called Private Eye and has been patented by the American engineer Allen Becker. It has been manufactured in the form of a prototype by Reflection Technology, a company close to the famous Massachusetts Institute of Technology.

Placed near to the eye, the screen forms an image comparable to that of a 30cm *11.8in* screen at a distance of 50cm *19.6in* from the spectator. It can be used for data processing, telecopying, graphic information, etc.

Video cameras and camcorders

From one year to the next video cameras make considerable advances in terms of weight, sensitivity and user friendliness.

Hypersensitive camera (1983)

New Saticon tubes for cameras (derived from professional models) and Newvicon (marketed by JVC since 1983) have a very high level of sensitivity. They allow filming to carry on in very bad light – to 7 lux, the equivalent of candlelight.

MOS integrated circuit camera (1984)

In **1984** the Japanese company **Hitachi** marketed the first amateur video camera equipped with an MOS (metal oxide semi-conductor) integrated circuit captor replacing the traditional analyser tube and improving performance. In 1983 several companies, in particular RCA, brought out prototypes of professional cameras of this type.

Betamovie (1982)

This was the first camcorder: a video camera and recorder combined. Presented at the Japan Electronic Show in Tokyo in October **1982** by **Sony**, it uses normal Betamax cassettes allowing up to 3h 35min recording time.

4mm video (1988)

In 1986 several Korean manufacturers, one of which was **Samsung**, caused great surprise by announcing the imminent perfection of a camcorder using the same 4mm cassette tapes as those used by DAT recorders. This machine could also be used by high performance recorders.

Super-VHS camcorder

Following Japan and the United States, S-VHS camcorders appeared in Europe at the end of 1988. In addition to significant improvements to the recorded image, certain models are also equipped with Hi-Fi stereo sound.

Sony's improved 8mm – the Hi Band – is that company's answer to the Super-VHS.

Professional camcorders

In addition to mass market camcorders, another generation of machines has emerged, reserved for professional use. They are as follows:

Betacam (1982)

Perfected by **Sony** and **Thomson**, this machine appeared in November **1982**. It uses Betamax cassettes and has now almost established itself worldwide as the standard machine for video reportage.

Quarter-cam (1983)

It was developed in **1983** by the German company **Bosch** and uses CVC format cassettes a quarter of an inch wide.

Hawk-Eye (1983)

It was launched in **1983** by the American company **RCA**. It operates with VHS format cassettes like the models launched by the Japanese companies Kegami and Matsushita.

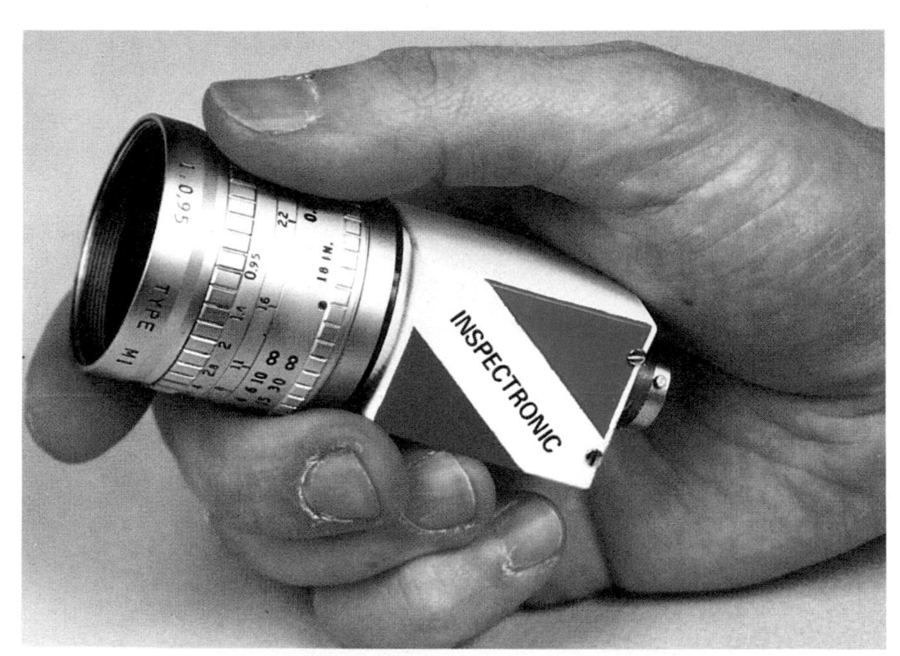

This tiny video camera was made in France for use by the armed forces and the medical service.

VIDEODISCS

Date	Inventor
1927	The Scot, John Logie Baird, carries out his first experiment using video signals on a disc to store images. His system is called phonovision.
1965	First attempt at a commercial venture on the part of the US firm Westinghouse (Phonovid method). Only 200 still images per disc can be recorded.
1970	AEG-Telefunken (W. Germany) and Decca (UK) come together to create the first commercial videodisc called the Teldec. Marketed 1974/5 and is a failure.
1972	Improvement of laser-read videodisc by Philips (Laservision) marketed in 1980 in the USA and in 1982 in Europe.
1974	The American company RCA opt for a capacitance reading system called Selectavision. Launched in March 1980 with a large advertising campaign, it proves to be a failure and in 1984 RCA are forced to abandon Selectavision.
1978	The Japanese company JVC develops a third videodisc system, the VHD (Very High Density), halfway between RCA's Selectavision and Philips Laservision.
1983	JVC develops the AHD (Audio High Density) system. Originally intended for audio-digital recording, the AHD procedure is completely compatible with the VHD on account of its decoder.
1984	The Laservision system receives strong support from two quarters: Sony and Hitachi. During this time Pioneer develops a compatible reader of compact dics and Laservision and makes the first Laservision videodisc with digital sound (similar to an audio compact disc). In Europe it seems the Laservision system is strictly for professional users.
1986	Improvement of the standard VHD by JVC. One version offers three-dimensional images. Introduction of Dolby surround method, used in the cinema.
1987	The videodisc finds applications such as in the form of back-up memory for microcomputers.
1988	Arrival of the Video CD, developed by Sony, Philips and Pioneer.

Video recorders

Video camera (1923)

The first video camera was the result of the work of the Russian-born American **Vladimir Kosma Zworykin** (1889–1982). Zworykin conceived of an electronic analysis procedure that led to the creation of the iconoscope: a camera tube that converts an optical image into electrical pulses. For this he gained the title 'father of television'.

After having applied for a patent for the procedure in December 1923, Zworykin performed an initial demonstration of his invention in the RCA (Radio Corporation of America) electronic research laboratories in 1930. The iconoscope was developed as of 1933. It was used for experimental broadcasts carried out by RCA and NBC from the top of the Empire State Building in New York in 1936.

Zworykin was also the inventor of a receiver tube called the cinescope, which appeared in 1929.

First video recording (1951)

The first company to carry out a demonstration of black and white video recording was **Mincom**, a branch of the M/Scotch company in the United States in **1951**.

In 1954 the RCA corporation built the first video recorder to be recognised as such. The following year, the BBC unveiled the VERA (Vision Electronic Recording Apparatus), a real monster which consumed 17km *10 miles 992yd* of tape per hour!

First retransmission of a recorded television programme (1956)

This took place on **30 November 1956**, on an Ampex VR 1000. That day the CBS studio at Television City in Hollywood recorded the *Douglas Edwards and the News* programme broadcast from New York, in order to retransmit it three hours later.

Magnetic video tape (1956)

The first video tape marketed was developed in **1956** by two researchers from the American firm 3M, **Mel Sater** and **Joe Mazzitello**. Working day and night the two men succeeded in presenting their invention the very day that Ampex Corporation placed the first video recorder on the market. This first Scotch 179 reel was 5cm *2in* wide, nearly 800m *875yd* long, and weighed 10kg *22lb 8oz*.

First colour video recorder (1958)

The first colour video recorder must also be credited to **Ampex**. It was presented in **1958**, two years after the first video recorder, under the name VR 1000 B. It was followed in 1963 by a transistor version, the VR 110.

Meanwhile the Japanese had been steadily working on video recorder technology:

in 1958 Toshiba announced the first single-head video recorder;

in 1959 JVC developed the first two-head video recorder, the KVI;

in 1962 Shiba Electric (now Hitachi), in co-operation with Asahi Broadcasting, presented a professional transistorised video recorder;

in 1964 Sony marketed the first video recorder for the general public;

in 1965 Shiba Electric marketed a small portable video recorder.

In Europe Philips launched their VR 650 in 1964.

Telecommunications

Post

Postal service (6th century BC)

Cyrus the Great (558–528 BC) the founder of the Persian empire, is said to have been the first to introduce a postal service. Cyrus had conquered a vast area and found, therefore, that messengers bearing missives and information were inadequate. This is the reason why the King organised a postal service with staging posts at regular intervals which would look after the horses after a reasonable day's journey.

The Romans copied this method of organisation and created, during the reign of Augustus (27 BC–AD 14), the *Cursus publicus*. Military routes were marked with *mutationes* which were staging posts providing rested horses and *mansiones* which were inns reserved for official travellers.

Monks' postal system

In the Middle Ages, Europe was speckled with monasteries and abbeys. In order to communicate with each other, a roll of parchment, called a *rotula*, was used on which the first abbey would write its message. During the course of the journey the abbeys visited would complete the message. The parchments could become very long; the Saint Vital Rotula announcing the death of its abbot measures 9.5m *31ft 2in* in length and 0.25m *9.85in* in width.

Envelope (1820)

A Brighton resident called **Brewer** stated in **1820** that he had invented the envelope. However, several envelopes dating back to 1615 are preserved in Geneva, Switzerland. At that time a letter would be folded and covered with silk thread with a wax stamp fixing the two ends together. Later the letter was wrapped in a folded white sheet on which the receiver's address would be written.

Telegraphy

Chappe's telegraph (1793)

After a preliminary demonstration in March 1791 a French engineer, **Claude Chappe**, sent his first telegraphic message along a distance of about 15km *9 miles* between Saint-Martin-du-Tertre and Paris on **12 July 1793**.

On 17 August 1794 the first line to Lille was opened.

Chappe's telegraph was a relay of semaphore signals from stations positioned about 12km *7.5 miles* apart at fairly high points so that someone with field-glasses could see the signals being made.

Directories (1785)

The first city directory was for the city and suburbs of Philadelphia, between 1st Street to the north and Maiden Street to the south, 10th Street to the west and Delaware Street to the east. Published by **John Macpherson** on **1 October 1785**, it included 6250 names and addresses.

Stamps (1834)

The Scot **James Chalmers** printed the first stamp in Dundee, in **1834**, but it was not until 1840 that stamps were used, following the British postal reform carried out by Sir Rowland Hill (1795–1879), who introduced penny postage.

The first adhesive stamp, which went into use in Great Britain on 6 May 1840, was the Penny Black which bore the profile of Queen Victoria on a dark background.

Perforated stamps (1854)

The first apparatus for separating stamps was invented in 1847 by the Englishman **Henry Archer**. It could only make slits, but its inventor perfected it one year later and the machine could then perforate a series of small holes. The first perforated stamp was the Penny Red, issued in February **1854**.

Postcard (1861)

The postcard was invented in Philadelphia in **1861** by **John P. Charlton**, who obtained a copyright for it and then sold his rights to a stationer named Harry L. Lipman. The latter

Nowadays, stamps are coated with a layer of glue.

published the cards with a picture and the words 'Lipman postcard, patent pending'.

The pre-stamped postcard was conceived by the Austrian Emmanuel Herrman, of the Neustadt Military Academy in Vienna. The first of these were made available on 1 October 1869.

Electric telegraph (1833)

In 1827 the German Steinheil discovered that a single earthed electric wire could be used as a transmission line. In **1833** the English physicist and chemist **Michael Faraday** (1791–1867) demonstrated that an electric current could be induced by moving a conductor within the field of influence of a magnet. If variants are used in accordance with a code common both to the sender and receiver, those two people will be able to send messages to each other.

It is on the basis of these principles that the Russian diplomat Pavel Schilling created an experimental telegraph in St Petersburg. His early death interrupted the experiment. The Englishman Sir William Cooke (1806–79), an Indian Army officer, and the great physicist Sir Charles Wheatstone (1802–75), continued Schilling's work and patented the first electric signal in June 1837.

With the help of a telegraphic message, the police were able to arrest a murderer, John Tawell, travelling on the 7.42 train from Paddington on 1 January 1845. This arrest increased the popularity of the telegraph in Britain.

Morse telegraph (1837)

On **28 September 1837** the American **Samuel F. B. Morse** (1791–1872) applied for a patent for electric telegraphy. However, his most original contribution was, without doubt, the invention of Morse code whereby letters are translated by a succession of dots and dashes.

On 30 December 1842, after a great deal of effort, he managed to obtain a grant of £30 000 to build an experimental line from Washington DC to Baltimore, which was opened on 24 May 1844. After a number of setbacks Samuel Morse, who saw his invention being contested, had his rights confirmed by a judgement of the Supreme Court. The telegraph subsequently experienced enormous growth and Morse became rich and famous.

Although its only purpose is to act as a support in the event of the radio system failing, the Morse code, 150 years after it was invented, is still taught to Navy radio controllers.

Underwater telegraphic cable

First cable under the Channel (1851)

In **1851** the first important telegraph cable was laid between Dover and Calais by the British steam ship *Blazer*. The British technician, **Jacob Brett**, was the main promoter of the operation which was made possible thanks to progress in many techniques. For example, the German engineer Werner von Siemens had developed a machine that could apply gutta-percha onto cables, thereby ensuring that they were insulated.

First transatlantic cable (1866)

On **27 July 1866** the first transatlantic telegraph cable was put in place. The American **Cyrus Field** (1819–92) had been the instigator of this operation which cost him almost his entire fortune. The first four attempts (one in 1857, two in 1858 and one in 1865) had ended in failure.

Telephone

Telephone (1876)

Candidates for the title of the inventor of the telephone are legion, with each country claiming its own. Before 1876 numerous researchers described some form of telephone system.

On **14 January 1876** two men, **Alexander Graham Bell** (1847–1922) and **Elisha Gray** (1835–1901) filed applications for patents at the Patents Office in New York. The first did so at 12pm, the second at 2pm. It is on this two hours' difference that the judges made a decision in Bell's favour after a long court case between the inventors.

Born in Edinburgh, Bell was sensitive to the problems surrounding deafness from a young age. His mother was deaf and his father was a specialist in the education of deaf children. The family emigrated to the United States and it was there that Bell became fascinated with telegraphy and later invented the telephone.

Bell invents the telephone by accident

In 1875 Bell worked with his assistant Watson towards improving the telegraph. All of a sudden, on 2 June, Watson made a mistake. The incorrect contact of a clamping screw which was too tight changed what should have been an intermittent transmission into a continuous current. Bell, who was at the other end of the wire, distinctly heard the sound of the contactor dropping. Bell spent the next winter making calculations and filed an application for a patent. However, it was not until 6 March 1876 that he succeeded in transmitting intelligible words to his colleague: 'Come here, Watson, I want you.'

The telephone had been born. The world was to learn about his invention when Bell presented it at the United States Centenary Exhibition held in Philadelphia in June 1876.

Alexander Graham Bell's experimental device, dating from 1875, led to the invention of the telephone. A false contact transformed an indirect current into a continuous one. The patent was filed on 14 January 1876.

PROFESSOR A. GRAHAM BELL'S TELEPHONE.—Fig. 1.

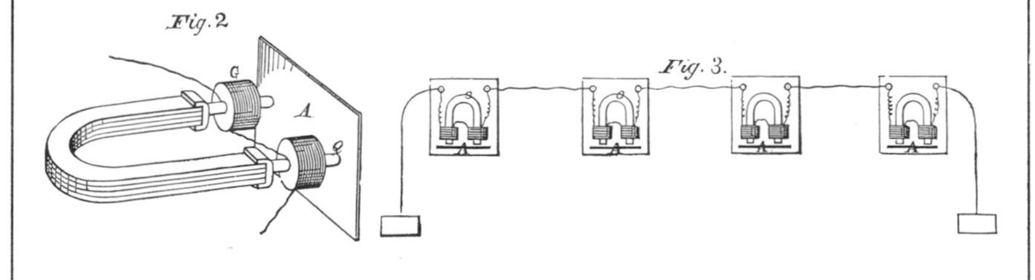

The group in Bell's study in Boston listen to a lecture given by the inventor in Salem, Massachusetts over the telephone. The audience at the lecture can also hear the conversation in Boston.

Radiotelephone (1900)

The ancestor of the radiotelephone is Chichester Bell and Charles Sumner Tainter's graphophone. The first demonstration took place on 15 February 1885. Radiotelephony is a means of communication by radio waves rather than along telephone wires.

The first proper radiotelephone was produced by **Reginald A. Fessenden**, who demonstrated the device in December **1906** at Cob Point, Maryland, USA.

The first transatlantic transmission was made by AT&T in 1915 between Virginia (USA) and the Eiffel Tower in Paris.

Portable radiotelephone (1979)

This apparatus, known as the cellular telephone, is causing a revolution in the field of communications.

It was in Sweden under the instigation of the **Ericsson** Company that it first appeared in **1979**. The area to be covered is divided into a certain number of small cells, each one served by a receiver. The unit is controlled by a data processing system. In 1986 some 200 000 Swedes, Danes and Finns already used a mobile telephone.

Automatic exchange (1891)

In **1891** an American undertaker from Kansas City, **Almon B. Strowger**, applied for a patent for the first automatic telephone exchange. Strowger had discovered that his competitor's wife, an operator at the local manual telephone exchange, was the first to learn of deaths in the city. Doubtless, she was directing the calls sent to Strowger's enterprise to her husband. It is thus easy to understand Strowger's interest in an automatic telephone system.

Telephone exchange (1878)

Manual exchange

In **1878** the first manual telephone exchange opened in New Haven, Connecticut (USA). It served 21 subscribers one of whom was the writer Mark Twain.

Hertz relays (1942)

Telephone transmission by ultra-short waves used in Hertz relays was perfected in the United States in the AT&T laboratories by **Harold T. Friis** around **1942**. They were a direct result of the war effort made by the Bell Laboratories.

Pay phones (1889)

William Gray from Hartford, Connecticut (USA) was granted the patent number 408709 on **13 August 1889** for a piece of apparatus that enabled telephones to be used with coins. The first telephone equipped in this way was installed in the bank at Hartford. In 1891 William Gray, Amos Whitney and Francis Pratt formed a company which installed pay phones in large stores.

Phototelegraphy (1904)

It appears that as far back as **1904** a German, **Professor Arthur Korn**, invented a procedure which was the forerunner of phototelegraphy. In 1906 he carried out a telegraphic transmission of a portrait of the Crown Prince from a distance of 1800km *1118 miles*.

In 1924 the Bell Laboratories carried out a preliminary demonstration of phototelegraphy. This was a transmission by telephone of photographs from the Chicago and Cleveland Conventions to New York when candidates for the Presidential Election were appointed. This method had been perfected by the engineers Ives and Gray from the Bell Laboratories.

Fax

The market for fax machines is now experiencing enormous growth: a million machines will probably be in operation by the end of 1990.

Photophone (1987)

Invented by two Californians **Gerald Cullen** and **David Monroe**, the photophone consists of a video camera and a receiver which can be plugged into any telephone. The device makes it possible to transmit still images and was marketed in **1987** by an American subsidiary of the Mitsubishi group.

Telephonic translator (1987)

Since October **1987** telephone subscribers in Italy can speak to another person in English or Arabic without knowing a word of these languages thanks to a simultaneous translation service developed by the company Italcable.

> **HOW TO FIND OUT WHO IS CALLING YOU**
>
> At the end of 1989 the American company Nynex launched a very practical gadget onto the market. Its name is Caller ID and it makes it possible to find out who is calling as the telephone number appears on a special screen. The number remains there for 30 seconds. Between 20 and 50 numbers can be stored in the memory with the date and time of the call.

This service operates from Italy and covers calls to all countries in the world.

Telex (1916)

The first teleprinter, making it possible to send written messages through telephone lines, was invented in **1916** by **Markrum Co.** of Chicago. The system became operational in 1928 and was extended at a national level by the Bell Laboratories in 1931 under the name telex from *tele*printer *ex*change.

Pocket telephone (1989)

Soon public telephones will be obsolete: all one will need is a cordless telephone not much larger than a packet of cigarettes.

Ticfir (1989)

The Ticfir was developed in 1986 by **Raoul Parienti**, a teacher of mathematics in France. This is a miniature version of a portable phone which will give access to the network through infra-red exchanges.

A product in competition with the Ticfir has been developed in Britain by British Telecom. The use of highly integrated electronic circuits makes it possible to reduce the size of this device to that of a cigarette packet weighing between 150g and 200g *5.29oz and 7.1oz* depending on the model. With this telephone, it is possible to call from a vehicle or outside so long as you are near a special terminal.

Vocal control telephone (1983)

The first telephone unit to be controlled vocally was invented in June **1983** by the American **Garth A. Clowes** and then marketed in 1984 under the name TTC 6012. This is a device which responds to the sound of the voice, either directly or through a memory capable of storing 80 numbers. The device is able to recognise the voice of three different people.

Recent developments on this principle are: a mobile telephone controlled by the human voice, developed by a Danish firm, Dancall, in co-operation with British Telecom in 1987; and an answer machine controlled by voice synthesis, the TD 9710, also launched in 1987 by Philips.

'Notebook' telephone (1987)

The number dials itself, thanks to a microcomputer, and everything you say is recorded on the screen on the page corresponding to the number you have called. It remains there, so you can consult your 'notes' next time, if by chance your memory fails you or if the person at the other end of the line challenges what was said. This invention was produced by AT&T Bell Laboratories, the American telephone giant.

Visiophone (1929)

It was around **1929** that American engineers experimented with visiophone, a device that enables two speakers to see each other on a screen while on the telephone. During the 1970s Bell Laboratories marketed the Picturephone in the United States but the exorbitant price led to its failure. Subsequently numerous prototypes have emerged throughout the world, in particular in Japan (the Scopephone by NTT, for example).

Videotex (around 1970)

Perfected by British and French engineers between 1970 and 1971, the videotex or viewdata developed at the same time as teletext. Videotex is an interactive videotext service that allows graphic images and texts to be transmitted from a 'feeder' system to a computer terminal.

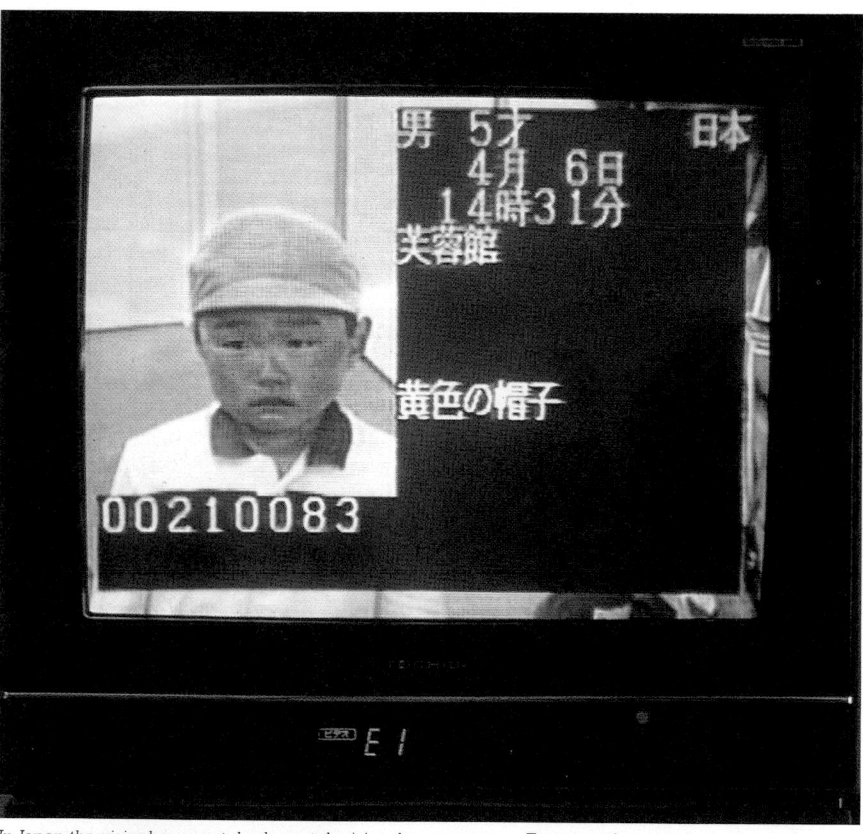

In Japan the visiophone, or telephone-television, has many uses. For example, parents can transmit photographs and details of their lost children.

> **THE RED TELEPHONE HAS BEEN SOLD**
>
> A Swiss company, Motor Columbus AG, has bought, through one of its American subsidiaries Tele Columbus USA, the famous red telephone that has provided a direct link between Washington and Moscow since 1963 and has been intended to reduce the risk of accidental war.
>
> On the American side, coded messages for the red telephone are transmitted from a small room in the Pentagon in the part of the building where computer terminals are fitted. The messages appear instantly on the Russian screens and can be printed at a speed of 1800 words per minute.
>
> In order to be sure that the lines are functioning properly, the Americans and Russians send tests to each other every hour.

Communication cables and satellites

Cable television (1927)

The first cable television transmission was carried out in the United States by the **Bell Telephone Company** in **1927**. The experiment took place between Washington and New York. A Nipkow scanning disc was used for the transmission and another for the reception.

This cable technique was then taken up again for the purposes of reaching those areas without access to traditional Hertzian transmission.

In 1949 a small town in Oregon in the United States had bad reception of programmes transmitted from Seattle on account of the mountains which surround it. It was decided that a large aerial would be installed on high ground. From there a cable network transmitted programmes, without any risk of parasitic oscillation.

Rapid growth of cable television (1960)

It was not until the 1960s that cable television experienced real growth in the United States and Canada. Today 22 million Americans are subscribers to different cable systems (paid by subscription), 2.5 million of which are Disney Channel subscribers.

The development of optical fibres in place of traditional coaxial cable (invented by the Americans Affel and Espensched in 1929) makes it possible to go from passive viewing of programmes to active audience participation, that is to say, users are able to choose their programmes and to participate directly in the contents of the programmes themselves (quick surveys, questionnaires, games etc.).

The fibre optic technique makes it possible to transmit through the same cables not only television programmes, but also radio programmes, telecommunications and data material.

Thanks to direct television satellites, Europe will be able to benefit from 120 channels in 1993.

First transatlantic telephone cable (1956)

On **26 September 1956** the first telephonic transatlantic link was made by cable. The cable made it possible to transmit 588 conversations, that is to say, more than all radiotelephone traffic in the previous ten days. In order to carry out this operation successfully, the American company AT&T, the British GPO and the Canadian company Canadian Overseas Telecommunications worked together.

First fibre optic transatlantic cable (1988)

The first fibre optic transatlantic cable, the TAT-8, has linked the United States with Great Britain and France since 1988. The cable is 6620km *4114 miles* long and carries television, telephone and data processing signals. The partners in this venture are DGT (France), American Telegraph and Telephone (AT&T), and British Telecom International. The total cost is estimated to be £220 million.

Vision, sound and data signals are transmitted along optical fibres.

Satellite television (1962)

There are two types of satellites used in television: direct television satellites (DBS) and telecommunications satellites (still called 'point-to-point' satellites), which are now very widespread. The first of these was Telstar which was launched in **1962**. They can transmit telephone conversations, computer data, television programmes, etc.

However, the line between these two types of satellite is becoming blurred.

Telecommunications satellite (1960)

In 1945 the science-fiction writer Arthur C. Clarke published the first theoretical analysis of an artificial-satellite system, in the radio buffs' magazine *Wireless World*.

On **12 August 1960 NASA** launched its first American telecommunications satellite called Echo 1. It was simply a 30m *98ft* diameter balloon, the metallic surface of which reflected radio signals without either magnifying or diverting them. Unfortunately, Echo 1 did not withstand meteorites for very long.

Telstar (1962)

On **10 July 1962 NASA** launched the first truly efficient civil telecommunications satellite on behalf of the American company AT&T: Telstar. It was put into orbit by a Delta rocket and contained 1064 transistors and 1464 diodes, fed by 3600 solar cells. Telstar cost $1 million and was able to transmit either 60 telephone calls or live television images.

ECS satellite (1983)

Under the auspices of the **Eutelsat** organisation, the first ECS satellite (European Communications Satellite) was sent into orbit by the Ariane rocket in **1983**. It was called ECS 1 and its main purpose was to re-transmit various television programmes (Sky Channel, Music Box, etc.) to European networks. The fourth satellite of this series, ECS 4, was launched in June 1986.

Since 12 March 1986 Europe has had access to the first entirely digital regional telephone network by satellite, thanks to Eutelsat.

Music

Instruments

It is virtually impossible to retrace the origins of most musical instruments as they are often the result of a long evolution and of traditions whose roots go far back into the history of mankind.

On the other hand, the efforts made by instrument makers to perfect and extend the possibilities of their instruments put them on a level with the greatest of inventors.

Traditional instruments

Harp (3000 BC)

The harp is one of the oldest of musical instruments, deriving from the primitive musical bow. It existed at the time of the Sumerians and of the Egyptians (**3000 BC**). The harp achieved its familiar form after various modifications which culminated in 1801 when the Frenchman Sébastien Erard put together the first double-action harps.

Flute (prehistory)

The origin of the flute lies in prehistory. In the Middle Ages a flute with a mouthpiece was generally used, and it was only from the 17th century that the transverse flute began to replace it. The change came about largely thanks to the Hotteterres, a family of flautists and instrument makers living in Paris at the end of the 17th century. In 1832 Theobald Böhm (1793–1881), flautist at the Chapel Royal in Munich, modified and improved the instrument (by covering the holes with keys), to such an extent that one can talk justifiably of a reinvention. It was at this time that wood was replaced by metal.

Oboe (2000 BC)

Instruments from the oboe family were played in Egypt around **2000 BC**. Like numerous instruments of this type, the oboe derives from the double-reed aulos whose invention the Greeks attributed to Minerva and even to Apollo. However, it was not until Frédéric Triébert's refinements in the 19th century that the oboe was perfected.

Trumpet (2000 BC)

A bronze trumpet dating from **2000 BC** has been discovered in Egypt. The form, for a long time straight, was bent into an S-shape in the 15th century. The art of rolling the tube into a loop was discovered in the 16th century. The valve trumpet, which increased the instrument's chromatic capacities considerably, appeared around 1815. The valve trumpet's invention is attributed to the German Stölzel.

Horn (prehistory)

The horn is undoubtedly one of the oldest of instruments. Hunters and warriors used primitive instruments carved from animals' horns. It is worth noting that the oliphant was originally made from an elephant's tusk. The horn was introduced into the orchestra in the 17th century, by which time it was made of metal. In 1815 the German Stölzel invented the chromatic valve horn, the only type used today.

Percussion (prehistory)

Percussion instruments were undoubtedly the very first musical instruments to be created by man. Drums, tambourines, maracas, linga, cymbals, etc are used in folk music all over the world. The richness and colour of the sounds has often earned these instruments a place in symphonic orchestras. Many composers have written for percussion instruments: the glockenspiel was first used by Handel in 1783. Bela Bartok wrote for the celesta which had been invented in 1868; Saint-Saëns introduced the xylophone into the orchestra in the *Danse Macabre* in 1874; and Ravel the whip in his *Concerto en sol* in 1931.

Harpsichord (3rd century BC)

According to legend, the harpsichord derived from Archimedes' monochord (278–212 BC). The word harpsichord first appeared in 1631.

Violin (17th century)

The violin evolved from medieval and Renaissance stringed bows. It was in **17th** and **18th century** Italy that the art of instrument makers such as **N. Amati** and **Antonio Stradivari** (c. 1644–1737) brought the violin to perfection. Stradivari and his family made more than 1100 instruments, half of which survive. It was he who also gave us the definitive form of the cello in 1680.

Clarinet (1700)

The German **Jean-Christophe Denner** invented the clarinet in Nuremburg around **1700**. However, its origins are ancient. After a series of modifications, the Böhm key system was applied to the clarinet and it reached its technical perfection.

Piano (1710)

The forerunner of the piano is the exchequer. But the inventor of the piano is the Italian **Bartolomeo Cristofori** who in 1698 created his first *cembalo a martelletti* (a clavichord with small hammers), and towards **1710** his first pianoforte. A number of improvements were later made: G. Silbermann perfected the system of hammers; J. A. Stein invented pedals (1789). But it is the Frenchman Sébastien Erard who, by inventing in 1822 the double escapement that allows a note to be repeated, can be considered the true creator of the modern piano.

Nikolaus Schimmel, nephew of the firm's founder, created this Perspex piano.

Tuning-fork (1711)

The tuning-fork was invented in **1711** by the English instrument maker, **John Shore**.

Music box (1796)

The music box was probably invented in Geneva in **1796** by the watch-maker **A. Favre**. It uses a pin-studded drum originally found in mechanical chimes, which existed in the 14th century.

Barrel organ (1800)

This instrument, probably invented around **1800** by **Barberi**, was built to accompany wandering musicians.

Metronome (1816)

The invention of the metronome is attributed to the German **Johann Maelzel** (1772–1832), a friend of Beethoven. Maelzel's device created an exact tempo for the musician to follow.

Harmonica (1821)

The harmonica was invented in **1821** by the German **C. F. L. Buschmann**. He was searching for something that would allow him to tune a piano more easily. In 1857, the German organ maker Mathias Hohner founded the first harmonica business.

Accordion (1829)

The accordion can claim a large number of ancestors, one of which is the Chinese sheng invented by a legendary queen, Nyu Wa, around 2500 BC. But we owe the invention of the accordion as we know it to the Austrian **Cyril Demian** who took out a patent on **6 May 1829**.

Saxophone (1846)

In his attempts to improve the bass clarinet a Belgian, **Adolphe Sax** (1814–94), invented a new instrument: the saxophone. Patented in **1846**, the saxophone first found success in military bands. It was later to become one of the leading instruments of the jazz era.

Electric and electronic instruments

Synthesiser (1965)

The synthesiser is considered by many to be the musical invention of the century. It is an electronic instrument which, in theory, is capable of recreating any sound imaginable.

It was in 1954 that the American engineer **Robert Moog**, together with composers **Herbert A. Deutsch** and **Walter Carlos**, started to look for a means of creating an universal electronic instrument. It was Moog who came up with the name 'synthesiser' and who invented the analogical control system that allowed him to define the basis of subtractive sound synthesis.

Production of the first model synthesisers began in **1965**. In 1970 the Minimoog appeared and its underlying principle is still used in the majority of synthesisers today.

Electric guitar (1935)

The principle behind the electric guitar was found in the United States at the beginning of the 1920s. Lloyd Loar invented the first microphone to be specially adapted for the guitar between 1920 and 1924. If one excepts the electric Dobros (1930) and the Hawaiian electric guitars (Frying pan, 1931), the first solid frame electric guitar was created by the American **Rickenbacher** who, in **1935**, designed the Electro Vibrola Spanish Guitar, the body of which was made of bakelite.

It wasn't until 1947 that Paul Bigsby (the inventor of the vibrato system) designed the first modern electric guitar at the request of guitarist Merle Travis. A veritable industry then began thanks to Leo Fender who, from

The saxophone was invented by Adolphe Sax in 1846.

1948, marketed the Broadcaster and, from 1950, the Telecaster. The latter is still popular today in its quasi-original form.

Electronic organ (1930)

The American Thaddeus Cahill developed an electromechanical organ in 1895. However, the pioneers of the electronic organ were the Frenchmen **Coupleux** and **Givelet** who, around **1930**, invented an organ whose sound quality resembled that of the classical organ. Unfortunately, the number of oscillators used (about 80) made it rather unstable. In 1943 Constant Martin introduced an improved model.

Electromagnetic organ (1934)

The inventor of the electromagnetic organ was **Laurens Hammond** (1895–1973), a former clockmaker from Chicago, who had been ruined in the Great Depression. Around 1934 Hammond decided to convert the unused cogwheels he had in stock to make an organ in which the keys set in motion wheels that released electric currents. The organ had two keyboards, electric tone generation and a wide variety of tone colours.

Digital organ (1971)

It was the American company **Allen** who, in **1971**, took up **Ralph Deutsch**'s patent for the use of digital synthesis in musical instruments. The Allen organ does not imitate the sound of an organ but faithfully recreates the sounds of numerous musical instruments from its memory bank. Some of the finest orchestras in the world now use this organ.

Sound reproduction

Phonograph (1877)

Thomas Alva Edison (1847–1931) invented the phonograph on **12 August 1877** and patented it on 17 February 1878. The machine was a revolving drum with spiralled grooves around its circumference. It brought Edison world-wide fame, but the sound quality was mediocre and the cylinders didn't last long. Edison preferred to move on to different lines of research and left the task of improving his machine to others. Two Americans, Chichester Bell (cousin of Alexander Graham Bell, the inventor of the telephone), and Charles Sumner Tainter took out a patent in 1886 for a piece of apparatus similar to the phonograph but which used wax cylinders. This was the graphophone.

Record (1887)

In **1887 Emile Berliner**, a German living in America, came up with the idea of replacing the cylinder of the phonograph with thin discs of zinc covered with a layer of wax into which fine grooves were cut. Having invented the record he went on to develop the machine on which it could be played: the gramophone. The record didn't replace the cylinder entirely and the two co-existed for many years.

In 1888 Berliner perfected a way of copying his records from matrices and on his return to Germany in 1898 he founded the Deutsche Grammophon Gesellschaft with his brother.

A STEREO BACTERIUM

In December 1988 the Japanese company Sony brought out a new audio headset. The fibre core of the MDR-10, created by a bacterium, the *Acetobacter aceti*, offers an acoustic quality equivalent to that of a concert hall. The result of biotechnical research, the MDR-10 has silk-insulated wiring and a gold-plated plug to avoid any loss of sound quality. It will be produced only to order.

Juke-box (1889)

The first juke-box (which worked with cylinders), was installed by the American **Louis Glas** at the Royal Palace in San Francisco on **23 November 1889**. It was followed by the public phonograph, the Automatic Entertainer produced by the Gabel Company in 1906, which offered a choice of music. Unfortunately, the sound quality was poor. The electric gramophone was introduced around 1925, and in 1926 a Swedish immigrant to the United States, J. P. Seeburg, invented the audiophone which offered a choice of eight records. It wasn't until 1950 and the introduction of machines that played 45s that the juke-box became truly popular. The most famous juke-box of all is the Wurlitzer 1015; 56 000 of them were produced between 1946 and 1947.

Microphone (1925)

In **1925** a team from the **Bell Laboratories**, directed by **Joseph Maxfield**, perfected an electrical system of recording. The microphone, by converting sounds into electric currents, replaced the huge horns that had been used until then. In the same year the first electrically engraved records were produced.

Stereo record (1933)

The first stereophonic records were produced in Britain by **EMI** (Electric and Musical Industries) in **1933**. The research, directed by the physicist **Alan Dower Blumlein**, culminated in the recording of stereo 78s. The work of Blumlein and EMI remained experimental until 1958 when the American company Audio Fidelity and the British companies Pye and Decca issued the first commercial stereo records thanks to numerous technical advances.

Long-playing record (1947)

Belgian René Snephvangers directed the CBS research team which in 1944 came up

This is a close-up of a stylus reading the grooves of a stereo record.

OLD RECORDINGS, NEW DISCS

In 1987 Sonic Solutions, an American company based in San Francisco, developed a process which has provoked a great deal of discussion. NoNoise is a process that removes the background noise and interference which tend to make old recordings difficult to hear. It enables such historic recordings as the *Bolero*, conducted by Ravel in person in 1932, to be revived and recorded on CD! It goes without saying that it is the computer that has made it possible to clean up recordings in this way. But given that the clean-up requires 53 million computer operations per second, NoNoise is a process which is at present proving quite costly.

with the first 33rpm record. Snephvangers was a friend of Arturo Toscanini, the famous conductor, for whom he created a HiFi well before the term was invented. The long-playing record was perfected in America by **Peter Goldenmark** for CBS in **1947** to replace 78s. The patent was taken out under the initials LP (long-playing). The first recordings on a long-player were Mendelssohn's violin concerto, Tchaikovsky's *Fourth Symphony* and the musical *South Pacific*.

Drum kit (1910)

The drum kit was born in New Orleans around **1910**. After a long evolution it took on its familiar form around 1950.

Organ (3rd century BC)

The earliest form of organ consisted of a sort of large set of panpipes fitted with two pumps which forced air through the pipes. The air pressure was created by pumping water by hand which is why it was known as a hydraulic organ. It was developed by **Ctesibius of Alexandria** around **270 BC**.

The largest organ ever built is in Atlantic City, New Jersey, USA. Completed in 1930, it has 33 122 pipes and seven keyboards.

The smallest organ in the world was created in 1984 by a Strasbourg carpenter, Hubert Molard. The Lilliput organ is 28mm *1.1in* wide and is played with a toothpick. It has a keyboard consisting of 23 keys, a pedal board of 12 pedals, a memory of 96 notes and a programme for eight pieces of music. Hubert Molard also invented the Smallest Musical Computer in 1983.

Stave (10th century)

A proper musical stave was introduced into written music by the Italian monk, **Guido of Arezzo** (c.990–c.1050). He advocated the use of four lines on which clefs and different colours provided the necessary reference points. Guido of Arezzo also invented the system known as solmisation, i.e. the naming of the notes of a scale by the syllables do, re, mi, fa, sol, la, si, which he devised from the first syllables of a Latin text.

Digital 'Sax' (1988)

This saxophone, the D11-100, presented at the Las Vegas Electronics Exhibition at the beginning of **1988**, was developed by the Japanese company, **Casio**. It is the first saxophone to be fitted with the MIDI interface (Musical Interface for Digital Instruments) and which can be connected to a synthesiser. Apart from producing the sound of the saxophone, the instrument can also reproduce the sounds of a trumpet, clarinet and flute.

Guitar (1850)

The origins of the guitar are much disputed. It takes its name from an Assyrian instrument of c.1000 BC called the *kettarah*, which had no neck, and it is supposed that the neck was added in the early years AD. After a long history, the guitar as we know it was developed in about **1850** by the Spanish instrument maker, **A. de Torres**. He concentrated particularly on enriching the tone, improving the upper soundboard and standardising the length of the strings.

Sampling (1980)

Based on the patent for number synthesis registered by **Ralph Deutsch** for Allen, the sampling technique enables a natural sound to be recorded, placed in the memory in the form of a binary code, and be reproduced in all its original purity of sound.

The Fairlight I was the first instrument of this type to be brought onto the market in **1980**. However, in 1985 the Japanese manufacturers Akai placed the system within the range of everyone's pocket with their Sampler S612.

MIDI interface (1981)

In **1981**, at the instigation of American synthesiser manufacturers, the MIDI interface (Musical Interface for Digital Instruments) was introduced. This made all synthesisers and their accessories (e.g. rhythm boxes) compatible, whatever their make or functioning method. One aspect of this system of intercommunication is that it has enabled the world of the synthesiser to be linked to that of the computer.

Guitar synthesiser (1978)

In **1978** a collaboration between Swedish instrument makers **Hagstrom** and American manufacturers **Ampeg** provided an opportunity to create a synthesiser based on a guitar. Modern technology has enabled synthesisers to be played using any guitar by means of the MIDI interface.

With the appearance of the SynthAxe in 1984 and the Steep in 1986, the British developed the idea of all-in-one instruments, i.e. guitars equipped with their own synthesising generators.

Electronic piano (1958)

The first instrument worthy of the name dates from **1958** when the American company, **Wurlitzer**, marketed one intended for use in music schools. It was very quickly adopted by the first rock musicians as it was easy to transport and had great potential for amplification. It was used for the recording of the Ray Charles classic 'What'd I Say?'.

In 1960 the German engineer, Zacharias, invented the Cembalet for the German company Hohner, an instrument which is inspired more by the harpsichord than the piano and which gave rise to the famous Clavinet as used by Stevie Wonder.

In 1963 the well-known Rhodes Fender piano made its appearance. Its particular tonal quality remains a feature of much modern music.

Electroacoustic piano (1978)

As electronic techniques developed, so an increasing number of instruments came onto the market, but none was able to convey the skill of a musician. The desire for realism, both in touch and sound, led manufacturers to develop the electroacoustic piano. This is a piano of almost traditional design in which the vibration of the strings is picked up by magnetic microphones. The first model to be widely used by professionals was the **Yamaha** CP70 which came onto the market in **1978**.

Remote control piano (1991)

NHK, the Japanese national television network, has developed a process which enables viewers to reproduce a recording of their favourite performer on their own piano, transmitted by satellite via their television. It is of course necessary to 'adapt' the piano so that it can receive and convert the numerical data of the sound. Once this has been done, the keys play on their own! At present, only the Yamaha U3A can be converted in this way, but conversions should become easier after **1991** with the launch of the Japanese television satellite.

Electronic drum kit (1980)

The first patent was registered by the English company **Simmons** in **1980**.

Electronic drumsticks (1986)

In **1986** the Japanese company **Casio** marketed the first electronic drumsticks. By

The decibel is a unit which measures the intensity of sound. It is equivalent to one tenth of a bel, a unit which is rarely used as the human ear is able to pick up sounds that are far weaker than the bel.

The unit of measurement is named after the famous Scottish-born inventor, Alexander Graham Bell (1847–1922). A specialist in phonetics, Bell directed his research towards the field of acoustics with the aim of enabling the deaf to hear. In 1847 he invented an artificial ear which was capable of registering sounds on a sheet of glass covered in lampblack. In 1876 his research led him to invent the telephone.

Sounds picked up by the human ear are contained within an interval of 130 decibels. The voice has an intensity of 55 decibels.

Light reflects on the surface of a compact disc.

striking any kind of surface, say a table, wall or saucepan, it is possible to obtain a similar sound to that produced by many percussion instruments. The drumsticks are connected to an electronic case.

Juke-box laser (1985)

The Lasergraph, invented by **Wurlitzer**, operates on the same principle as the juke-box except that it works with laser videodiscs.

Pulse code modulation (1926)

Pulse code modulation appears to have been invented by **Paul Rainey** in **1926**, reinvented in 1939 by the American inventor H. A. Reeves, and rediscovered during the Second World War by Bell Laboratories, USA, to fulfil the need for secrecy in telephone conversations. The process enables a continuous signal, such as a telephonic signal, to be sampled. The value of each sample is then quantified and converted by coding into a digital signal.

Photographic sound recordings (1901)

The photographic recording of sound using a sensitised tape was developed simultaneously by the German **Ruhmer** and the English engineer **Duddell** in **1901**. They had used the phenomenon of the singing arc which had been discovered in 1892 and studied by Thomson and Simon.

Compact disc (1979)

The compact disc was developed in **1979** by the Dutch company **Philips** and the Japanese company **Sony** under a joint licensing agreement. On this type of disc, a process of digital recording is used rather than the analogue recording process used for the microgroove. The signal is coded in binary form, using the series 0 and 1. The conventional groove has therefore disappeared and has been replaced by millions of microcells known as pits: about four million per second. The compact disc has a diameter of 12cm *5in* and can hold up to an hour of music or sound on one side.

The sound is reproduced by a laser beam. The invention of the compact disc was the direct result of research carried out on the video disc, also invented by Philips.

The compact disc was first marketed in Europe in March 1983, and in 1988 nearly 26 million compact discs were sold! In the space of a few years, the compact disc has enjoyed incredible success and its applications are many and varied.

From 1984, Matsushita and Philips brought out the prototypes of decoders which enabled fixed images, which had been stored on compact disc alongside an audio signal, to be viewed on a television. In 1985 the extensive storage capacity of the CD was applied to computers.

Compact videodisc (1987)

In **1987** there was a new development. The combi, still referred to as the CVD, the compact videodisc, was brought out by **Philips** and developed in conjunction with **Sony**. It enables video pictures to be shown on a television screen while laser quality sound is produced simultaneously on stereo. The new readers can reproduce both sound and picture. They will read standard compact discs, the gold CVDs of the same format which play for longer and reproduce pictures and sound as well as 20 minutes of music, the 20cm *8in* CVDs which offer 40 minutes of pictures and sound, and the 30cm *12in* CVDs which last for a maximum of two hours, and which give additional backing to films and operas. The CVD is one of the modern answers to the competition offered by DAT (Digital Audio Tape), an audio-digital cassette.

CD single (1988)

The 45 is still very much alive! Since the beginning of **1988**, **Sony** and **Philips** have been marketing their latest invention, the CD single, the audio-digital equivalent of the microgroove 45. This new disc which holds approximately 20 minutes' worth of music was invented to supply a demand created primarily by young people. On a similar basis, Sony presented the Pocket Discman, an extra-light reader for the CD single, at the 1988 Music and Sound Exhibition.

Re-recordable CD (1991)

The next stage is the CD which can be wiped clean and re-recorded: the MOD or Magneto Optical Disc. The first prototype was developed by **Thomson** of West Germany, and in 1987 they won the German award for the best economic invention, but it is not expected to be marketed before **1991**. The American company Tandy Corporation is also preparing to market a CD which can be wiped clean, re-recorded and will last almost indefinitely.

Intelligent tape recorder (1985)

In **1985** a French engineer, **Louis Kanny**, developed a very sophisticated tape recorder: the Lea box can read and record automatically and, among other things, is voice activated. The Lea Box has been on the market since January 1989.

In 1986 Philips brought out a small tape recorder which is sound-activated and which stops after four seconds' silence. It is ideal for recording enthusiasts and reporters.

Tape recorder (1888)

The principle of the tape recorder was worked out theoretically in **1888** by the Englishman **Oberlin Smith**. Ten years later, the 20-year-old Dane Valdemar Poulsen put the theory into practice. However, his presentation of the new machine at the 1900 Paris Exhibition didn't raise much interest. It was not until 1935 that two Germany companies, AEG Telefunken and I. G. Farben, made a device based on Poulsen's principle with a plastic tape which ran at 7.6m *25ft* per second.

Magnetic recording tape (1928)

Magnetic recording tape was patented in **1928** by the German **Fritz Pfleumer**. Back in 1888, Oberlin Smith had already proposed using strips of fabric covered with iron filings.

AEG became interested in Pfleumer's invention; they concentrated on the development of the tape recorder and passed the work of improving the tape on to I. G. Farben. The first tests were carried out by the two firms in 1932 and in 1934 I.G. Farben was able to produce 50 000m *54 700yd* of tape.

Compil box (1989)

Patented in **1989** by the Swiss inventor, **Pierre Schwab**, the Compil Box is a complete audio and video recording studio in a suitcase. Developed using Sony equipment, it works on a battery, a car cigarette lighter socket or standard electric current. It contains no less than two speakers, headphones, a TCD-D10 portable DAT, a CD reader, a walkman video, a camera input point for Betacam, etc.

Tape cassette (1961)

It was in **1961** that the Dutch company **Philips** developed the first mini tape cassette which was 100mm *3.9in* long and designed for stereo and mono recordings. This cassette, along with the first cassette recorder, was presented in Berlin in 1963. Philips decided to allow manufacturers to use its patent free of charge so as to encourage the spread of the system throughout the world.

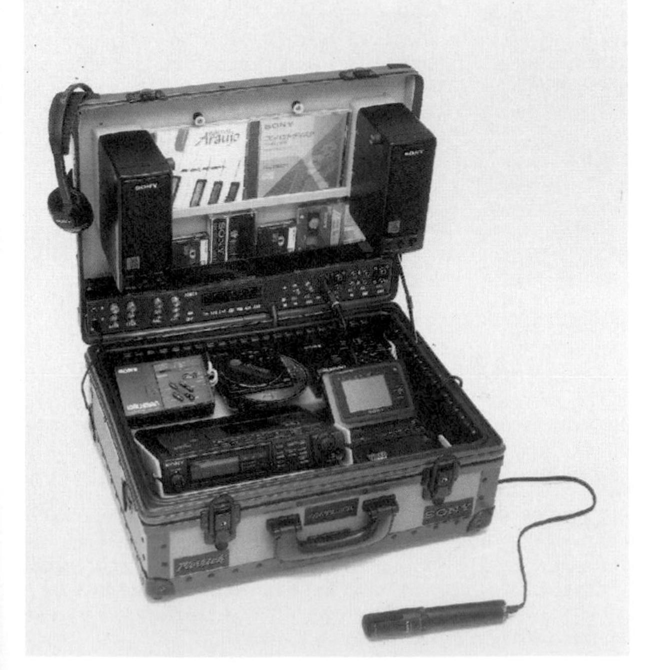

Pierre Schwab's Compil Box is a mini audio and video recording studio.

Micro cassette (1976)

The micro cassette, patented in **1976** by the Japanese company **Olympus**, is another type of audio cassette, smaller than the Philips mini cassette.

It operates at a greatly reduced speed and is mainly used in pocket tape recorders which act as audio notepads.

Pico cassette (1986)

In April 1985 the Japanese company **JVC**, in conjunction with the American company **Dictaphone**, brought out a new format audio cassette, the pico cassette, and a new dictation machine based on the same model. The pico cassette contains a tape 2mm *.08in* wide, and has a maximum recording time of one hour at a speed of 0.9cm *.35in* per second.

Digital audio tape (1987)

From 1980 several Japanese manufacturers researched the possibility of an audio cassette reader which used digital recordings (*D*igital *A*udio *T*ape or DAT) and offered a quality of sound equivalent to that of the compact disc.

There were two rival models: the JVC fixed-head S-DAT, and the R-DAT developed by Matsushita and Sony which had rotating heads and has since been adopted by the majority of manufacturers.

Aiwa was the first company to market the DAT in Japan in February **1987**.

Personal cassette (1989)

This is an invention by Californian, **Charles Garvin**, which, with the help of a computer, enables you to record all your favourite songs on one cassette in less than five minutes. Using a catalogue of some 3000 or so titles, customers can listen, make their choice and then give the list to an assistant who makes the recording. The result is 90 minutes of personally selected music.

Speaker (1877)

The first patents relating to the coil-driven speaker, which is virtually the only type in use today, date back to the 19th century. A patent was registered by **Ernst Wermer** for the German company Siemens on **14 December 1877**, and by the British physicist Sir Oliver Lodge on 27 April 1898. But at the time, there was no electrical source that would have enabled the mechanism to operate.

In 1924 Chester W. Rice and Edward Kellogg, both of the American company General Electric, registered a patent for a voice coil speaker as well as constructing an amplifier capable of providing power of 1W for their device. The speaker, known as the Radiola Model 104, had a built-in amplifier and came onto the market the following year at a price of $250.

Ribbon speaker (1925)

The technique of the ribbon speaker was invented in **1925**. The standard cone was replaced by a very fine aluminium ribbon which was concertinaed and exposed to a magnetic field. The first models were brought out in America in the 1940s but in limited numbers. Today three American companies – Apogee Industries, Magnepan Inc. and VMPS – control their distribution.

Musical cushion (1988)

Brought onto the market in **1988** by the Japanese company, **Tokyo Tire Rubber Lance**, this relaxing musical cushion contains a network of microphones which transmit music as soon as you sit down. It is reminiscent of the Boodo Khan walkman, named after a famous concert hall in Tokyo, invented by Sony in 1987 and which connected a headset to a cushion. It enabled the tone to be adjusted and reproduced the acoustics of a concert hall. These cushions transmit low frequency vibrations.

Visual speaker (1989)

The visual speaker had its world premiere in Japan in **1989**. It is omnidirectional and based on the vibrations of an arc of light. The system, known as the sound arc, was developed by **Mitsubishi Heavy Industries** and is intended

AN IDEAL ACOUSTIC SYSTEM

Any theatre can become a concert hall with an ideal acoustic system thanks to the Acoustic Control System, invented by the Dutch company, Griffioen. On 23 November 1988 it held a world premiere to demonstrate the system which consists of microphones, speakers and a computer. Griffioen has already received an order from the National Theater of Toronto, Canada.

mainly for cultural and commercial purposes in large areas.

Walkman (1979)

The Walkman is a portable, stereophonic cassette reading unit equipped with lightweight headphones. It was devised by **Akio Morita**, President of Sony, at Shibuara in April 1987. Morita, a keen golfer and music lover, wanted a lightweight and compact device which would allow him to pursue both his hobbies at the same time. The result was the Walkman. The first model was known as the TPS 12.

By 1988, 30 million Walkmans had been sold worldwide.

Wireless Walkman (1988)

In May **1988 Sony** brought onto the Japanese market the first wireless Walkman. It weighs 210g *7.4oz* and consists of a reader, a radio receiver and a micro transmitter which relays the selected programme to the headphones. The connection between the cassette reader and the headphones is effected by radio frequency.

Dolby (1967)

The first noise reducer, designed to improve the signal/noise ratio, was the work of the American **Ray Dolby** in **1967**. (Dolby had begun his career working on the tape recorder.) Dolby A was intended for professional use; a few years later, Dolby brought out a simplified system for general use called Dolby B. Dolby stereo, as used in the cinema, is produced from four sources which improve the sound quality and gives the audience a sense of depth and relief.

Boxed-in loud-speakers (1958)

It was only later that someone had the idea of putting one or more speakers in a wooden or plaster case to form the loud-speakers we know today. In **1958** the French firm **Cabasse** built the first speakers with an incorporated amplifier.

Musical composition and performance

Composition (17th century)

The Italian-born French composer **Jean Baptiste Lully** (1632–87) was superintendent of music at the court of Louis XIV (1638–1715), the Sun King. Lully insisted that the violinists play exactly what was written. He was thus the first composer in the modern sense of the word. Indeed, two identical manuscripts of the same work do not exist from the Middle Ages. Like jazz musicians, performers then would change the piece slightly every time they played.

Opera (1598)

In **1598 Jacopo Peri** (1561–1633), court singer to the Medici family, presented the first dramatic entertainment with continuous music: *La Dafne*. It was performed at Carnival time at the Palazzo Corsi in Florence.

In 1660 Peri put to music a text by Ottavio Rinuccini (1563–1621), *Euridice*, which was performed at the marriage of Maria de' Medici to Henry IV of France at the Palazzo Pitti.

The first long work was **Claudio Monteverdi**'s (1567–1643) *L'Orfeo*, which was performed in 1607.

Concert (1672)

The first paid concert was given before an audience of 42 people and organised in **1672** in London by an Englishman called **Bannister**.

Jazz (c.1914)

Profoundly influenced by the blues and the secular and spiritual songs of the American blacks, jazz was born around **1914** in New Orleans, Louisiana. The first jazz recording took place in 1917 on the other side of America in Chicago, by the Original Dixieland Jazz Band. Since then jazz has been through many changes of style from bebop to free jazz.

Blues (1860)

Blues, sung by the black slaves in the South, became known to a wider audience around the time of the American Civil War (1861–65) as it spread to the North of the country. The music owes its name to the fact that it uses many blue notes: the flattened third and seventh notes.

Dance

Tango (15th century)

A dance similar to the tango can be traced to southern Spain at the beginning of the **15th century**, at the time that the country was under

Waslaw Nijinsky, the great Russian dancer and choreographer, features in this design for the ballet L'apres-midi d'un faun *in 1912 with costumes by Léon Bakst.*

Arab domination. Carried to Latin America, the initial rhythm of two beats in the bar underwent numerous changes and gave rise to dances like the Cuban habanera. The tango is characterised by its gliding steps and abrupt pauses with a duple rhythm.

Ballet (1581)

The first real ballet was the *Ballet comique de la reine* which was commissioned by **Catherine de' Medici** (1519–89) in **1581**.

The modern trappings of ballet appeared centuries later. It was not until about 1826 that Maria Taglioni began to dance on her points thanks largely to the developments in dance shoes. The tutu was designed by the painter Eugène Lamy for the performance of *La Sylphide* at the Paris Opera in 1832. The tutu was made from white muslin and reached mid-calf. The short, flared tutu was first worn by an Italian dancer, Virginie Zucchi, at the Imperial Theatre at Saint Petersburg in 1885. The name tutu derives from the childish French *cucu* meaning 'little bottom'.

Choreography (c.1671)

Charles Louis Beauchamps (1636–1719), who was the ballet master at the court of the Sun King (Louis XIV), invented a method of notation around **1671** that allowed dance steps to be recorded on paper.

Waltz (18th century)

The origins of this popular dance are obscure. However, it seems to have come from Bavaria at the end of the **18th century**. The name of the dance comes from *walzen*: Middle High German meaning to roll.

The Austrians Johann Strauss (1804–49) and his son (1825–99) wrote many popular waltzes; the younger Strauss was known as the waltz king.

Photography

Origins

Photography results from the combination of two disciplines: optics on the one hand and photo chemistry on the other. Each followed separate paths through history: the former dating back to the middle of the 16th century, the latter to the beginning of the 18th century. Once they had come together, lenses and films progressed along parallel roads.

It was the Englishman Sir John Frederick William Herschel (1792–1871) who, in 1836, coined the word photography (from the Greek *photos* light and *graphein* to write) to describe the action of light on certain sensitive surfaces.

Camera obscura (16th century)

The Chinese probably knew the principle of the dark room – the *camera obscura* – as early as the 4th century BC. The Arab scholar Alhazen (965–1038) used it to observe solar eclipses and Leonardo da Vinci (1452–1519) described the phenomenon very clearly: a beam of light entering a darkened room through a small hole projects on the opposite wall the reverse image of the outside scene. The Venetian Daniele Barbaro (1513–70), in 1568, placed a lens over the hole and this gave a sharper image.

However, it seems to have been the Neapolitan dramatist and writer on natural magic Gianbattista della Porta (1538–1615) who first advocated the use of the *camera obscura* to reproduce images (by drawing them).

In the 17th century several inventors including Johann Zahn and Athanase Kircher made portable 'dark rooms'.

Photographic lenses (1560)

When Gerolamo Cardano put a biconvex lens in front of the hole of a *camera obscura*, he invented the first photographic lens. But it was the French philsopher, **René Descartes** (1596–1650) who laid the foundations of modern optics in *Dioptrics* (1637). The photographic lens corrects the distortion of the image that forms on the light-sensitive surface when the light has come through a simple lens.

The earliest positive image captured on a plate, taken by Nicéphore Niepce from the window of his home in 1826.

The performance of the first photographic lens (single lens) was improved by a series of inventions: diaphragm (Barbaro, 1568), moveable mirror (Egnazio Danti, 1573), meniscus (Wollaston, 1812).

Chevalier's invention (the compound lens, 1840), improved by the Austrian J. Petzval, allowed for a larger aperture which meant a shorter exposure time and made daguerreotype portraits possible.

Birth of modern camera lenses

The rapid rectilinear lens (Thomas Dallmayer, Germany, 1865) gives a sharper picture, the aplanat (Dallmayer and Steinheil, 1866) does away with image distortion, and finally the anastigmatic lens (Doctor Schott, Germany, 1888) gives a sharp picture round the edges. H. D. Taylor's Triplet lens (Great Britain, 1893) was the first of a long series of rapid lenses such as Doctor Rudolph's Zeiss Planar (Germany, 1896) and the most famous of them all, the Zeiss Tessar from the same inventor (1902).

Diaphragm (1816)

The diaphragm regulates the amount of light passing through the objective lens. The iris diaphragm used in all modern cameras and invented by **Nicéphore Niepce** (1765–1833) around **1816**, only came into general use around 1880.

The valve diaphragm was created in 1858 for rapid rectilinear lenses or doublets.

The automatic control of the diaphragm's aperture according to the amount of light available was the work of the Frenchman Alphonse Martin who patented his invention in 1939.

Lens coating (1904)

Lens coating reduced reflections from the glass surfaces within the lens. In **1904**, **Taylor** patented a method for artificially tarnishing lenses. In 1935, Carl Zeiss patented a method of coating the lens surfaces with fluoride of magnesium.

Since 1945, all lenses have had coated surfaces, improving light transmission and contrast, and making it possible to produce multiple-element lenses.

Wide angle lenses (1859)

Their ancestor, the Englishman **Thomas Sutton**'s panoramic lenses (**1859**), covered a 120° field at f/12. To reduce refraction, the spaces between the lenses were filled with water.

The first wide angle lens was von Hoëgh's Hypergon, in 1900. Robert Hill's disymetric objective lens (1924) gives an even wider field. The modern 'fish-eye' lenses were derived from it.

Telephoto lens (1891)

The telephoto lens is an objective lens in which the focal length is greater than the negative image's diagonal. The first one was used by **T. R. Dallmayer** in **1891**. Some telephoto lenses with a very long focal length use curved mirrors: they are the catadioptric telephoto lenses perfected by B. Schmidt in 1931.

Zoom lenses (1945)

Zoom lenses are objective lenses with a variable focal length. In 1896, T. R. Dallmayer and J. S. Bergheim perfected a rudimentary lens of this type. It can be considered to be the distant ancestor of the modern zoom lenses. The idea was taken up again in **1945** with **Frank Back**'s Zoomar (USA), a combination of two sets of mobile lenses. But it was with Roger Cuvillier's Pancinor (France, 1949) that the principle of the variable focal length was finally adopted.

The largest zoom lens in the world was built by the French company Angénieux in 1987: 3m long, × 300 focal length variation, germanium lenses. It was designed for the American Army's observations under infrared light.

Space objective lenses (1962)

The first space photographs were taken on **3 October 1962** during the Mercury mission with a German **Carl Zeiss** objective lens (and a Swedish **Hasselblad** 500C camera). It was also a Carl Zeiss lens that recorded the first steps of the first man on the moon, Neil Armstrong, during the Apollo II mission (16–24 July 1969). During the same period NASA also used the French Angénieux lenses, which can be used in extreme light conditions, to equip the Ranger 7, 8, 9 missions. The Angénieux lenses also went on the first trip to the moon.

The most commonly used objective lenses are the 1.2/55mm outside and 1.4/35mm inside. NASA certainly puts photographic materials to the test and therefore acts as a gigantic laboratory which helps the progress of the photographic industry.

On the 1984 Challenger mission, pictures of an exceptional quality were taken at a speed of 28 000km/h *17 400mph* and without blur! In 1986 the space probe Giotto, after travelling 685 million km *425 659 000 miles*, photographed the nucleus of Halley's comet for 45 minutes and sent back the pictures in eight minutes!

First light sensitive surfaces (1727)

As early as **1727**, the German physicist **Johann Heinrich Schulze** noticed that silver nitrate turned black when exposed to the light. In 1777, the Swedish chemist C. W. Scheele made an extensive study of the effect of light on paper with silver chloride. In 1802, the Englishman Thomas Wedgwood (son of Josiah Wedgwood, the potter) went further and obtained contact images on silver-nitrate coated paper of objects and drawings under glass. But as they were not fixed, these images disappeared when exposed to light.

Photography (1826)

The first attempts by the German J. W. Ritter and Englishmen Thomas Wedgwood and Humphry Davy between 1801 and 1802 to fix an image on light sensitive paper in a *camera obscura* were unsuccessful.

A Frenchman, **Nicéphore Niepce**, was looking for a means of reproducing his son's drawings mechanically. Having thought about the effect of light on some chemicals, in 1816, after many unsuccessful attempts, he managed to capture a picture on silver-chloride coated paper, at the back of a *camera obscura*. But the image was negative and short-lived. He then tried a different process: Jew's pitch, which turns white and hardens when exposed to light. In 1822, he reproduced translucent pictures by contact on a pitch-covered plate. These 'heliographs' were the ancestors of photogravures. In **1826**, carrying on his research in the *camera obscura*, he captured on the same plate, after eight hours' exposure, the first positive image. Photography had been invented.

Daguerreotype (1835)

It was in **1835** that **Jacques Louis Mandé Daguerre** (1787–1851), Niepce's partner, perfected a new process to capture the image of an object on a metal plate. In spite of its faults (it was fragile, could be reproduced only by rephotographing and was reversed left to right), the daguerreotype produced a picture of an exceptional quality for the time. On 19 August 1839, Arago, permanent secretary at the French Academy of Science, revealed how daguerreotypes were obtained. It was an immediate success. 'Daguerreotypomania' spread like wildfire all over the world: 500 000 plates sold in Paris in 1846, and more than three million in the United States in 1853.

Photography on paper (1839)

The daguerreotype might be perfect from the artistic point of view but on the technical side

A daguerreotype of the man who perfected the technique: Jacques Louis Mandé Daguerre.

it was not the ancestor of contemporary photography. Modern photography has its roots in the work of two men who had no knowledge of the work of the other: the Englishman **William Henry Fox Talbot** (1800–77) and the Frenchman **Hippolyte Bayard** (1801–87).

As early as 1833, Talbot tried to fix on paper sensitised with silver chloride the image obtained at the back of a *camera obscura*. The following year, he obtained 'photogenic drawings': negative images of the objects placed on the paper. He fixed the images with cooking salt. Improving his technique, in 1835 – after several hours' exposure and using a wooden box at the back of which he had placed some sensitised paper (a box his wife nicknamed the mousetrap) – he managed to obtain a faint negative image: the first negative on paper.

In **1839**, when Daguerre's process was revealed, Talbot started his work again and discovered that by using this negative image as 'an object to be copied' he could obtain a positive image on sensitised paper. The negative/positive process had been discovered.

Positive images on paper (1939)

In that same year of **1839**, in Paris, **Hippolyte Bayard** invented a process which enabled him to obtain positive images directly on paper. They were of such good quality that they were shown at the first photographic exhibition on **24 June 1839**. It is undeniable that these unique pictures, obtained after a 15 minute exposure, came first. However, Arago asked Bayard not to reveal that fact so as to avoid upsetting Daguerre.

Calotype (1840)

In **1840 William Henry Fox Talbot** revealed what Bayard had earlier discovered: that it was possible to make the latent negative image formed on paper treated with silver iodide that had been subjected to strong light appear by putting it in a developer. This process meant that a pose could be held for less than 30 seconds, but only Talbot dreamed of using the negative to obtain a positive image through contact. This development led photography into the era of endless reproduction. Talbot called it a Calotype from the Greek *kalos* meaning beauty.

Wet collodion plate (1851)

In 1849, the Frenchman Le Gray light-sensitised his negative materials with a collodion based preparation, that is, one made of gun-cotton dissolved in ether. Nonetheless, it was the Englishman **Frederick Scott Archer** who first promoted the use of this process in **1851**.

Fifteen times faster than the daguerreotype, the wet-collodion process was a resounding success because of the quality of reproduction, and was adopted by all the great photographers including Roger Fenton and Matthew Brady.

Dry plates (1871)

In **1871** an English doctor, **Richard L. Maddox**, published the first description of a process using really efficient dry plates. Improved by Charles Harper Bennett in 1876, gelatine bromide plates revolutionised the world of photography. And suddenly, industry took over and quickly popularised photography.

In 1879 a young American amateur photographer, George Eastman (1854–1932) invented a machine that could produce dry plates in large quantities. On 1 January 1881 he founded the Eastman Dry Plate Company with Henry A. Strong, a horse-drawn carriage whip manufacturer. This small company, based in Rochester, New York, went through a spectacular expansion and is today universally known under the name of Kodak.

The art of photography was soon to capture the imagination of travellers; this daguerreotype of Siberian villagers was made in 1847.

The Kodak camera (1888)

In **1888** the American **George Eastman** revolutionised the photographic market by creating a simple, light and cheap camera: the Kodak camera. The small 82 × 95 × 165mm *3.2 × 3.7 × 6.5in* box, with its F/9 objective lens and a shutter with a 1/20 of a second speed, was a resounding success. It produced 100 round photos measuring 63mm *2.48in* in diameter. Once all the photos had been taken, the camera was sent back to the factory who took the film out, put a new one in, developed the photos and sent the whole lot back to its owner within ten days. 'You press the button, we do the rest!': the slogan was launched and 90 000 cameras were sold in the first year.

Why the name Kodak? Because it was short, easily remembered and could be pronounced throughout the world.

Celluloid roll film (1889)

The first celluloid roll film to be used commercially was developed by **Henri Reichenbach** for **George Eastman** and marketed in **1889**. Already in 1884 Eastman and William H. Walker had invented a container for rolls of negative paper. In 1885 they developed the film Eastman America. Unlike negative paper, it was a thin film that used paper only as a temporary support for the emulsion. The paper was eliminated after development and a thin negative film remained, which was then mounted on glass for the production of prints.

Flash (1850)

Towards **1850**, it was discovered that the combustion of magnesium wires produced an extremely brilliant light, accompanied, however, by a thick cloud of white smoke. In 1887 Germans **Adolf Mietke** and **Johannes Gaedicke** invented flash powder (magnesium powder), an explosive mixture with a base of magnesium, potassium chlorate and antimony sulphate. This process won over photographers as a whole, despite the dangers involved in its use.

Flash bulbs (1929)

In 1925 **Paul Vierkotter** took out a patent for a new flash method whereby the inflammable mixture was contained in a vacuum lamp and set alight by means of an electric current.

But it was in **1929** that the first real flash bulb appeared, thanks to the German **Ostermeier** who perfected the system and created the Vacublitz bulb. This had the advantage of being both silent and smokeless. Today's flash cubes are miniaturised versions of the flash bulb.

Electronic flash (1931)

The principle goes back to the very beginning of photography. In 1851 Fox Talbot succeeded in obtaining an image of a newspaper fixed to a rotating wheel thanks to a spark supplied by means of a condenser battery.

In **1931 Harold Edgerton** of the Massachusetts Institute of Technology (MIT) invented the electronic flash; the current accumulated at high voltage in a condenser explodes in a tube filled with rare gases.

Underwater photographic observations (1856)

The first underwater photograph was presented by **William Thompson** to the London Art Society in **1856**. The photographic apparatus had been placed at the bottom of Weymouth Bay at a depth of 6m *19.6ft*. The lens, controlled from the surface, stayed open for ten minutes.

In 1889, the French zoology professor, Louis Boutan, took photos at a depth of 50m *164ft*. The photographs thus obtained were published in 1900 in an album which was the first of its kind.

In 1935 the Englishman William Beebe took photographs from the porthole of a bathyscape at a depth of 900m *2952ft*.

Ten years later, an American, Maurice Ewing, developed an automatic underwater camera to photograph geological formations at depths between 400 and 700m *1312* and *2296ft*.

In addition to these purely photographic techniques, other means of looking at the sea depths have come into existence. One of the latest is the Sea Beam, from the French company, Thomson, which takes radar images (viewed on a TV screen) with an apparatus mounted on a ship. This was used during the exploration of the wreck of the *Titanic*.

The first underwater shots were taken in 1856. Since then, the technique has improved dramatically.

Photobooth (1924)

The Photobooth is the invention of the Hungarian **Anatol Marco Josepho** who registered his patent in Germany on **13 January 1924**. The machine was marketed in England in 1928. The attendant was replaced by a coin slot in 1968 and in the same year colour was introduced.

Colour photography

First attempts (1891)

As soon as photography was invented the question of colour attracted inventors (coloured daguerreotypes 1840; photographic glazes by Lafon de Carmasac in 1854).

In 1848, Edmond Becquerel (1820–91) managed to photograph colour prints but could not fix the images. It was **Gabriel Lippmann** who, in **1891**, obtained the first direct colour pictures by the interferential process (the fixation of the luminous vibration traces). That success won its inventor the Nobel Prize for Physics in 1908. The only problem was that his work could only be seen at a certain angle and could not be copied.

Autochrome process (1903)

It was by applying the principles of additive

One of the earliest Photobooths in use in Oslo in 1929.

This 1925 autochrome is the first colour photograph of a music hall scene. However, the process had been developed as early as 1903.

colour synthesis that the **Lumière** brothers, in **1903**, perfected the autochrome plate process. Each plate was sprinkled with a mixture of potato starch grains dyed green, red and blue. A single 9 × 12 plate carried nearly 90 million! This process remained popular until 1932 when it was replaced by Lumicolor and Filmcolor, also Lumière processes. It was subsequently abandoned altogether.

The autochrome technique produced some masterpieces which were very close to paintings in their texture and rich colours.

Additive process (1855)

This process is based on the discovery made in **1855** by the Scottish physicist **James Clerk Maxwell** (1831–79). The desired colours can be created by adding together the three fundamental colours, red, green and blue, in appropriate proportions.

Subtractive synthesis (1869)

It is the process from which all modern colour photography processes are derived. The announcement of its discovery was marked by a dramatic turn of events: on **7 May 1869** two men who knew nothing of each other's work presented similar conclusions to the French Photographic Society. **Charles Cros** and **Louis Ducos du Hauron** (who had registered a patent in November 1868) had discovered trichromatic synthesis.

Modern films

Kodachrome (1935)

In 1911/2 the German chemist Rudolf Fischer discovered that dyes could be obtained by oxidation or coupling with other chemical substances. All modern colour films follow this general principle.

It was not until **1935** that subtractive colour synthesis found its first commercial application. Two American musicians **Leopold Mannes** and **Leopold Godowsky**, with **Kodak**'s financial backing, invented Kodachrome.

Agfacolor (1936)

In **1936** the German company **Agfa**'s first reversal film, Agfacolor, came out. It had colour couplers present in the emulsion. Subtractive colour synthesis then spread all over the world and improvements were made in rapid succession: in 1939, the Agfacolor negative/positive film appeared, a process which was immediately used by Hitler's propaganda machine (e.g. in the making of prestige films such as the *Adventures of Baron Münchhausen* which created a sensation at the time).

In 1942 the United States' reply came in the shape of Kodacolor. The reversal and negative-positive processes continued to evolve in parallel: the former producing the Ektachrome (1945), the Anscochrome (1955) and the Fujicolor (1948); the latter the Ektacolor (1947) and the Gevacolor. In 1949 a new generation of Kodacolor films and in 1953 of Agfacolor, applied a technique using colour 'masks' which gave better colour saturation.

In addition to its Kodacolor range, at the end of 1988, Kodak introduced the Ektar 25 and 1000, two types of negative films for print making, aimed at the amateur wanting very high-quality pictures. These can only be used in 24 × 36 reflex cameras.

Contemporary colour emulsions

The perfecting of the renowned 'T' grain by Kodak (1983) meant an amazing improvement in sensitivity. This grain makes it possible, among other things, to raise the sensitivity of Kodacolor negative films to ISO 1000. There is also a fantastic improvement for reversal films and the new Ektachrome comes with ISO 400, 600, 1600 even 3200, which makes it the most sensitive 'daylight' film to date. Other companies have researched the same areas: Agfa (use of structured twin crystals) and Fuji (double structure grain SG and Fujicolor that can be used at ISO 3200).

Films with enzymes (1987)

The biotechnological revolution has reached photography with the process perfected by **Canon**. The enzyme film uses amylase (an enzyme which attacks starch) emulsions on a mixture of starch and colour pigments. Light deactivates the enzyme. After reaction and drying, the black and white photo obtained is identical to standard film ones.

Polacolor (1963)

The announcement in **1963** by **Edwin H. Land** and his Polaroid Corporation that they had created a one-minute colour film, Polacolor, caused a sensation. Extremely sophisticated in its conception, this film allowed a colour print to be made on paper in only 60 seconds. The emulsion consisted of a negative part half the thickness of a hair, comprising nine distinct layers, and a positive part of four layers, as well as one last layer enclosing a capsule containing an alkaline solution which triggered off treatment. In total, 14 stratifications in a single film!

In 1982 Polaroid launched a process for producing instant colour or black and white transparencies.

3-D

Stereoscopy (1838)

In **1838** the Englishman **Charles Wheatstone** made the first stereoscope, an apparatus which permitted geometric patterns to be seen in three dimensions. Each eye receives an image of the same object from a slightly different angle and the two images are synthesised by the brain. This gives us the illusion of three-dimensional perception.

In 1849, the Scottish physicist Sir David Brewster built a simple and practical stereoscope (a small closed box with two viewing holes) which, factory produced by the Frenchman Dubosq, was all the rage in London at the Great Exhibition of 1851.

The stereoscopic revolution was all the more important as, because of the lenses' short focal length, exposure time was reduced to about a quarter of a second which made the first snapshots possible.

Anaglyphs (1891)

In **1891 Louis Ducos du Hauron** discovered a new process, anaglyphs, which consisted in superimposing two separate images of the same picture after dying one violet-blue and the other red and then looking at them through glasses with one lens each of the same two colours. Each eye saw only one of the images and the brain combined the two to give the illusion of a single three-dimensional picture.

Current cameras and processes

Automatic focus (1945)

The first automatic focus device appeared in **1945**. It was the Optar, invented by the German **Doctor Kaulmann**. However, the device was too bulky to be fitted on a camera. Automatic focus is now a common feature and the device is an integral part of the camera. There

One of the pioneers of colour photography was Louis Ducos du Hauron. Here is an example of his work.

are various principles: the CCD (Charge-Coupled Device), the Sonar by Polaroid (which uses ultrasound), and a system exclusive to Minolta (1983) using an infra-red beam. Apart from the auto-focus cameras, the Minolta 7000 was the first reflex camera to use this system. The Minolta 9000 (1986) applies this automatic focusing system to all its range.

Compact auto-focus camera (1976)

The first compact auto-focus camera was the Japanese Konica in **1976**. Since then, cameras have become more and more sophisticated, such as the Konica MR 70 or the Fuji TW 300. The culmination came in 1985 with the Minolta AFT which seems to have made everything it could automatic: its two objective lenses, focusing, exposure, loading, film winding and DX selection.

Kodak produced the first small, popular cameras in 1888. Backed by catchy slogans and attractive posters, the firm is still a worldwide success over a century later.

Instant photography (1948)

In November **1948** the American **Edwin Herbert Land** launched on the American market the first instant picture camera, the Polaroid 95. At the age of 28, Land had founded the Polaroid Corporation which specialised in the manufacture of sunglasses and polarising filters.

In 1947, Kodak had shown no interest in Land's invention. But on seeing how popular Polaroid cameras were, in 1976 Kodak decided to launch their own instant picture cameras. However, Polaroid sued them for patent violation and after proceedings lasting ten years won the case in October 1985. This gave back to Polaroid exclusive use of the process and a minimum of $1000 million damages!

Edwin H. Land holds the second largest number of registered patents in the United States: 533.

Onyx (1987)

Onyx is the first transparent camera. A **Polaroid** invention which makes it possible to see the devices that operate the focus and allow a shutter speed of 50 milliseconds.

Magnetic photography (1981)

The first in this category came from the Japanese firm **Sony** which introduced its Mavica-Magnetic Video Camera on **24 August 1981**. This camera, which uses a magnetic disc, represents a revolutionary invention in photography: an electromagnetic system now replaces chemicals; it does not use a film, therefore there is no processing or printing as the pictures can be seen on a screen.

In 1984 Panasonic introduced a camera of this type, but that recorded colour pictures on a video floppy disc with a 50-frame capacity. Although they were the pioneers in this domain, Sony only marketed its Mavica (the Mavica MVC 1, a greatly improved version of the one introduced in 1981) at the end of 1988 in Japan.

First application (1984)

World premiere at the Los Angeles Olympics in July **1984**; a camera which looked like a 24 × 36 motorised reflex, the **Canon** Still Video System D413 took magnetic colour photographs which were transmitted over the telephone to a newspaper and were viewed on a television monitor. The camera had interchangeable lenses and used a 4.7cm *1.85in* diameter disc which could hold 50 pictures.

Fujix TV-Photo (1984)

The Fujix TV-Photo system, invented by the Japanese **Fuji** and introduced in September **1984**, allows you to view your favourite pictures (prints, negatives, slides) on your colour television set. The system comprises a compact and light reader and an 8g *.28oz* video diskette on which up to 48 photographs have been electronically recorded (with a reader-recorder). The diskette can be inserted into the reader as easily as a cassette in a video recorder.

Kodak disc-film (1982)

On **3 February 1982 Kodak** brought out a new type of photographic base: the disc-film. The flat plastic cartridge holds a disc-film which rotates round a hub in front of an exposure window which is behind the objective lens. Thicker than that of ordinary films, the base of the Kodacolor HR Disc film is made of Estar, a thick and rigid material.

It was the German engineer Oskar Barnack between 1911 and 1913. A mountain-climbing and photography fanatic, he dreamt of making his mountain photographic equipment lighter. The first Leica came on the market in 1925. It marked the beginning of photojournalism as we know it today.

DX coding (1984)

The 24 × 36 film cassettes have been coded since **1984**; initially perfected by **Kodak**, the coding system tells the camera the type of film used, its light sensitivity and the number of exposures. Absent-minded photographers (many devices have been geared to them!) will no longer run the risk of making dreadful mistakes. All the films on the market at present have such coding and most modern cameras are fitted with the system.

Quicksnap (1986)

In **1986 Fuji** invented a negative colour film wrapped in a box which includes the lens and shutter. After the 24 pictures have been taken, the whole lot is sent to a laboratory for processing. No adjusting required, this film-camera is incredibly easy to use and should be made available from automatic dispensers. In 1987 it was Kodak's turn to put a camera of that type on the market, the Fling, which joined the Quicksnap, a second generation Fuji disposable camera. In 1988 Fuji brought out a Quicksnap with incorporated electronic flash.

Cinema

Origins (1654)

The ancestors of modern cinema are the magic lantern invented in **1654** by a German priest, **Athanase Kircher**, and the fantasmagoria developed by the Frenchman Gaspard Robert in 1798. These were both projection techniques with painted backgrounds mounted in front of a light source.

The Belgian Joseph Plateau demonstrated in 1828 the principle of the persistence of luminous images on the retina, thereby formulating the basis of cinema. He invented a *phénakistiscope* in 1832 that, for the first time, produced a moving image. A cardboard disc carried a sequence of pictures of a subject such as a dancer. As it revolved, the disc created the impression that the figure was moving.

The stroboscope, which worked on a similar principle, was invented by the Austrian Simon von Stampfer in 1833. It is used today to give the appearance of immobility or to slow down the image of something moving quickly. It has industrial applications but is also to be found in discos.

A scene from a magic lantern show dating from the end of the 18th century. By then the technique had been known for over a hundred years.

Pictures of movement (1878)

After years of work, the English photographer **Eadweard Muybridge** succeeded in demonstrating, in **1878**, that a galloping horse does indeed lift all four legs completely off the ground. (The horse set off a series of cameras as it galloped along the track.)

Photographic gun (1882)

In 1882, the French scientist **Jules Marey** developed 'chronophotography' by constructing a repetitive-action 'photographic gun' that allowed him to take successive shots of a bird in flight. At the sea-shore in Naples, Marey chronophotographed the flight of sea gulls. Peasants who saw him spoke about a crazy man armed with a gun who aimed at birds without ever shooting and seemed delighted to come back from hunting empty-handed.

The invention of chronophotography helped the development of the cinema, even if Italian peasants thought the inventor was mad.

Studio

W. K. L. Dickson, a collaborator of Thomas Edison, shot the first films for him in the world's first studio. It was more or less a shed lit by large windows and constructed on a pivot. The whole thing turned, following the path of the sun; thus the scenes, played inside, received a constant flow of light. The first films shot in this studio were shown on the Kinetoscope, a machine which contained an endless strip of film.

Lumière cinematograph

On **13 February 1895** the **Lumière** brothers, **Louis** (1864–1948) and **Auguste** (1862–1954), took out a patent for 'a device for obtaining and viewing chronophotographic prints.' They baptised it the *cinematographe*.

Public showing

The first public showing was held on

28 December 1895 in Paris. This new form of entertainment attracted crowds, and its success led to the growth and development of the motion picture industry. One might add to the Lumière brothers' credit by mentioning that their films, which were the first moving picture films, had real artistic value: the very first, the entry of a train into the La Ciotat station, was taken at such an angle that the spectators were terrified.

Movie projector (1894)

Credit must be given to the American **Le Roy** for having invented the first projector that brought together the essential elements of our current models. On **5 February 1894**, at 16 Beekman Street, New York, he publicly projected two of Thomas Edison's Kinetoscope films.

Movie camera (1897)

The Frenchman **Charles Pathé** (1867–1957) broke down the Lumière brothers *cinematographe* into two distinct elements, the camera and the projector, and made them independent.

Around 1904, camera speed was made variable in order to allow speeded-up and slow-motion films. Sixteen images per second were used for ordinary shooting, but 24 per second were necessary for sound films.

Films with sound (1896)

Voices and music were added to **Thomas Edison**'s films thanks to the Kinetophone in **1896**.

In 1902 the Frenchmen Baron and Glaumont built an automatic electric synchronisation system using the phonograph and the projector, but the phonographs were not powerful enough. Sound amplification was a difficulty resolved in 1910 by the Frenchmen Decaux and Laudet for the Gaumont company. They modulated a flow of compressed air. In 1912 this process was used in Gaumont's large cinema in Paris.

Talkie

The first successful 'talkie' was *The Jazz Singer*, an American film produced by Warner Brothers in 1927. Several of its sequences had sound added by means of 33rpm records.

Optical sound track (1929)

In **1929** the first truly talking film came out: *Hallelujah*, an American film produced by **Metro-Goldwyn-Mayer** and directed by **King Vidor**. On the film itself, next to the images, an optical soundtrack was engraved through photoelectric processes. This process had been demonstrated by Lee De Forest in 1923.

Slow motion (1904)

Two students of Jules Marey, the Frenchmen **Lucien Bull** and **Henri Nogues**, invented slow-motion filming in **1904**. They understood that by shooting more frames per second than one normally would shoot, one obtains a longer film. Then, projecting this longer film at normal speed, events are seen to take place less rapidly than they do in reality. That's slow motion.

Colour films (1900)

The first colour films were hand painted, frame by frame. Then from **1900 Méliès** and **Pathé** used a mechanical technique: the stencil. Despite the years, the few films that still exist from the era retain their fresh colour. Subsequently, progress in the development of moving film took place hand in hand with that of photography.

Technicolor (1881)

Technicolor had been imagined by the American **Warnake** in **1881**, but it only appeared in 1917. This was in a two-colour film, *The Gulf Between* with Grace Darmond and Niles Welch. In 1932 a three-colour film, *Flowers and Trees*, was produced by Walt Disney.

The first full length film shot in three-colour and in Technicolor was *Becky Sharp*, directed by Rouben Mamoulian in 1935.

The subtractive process is still used today, particularly for taking copies of films, because of the fine quality it produces.

Colouring of the great classics (1986)

The Canadian **Wilson Markle**'s invention revolutionised the American world of cinema. This ex-NASA engineer has perfected a computer system which makes it possible to colour films shot in black and white. The main American television networks rushed out to buy film rights and have films coloured, ranging from Laurel and Hardy to John Huston's *Maltese Falcon* and *The Hunchback of Notre Dame*. In spite of protests from the traditionalists – headed by Woody Allen – this process, used by Vidcolor Inc. in Toronto and Color System Technologies in Los Angeles, appears to have a promising future.

Modern cameras

Steady Cam (1970)

Perfected by the American technician **Garrett Brown**, Steady Cam allows camera move-

Wilson Markle developed the process that allows black and white films to be viewed in colour.

ments without a dolly with crane and rails. The camera is strapped onto the cameraman's body which means he needs only one hand to hold it. Thanks to a series of controls and balancing devices the cameraman is completely autonomous and can move about and even run. Dustin Hoffman's run in John Schlesinger's *Marathon Man* and the little boy's flight through the snowy maze in Stanley Kubrick's *The Shining* were filmed using a Steady Cam.

Sky Cam (1984)

Since then **Garrett Brown** has perfected two more revolutionary cameras: the Sky Cam (first used in Alan Parker's *Birdy*, 1984), and more recently, the Cable Cam. These two cameras, fitted with video-relays, are fixed on cables stretched at very high altitudes and can glide along them at varying speeds. The result is an amazing sequence of shots that cannot be equalled even from a helicopter.

Shaky Cam (1982)

Devised by **Sam Raini**, a young American director, on the set of his film *Evil Dead*, the Shaky Cam is a motorised version of the Steady Cam. The camera is fitted on to a motorbike and protected from bumps by a system of shock absorbers. This allows for very smooth and fast runs at ground level. Raini used the Shaky Cam to give a vision of things from the point of view of a crawling creature crossing a forest at high speed and brushing against the trees.

Wide screens and 3-D

Pan-and-Scan (c.1980)

A process perfected by Warner Studios which allows the recentering of a 'big screen' film picture so that it takes up the whole area of the television screen. Pan-and-Scan (from *Pan*avision and *Scan*ner) is used by the video industry and television networks.

OMNIMAX cinema (1971)

It was a Canadian engineer **Graeme Ferguson** who, in **1971** in Toronto, invented the cinematographic process called OMNIMAX. This process uses large-size films, each frame being 70mm *2.76in* wide and 50mm *1.97in* high, with 15 perforations as opposed to the five used on a traditional 35mm film. The film moves horizontally, projected onto a semi-hemispherical screen by a special projector. The film is shot with a special camera. Ferguson is also the inventor of IMAX, a similar system but projected onto a flat screen. In 1967, Ferguson, with his partner Robert Kerr, shot his first multi-screen film, *Polar Life*. At present very few cinemas in the world can show OMNIMAX films: about 20 in all.

3-D (1937)

Starting as early as 1894, attempts were made in the direction of the 3-D cinema. Efforts at developing a lenticular system were very quickly abandoned, except in the Soviet Union, where experiments were tried out in the 1940s.

It was in the 1930s that the first commercial 3-D films were produced, in the form of shorts.

But it was the Polaroid system, based on the principle of light filtering, that sparked off the great epoch of the 3-D cinema. The first full-length film was released in Germany in **1937**. The Gunzberg brothers were to import into the United States their application of the process developed by Edwin H. Land which permitted producer Arch Oboler to shoot *Bwana* in 1952. The film was shot with two cameras; the two reels had to be projected by two machines that were strictly synchronised with each other. And, of course, the spectators had to wear Polaroid glasses. A good hundred films were produced in this way.

In 1966, Arch Oboler shot *The Bubble* with the use of a special lens, the polarisator. This device did away with the inconveniences of double projection.

Showscan (1983)

The result of several years' research by **Douglas Trumbull**, the special effects supervisor for Steven Spielberg's *Close Encounters of the Third Kind* (1977) and Ridley Scott's *Blade Runner* (1982), Showscan is a new process which gives the audience an almost tangible illusion of reality, a greatly increased three-dimensional impression.

It uses sophisticated equipment (70mm film, 60 frames/sec., Dolby stereo magnetic sound track). A demonstration film, made by Trumbull himself, is being shown at present in Disneyworld, Florida (USA). A network of appropriate cinemas is growing worldwide.

Holographic film (1985)

It was at the Franco-German research institute in Saint-Louis (Haut-Rhin) that the first convincing *cinéhologrammes* were made. Previous ones had been produced in the USSR (V. G. Komar, 1977) and in the United States (A. J. Decker, 1982).

The holographic cinema principle: to record on film the light reflected by an object or a person lit by laser and at the same time the light coming from the laser itself. From the meeting of the two lights a 3-D picture is born.

Holographic cinema is of particular interest in the study of four-dimensional physical phenomena. The branches of industry which are interested in it are: aeronautics, medicine, new materials' inventors, etc.

Underwater cinema

Origins

It was probably **Jacques-Yves Cousteau** who invented underwater cinema. However, before him, a few attempts had been made using some rule of thumb techniques: for instance, during the shooting of Buster Keaton's *The Navigator* (1926), in which there was an underwater scene. The camerman had been lowered into the water in a glass cage but, as he breathed, the glass walls steamed up and he could not film through them any longer. The problem was remedied by filling the cage with blocks of ice.

The Silent World (1956)

It was **Jacques Cousteau**'s film *The Silent World*, produced in **1956**, that made him famous the world over. But he also has a lot of inventions to his credit: in 1952, the first underwater industrial television equipment with engineer André Laban and Thomson's co-operation; in 1954, a first 'sledge' for underwater photography using Professor Harold E. Edgerton's equipment; in 1955, the first 35mm underwater camera with Armand Davso. In that same year deep sea photography was also attempted, using Professor Edgerton's equipment; in 1956, first survey of the well-known Rift Valley in the Atlantic Ocean; in 1957 final perfecting of the photographic

Walt Disney poses with one of his most popular creations: Mickey Mouse.

'sledges', called *troïkas*, to be used in deep waters; on 15 June, first undersea television programme in the Marseilles area in France, showing the *Calypso* team at work; in 1967, first 16mm underwater camera with Armand Davso.

Cartoons

Origins (1908)

The forerunner of cartoons was, in 1893, Emile Reynaud's optical theatre. The scenery was projected behind the screen by a magic lantern, the characters painted on a white background and their successive positions transferred onto a perforated translucent strip. In **1908 J. Stuart Blackton** and **Emile Courtet**, known as Emile Cohl, invented filmstrip cartoons. After them, the American Earl Hurd perfected, in 1914, a system which made it unnecessary to draw in the background each time.

Rotoscope (1915)

Invented by the **Fleischer** brothers, pioneers of the American cartoon who created *Betty Boop* and *Popeye*, the rotoscope is a technique which consists in filming a real actor and tracing this film so as to give the cartoon character the same ease of movement. The Fleischer brothers used that system in a feature-length cartoon which has remained famous: *Gulliver's Travels* (1939). The American Ralph Bakshi, a leader in cartoons for adults, used the technique again in his film *Fritz the Cat* (1970) as well as in *Lord of the Rings*. The rotoscopy principle is also used to make a 'mask' to hide unwanted elements such as the models' stands or even the technicians' heads showing behind miniature scenery. This indispensable process is still in use.

Cartoons by superimposition (c.1915)

We owe the first superimposition of cartoon characters on to live action films to the **Fleischer** brothers who created *Betty Boop* and *Popeye*.

In *Out Inkle* (**1915**) a small cartoon character was seen coming out of a real inkwell and walking on Max Fleischer's drawing board. Due to the transparency and blue screen processes, the technique has continued to improve. Walt Disney productions were the first to extend the process to feature-length films. The most famous are: *The Three Caballeros* (1945) in which Donald Duck seduces a Mexican dancer; *Mary Poppins* (1964) and *Peter and Eliott the Dragon* (1987).

The technique, enhanced by the use of computer-controlled cameras, found its second wind in Bob Zemeckis's *Who Framed Roger Rabbit?*, co-produced by Steven Spielberg and the Disney Studios. In it, actor Bob Hoskins does battle with a band of cartoon characters referred to as 'toons'.

Disney's Multiplane camera (1941)

It was in a short film starring Mickey Mouse, *The Little Whirlwind*, that **Disney Studios** tested a revolutionary camera: the Multiplane. Meant to re-establish the laws of perspective in cartoon films and allowing shots on several planes, the Multiplane camera has accentuated the realism of scenery and the three-dimensional impression.

Bambi (1942) was the first feature-length Disney cartoon to use Multiplane.

Special effects and trick photography

Origins

Since the beginning of the cinema, filmmakers have had to use special effects and trick photography to recreate dreams and imaginings.

There have been so many different techniques that it is impossible to name them all. A few names and a few films, however, are landmarks in the history of the cinema. Among these the first and greatest is without a doubt the Frenchman **Georges Méliès** (1861–1938), a true magician of the cinema who produced numerous films.

Single frame animation (1897)

This special effects technique, which is used to give the impression that inanimate objects are moving, originated at the end of the last century. The Frenchman **Georges Méliès** and the American production house **Vitagraph** (when they made the short film *Humpty Dumpty Circus* in 1897) were doubtless the first to attempt to make an object move by breaking up this motion into as many positions as there are frames to record.

In 1913 Willis O'Brien thought of filming a clay dinosaur with a wooden skeleton one frame at a time in *The Dinosaur and The Missing Link*. He had invented Stop Motion, a technique that he used again in 1933 in *King Kong*, this time using transparencies which enabled him to overlay the puppet on the real film.

Ray Harryhausen, Willis O'Brien's pupil, improved his techniques and obtained fabulous results in Don Chaffey's *Jason and the Argonauts* (1963) with the famous sequence of the fight with the skeletons; and in Gordon

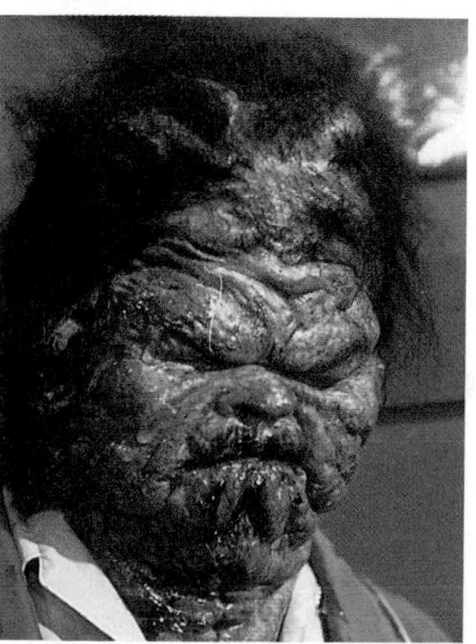

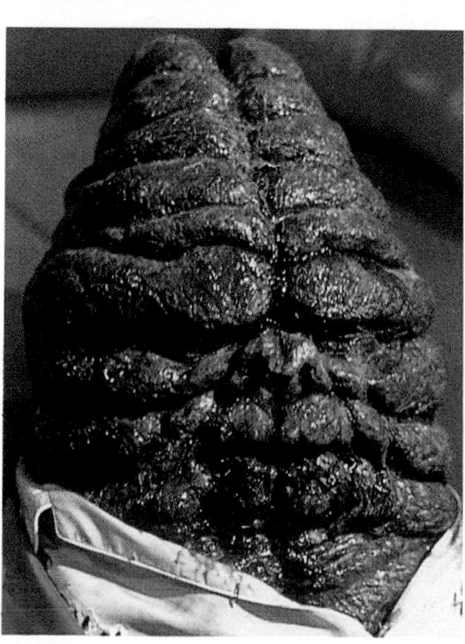

In many horror, fantasy and science-fiction films, the work of the make-up artist is vital. Here are four stages from the transformation of a man into a troll, created by John Buechler in 1986.

Hessler's *The Golden Voyage of Sinbad* (1973).

During the shooting of *Dragonslayer* in 1981, Denis Murren, a young American technician, introduced a new frame-by-frame animation technique: Go Motion. The puppets' movements were computer controlled which saved time and improved smoothness.

Blue screen (c.1950)

Discovered in the **1950s** and put to magnificent use by the technician **L. B. Abbott** for the most famous crossing of the Red Sea in Cecil B. DeMille's *Ten Commandments* (1955), the blue screen principle is one of the most common principles of illusion in the cinema and in television (video shot). A subject filmed against a background of a special blue is isolated on a separate film so as to be inserted in a more complex scene. The blue is identified and eliminated in a laboratory process called 'separation'. This means that the subject, Superman for instance, will be on his own on a transparent film. He can then be overlaid on a shot of New York taken from a helicopter by the matte/counter matte process. The confrontation in one shot of the giant ectoplasms of *Ghostbusters* (Ivan Reitman, 1984) and the ghostbusters themselves was obtained using the same technique.

Zoptic (1978)

During the shooting of Richard Donner's *Superman* (1978), Hungarian born **Zoran Perisic** used his Zoptic for the first time. Combining front-projection (the projection of a moving set on a giant screen), the pole-arms technique, which makes it possible to have actors hanging in the air, and the use of very mobile cameras, Perisic managed to create the illusion of light and motion through the air. His technique, even more refined in Richard Lester's *Superman II* (1980) reached its peak in *Sky Bandits* (1987) which he directed himself.

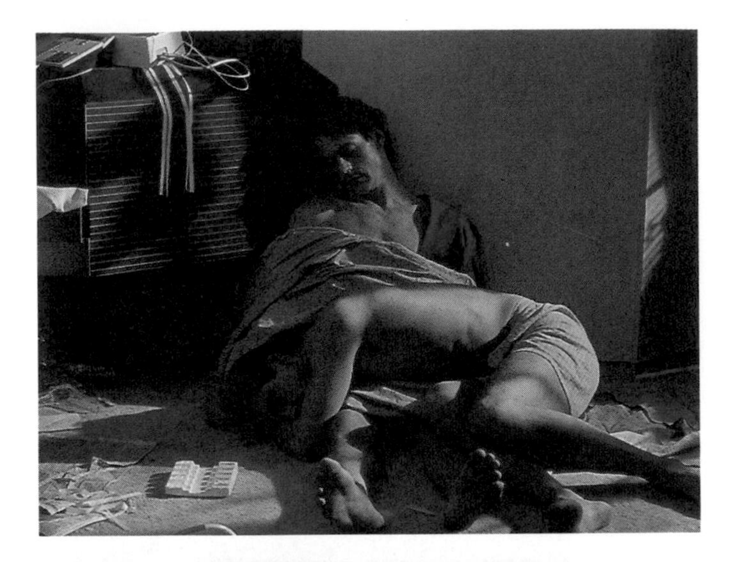

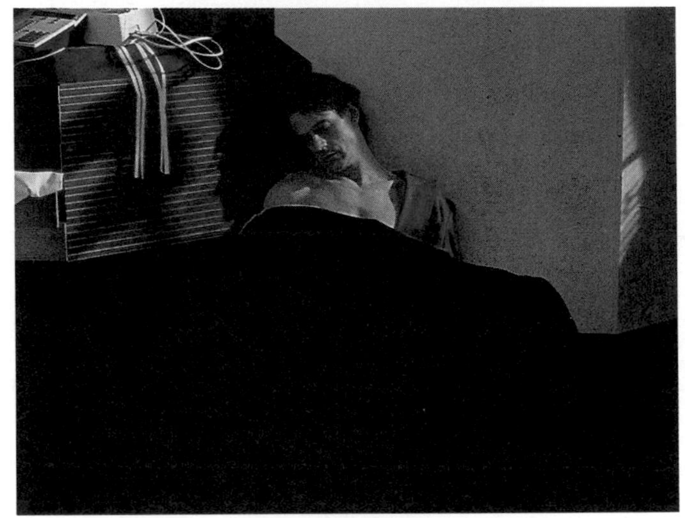

In Dead Ringers *by David Cronenberg, Jeremy Irons plays his own twin brother. In scenes where they appear together, a computer-controlled camera shoots the action once, then reproduces the timing and movements exactly as the actor plays the second part.*

THEY CHANGED OUR LIVES

André Marie Ampère
(1775–1836)

A romantic figure, Ampère was self-taught and possessed a truly encyclopaedic mind. Despite the terrible personal suffering he endured in the French Revolution, when his father was guillotined, he went on to become Professor of Philosophy in the Faculty of Arts, of Physics at the Collège de France, and of Mathematics at the Polytechnique. In 1820 he pioneered the study of electrodynamics, based on Oersted's discovery of the relationship between electricity and magnetism. Ampère was so involved in his research and university studies that he had little concern for material considerations and died virtually penniless. His name is commemorated in the basic SI unit of electric current.

Archimedes
(c.287–212 BC)

Archimedes, son of the astronomer Pheidos, was born in Syracuse and was a pupil of Euclid in Alexandria. He became a renowned engineer and was put in charge of the work on the port of Syracuse, commissioned by King Hieron. His fascination with finding solutions to practical problems led him to study mathematics and geometry. He was the first to make an accurate calculation of π, he developed Archimedes' Principle and the Archimedean screw, a device for lifting water. He was killed during the seige of Syracuse despite the orders of the general, Marcellus, that his life should be spared.

John Logie Baird
(1888–1946)

The Scottish engineer was one of the pioneers of television. In 1926 he gave a public demonstration of what he referred to as 'seeing by wireless'. In 1928 he invented the colour television, and a year later he began a daily black and white broadcast using a BBC

Ampère made important progress in the fields of electricity and magnetism.

John Logie Baird adjusts the transmitter for his television.

transmitter. But, for technical reasons, Baird's 240-line mechanically scanned system was abandoned by the BBC in 1935 in favour of the electronic system developed by Marconi and EMI which produced a much clearer image.

John Bardeen
(born 1908)
Walter Brattain
(1902–87)
William Shockley
(1910–89)

These three American engineers are the inventors of the transistor. They met at the Bell Laboratories and worked together from 1941. At the time their research concentrated on semi-conductors, which were generally considered to be of no practical use. On 23 December 1947, after a great deal of experimentation and trial and error, the three researchers presented the 'persistor', a name chosen as a joke in view of their persistent efforts. After various stages of development, the final version of the transistor was developed in 1950, since when it has revolutionised our way of life. In 1956 the three researchers received the Nobel Prize for Physics for their work.

They then went their separate ways. John Bardeen received a second Nobel Prize in 1972 for his work on superconductors, which seem as promising as the semiconductor. Walter Brattain retired from Bell Laboratories in 1967 and taught at the Washington State University. William Shockley became involved in controversial genetic research which earned him a degree of unpopularity.

Christiaan Barnard
(born 1923)

Christiaan Barnard was born in Beaufort in Cape Province, South Africa. He studied at the University of Cape Town and became a doctor in 1946, specialising in heart surgery. He introduced open-heart surgery into South Africa having learned the technique in the United States while on a scholarship there. He became famous in 1967 when he performed the first successful heart transplant. In 1974 he performed a successful double transplant. His youth and playboy image made him a symbol of social success.

Antoine Henri Becquerel
(1852–1908)

The physicist Henri Becquerel was particularly interested in problems related to phosphorescent materials. After the discovery of the X-ray in 1895, he studied the relationship between these two phenomena which led him to discover radioactivity. He presented his findings to the Academy of Science in Paris in 1896. He then worked with Pierre and Marie Curie on polonium and radium, for their efforts all three were awarded the Nobel Prize for Physics in 1903.

Alexander Graham Bell
(1847–1922)

Alexander Graham Bell, a Scottish-born American, took an early interest in acoustics. He was Professor of Vocal Physiology in Boston when he patented his 'electrical speech' machine in 1876. Later that same year, he gave a public demonstration of his machine at the World Fair in Philadelphia. This machine, better known as the telephone, was extremely successful and Bell became a rich man. He was able to devote the remaining 45 years of his life to the education of deaf and dumb children.

Karl Benz
(1844–1929)
Gottlieb Daimler
(1834–1900)

Inspired by an engine invented by the Franco-Belgian engineer Lenoir, Benz invested all his savings to develop a prototype motorcar in 1885 which was the first to have an internal combustion engine. A year later, working on the same lines, Daimler successfully developed a motorcar which reached a speed of 17km/h *10.6mph*. As a result of improvements made by his assistant Wilhelm Maybach (who invented the carburettor in 1893 and the radiator in 1897), Daimler produced the first Mercedes,

Alexander Graham Bell demonstrates the telephone to an admiring audience.

named after the daughter of his sponsor. In 1926 the companies of Benz and Daimler were amalgamated and manufactured the Mercedes-Benz.

Emil Berliner
(1851–1929)

In 1888, while living in the United States, Berliner developed the flat gramophone record. It replaced the cylinder invented by Thomas Edison in 1877 where the needle moved vertically over a rotating cylinder. In 1898 Berliner returned to Germany where, with his brother Joseph, the director of a telephone factory in Hanover, he founded the Deutsche Grammophon Gesellschaft. By 1901 the company had a catalogue of some 5000 recordings.

Clarence Birdseye
(1886–1956)

Birdseye, the inventor of the deep-freeze method, developed a process which made it possible to commercialise the preservation of food by freezing. He said that the idea was the result of a journey to Labrador with his wife and his five-week-old son. There, Birdseye became interested in deep-freezing, a method used by the Inuit. He founded the Birds Eye company, which, in 1925, started producing frozen foods as we know them today. From 1945 he also began to utilise the process for dried foods invented by Howard.

Laszlo Biro
(1900–85)

Laszlo Biro, a Hungarian artist and journalist from Budapest, was struck by the speed with which printing ink dried. In the 1930s he invented what was to become the ballpoint pen and requested a patent in 1939. But war broke out and he had to flee from the Nazis. In Buenos Aires in 1940 he began production of his pen which proved extremely successful and was supplied to British and American troops. In 1944 Biro sold his shares to one of his partners. The patent was bought a little later by the French company Bic, and Biro did not gain very much from his amazingly successful invention.

Louis Braille
(1809–52)

Braille was blinded at the age of three by one of his father's shoemaking tools. In 1819 he won a scholarship from the Institute for the Blind in Paris. Having attended a demonstration of Captain Barbier's 'night writing', he developed the idea for the blind. The system was published in 1829 but was not adopted on an international scale until 1932.

Sadi Carnot
(1796–1832)

Sadi Carnot was a brilliant student, graduating in first place from the Polytechnique at the age of 17½. In 1824 he laid the foundations of thermodynamics but his work did not become widely known until publicised by Lord Kelvin in 1844. Carnot died of cholera at the age of 36, just as he was intending to study the effects of the disease which was devastating Paris.

Wallace H. Carothers
(1896–1937)

In 1935 the American chemist Wallace Carothers was Director of Research at the Du Pont chemical company. He successfully produced a synthetic fibre which appeared to have the same characteristics as silk. Nobody had ever managed to develop an entirely synthetic fibre, artificial silk (1889) and rayon (1895) were in fact both derived from cellulose. Nylon, which was used initially for toothbrushes, was soon successfully applied in other areas.

Sir George Cayley
(1773–1857)

Born into an aristocratic Yorkshire family, George Cayley was one of the most brilliant pioneers of aviation. In 1796 he repeated the helicopter experiments of the French engineers, Launoy and Bienvenüe, and went on to clearly expound the principles of the aeroplane. He recommended the use of the propeller and the internal combustion engine, in this way defining the characteristics of the modern aeroplane. In 1816 he developed the use of the airship for long distance voyages. In 1853 he built the first glider to be flown by man.

Claude Chappe
(1763–1805)

Chappe took an early interest in physics and mechanics, inventing semaphore signalling which he presented to the French National Convention in 1793. In 1794 the first transmission to be made via the system was the news of the Austrian defeat and the capture of the town of Le Quesnoy. The first telegraph wire was erected between Paris and Lille. Chappe's claim to the invention was disputed and he committed suicide. He is still considered the inventor of this important invention which was replaced half a century later by the electric telegraph.

The Curies

The names of Pierre and Marie Curie are always associated in the field of science and they were the joint winners of several Nobel Prizes. Pierre (1859–1906) and Marie (1867–1934) met while working as laboratory assistants at the Sorbonne in Paris and they married in 1895. They carried out joint research on radioactivity with their friend, Becquerel, and in 1903 all three were awarded the Nobel Prize for their

Marie and Pierre Curie in a cartoon by Imp that appeared in Vanity Fair *in 1904.*

discovery of polonium and radium. Pierre died in 1906 when he was run over by a lorry. Marie then took up a post at the Sorbonne and received the Nobel Prize for Chemistry in 1911. Their work was carried on by their daughter Irène (1897–1956) who, with her husband Frédéric Joliot (1900–58), invented artificial radioactivity.

Jacques Daguerre
(1787–1851)

Daguerre was a scenery painter who was extremely interested in the developing techniques of the dark room. In 1829, in association with Niepce, he produced the first permanent photograph. He improved upon the rather slow process invented by Niepce, and by 1838 he could produce an image in 20 minutes. His invention, known as the daguerreotype, was a great success and established the fashion for the portrait.

Humphry Davy
(1778–1829)

After a mediocre career as a student, Davy became apprenticed to an apothecary surgeon and quickly became a renowned chemist. In 1801 he obtained a post at the recently-founded Royal Institute in London, and won the Academy of Science award for his discoveries in electrochemistry. In 1806 he discovered the properties of chlorine. On his return to England after travelling in France in 1815, he developed the safety lamp which was named after him. His invention made mining much safer, because his light would not ignite firedamp, the explosive mixture of methane and air which caused so many underground disasters.

Rudolf Diesel
(1858–1913)

On finishing his studies at the University of Munich, Diesel became an engineer and developed a keen interest in the theory of engines. He wanted to construct a high-performance engine based on the Carnot cycle, and in 1893 tested a single-cylinder engine which demonstrated the validity of the principle of ignition by compression. Diesel achieved fame with his invention and sold his patents for huge sums which he frittered away. He tried to revive his fortunes on the Stock Exchange, but lost heavily. He was drowned during a voyage in 1913, and it was assumed that he had committed suicide.

The bicycle became very popular thanks to Dunlop's invention.

John Boyd Dunlop
(1840–1921)

Dunlop was a Scottish-born veterinary surgeon practising in Dublin, Ireland. In 1888 he produced tyres with inflated inner tubes for his son's tricycle and patented his invention later that year. Commercial production was started in 1890, and the Dunlop company was formed at the same time. The Dunlop pneumatic tyre played an important part in the increased popularity of the bicycle, and subsequently in the development of the car industry.

George Eastman
(1854–1932)

While on holiday in Michigan, USA in 1877, Eastman, who was at the time a bank clerk, took some holiday photographs but found the preparation too long. In order to make photography easier for amateurs like himself, he invented roll film in 1885. Three years later he produced a portable box camera, the Kodak, which made him a multimillionaire. In spite of his success, he grew disillusioned and committed suicide in 1932.

Thomas A. Edison
(1847–1931)

Edison was born in small village in Ohio, USA, and at the age of 12 began work as a newsboy on a train. He went on to found his own newspaper using an old printing press based in the baggage car. He was interested in chemistry, mechanics and physics, but was also creative and had sound business sense. In 1869 he invented the teleprinter, the first of which he sold to Western Union for $40 000. He patented more than 1 200 inventions of all kinds, from the gramophone to the incandescent lamp.

Alexander Fleming
(1881–1955)

Fleming, a graduate from the School of Medicine at Saint Mary's Hospital in London, carried out research into bactericidal substances. In 1928 he noticed a total absence of staphylococci around a group of mycelia, after accidentally contaminating the middle of the culture. In this way he discovered penicillin which paved the way for cures of many infectious diseases. Fleming was awarded the

Thomas Edison is pictured here with a few of his many inventions.

Gramme at work trying to perfect his dynamo.

Robert H. Goddard
(1882–1945)

During the 1920s, the physicist Robert H. Goddard claimed that it would be possible to reach the Moon using a rocket of the type he had patented in 1914. His statement was greeted with scepticism, although Goddard had been a professor at Clark University. In 1926 he launched a rocket which reached a height of 12.5m *41ft*. He registered more than 200 rocketry patents and constructed a liquid fuel motor with a thrust of 300kg *661lb*. The importance of his work was not recognised until after his death.

Gordon Gould
(born 1920)

Gordon Gould, born in New York, is the physicist who invented the laser. He was a professor at the University of Columbia when he made this important discovery in 1957, although he did not register the patents until 1959 when two of his colleagues, Charles H. Townes and Arthur L. Schawlow, had started to follow the same line of research. He became involved in a legal battle which he eventually won in 1987. He has recovered most of his financial rights, but he is reported as saying: 'Money alone cannot compensate for all those years of lost recognition.'

Zénobe Gramme
(1826–1901)

Gramme came from a modest background and did not prove a very able student, but he showed considerable ingenuity and manual dexterity. He became a carpenter in Liège, Belgium and then moved to Paris in 1857. He was greatly interested in electrical phenomena and took out patents for several devices in 1867. In 1869 he perfected the famous dynamo which was of immediate interest in the fields of science and industry. He founded a company for the manufacture of Gramme machinery which was subsequently considerably extended.

Gutenberg
(1400–76)

Johann Gensfleisch, born in Mainz, Germany, where his father was an officer of the Mint, adopted his mother's name. He first became a goldsmith in Strasbourg and then, in about 1445, published his first books in Mainz with the help of the banker, Johann Fust. The famous Bible appeared in 1445. A few years later he was involved in a lawsuit with Fust and founded a new workshop. Contrary to popular belief, Gutenberg did not invent the printing process, but developed the cast of the characters and applied his genius to the research carried out by his predecessors. He was awarded a title – and in later years a pension – by the Archbishop-Elector of Mainz.

Copies of Gutenberg's books are extremely valuable.

Christiaan Huygens
(1629–95)

Huygens was initially educated by his father, who was a poet and a mathematician, and then went on to study law at the Dutch University of Leiden. He wrote many works on physics and invented, among other things, the pendulum clock. In the field of optics he developed the theory which was named after him as well as the theory of the vibration of light. Finally, in astronomy, he discovered the ring around Saturn. He travelled in England, where he met Newton, and in France where he founded the Academy of Science in Paris.

Joseph Marie Jacquard
(1752–1834)

Jacquard, like his father, was a weaver. From 1790 he tried to develop a mechanism that would automatically lift the threads of the weft which were then still lifted manually. He constructed his first machine in 1801, which he improved and completed in 1806. Jacquard's loom, which was programmed by perforated cards rather like those used in the first computers, controlled the needles and selected and lifted the threads of the weft. It did the work of five people.

Jack St Clair Kilby
(born 1923)

Kilby was born in Jefferson, USA and became an engineer in 1950. He had been working for Texas

Louis Lumière and his brother pioneered the idea of the cinema.

Instruments for two months when he developed the first integrated circuit, the future microchip, in 1959. His invention was to have considerable repercussions. It revolutionised the field of electronics by the sheer range of its different applications which affected all fields: medicine, aeronautics, computers etc. Everyday life also benefited from another of his inventions, the electronic pocket calculator, which first appeared in 1972 from Texas Instruments. As a reward for his work he became one of their directors.

Louis Lumière
(1864–1948)

Due to his family background – his father was a photographer in Besançon, France – Louis and his brother, Auguste, became interested in moving pictures at a very early age. They were both familiar with Plateau's research on the persistence of images on the retina, Marey's work on chronophotography and the inventions of Edison. Auguste said that the idea of the cinematograph 'came overnight'. After three months of experiments, the first film was projected in Lyon in 1895. Louis also worked on developing the technique of colour photography, producing the autochrome photograph which became available in 1907 and contrasting colours in 1920.

Guglielmo Marconi
(1874–1937)

The young Marconi was handsome, gifted, inventive and had a good head for business. He studied magnetic waves and, after a year of experiments in the attic of the family villa in Bologna, Italy, he successfully transmitted the first messages through space in 1895. However, the Italian authorities refused the patent that Marconi offered them. His Irish mother had connections in England and it was there that he raised the funds to found a company in 1897. The following year, he introduced the sports commentary. He accumulated both honours and wealth and, unlike many inventors, died rich and well-respected.

Akio Morita
(born 1921)

Akio Morita was born into a large family of sake brewers, but decided to break with family tradition and became a physicist and engineer. In 1946 he and a friend founded Tokyo Tsuching Kogyo which was renamed Sony shortly afterwards. In 1949 he developed the first Japanese tape recorder which weighed 50kg *110lb*! From then on the motto of Sony was 'Small is beautiful'. In 1955 the company produced the first transistor radio, followed in 1957 by the first pocket radio and in 1979 by the Walkman. This last invention was virtually for his own personal use as it made it possible for him to indulge in his two favourite hobbies at the same time: golf and music. He says of himself: 'I am constantly thinking, so I think my inventiveness is almost permanent.'

Samuel Morse
(1791–1872)

Morse was a painter and Professor of Design at New York University. The early stages of his career were difficult, so much so that he invented a fire engine in order to earn some money. He was returning from Europe by sea in 1832 when he had the idea that was to make him famous: the electric telegraph. In 1834 his application for the post of official painter was rejected by the American government. Disillusioned with his career as an artist, he devoted himself to being an inventor. Finally, in 1842, the telegraph line from Washington to Baltimore was officially opened and Morse, after a few more setbacks, achieved fame and fortune. He developed the Morse code to further ease communications.

Thomas Newcomen
(1663–1729)

Newcomen was a blacksmith from Dartmouth. In 1712 he constructed an atmospheric engine to pump water which was the first

Marconi in the wireless room of his yacht in 1920.

steam engine to be used in industry. It was extremely successful both in England and abroad. Along with Thomas Savery and James Watt, Newcomen's invention made a major contribution to the Industrial Revolution.

Isaac Newton
(1642–1727)

Newton studied at Cambridge where he then became a professor. In 1665 he returned home to spend two years with his family. It is probably during this period that he made the discoveries which were eventually published in 1687, and then only on the insistence of the famous astronomer Edmund Halley. In his work *Philosophiae Naturalis Principia Matematica*, he states the basic laws of mechanics and universal attraction. Newton was extremely interested in optics, laying the foundations of spectroscopy, and he invented the telescope in 1672.

Nicéphore Niepce
(1765–1833)

Niepce is today considered to be the true inventor of photography. As early as 1813 he became interested in lithography and then tried to find a method of producing an image of reality using light. In 1822 he discovered the process which enabled him to obtain the first heliograph, a still life. In 1829 Niepce signed a partnership agreement with Daguerre who wanted to discover his secrets. Daguerre was the only one to profit from the partnership.

Alfred Nobel
(1833–96)

When the family nitroglycerine factory blew up in 1863, Alfred Nobel was left to pursue his experiments on this dangerous substance on his own. In 1867 he invented dynamite, the first effective explosive which could be handled safely. A somewhat eccentric bachelor, he led an extravagant and wandering life. When he died, the conditions of his will were surprising. He was opposed to the idea of inheritance and so left his fortune to a foundation entrusted with the task of making an annual award to people who had made a major contribution to humanity.

Denis Papin
(1647–1714)

In 1675 Papin witnessed the experiments carried out by

Samuel Morse, who made important improvements in communications.

Le Petit Journal

Le Petit Journal
CHAQUE JOUR 5 CENTIMES
Le Supplément illustré
CHAQUE SEMAINE 5 CENTIMES

SUPPLÉMENT ILLUSTRÉ
Huit pages : CINQ centimes

ABONNEMENTS

Sixième année — DIMANCHE 13 OCTOBRE 1895 — Numéro 256

A LOUIS PASTEUR

Le Petit Journal *mourns the death of Louis Pasteur in October 1895.*

Huygens in London into vacuums. He remained in London where he invented the 'steam digestor' and the safety valve. In 1688, while he was Professor of Mathematics at the University of Marburg in Germany, he repeated Huygens' experiment and invented the principle of the steam engine, but the model he invented did not get beyond the experimental stage. He could think of many possible applications, but did not have enough capital and was forced to abandon his project to construct a steam ship, and he died in poverty.

Blaise Pascal
(1623–62)

Pascal was a child genius. At the age of 11 he wrote a treatise on sound propagation and at 16 a treatise on conic sections. In 1639, to help his father, he made an adding machine. In 1648 he carried out the famous demonstration of atmospheric pressure at the top of the Puy-de-Dôme in France. He was converted to Jansenism and, suffering from a serious illness, retired to Port-Royal. He spent the last four years of his life writing *An Apologia for the Christian Religion*, better known under the French title of *Pensées*.

Louis Pasteur
(1822–95)

Pasteur was born in the French Jura Mountains and studied at the college of Besançon before being accepted into the Ecole Normale in Paris in 1843. He was a doctor of physics and chemistry and became famous with the publication of his dissertation on crystallography. From 1862 he devoted himself to the study of micro-organisms, and his research on rabies, and how to prevent it, gave rise to violent polemics but led to the development of the vaccine for which he is remembered.

Wilhelm C. Röntgen
(1845–1923)

Röntgen was Professor of Physics at the University of Würzburg in Bavaria when, by chance, he discovered X-rays in 1895. While carrying out an experiment with a cathode-ray tube in a darkened room, he noticed that he could see the bones of his hand. He called these unknown rays X-rays. His discovery, for which he was awarded the Nobel Prize in 1901, revolutionised medicine by making it possible to see inside the body. Röntgen refused to patent the X-ray and died in poverty in 1923, a victim of the economic crisis.

Jonas E. Salk
(born 1914)

Having studied medicine and bacteriology in New York, Salk began his research into the 'flu virus in 1942. Later, he began to work on discovering a serum against poliomyelitis and, in 1953, announced that his experiments had been successful. His anti-polio vaccine was used extensively and produced excellent results, as did his 'flu vaccine which gives protection for a period of more than two years. Today, Salk is working on the Aids virus, with very encouraging results which he presented to a recent congress in Montreal.

Ignaz Philipp Semmelweis
(1818–65)

In 1844 the Hungarian Ignaz Semmelweis was working as an assistant doctor in the maternity section of the Vienna Hospital when he discovered the infection process. To prevent this problem he invented rules for medical and surgical hygiene. As a result of a rigorously enforced asepsis in his clinic, the mortality rate fell from 18 per cent to 1.2 per cent. In spite of the evidence of these results, he was confronted by the indifference and even the hostility of the medical world. He wrote and published a great deal in defence of his theory, but, by a cruel stroke of fate, he died of an infection in 1865.

Alessandro Volta
(1745–1827)

Born in Como in 1745, where he subsequently became Professor of Physics, Volta devoted himself to research into electricity. In 1793 he had the idea of developing the first electric cell, following work by Galvani on the contractions of the muscles of a frog on contact with two different types of metal, which suggested an electric current. The first successful demonstration of his cell was in 1800.

Wilhelm Röntgen, the discoverer of X-rays.

James Watt
(1736–1819)

In 1764 Watt was responsible for the maintenance of physics equipment at the University of Glasgow. While repairing a Newcomen engine, he noticed that it had a major defect and developed an improved model. In 1765 he invented the condenser and, after making several improvements, constructed the double-acting engine. He went into partnership with the owner of a factory and enjoyed immediate success in all areas of industry.

Eli Whitney
(1765–1825)

In 1792, while staying in Georgia, USA, Whitney had the idea of a machine to gin cotton which would considerably speed up the grading of the fibre. The machine, which played a major part in the development of the cotton industry, did not make Whitney's fortune. Although his machine was patented, it was widely copied and he was forced to start proceedings many times in order to protect his rights. However, he did make his fortune by manufacturing rifles with parts that were standardised and therefore interchangeable following an order from the army for 10000 which was fulfilled in 1801.

Wilbur Wright
(1867–1912)
Orville Wright
(1871–1948)

Orville and Wilbur Wright did not complete their secondary education, but in 1892 they opened a workshop for the repair and manufacture of bicycles. It was a bicycle, to which they had attached wings and an engine, that provided the model for a 12 hp biplane in which Orville made the first flight in December 1903. They then set about making an improved machine, the *Flyer*, that they exhibited in France in 1908. The next year they received an order from the American Army for the first military aircraft. In 1912 Orville only just survived an attack of typhoid which killed his brother Wilbur.

Vladimir Zworykin
(1889–1982)

After serving in the Russian Army as a radio officer during the First World War, Zworykin emigrated to the United States where he patented the iconoscope in 1923. The American company RCA, which appointed him a director in 1929, invested several million dollars in developing this invention which enables a quicker and clearer transmission of television pictures. Zworykin also developed the electron microscope and contributed to the development of colour television.

Orville Wright in 1922 on an aeroplane named after his brother Wilbur.

INDEX OF INVENTORS

A

Aaron, Danielle, 179
Abbott, Bob, 121
Abbott Laboratories, 91
Abbott, Scott, 206, 207
Abel, Niels Henrik, 53
Abelson, P. H., 64
Abulcasis of Cordoba, 111
Achenwall, G., 53
Ackeret, 162
Ackroyd-Stuart, H., 159, 161
ACS, 7
Adamadama, Randy, 225
Ader, Clément, 17, 236
Adobe Systems, 117
A.E., 3
AEG-Telefunken, 239
Aérospatiale, 18, 19, 22, 47
Aganessian, Y., 63
Agfa, 256
Agricola, 189
Agriculture Canada, 130
Agusta, 47
Aiken, H. H., 115
Ainsworth, Gregory, 225
Airship Industrie, 17
Akutsu, Fumiko, 198
Alan Leibert Associates, 201
Albaret, 128
Aldus, 119
Alexopoulos, K., 150
Alfa-Laval, 133
Alfvén, H., 156
Allen, Paul, 118
Allison, D. J., 211
Alter, Hobart L., 213
Amdahl, Gene, 116
American Express, 200
Ampeg, 248
Ampère, A. M., 56, 61, 263
Ampex, 238, 240
Amsterdam, Diana, 226
Amstrad, 122
Amyan, Claudius, 98
Amynthas of Alexandria, 98
Analytical Instruments Ltd, 201
Analytics Inc., 126
Anaximander of Miletus, 147, 148
Anderson, C., 238
Anderson, C. D., 65
Anderson, Gary, 132
Andrews, Thomas, 59
Angel, Rob, 206, 207
Anschutz, 23
Apolonius of Perga, 51
Appert, Léon, 174
Appert, Nicolas, 156
Apple, 122
Appleby, John F., 128
Arad-Manai, 97
Arago, F., 61
Archer, Henry, 240
Archer, Scott, 254
Archigenes, 112
Archimedes, 54, 56, 57, 153, 167, 168, 263
Archytas of Tarentum, 168
Argand, A., 189
Argand, J.-R., 51
Aristaeus, 133
Aristarchus of Samos, 50, 55
Aristophanes, 228
Aristotle, 147, 149
Arlandes, Marquis of, 16
Armati, Salvino degli, 109
Armengaud, 162
Armstrong, Edwin H., 236, 237
Aron-Rosa, Danièle, 109
Artsimovitch, L. A., 165
Arzco Medical Electronics, 96
Asahi Glass Co., 144
Ashton-Tate, 118
Aspdin, J., 174
Aston, W. F., 70, 164
AT&T, 242, 243, 244
Atari, 122, 206
Ateliers et Chantiers du Havre, 26
ATG, 125
Athotis, 97
Atkinson, Bill, 118
Auenbrügger, L., 89
Auth, David, 95
Avery, O. T., 71
Avogadro, Count, 62, 63
Avro Research Laboratories, 156

B

Babbage, Charles, 114
Babington, B., 92
Babolat, Pierre, 220
Back, Frank, 253
Baekeland, H., 69
Bailey, Jeremiah, 128
Bain, A., 175, 229
Baird, John Logie, 237, 239, 263
Baldet, Patrick, 130
Bally, Timothy, 178
Barber, John, 162
Barberi, 246
Barbier, Claudio, 221
Bardeen, J., 119, 237, 264
Barker, 69
Barnack, Oskar, 257
Barnard, Christiaan, 99, 264
Barnett, William, 163
Barraquer, Prof., 109
Barron, Ian, 121
Barron, Robert, 192
Barton, John, 154
Battelle Institute, 127
Baxter Laboratories, 92
Bayard, Hippolyte, 254
Bayer, Otto, 69
BBC, 237
Beach, Hamilton, 198
Beart, Robert, 157
Beaumarchais, 175
Beck, Léon, 135
Becker, Allen, 238
Becquerel, Antoine, 167, 264
Becquerel, Henri, 63
Bedford, B., 3
Bednorz, J. G., 62
Beechcraft, 21
Behring, E. von, 94
Beidler, G. C., 203
Beiersdorf, Paul, 194
Beijerinck, M. W., 73
Békésy, Georg von, 110
Belairbus, 19
Bell, Alexander Graham, 241, 248, 264
Bell Aircraft Corp., 22, 23, 42, 43
Bell, Chichester, 242
Bell, J., 79
Bell Laboratories, 61, 62, 123, 124, 247
Bell Telephone Co., 243
Benedictus, E., 5, 171
Benitz, Earl, 199
Bennett, Bill, 215
Bentall, E., 213
Benton, Stephen, 179
Benz, Karl, 158, 159, 161, 163, 264
Berger, Franka, 194
Berger, Hans, 90
Bergès, Aristide, 155
Bergheim, J. S., 253
Berliner, Emil, 247, 265
Bernard, Noël, 136
Bernoulli, Jean, 54
Berson, Salomon, 90
Berthollet, C. L., 191
Berzelius, J. J., 68
Bessel, F., 79
Bessemer, Sir Henry, 175, 223
Best, 130
Bethe, Hans, 80, 165
Betti, Oswaldo, 98
Bettosini, Ambrosio, 219
Bevan, Edward J., 69
Beyl, Jean, 219
Bic, 194
Bich, Baron, 202, 230
Bickford, William, 34
Biesenberger, H., 98
Bigsby, Paul, 246
Binet, J. P., 99
Binning, Gerd, 90
Biotronik, 96
Birdseye, C., 184, 202, 265
Biro, Laszlo, 230, 265
Bismuth, H., 99
Bissel, Melville R., 190
Bissell, G. H., 158
Black, S. D., 199
Blackton, J. Stuart, 261
Blaupunkt, 236
Blech, Ilon, 67
Blenkinsop, John, 11
Blickensderfer, C. C., 229
Bloch, F., 61, 64
Blumlein, A. D., 247
Blyth & Co., 206
BMW, 10
Bocage, André, 91
Boeing, 18, 19, 20
Boge, 4
Bohr, Niels, 56, 63
Bollée, Amédée, 153
Bolyai, Janos, 52
Bombardier, J. A., 9
Bond, Alan, 81
Bonnet, J. P., 235
Bonneville, Louis, 4
Boole, George, 53, 115
Booth, A. D., 125
Booth, Hubert C., 190
Booth, Oliver, 216
Borden, Gail, 186
Borel, E., 54
Borel, J.-F., 98
Borriello, Vincenzo, 224
Bosch, 4, 10, 239
Bouet, 104
Boulanger, Pierre, 1
Bouton, G., 161
Boxer, 30
Boyle, David, 185
Boyle, R., 58, 106, 166, 190
Braham, R. R., 146
Brahe, Tycho, 57
Braille, L., 235, 265
Bramah, Joseph, 27, 172, 192
Bramani, Vittorio, 221
Branderburger, J. E., 69
Brandt, Edgard, 32
Brandt, Georg, 67
Branemark, P. I., 111
Branly, Édouard, 236
Brattain, Walter, 119, 237
Braun, K. F., 61, 237
Braun, W. von, 36, 161
Brayton, 160
Brearley, Harry, 173
Breguet, Louis, 158
Brendel, Walter, 99
Brett, Jacob, 241
Breuner, Don, 119
Brewer, 240
Brewster, Sir David, 256
Bréziers, M., 138
Bricard, A. and J., 192
Bricard, Eugène, 192
Bricklin, Dan, 118
Briggs, Henry, 54
British Aerospace, 19, 85
British Rail, 14
Broglie, Louis de, 56, 63
Brongniart, A., 67
Brown, Garrett, 256, 260
Brown, Robert, 71
Brown, Samuel, 159
Browning, J. M., 31
Brownrigg, William, 184
Brucë, D., 93
Brunel, Isambard Kingdom, 11, 169
Brunet, A., 172
Brunet, Jean-Louis, 106
Bruno, Giordano, 57
Brunot, James, 207
Buchanan, B. G., 119
Büchi, A., 159
Buckett, F. M., 173
Buckley, John C., 25
Budding, 135
Budin, 103
Buecherl, E. S., 95
Buehler, William, 177

Bull, Lucien, 259
Burke, John, 99
Burton, Luke, 178
Busbecq, Augier de, 136
Busch, Hans, 90
Buschmann, C. F. L., 246
Bushnell, David, 37
Bushnell, Nolan, 206
Byers, M. H., 146

C

Cabasse, 251
Cabrera, Blas, 61
Cadolle, Herminie, 197
Cahill, T., 229
Cahn, John, 68
Cailler, F.-L., 183
Cailletet, L.-P., 59
Cai Lun, 228
Callinicus, 34
Camm, Sir Sidney, 43
Canon, 125, 203, 256, 257
Cantor, Georg, 53
Capek, Karel, 127
Caquot, Albert, 168
Cardan, Jérôme, 52
Carlier, François, 191
Carlos, W., 246
Carlson, Chester, 203
Carnot, Sadi, 59, 265
Carothers, W., 69, 195, 265
Carpentier, A., 99
Carré, Edmond, 185
Carrel, Alexis, 72
Carrier, Denis, 133
Carrier, Willis, 188
Carswell, 106
Cartan, Henri, 54
CASA, 19
Case, 128
Casio, 230, 248
Cavanilles, 136
Cavendish, Henry, 58, 166
Caventou, J., 101, 108
Caxton, William, 230, 232
Cayetty, Joseph, 193
Cayley, Arthur, 53
Cayley, Sir George, 117, 265
CBS, 247
CdF Chimie, 141
Celsus, 108, 111
Chadwick, James, 65
Chaillou, A., 94
Chalmers, James, 240
Chamberlen, P., 103
Chambers, Arthur, 216
Champollion, J. F., 234
Chang Heng, 150
Chaplin, D. M., 167
Chaplin, G. B. B., 178
Chapman, Dallas, 187
Chappe, Claude, 240, 265
Charles, Jacques, 16, 59
Charles II, 213
Charlton, John P., 240
Charron, G. V., 33
Chaussy, Christian, 99
Chen, Steve, 123
Chevalier, 252
Chilvers, Peter, 218
Christiansen, O. K., 206, 207
Chu, Paul, 62
Chukov, V., 157
Ciba Geigy, 104, 107
Cierva, Juan de la, 22
Citroën, 4
Clair brothers, 31
Clapeyron, Émile, 59
Clark, A. R., 79
Clark, Marione, 7
Clark, Steven, 105
Claude, Georges, 59, 61, 189
Clément, 193
Clerck, Dugald, 160
Clifford, Thomas, 199
Clowes, Garth A., 13
CNRS, 132, 178
Cochot, J. J., 172
Cockerell, Sir Christopher, 25
Cohen, Samuel, 34
Colburn, Irving, 174
Collen, 94
Colmerauer, Alain, 118
Colt, Samuel, 31
Colvin, 133
Compton, A. H., 34, 66
Congreve, William, 160
Conté, J.-N., 229
Coolidge, William D., 189
Cooper, 220
Cooper, Geoffrey, 73
Cooper, I. S., 99
Copernicus, Nicolaus, 55, 57
Cor, John, 187
Corbato, F., 116
Cormery, 9
Corning, J. L., 97
Corning Glass, 174
Cornu, Paul, 22
Cort, Henry, 176
Coulomb, Ch. de, 60
Coupleux, 247
Courtet, Émile, 261
Cousteau, J.-Y., 149, 260
Cowan, G. A., 65
Craig, Eliza, 388
Crampton, T. R., 11
Crane, 178
Cray, Seymour, 123
Creative Programming Inc., 224
Crick, F. H. C., 71
Cristofori, B., 245
Crompton, R. E. Bell, 192
Crompton, Samuel, 77
Crookes, William, 61
Croskill, William, 128
Cross, Charles, 69, 70, 256
Crow, Bruce, 150
Ctesibius of Alexandria, 248
Cugnot, J.-N., 102
Cullen, Gerald, 242
Cullen, William, 185
Curie, P. and M., 63, 265
Cushing, Harvey, 98
Cuthbert, J. A., 199
Cutte, H., 175
Cuvillier, R., 253
Cyberg, 127
Cyrus the Great, 241

D

Daguerre, J. L. M., 253, 266
Daikin, 121
Daimler, Gottlieb, 1, 8, 161
Daimler-Benz, 4
Dallmayer, Thomas, 252, 253
Dalton, John, 62
Damadian, R., 92
Darby, Abraham, 170, 175
Darby, Newman, 218
Darrow, Charles, 208
Darwin, 72
Dassler, 210
Daughter, Charles, 96
Dausset, Jean, 73
Davenport, F. S., 128
David, Loïc, 219
Daviel, J., 109
Davies, K. H., 124
Davis, 33
Davy, Sir Humphry, 96, 176, 189, 266
Dawson, 178
Decaux, 259
Decca, 239
Decker, Alonso G., 199
Decroix, 178
Deep Ocean Engineering Co., 150
Deere, John, 128
De Forest, Lee, 236
De Havilland, 18
Delamare-Debouteville, E., 160, 161
Delambre, J.-B., 54
Delens, Gérard, 70
Delessert, Benjamin, 187
Della Porta, G., 252
Delvigne, Henri-Gustave, 29
Demian, Cyril, 246
Demling, Ludwig, 99
Denayrousse, A., 149
Denis, J., 94
Denner, J.-Ch., 245
Deprez, Marcel, 156
Desaguliers, Jean, 60
Descartes, René, 51, 110, 252
Desgranges, Henri, 261
Desmaret, Nicholas, 171
Deutsch, Ralph, 246, 247, 248
Deutsche Bundesbahn, 14
Deutsches Erdöl, 158
Deverell, R. S., 168
Dewar, Sir James, 59, 187
Diablo, 125
Diagnostic Pasteur, 90
Dickson, William, 258
Dictaphone, 250
Diesel, Rudolf, 162, 266
Dion, Albert de, 8, 161
Dirac, Paul, 56, 61, 65
Dolby, Ray, 238, 251
Donkin, Bryan, 181, 229
Doppler, Christian, 147
Dorn, Ernst, 58
Dornberger, Gen. Walter, 36
Doubleday, Abner, 211
Douez, Françoise, 185
Douglas, 18
Dowsing, H. J., 192
Dowson, Emerson, 5
Doyle, Mike, 218
Drais von Sauerbronn, Karl von, 6
Drake, Alfred, 163
Drake, Jim, 218
Drebbel, C., 36
Drela, Mark, 20
Drew, Dick, 202
Dreyse, J. N. von, 30
Drinken, Philip, 96
Dubois de Chemant, 110
Dubosq, 256
Dubus-Bonnel, I., 176
Ducos du Hauron, Louis, 256
Duddell, 249
Dudley, H. W., 124
Dumas, 207
Dumont, Ben, 216
Dumont, Frank, 216
Dunlop, John Boyd, 5, 267
Dunwoody, 236
Du Pont, 67, 69, 174, 194
Dupuy de Lôme, St., 36, 62
Durand, Pierre, 181
Dutrochet, René, 71
Dynes, Robert, 62

E

Eastman, George, 254, 267
Eckert, J. P., 115, 116
EDF, 143
Edgerton, H., 254
Edison, Thomas, 155, 156, 176, 189, 236, 247, 259, 267
Edoux, Léon, 190
Edwards, R., 103
Egberts, E., 178
Eiffel, Gustave, 173
Einstein, A., 55, 56, 57, 66
Electrolux, 143
Elf, 10
Elf-Aquitaine, 141, 157
Ellis, W. W., 211
Elshots, 106
EMI, 247
Empedocles Agrigentum, 150
Encil, George, 219
Engelbart, Doug, 124
Eötvös, Lorand, 57
Erasistratus, 96, 111
Eratosthenes, 57, 148
Ericsson, John, 27, 160
Ericsson Co., 242
Eriksen, Stein, 219
ESA, 78, 84, 87, 88
Espitallier, George, 173
Esselte, 201
Essex, Max, 93
Estridge, Philip, 122
Ethyl Corp., 68
Eton, William, 97
Euclid, 50
Eudosius, 55
Euler, 54
Eutelsat, 244
Evans, Olivier, 185
Evinrude, Ole, 27
Eye Dentify, 200, 201

F

Fage, J.-M., 147
Fahlberg, Constantin, 187
Falabella, Patrick, 132
Fallopia, Gabriele, 102

Faraday, Michael, 59, 61, 156, 241
Farina family, 194
Faucher, Alfred, 5
Favre, A., 246
Fawcett, 70
Feigenbaum, E. A., 119
Fenaille, Pierre, 4
Ferguson, Graeme, 260
Ferguson, H., 3
Fermat, P. de, 51, 52
Fermi, Enrico, 34, 56, 164
Ferrabee, 135
Ferraris, Galileo, 160
Ferro, Scipione del, 52
Ferruzi, 142
Fessenden, R. A., 242
Feynmann, Richard, 56
Fichet, Alexandre, 192, 201
Fidelity Electronics, 205
Field, Cyrus, 241
Finlay, C., 93
Fiocco, G., 147
Fischbach, 56
Fischer, 33
Fischer, B., 52
Fischer, Rudolf, 256
Fisher, Alva J., 192
Fisher, Gary, 222
Fisher Research Laboratory, 198
Fitch, John, 24
Fitzroy, Admiral, 146
Fleischel, Gaston, 4
Fleischer, J. F., 68, 261
Fleischmann, Martin, 165
Fleming, Alexander, 104, 267
Fleming, John A., 236
Flobert, 30
Flosdorff, E. W., 188
Flymo, 135
Focq, N., 172
Fokker, 20, 47
Foldes, Peter, 126
Foljambe, Joseph, 128
Folli, Francesco, 94
Fondeur, 128
Fontana, Niccolo, 52
Ford, 145
Ford, Henry, 1
Forest, Fernand, 161
Forster, Max von, 33
Forsyth, Alexander J., 29
Fosberry, G. V., 31
Fosbury, Richard, 210
Foster, Len, 131
Föttinger, 2
Foucault, Émile, 174
Foucault, Léon, 23, 61
Fowler, John, 128
Fox, Uffa, 213
France, P., 105
Francis, James, 155
Franklin, Benjamin, 60, 109, 268
Frankston, Bob, 118
Fréchet, Maurice, 54
French Commission for Atomic Energy, 127
Frenkel, Michel, 145
Freud, Sigmund, 108
Frézier, François, 138
Friedel, 68
Friis, Harold T., 242
Frost, George, 237
Froude, William, 57
Fry, Arthur, 202
Fuji, 257
Fujinaga, 134
Fuller, C. S., 167
Fulton, Robert, 24, 37
Funakoshi, Gichin, 216
Funakubo, Hiroyasu, 127
Funk, C., 109
Furness, Tom, 126

G

Gaedicke, Johannes, 254
Gagarin, Yuri A., 83
Gaggia, 182
Galileo, 55, 56, 57, 74, 268
Galton, Sir Francis, 201
Galvani, Luigi, 60
Gamblin, Rodger L., 229
Garcia, Rafael, 86
Gardena, 135
Garnerin, A.-J., 24
Garros, Roland, 40
Garvin, Charles, 250
Gates, Bill, 117, 118
Gaulard, Lucien, 156
Gauss, J. K. F., 52
Gautheret, Roger, 72
Gautier, Daniel, 87
Gavinsky, Robert M., 20
Gay-Lussac, J.-L., 59, 62
Geiger, Hans, 62, 63
Geissler, Heinrich, 61
Gell-Mann, M., 65
Genentech, 94
General Dynamics, 36, 46
General Electric, 32, 67, 68, 126, 184
Geobra-Brandstätter, 209
Gerhardt, Charles, 104
Gernelle, François, 122
Gibbs, C., 107
Gibbs, J. D., 156
Gibson, 70
Gienenberg, 163
Giffard, Henri, 16
Gigax, Gary, 206
Gilbert, William, 60
Gillette, 230
Gillette, King Camp, 194
Gindrey, 99
Giova, Flavio, 26
Givelet, 247
Glark, 94
Glas, Louis, 247
Glashow, Sheldon, 56
Gliden, Joseph, 132
Glock, Gaston, 32
Glouchko, Valentin, 268
Goddard, R. H., 80, 160, 271
Godowsky, Leopold, 256
Golden, Rob, 112
Goldenmark, Peter, 248
Goldstein, E., 65
Gollan, 94
Gondard, Maurice, 163
Goode, M., 221
Goodyear, Charles, 69
Gorrie, John, 185
Gould, R. Gordon, 179, 271
Grafenberg, Ernst, 103
Graham, George, 175
Gramme, Zénobe, 155, 271
Granhed, Magnus, 219
Grassmann, H., 51
Grataroli, 221
Gratias, Denis, 67
Gray, Elisha, 241
Gray, Stephen, 60
Gray, William, 242
Green, Howard, 99
Green, M., 56
Greenberg, 65
Grégoire, Jean, 4
Grégoire, Marc, 69, 186
Griess, R., 52
Griffloen, 250
Grotefend, G. F., 234
Grumman, 43, 85
Grundig, 238
Grüntzig, Andreas, 96
GTE, 35, 179
Guericke, Otto von, 58, 60, 153
Guido of Arezzo, 248
Gunson Ltd, 158
Gurevich, G., 41
Gurney, Sir Goldsworthy, 154
Gutenberg, Beno, 150
Gutenberg, Johann G., 230, 233, 271
Gutte, 73

H

Haarman, W., 188
Haber, Fritz, 66
Hadley, John, 27
Haensel, V., 157
Hageman, Geert, 137
Hagstrom, 248
Hahnemann, C. S., 105
Hainsworth, M., 234
Haldane, T. G. N., 167
Hale, G. H., 75
Halkett, 128
Hall, Charles Martin, 66, 176
Hall, John, 181
Halley, Edmund, 149
Halpern, 104
Halsted, 97
Hamilton, Thomas S., 20
Hammond, Laurens, 247
Han Si, General, 40, 205
Hancock, Walter, 15
Haney, Chris, 206, 207
Haney, John, 206, 207
Hansen, Malling, 229
Harel, Marie, 184
Harington, Sir John, 191
Harrington, 112
Harrison, James, 185
Harrison, John, 175
Harrison, R. G., 72
Harryhausen, Ray, 261
Harvill, L., 124
Harwood, J., 175
Hary, J. D., 99
Hasegawa, Goro, 205
Hass, Earl, 102
Hayakawa, Tokuji, 230
Heathcoat, John, 128
Heathkit, 127
Heinkel, Ernst, 42
Heinz, Henry, 186
Heisenberg, W., 63
Helmont, J. B. Van, 57
Helvig, 206
Hennebique, F., 174
Herbin, Jean-Pascal, 127
Hero of Alexandria, 153
Herophilus, 96
Héroult, Paul-Louis, 66, 176
Herreshof, F., 213, 215
Herschel, W., 75, 252
Hertz, H. R., 61, 236
Heusquin, M., 135
Hevelius, J., 75
Higgs, Thomas, 177
Higonnet, René, 233
Hilbert, David, 54
Hildebrand, H. and W., 8
Hill, Henry, 229
Hill, Robert, 253
Hiller, M., 125
Hipparchus of Nicea, 50
Hippocrates, 72
Hitachi, 239
Hoe, R., 230
Hoëgh, 253
Hoff, Marcian E., 119
Hohmann, W., 87
Holden, Colonel, 9
Hollerith, H., 114, 115
Hollister, Donald, 189
Holmuth, Fred, 152
Holzwarth, H., 162
Home Diagnostics Inc., 91
Homma, Dai, 178
Honda, 9, 10
Honold, Gottlieb, 163
Hooke, Robert, 71
Hornby, Frank, 206, 207
Hounsfield, Sir G. H., 91
Hovelian, Krikor, 125
Howard, 184
Hoyle, E., 204
Hulls, Jonathan, 24
Hulse, 80
Hume, Ian, 224
Humphrey, Cliff, 143
Hunley, Horace L., 38
Hunt, A., 94
Hunt, Seth, 195
Hunt, Walter, 192, 195
Hutchinson, M. Reese, 110
Huygens, Christian, 53, 159, 175, 271
Hyatt Bros., 69
Hystar Aerospace, 17

I

IBM, 62, 116, 123, 125, 230
IBMP, 132
Ichbiach, Jean, 118
ICOT, 123
Identix, 201
IFREMER, 149
I Hing, 175
Illmensée, 71
ImageData, 229
Imhotep, 168
Indian, 10
Inmos, 121
Intel, 120, 121
Inter, 121
International Hard Suits, 149
International Harvester, 130
Ipp Wing, 87

Isaacs, A., 105
Isaacson, M., 125
ISAS, 78
Issigonis, Sir Alex, 1
Italcable, 242
Ivanovsky, D., 92

J

Jacobi, M. H. von, 159
Jacquard, Joseph-Marie, 114, 124, 271
Jacques, John, 221
Jakob, Rainer, 191
Jansk, Karl, 76, 94
Janssen, Hans, 89
Japanese National Railways, 14
Jarvik, Robert, 95
Jas, Raveno, 194
Jayaraman, G., 105
Jeffreys, Alec J., 72
Jenkins, 97
Jenner, Edward, 93
Jessop, William, 11
Jobs, Steve, 119, 122, 123
John Shearer and Sons, 128
Johnson, H., 128
Joliot, Frédéric, 64, 108
Joliot-Curie, Irène, 108
Jolly, J.-B., 190
Josepho, A. M., 255
Joule, J. P., 60, 61
Judson, Whitecomb, 196
Julliot, Henri, 40
JVC, 238, 239, 250

K

Kamerlingh, Onnes Heike, 59, 60, 62
Kamo, Kosuke, 152
Kanny, Louis, 249
Kano, Jigoro, 216
Kaplan, Viktor, 155
Kashiwase, Toshio, 127
Kaulmann, Dr, 256
Kawase, Taro, 127
Kay, John, 177
KB-Pillow Co., 225
Kégresse, Adolphe, 3
Keinath, Gerald, 5
Kellogg, Will K., 183
Kelman, Charles, 109
Kelvin (Lord), 59, 60, 167
Kemeny, John G., 118
Kennedy, J. W., 64
Kenner, 207
Kepler, Johannes, 57
Kessis, Jean-Jacques, 127
Khorama, H. G., 72
Kilbourne, Dave, 215
Kilby, Jack St Clair, 120, 122, 271
Kildall, Gary, 116
Kim, Dave, 226
Kimberley-Clark, 194
Kindermann, F., 160
Kinley, K. T., 158
Kircher, Athanase, 258
Kitasato, S., 94
Klatte, 69
Klein, Christian F., 52
Kleist, D. J. von, 60
Klietsch, Karl, 231
Klingert, 149
Knerr, Richard P., 205
Knopoff, Leon, 150
Knopp, 130
Koch, R., 92, 101
Kodak, 124, 219, 256, 257
Kohler, George, 104
Koken, 127
Kolegov, Alexandre, 228
Kolff, Willem, 95
Köller, K., 97
Konica, 257
Korn, Arthur, 242
Kornberg, Arthur, 73
Korolev, 84
Kremer, Gerhard, 148
Kugelfischer, 220
Kulyeziski, 183
Kurtz, Thomas E., 118
Kwolek, Stéphanie, 69

L

La Condamine, C. M. de, 69
Lacoste, René, 221
Laennec, R., 89
Lafuma, 197
Lakeshore Industrial, 136
Lamarck, J.-B. de, 72
Lambert, J.-H., 52
Lanchester, 3
Land, E. H., 109, 256, 257
Landmark Entertainment Group, 206
Landsteiner, K., 94
Lane, 130
Lang, Franz, 159
Lange, 219
Langen, Eugen, 160
Langevin, Paul, 35
Langmuir, Irving, 143
Lansing, Lawrence A., 222
Lanston, Tolbert, 233
Laposky, Ben F., 125
Larderel, F. de, 166
Larsen, Bent, 150
Latil, Georges, 3
Laubeuf, Maxime, 38
Laudet, 259
Lauterbur, P. C., 92
Laval, G. de, 133, 162
Laveran, A., 93, 101
Lavoisier, A.-L. de, 58, 167
Lawrence, 133
Lawrence Livermore National Laboratory, 179
Lawson, John, 165
Le Baron Jenney, William, 170
Leder, Philip, 73
Lederberg, J., 119
Lee, William, 178
Leeuwenhoek, A. Van, 71, 90
Lefèbvre, André, 1, 4
Leibnitz, G. W. von, 53, 54, 113
Lelièvre, René, 197
Lemaire, Jean, 99
Lemoine, Louis, 172
Lemoine, Roger, 197
Lenoir, Étienne, 160, 163
Lenthéric, 198
Leonardo da Vinci, 110, 252
Lepivert, Patrick, 199
Le Prieur, 149
Le Roy, 259
Lethbridge, John, 149
Letourneur, 99
Le Verrier, Urbain, 146
Levine, 94
Levine, Alexandra, 107
Lewis, Gilbert N., 67
Lexcen, Ben, 28
Libby, Willard F., 64, 175
Liebau, Gustave D., 4
Lilly, Simon, 80
Linde, Karl von, 59, 185
Lindermann, J., 105
Lindt, James, 137
Lippmann, Gabriel, 255
Little, C. G., 147
Loar, Lloyd, 246
Lobatchevski, N. I., 52
Lockheed Aircraft Corp., 43, 44
Lodge, Sir Oliver, 236
Loewy, Raymond, 85, 86
Long, C. W., 96
L'Orange, Prosper, 159
L'Oréal, 198
Lorin, 161
Lorito, 199
Lotus, 118, 119
Lovelace, Ada, Countess, 114, 115, 118
Lower, R., 94
Lumière, A. and L., 256, 258, 272
Lunn, Arnold, 218
Luppis, Captain, 39
Luria, Salvador, 72
Lwoff, André, 73
Lynch, Bernard, 24

M

MacGhee, Robert, 127
Mach, Ernst, 57
Mackintosh, C., 196
Macmillan, Kirkpatrick, 7
MacPherson, John, 240
Maddox, Richard L., 254
Madsen, 32
Maelzel, Johann, 209, 246
Magalhaens, 230
Magee, Carlton, 6
Mahler, Ernest, 102
Maiman, T. H., 179
Majima, Hiroshi, 225
Malandin, L., 160, 161
Malewicki, D. J., 220
Malske, Johan Georg, 13
Manby, George, 191
Maness, W., 112
Mannes, Leopold, 256
Manning, William, 128
Manson, P., 93
Man Te Seol, 7
Marbach, André, 141
Marchetti, Charles, 22
Marchoux, E., 94
Marconi, Guglielmo, 236, 272
Marey, Jules, 258
Mariotte, Edme, 58
Marius, Armand, 219
Markle, Wilson, 259
Markrum Co., 243
Marrison, Warren Alvin, 175
Martin, Alphonse, 252
Martin, Gardner, 212
Martin, J., 73
Martin, L., 94
Marty, Raul Espinosa, 193
Massey, Edward, 27
Massey-Harris, 130
Masson, Antoine, 158
Masson, Claudine, 70
Masuda, Shigenori, 227
Mathijsen, Antonius, 97
Matsushita, 225, 230, 237, 238
Mattel, 206, 209
Mauchly, J. W., 115, 116
Maudslay, Henry, 172
Maurer, Edouard, 173
Mauser, 32
Mauser, Paul, 30
Maxfield, Joseph, 247
Maxim, Hiram P., 31, 32
Max-Planck Institut, 168
Maxson, 184
Maxwell, James Clerk, 56, 61, 236, 256
Maybach, Wilhelm, 1, 8, 158, 159, 161
Mayer, Julius Robert von, 59
Mayor, 94
Mazzitello, Joe, 240
McAdam, John, 6
McAllister, L. G., 147
McCarthy, J., 118
McCormick, Cyrus, 130
McCready, Paul, 19
McDonald, M. and R., 186
McDonnell Douglas, 36, 46, 47, 163, 179
McMillan, E. M., 64
Méchain, Pierre, 54
Mège-Mouriès, H., 188
Melchers, Georg, 138
Méliès, Georges, 259, 261
Melvin, Arthur K., 205
Mendel, Gregor, 71, 72
Mendeleïev, Dimitri, 63
Mensiga, 102
Mercator, 148
Merck and Co., 107
Merganthaler, O., 233
Merino Wool Harvesting, 127
Merlin, Joseph, 218
Merrifield, Bruce, 73
Merrill, J. P., 99
Merryman, J. D., 122
Mesmer, F. A., 108
Messerschmidt-Bölkow-Blohm, 47
Messier, Georges, 3
Messinger, Nicholas, 138
Mestral, Georges de, 197
Mésué, Major, 111
Metchnikoff, I., 73, 94
Metropolitan Vickers, 162
Michaux, Pierre Ernest, 7
Michelin, 5
Michelson, 56
Micropro, 118
Microsoft, 116, 117, 206
MicroSolutions, 123
Mietke, Adolf, 254
Mikimoto, Kokichi, 134

Mikoyan, A., 41
Miller, Patrick, 215
Millet, 160
Milstein, César, 104
Mincom, 240
Minié, Claude-Etienne, 29
Minolta, 257
Mintrop, D. L., 157
Mir International, 6
Mirovski, 96
MIT, 90, 126, 179, 238
Mita Tsutomu, 127
Mitsubishi, 46, 150, 201, 250
MMB, 19
Mock, E., 202
Mohr, L., 103
Moloney, Richard T., 205
Molun, 7
Mondragon Général, 31
Monod, Robert, 97
Monroe, David, 242
Montagu, John, 188
Montgolfier, E. and J. de, 16, 153
Moog, Robert, 246
Morel, G., 73
Morey, Tom, 217
Morgan, T. H., 71, 72
Morgan, William G., 211
Morita, Akio, 251, 272
Morley, 56
Morris and Salom, 2
Morrisson, Fred, 205
Morse, Samuel F. B., 24, 272
Motorola, 121
Mouillé, René, 22
Moyroud, Louis, 233
Müller, J., 202
Müller, K. A., 62
Müller, Richard, 158
Murdoch, Thomas, 150
Musschenbroek, Petrus Van, 60
Muybridge, E., 258

N

Nag-Hyun, Kim, 4
Nairne, Gerald, 185
Naismith, James, 211
Naito, Ryochi, 94
Nakamats, Yoshiro, 124, 226
Nambu, 56
Napier, John, 54, 115
NASA, 77, 78, 86, 88, 141, 143, 244
Nasmyth, James, 172
Natta, G., 70
Nau, Christian, 217
Neddemeyer, 65
Nelson, Alan, 90
Nelson, C. K., 184
Nestlé, 182, 198
Neumann, John von, 115
Neveu, A., 56
Newcomen, T., 154, 159, 272
Newell, A., 118
New Holland, 130
Newton, Sir Isaac, 54, 55, 56, 57, 75, 273
Next, 123
Nicholson, Charles E., 215
Nicot, Jean, 68
Niepce, N., 231, 252, 253, 273
Nikon, 110
Nipkow, Paul, 238
Nippon Kokan, 25
Nippon Telegraph and Telephone Corp., 237
NipponZeon, 177
Nissan, 3, 5
Nobécourt, Pierre, 72
Nobel, Alfred, 34, 273
Noël, Suzanne, 98
Nogues, Henri, 259
Nomicos, K., 150
Nynex, 243

O

Oberth, Hermann, 36, 87, 161
O'Brien, Willis, 261
Odetics, 127
Oehmichen, Etienne, 22
Oersted, Christian, 56, 61, 66
Offenstadt, Eric, 10
Ohlhausen, Howard, 5
Ohm, Georg Simon, 60
Olivetti, 230
Olovson, Gudmar, 195
Olsen, S., 108
Olympus, 250
Oppenheimer, Robert, 34
Optonics, 211
Orata, Sergius, 134
Osterhoudt, J., 181
Ostermeier, 254
Osti, Massimo, 222
Otis, Elish Graves, 15, 190
Otto, Nicolas, 160
Outram, John, 14
Ozun, Gilbert, 99

P

Panasonic, 203, 238
Papin, Denis, 153, 273
Paquette, Daniel, 226
Paracelcus, 58
Parc, Ambroise, 193
Parienti, Raoul, 243
Parissot, Pierre, 201
Parsons, Charles A., 162
Parsons, William, 75
Pascal, Blaise, 14, 57, 113, 274
Paschen, Friedrich, 61
Pasteur, Louis, 73, 93, 274
Pathé, Charles, 259
Paulesco, 106
Pauli, Wolfgang, 65
Paulos, L., 105
Peano, Giuseppe, 54
Pearson, G. L., 167
Péchiney, 7
Pecqueur, O., 3
Pedan, Bob, 150
Peddle, Chuck, 120
Pelletier, J., 101
Pelton, Lester Allen, 155
Pemberton, John, 183
Penguin, 232
Pentel, 230
Penzias, A., 80
Percy, Pierre-François, 97
Peri, Jacopo, 251
Pérignon, Pierre, 183
Perisic, Zoran, 262
Perl, Martin L., 65
Perraud, M., 230
Perrelet, A.-L., 175
Perrin, Jean, 61, 62
Petersson, Bengt, 221
Petty, William, 213, 229
Peugeot, 7
Pfleumer, Fritz, 250
Philippe, A., 175
Philippe, Odette, 138
Philips, 124, 125, 189, 239, 249, 250
Pickard, 236
Pictet, Raoul, 185
Pictet, Raoul-Pierre, 59
Pierce, William, 95
Pilkington, 110
Pilkington, Sir Alistair, 174
Pincus, Gregory, 103
Pioneer, 232, 239
Planck, Max, 54, 55
Planté, Gaston, 155
Plant Genetic System, 131
Plateario, Giovanni, 111
Plateau, Joseph, 258
Plenciz, M. A., 92
Plessinger, John, 220
Plessner, 31
Pliny, 96, 97, 130
Plunkett, Roy J., 69
Plus Development Corporation, 124
Pohlman, Bill, 120
Poincaré, Henri, 54
Poiseuille, J.-L., 89
Polch, 133
Pons, Stan, 165
Popov, Aleksandr, 236
Porsche, Ferdinand, 1
Posselt, R., 68
Poulet, Pedro P., 80
Poulsen, V., 250
Premack, David, 228
Progin, Xavier, 229
Pternitis, C., 90
Ptolemy, Claude, 55, 148
Pullman, George M., 12
Purcell, E. M., 64
Puricelli, Piero, 6
Pythagoras, 50

Q

Quant, Mary, 196
Queensbury, 216
Qui-Lim-Choo, 90

R

Racine Universal Motor, 198
Raichie, 219
Raimi, Sam, 260
Rainey, Paul M., 249
Raman, C., 147
Ramon, Gaston, 93
Ransome, Robert, 128
Ratcliff, C. Wayne, 118
Rateau, A., 160, 162
Rattner, Dan, 132
Rausing, Ruben, 181
Ravenscroft, George, 174
Rawson, 133
Rayleigh, J. W., 147
Raymond, P.-A., 195
RCA, 147, 239
Réard, Louis, 195
Reber, Grote, 76
Redier, Antoine, 175
Reed, Ezekiel, 199
Reichenbach, H., 254
Reid, 52
Reimann, R., 68
Reines, 65
Remington Rand, 125
Renault, Louis, 2
Reno, Jesse W., 191
Revelli, Abiel, 31
Reynolds, 10, 230
Reynolds, Osborne, 57
Ricardo, Henry, 159
Ricardo Consulting Engineers, 2
Richter, Lieutenant, 24
Richter, Charles, 150
Rickenbacher, 246
Rickover, Admiral Hymar G., 38
Ridley, Harold, 109
Riemann, Bernhard, 52, 54
Riesz, Frédéric, 54
Riggenbach, Niklaus, 12
Ritchie, Dennis, 117, 118
Ritty, James J., 202
Robert, Nicolas, 229
Roberts, 33
Roberts, H. Edward, 122
Rock, John, 103
Roe, 98
Roemer, Stephen von, 190
Rogallo, F. M., 215
Rogers, Moses, 25
Rohm, 69
Rohrer, Heinrich, 90
Rolex, 175
Rolls Royce, 43, 163
Röntgen, W. C., 64, 91, 274
Rosen, Arnold, 6
Rosen, G., 105
Rosenberg, S., 105
Rosnay, Arnaud de, 217
Ross, R., 93
Rotheim, Erik, 188
Rouquayrol, Benoît, 149
Roux, E., 94
Rover, 2
Rowland, Thomas F., 157
Rowntree, 182
Rozier, Pilâtre de, 16
Rubbia, Carlo, 56, 66
Rubel, W., 231
Rubik, Erno, 206, 209
Rubin, Colonel, 30
Ruder, Ed., 119
Rudge, Whitworth, 10
Rudolph, 252
Ruhmer, 249
Rumsey, James, 24
Rundölf, 161
Rutan, Burt, 20, 21

Rutherford, Ernest, 62, 63
Ryle, Sir Martin, 77

S

SAAB, 46, 160
Sabine, Thierry, 220
Saccheri, G. G., 52
Sachs, F. G., 130
Sachs, Jonathan, 118
Saft, 156
Saint-Laurent, Yves, 195
Saji, Yoshiro, 26
Salam, Abdus, 56
Salimbeni, A., 94
Salk, Jonas, 107, 274
Salmon, Robert, 128
Saloman, Steen, 150
Salomon, 219
Samuelson, Ralph, 218
Sänger, E., 82, 160
Sanger, F., 73
Santorio, 89
Sanyo, 167
Sarich, R., 158
Satellite Wheel Safety Light, 7
Sater, Mel, 240
Sato, T., 109
Saulnier, Alain, 198
Saulnier, Raymond, 40
Sauria, Charles, 190
Sauvage, Nicolas, 15
Savery, Thomas, 153
Savonarola, G., 111
Sax, Adolphe, 246
Schank, 213
Schechtmann, Dan, 67
Scheele, C. W., 68
Scherk, J., 56
Schick, Jacob, 194
Schickard, Wilhelm, 113
Schleyer, J., 228
Schlimok, Gunther, 91
Schlumberger, 158
Schmidt, Paul, 163
Schmied, Egbert, 99
Schneider, 30
Schneider, Ralph, 200
Schönbein, C. F., 34, 69, 144
Schott, 252
Schrauser, Gerhard, 193
Schröder, 54
Schrödinger, Erwin, 63
Schueller, Eugène, 198
Schüler, 23
Schulze, J. H., 253
Schwab, Pierre M., 250
Schwarz, J., 56
Schwarzmann, E., 98
Schweitzer, 217, 218
Schweppes, 186
Schwinger, Julian, 56
Sclaco, A., 94
Scribonius Largus, 111
Seaborg, G. T., 64
Sear, John, 199
Searle Laboratory, 187
Seebeck, Thomas, 155
Seeley, Henry W., 192
SEGA, 206
Ségalas, P. S., 92
Seifert, Lieutenant, 24
Seiko, 175
Semmelweis, I. P., 97, 274
Semple, William F., 188
Senefelder, A., 230–231
Senning, Ake, 96
SEP, 174
Servier Laboratories, 107
Shakespeare, Scott, 105
Shaw, D., 118
Shearer, John, 128
Shell, 131
Shillibeer, George, 15
Shockley, William, 119, 237
Sholes, C. L., 229
Shore, John, 246
Shrapnel, Henry, 32
Shuler, Kurt, 193
Siebe, 149
Siemens, Werner von, 12
Siemens & Halske, 190
Sikorsky, Igor, 22, 46, 47
Sillam, Bernard, 179
Silliman, B., 156
Simmons, 248
Simms, F. R., 58
Simon, F., 118
Singer, Isaac, 192
Sinsheimer, R. L., 73
Sivrac, Comte de, 6
Skinner, 69
Slagle, J. R., 119
Slaviter, Henry A., 94
Smeaton, John, 174
Smith, Capt., 24
Smith, Kline & French, 107
Smith, Oberlin, 250
Smith, William Henry, 200
Smith, Wilson, 218
Smullin, L. D., 147
Soddy, Frederick, 63, 70
Sogetic, 215
Solaro. Ascanio, 34
Sommelier, Germain, 172
Sommerfeld, Arnold, 63
Sonex Research Inc., 3
Sonic Solutions, 248
Sony, 121, 124, 125, 238, 239, 247, 249, 251, 257
Sozio, 110
Spallanzani, Lazzaro, 132
Spencer, Percy Le Baron, 184
Sperry, Elmer and Lawrence, 23
Spielman, R. J., 211
Spilbury, John, 207
Spill, Bill, 199
Sprague, Frank, 15
Spratt, James, 135
Stair, 138
Staite, W. E., 189
Stampfer, Simon von, 258
Stanley, William, 156
Stanley, W. M., 73
Starley, J. K., 7
Staub, 104
Stein, Claude, 141
Stephenson, George, 11
Stephenson, Robert, 11
Steptoe, Patrick, 103
Sterlin, Charles-Louis, 192
Stevens, John C., 213
Stevin, Simon, 57
Stewart, Timothy, 73
Stewart Automotive Ltd, 6
Stihl, 143
Stimula, 10
Stirling, Robert, 159
Stoffel, Wilhelm, 106
Stokes, 32
Stolze, 162
Stone, J. McW., 236
Stoney, George, 61
Strauss, Benno, 173
Strohmeyer, F., 66
Strowger, Almon B., 242
Strumienski, 168
Sugar, Alan, 122
Sugar Test Instrumental Co., 219
Sulzer, 13
Sun, Simao, 33
Sutherland, I. E., 124
Sutton, Thomas, 252
Svenska, E., 109
Swan, Sir Joseph, 189
Sylver, Spencer, 202
Szilard, Léo, 34

T

Takayanagi, Kenjiro, 237
Talbot, Henry Fox, 254
Tanaka, Kogyo, 199
Tarrantini, 179
Tartaglia (N. Fontana), 52
Taurinus, 52
Tavrison, A. A., 157
Taylor, 80, 252
Tchechett, Victor, 215
Teller, Edward, 34
Tellier, Charles, 184
Temin, H., 93
Terrier, 97
Terrillon, 97
Terrot, 10
Tesei, 40
Texas Instruments, 119, 121, 122, 125, 211
Thal, M.-F., 70
Thalus of Miletus, 50, 60, 147
Thévenot, M., 168
Thimonnier, B., 192
Thomke, E., 202
Thompson, Ken, 117
Thompson, William, 255
Thomson, 26, 239, 249, 255
Thomson, Elihu, 176
Thomson, Sir Joseph, 61, 62, 65, 70
Thomson, R. G., 230
Thomson, William, 59, 60
Thomson-CSF, 35
Thorpe, J., 177
Tiemann, Ferdinand, 188
Tokyo Tire Rubber Lance, 250
Torres, A. de, 248
Torricelli, E., 58, 146
Torschi, 40
Toshiba, 147
Townsend, John, 61
Toyota, 4, 145
Trevithick, Richard, 11, 154
Trouvé, G., 23
Truffault, 10
Trumbull, Douglas, 260
Tukey, John, 116
Tupolev, 19, 21
Tupper, Earl W., 187
Turck, Fred W., 108
Turing, Alan M., 114
Turri, P., 229
Tyndall, John, 147

U

Ueshiba, Morihei, 216
Ugelstad, John, 69
Umpleby, 9
United Service Garages, 3
Urey, H. Clayton, 67

Vadasz, Les, 121
Vaillard, 94
Valeriano, Piero, 138
Valzani, Giovanni, 215
Van Allen, Joseph, 78
Van Breems, M., 215
Van de Haar, A., 193
Van der Meer, Simon, 56, 66
Van Houten, C. J., 181
Van Marum, M., 144
Van Tassel, J. H., 122
Varostos, P., 150
Vaucanson, J. de, 126, 172
Verhoeven, A., 232
Versailles, Lambert de, 25
Vickers-Armstrong, 18
Vierkotter, Paul, 254
Vink, Henk, 9
VISA, 179
Vistakon, 110
Vitagraph, 261
Volkswagen, 1, 3, 146
Volta, A., 60, 166, 275
Volvo, 3, 45, 160
Von Hofmann, A. W., 68
Vries, Hugo de, 73
Vries, R. C. de, 68
Vuillemin, 104

Waaler, Johann, 203
Wahl, C. C., 64
Waldeyer, G., 71
Wales, Nathaniel, 184
Walker, Fred, 218
Wall, William, 60
Waller, A. D., 90
Walsh, 79
Walter, Helmuth, 161
Walton, Patrick, 134
Wampler, R. K., 95
Wankel, Felix, 162
Warnake, 259
Warner Bros., 260
Waterman, Lewis E., 230
Watson, J. D., 71
Watson-Watt, Sir R., 35
Watt, James, 154, 191, 275

Weaver, W., 125
Wedgwood, R., 203
Wedgwood, Thomas, 253
Wegener, Alfred, 148
Wegmuller, William, 219
Weinberg, Robert, 73
Weinberg, Steven, 56
Wermer, Ernst, 250
Werner, Eugene and Michel, 8, 9
Wessel, Caspar, 51
West, 194
Westinghouse, G., 12, 239
Westland, 47
Weymann, 79
Wheatstone, Sir Charles, 175, 256
Wheeler, George H., 191
White, Philip, 72
Whitney, Eli, 172, 275
Whittle, Sir Frank, 42, 162
Whitworth, John, 199
Whitworth, Joseph, 172
Wiencziers, 24
Wiener, 94
Wiener, Norbert, 116
Wigginton, Randy, 119
Wigler, Michael, 73
Wilcox, Howard A., 134
Wilkes, M. V., 118
Wilkinson, John, 172
Willer, Claude, 91
Williams, F. C., 121
Williams, Harry, 205
Wilsdorf, H., 175
Wilson, Major, 2
Wilson, R., 86
Winchester, Oliver, 30
Wingfield, Walter C., 220
Winkler, Clemens, 67
Wirth, Niklaus, 118
Wöhler, Friedrich, 66, 68
Wolfmüller, Alois, 8
Wolkovitch, Julian, 20
Wood, Kenneth, 188
Wood, Walter, 130
Woolf, Arthur, 154
World Promotions, 5
Wozniak, Stephen, 120, 122
Wren, C., 106
Wright, Wilbur and Orville, 17, 275
Wurlitzer, 248, 249
Wynne, Arthur, 207

Y

Yablochkov, P. N., 189
Yacoub, M., 99
Yale, Linus, 192
Yalow, Rosalyn, 190
Yamada, 99
Yamaguchi, Toru, 127
Yamaha, 9, 248
Yannas, Ioanis, 99
Yersin, A., 94, 100
Young, James, 157
Yu Chang Yang, 105
Yukawa, Hideki, 56, 65

Z

Zamenhof, L. L., 228
Zeiss, Carl, 174, 253
Zeppelin, F. von, 16
Zero Products, 6
Ziegenspeck, Fritz, 27
Zimmerman, Thomas, 124
Zipfel, Heinz, 135
Zuse, Konrad, 114
Zweig, George, 65
Zwicky, Fritz, 80
Zworykin, V. K., 237, 240, 275

INDEX OF INVENTIONS

1-2-3 software, 118
6502 microprocessor, 120
8086 microprocessor, 121
68000 microprocessor, 121
80386 microprocessor, 121

A

A-12 fighter plane, 46
Abacus, 113
Accordion, 246
Acid rain, 145
Acoustic Control System, 250
Acupuncture, 103
Ada, 118
Additive process, 256
Adidas shoes, 210
Advanced Tactical Fighter (ATF), 46
Adventure book-game, 207
ADX, 91
Aerial, 236
Aerosol, 188
Agenda, 119
Agfacolor, 256
AGLAE, 66
AGM-129/A, 36
Agricultural machines, 128
AIDS, 106, 107
Aikido, 216
Airbag, 4
Airborne telescopes, 77
Air bottle, underwater, 149
Air brakes, trains, 12
Airbus, 19
Air conditioning, 188
Aircraft carriers, 38
  Russian, 39
Aircraft to prevent nuclear disasters, 164
Air Cushion Vehicle (ACV), 25
Air hostess, 23
Airship, 16
Air, liquid, 59
Air pollution, 143
Air reserve chamber, 159
Airship, military, 40
Alarm clock, 175
Alchemy, 176
Alcohol-free beer, 181
Alexandria library, 232
Alfa Feed, 133
Algebra, 52
Algometer, 91
Algorithm, 115
Alkaline storage batteries, 155
All-aluminium bicycle, 7
Alloys, shape-memory, 177
All-terrain bike, 212
All-terrain vehicle (ATV), 220
Altair 8800, 122
Alphabets, 235
Alternator, 155
Aluminium, 66
Alyoundi, 228
American football, 211
Ammonia, 66
Amorton technology, 167
Ampex video recorder, 238
Amstrad PC1512, 122
Anaesthesia, 96
Anaesthetic, oral, 111
Anaglyphs, 256
Analysis, 54
Analytical engine, 114
Anchor, 26
Androids, 126
Angioplasty, 96
Angioscope, 92
Animal husbandry, 132
Anti-asphyxia alarm, 193
Antibiotics, 104
Anti-Blocking System (ABS), 4
  for motorcycles, 10
Antibodies, monoclonal, 104
Anti-cholesterol machine, 106
Anti-explosives device, neutron, 34
Anti-fraud felt-tip pen, 230
Antihistamine, 104
Anti-laser material, 179
Anti-noise noise, 178
Anti-pregnancy vaccination, 103
Antipyretic agent, 104
Anti-rain spectacles, 110
Antisepsis, 97
Anti-ship missiles, 36
Anti-smoking device, 107
Anti-sugar medication, 107
Antitank Ant, 33
Anti-tank gun, 33
Anti-theft
  device, 6
  label, 201
Antonov-225, 44
Aortic valve transplant, 99
Apache, 47
Appendectomy, 98
Apple II, 122
Apple Macintosh, 122
  PC - Macintosh interface, 123
Applications software, 118
Apricot, 137
Aquaculture, 134
Aqualung, 149
Aquarium ties, 222
Aquarius underwater station, 150
Aqueduct, 169
Arabic alphabet, 235
Archery, 215
Archimedes
  Principles, 56
  screw, 153
Architecture, 168
Argot, 228
Ariane, 80
ARI radio control system, 236
Arithmetic machine, 113
Armoured vehicles, 33
Armour-plated bullet, 30
Arteries, clearing, 95
Artificial flavouring, 189
Artificial insemination, 132
Artificial intelligence, 119
Artificial intelligence programming languages, 118
Artificial languages, 228
Artificial rubber, 69
Artificial snow, 219
Artificial ventilation, 189
Artillery, 32
Artistic skiing, 219
Asepsis, 97
Ashtray, pocket, 191
Asparagus, 137
Aspartam, 187
Aspirin, 104
Assault rifle, 31
Assembler, 118
Association football, 211
Astro-C satellite, 78
Astronomical lens, Galileo and, 74
Astronomic satellites, 78
Astronomy, 74
Astrophysics, 79
AS-X-19, 36
Athletics, 209
Atlantic locomotive, 13
Atlases, 148
Atmospheric engine, 153
Atmospheric pressure, 58
Atomic bomb, 34
Atomic clock, 175
Atomic energy, 164
Atomic physics, atomic theories, 62
Atomic pile, 164
Atomic-powered ships, 26
Atom in wave mechanics, 63
Audion, 236
Auscultation, 89
Austin Mini, 1
Australia's keel, 28
Autochrome process, 255
Antechamber, 159
Automata, 126
Autogiro, 22
Automatic door, 193
Automatic focus, 256
Automatic Layshaft Transmission (ALT), 2
Automatic pilot, 23
Automatic telephone exchange, 242
Automatic transmission, 2, 125
Autopsy, 96
Avogardro's number, 63
AWACS, 40
AZT, 106

B

B-2 bomber, 43, 44
Back axle with integral trailing arm, 3
Back Sensor, 5
Bacteriological weapons, 34
Bacteriology, 73
  medical, 92
Babyliss, 197
Badminton, 221
Bakelite, 69
Balance wheel, 175
Ballet, 252
Ball games, 210
Ballistic missiles, 35
Balloons, 16
Ballpoint pen, 230
Bandages, 97
Bank notes, 201
  holographic, 180
Barbed wire, 132
Barbie doll, 209
Barometers, 146
Barrel organ, 246
Baseball, 211
Basic, 118
Basketball, 211
Bath, 193
Bathroom scales, 194
Batteries, 60, 155
Battleship, 36
Bayonet, 29
Bazooka, 33
Beaches, pollution, 142
Beekeeping, 133
Beer, 181
Beetle, 1
Bell X-1, 42
Bessemer converter, 175
Betacam, 239
Betamax, 238
Betamovie, 239
Bethe Cycle, 165
Bic, 230
Bicycle, 6
  with propeller shaft, 7
  mountain bike, 212
Bifocals, 109
Big Bang to Big Crunch, 80
BIK-COM-2, 39
Bikini, 195
Bilharzia, 101
Billiards, 204
Bill of exchange, 201
Binary computer, 114
Binary logic, 115
Biology and genetics, 71
Biomass, energy source, 166
Biometal, 178
Biometry, 200
Biotextiles, 177
Biped robots, 127
Birmingham, USS, 38
Biscuit for koala bears, 224
Bissel carpet-sweeper, 190
Bit, 116
Black holes, 79
Black-ice detector, 6
Black tulip, 137
Blast furnace, 175
Bleach, 191
Blenkinsop locomotive, 11

Blood,
artificial, 94
circulation, 94
hormone for the red corpuscles, 94
transfusion, 94
Blood pressure measurement, 89
Blue Ribband, 214
Blues, 251
Blue screen, 262
Boeing,
247, 707, 18
747, 19
767, 20
Bohr's atom, 63
Boiler, water-tube, 154
Bombers, 41, 42
Bone transplant, 99
Bonsai, 136
Book, 232
Book publishing, 231
Boolean logic, 115
Bosons, 66
Bourane, space shuttle, 82
Bourbon, 188
Boxing, 216
Bra, 197
Brace, 199
Braille, 235
Brain, European neurocomputer programme, 116
Brain lesions, treatment of, 98
Brakes, 3
ABS, 4
Brakes, air, trains, 12
Bread, wheatless, 188
Breech-loading, 30
Breeder reactor, 165
Bridges, 169, 204
Bridges, dental, 110
Browning pistol, 31
Bullets, 30
Bullet trains, Japan, 14
Bullpup rifle, 31
Bumpers, 5
Bus, 15
Button, 196
screw-on, 195
Byte, 116

C

Cable car, 16
Cable television, 243
Cadmium, 66
Caesarean section, 193
Calculator clock, 113
Calculus, 54
Caller ID, 243
Calotype, 254
Camcorders, 239
Camembert, 184
Camera lenses, 252
Camera, movie, 259
Camera obscura, 252
Cameras and processes, current, 256
Canal lock, 26
Cancer and immunotherapy, 105
Cannon, 32
Canoe, 212
Canvas, 233
Car, 1
armoured, 33
motor sports, 220
Car hire, 201
Car-sleepers, 12
Car that walks like a crab, 224
Car umbrella-holder, 223
Caravelle, 18
Carbon 14, 64
Carbon-carbon, 174
Carbon dating, 64
Carbon fibres, 176
Carbon paper, 203
Carburettor, 158
Cards for shy people, 226
Cargo ships, wind-powered, 25
Carnot principle, 59
Carpet-sweeper, 190
Carriages, railway, 11
Cars, model, 208
Carton, 181
Cartoon films, 261
Cartridge, 29
Cash register, 203
Cassettes giving off scents, 225
Cassini project, 87
Catalytic exhaust, 145
Catalytic reformation, 157
Catamaran, 213
Cataract operation, 109
Cathode rays, cathode-ray oscilloscope, 61
Cathode-ray tube, 237
Cats, domestication, 134
yoga for, 227
CCS suspension, 3
Celeripede, 6
Cell,
electric, 155, 156
photovoltaic, 167
thermochemical, 168
Cellophane, 69
Celluloid, 69
Celluloid roll film, 254
Cellulose acetate, 69
Cement, 174
Centaur II, 127
Central engine in competition cars, 220
Central heating, 191
Centreboard, 213
Centre priming, 30
Cepheid stars, 79
Ceramic-ceramic, 174
Ceramics, 67
Cereals, breakfast, 183
Cerebral radiotherapy, 98
CFCs, 143, 144
Chain drive, motorcycles, 10
Chainsaw, environment-friendly, 143
Chair, dentist's, 111
Champagne, 182
Chappe's telegraph, 240
Channel Tunnel, 171
Characteristic numbers, 57
Charles' Law, 59
Chemical waste, destruction, 143
Chemical weapons, 34
Chemistry,
inorganic, 66
organic, 68
Chemotherapy, 104
Cheques, 200
Chess, 205
Chewing-gum, 188
for smokers, 104
Chicken, domestication of, 133
Chicory, 138
Chimpanzees, communicating with, 228
Chip, silicon, 120
Choc-ice, 184
Chocolate, 183
Cholera, 101
Cholesterol checks, 91
Choreography, 252
Chromosomes, 71
Chronometer, 175
Chrysalis biplane, 20
Cigarette, 199
Cinema, 258
computer applications, 126
Cinema, air-conditioned, 190
Cinematograph, 258
Circle, impossibility of squaring, 50
Citroën 2 CV, 1
Citroën 7, 4
City of Ragusa, 25
Clarinet, 245
Clean Air Act, 143
Cleaner fluid for microchips, 121
Clean ink, 229
Clermont, 24
Clocks and watches, 175
Clod-crusher, 128
Clones, 71, 72
Clothes dryer, pocket, 192
Clutch, 2
Cobalt, 67
Cobol, 118
Coca-Cola, 183
Cochlear implant, 110
Cocoa powder, 182
Coding, 115
Coffee, 182
Coherer, 236
Coil, contraceptive, 102
Coining, 200
Coke, 175
Colossus, 114
Colour films, 259
Colouring of classic films, 259
Colour photocopies, 203
Colour photography, 255
Colourtune spark plug, 158
Colt, 31
Comb, 196
Combine harvester, 130
Combustion chamber, separate, 159
Comet, De Havilland, 18
Comma bacillus, 101
Communicating with chimpanzees, 228
Communication cables, 243
Communication satellites, 244
Compact auto-focus camera, 257
Compact disc, videodisc, CD single, re-recordable CD, 249
CD-I, 125
CD-ROM, 124
Compass, 26, 28
Compil Box, 250
Complex numbers, 48
Composite materials, 174
Composite particles, 65
Composition, musical, 251
Computer-aided design (CAD), computer-aided drawing, 126
Computer chess, 205
Computer cushion, 225
Computer games, 205
Computer golf course, 211
Computers, 114
fifth generation, 123
graphics, 125
languages, 117
memories, 121
operating systems, 116
peripherals, 123
programs, 114, 115
underwater, 150
viruses, 117
volcanic eruption forecasting, 152
Concert, 251
Concorde, Super Concorde, 19
Concrete, 174
Condensed milk, 186
Conductors, 60
Conic sections, 51
Connecting-rod system, 153
Connector, 223
Construction, 169
Contact lenses, 110
disposable, 110
Continental drift, 148
Contraception, 102
Convair XFY-1, 43
Cookery book, 231
Cooling rocket engines, 160
Co-ordinates, 51
Copernican system, 55
Core sampling, 158
Corkscrew, 187
Corneal graft, corneal incision, 109
Cornea transplant, 99
COS-B satellite, 78
Cosmic Background Explorer (Cobe), 88
Cosmic exploration, 86
Cosmological model (antiquity), 55
Cosmos satellites, 85
Coulomb Laws, 60
Cousteau aqualung, 149
CP/M, 116
Crab Motor, 224
Cracking process, 157
Crampons, 221
Crampton locomotive, 11
Cranes, 168
Cravat, 196
Cray 1, Cray X-MP, Cray Y-MP, 123
Cream separator, 133
Credit card, 200
holographic, 179
Cricket, 210
Croissant, 183
Crosswords, 207
Crowns, dental, 110
Cruise missiles, 36
Cruiser, 39
Cryoanaesthesia, 97
Cryogenic probes, 99
Cryosurgery, 99
Crystal-glass, 174
Crystalline lens implantation, 109
Crystal set, 236
Crystals, liquid, 68
Cultured pearls, 134

Cuneiform, 234
Curling tongs, electric, 197
Cybernetics, 116
Cycletouring, 212
Cycling, 211
Cyclosporin-A, 98
Cyrillic alphabet, 235

D

Daedalus aircraft, 20
Daguerreotype, 253
Dahlia, 136
Dahon pocket-size bicycle, 7
Daisy wheel printer, 125
Dam, 155
biggest, 153
Dance, 251
Darwinism, 72
Dataglove, 124
David, 38
dBase II, 118
Decibels, 248
De Dion three-wheeler, 8
Deep-freezing, 184
Deep Rover, 150
Defibrillator, 96
De Havilland Comet, 18
Deltabox frame, 11
DENDRAL project, 119
Dental equipment, 111
Dental implants, dental extraction, 111
Dental scanner, 112
Dental surgery, 110
Dental transplants, 111
Dentures, 110
Deoxyribo-nucleic acid (DNA), 71
Department stores, 201
Dérailleur gears, 7
Derby Flex, 219
Desert, combating its advance, 143
Design of space vehicles, 85
Detection systems, 35
Detector, electronic, for reversing, 5
Detergent, 191
Diagnostic robots, 127
Diamonds, synthetic, 68
Diaphragm, 252
contraception, 102
Dibbles, 128
Dictionary, 231
Diesel engines and the environment, 146
Diesel locomotive, 13
Diesel lorries, turbocharged, 160
Differential, 3
Digital audio tape (DAT), 250
Digital organ, 247
Digital paper, 229
Digital 'Sax', 248
Dinosaurs, what happened?, 143
Diode, 236
Direct fuel injection, 220
Directories, 240
Direct transmission, 2
Disabled driver controls, 4
Disc brake, 3
Disc-film, 257
Dishwasher, 192
Diskettes, non-degradable, 124
Disks, computer, 124
Disneyland, 221
Disney's Multiplane camera, 261
Dispersant, oil spillage, 141
Disposable razor, 202
Distance alarm, 133
Distillation,
petroleum, 156
continuous, 157
whisky, 187
Diver, robot, 150
Diving equipment, 149
DNA, 93
Dog, electronic, anti-theft, 193
Dog, domestication of, 135
Dolby, 251
Dolls, 209
Domestic animals, 134
Dominoes, 204
Door, automatic, 193
Doppler sonar, 147
Dot matrix printer, 125
Double glazing in cars, 5
Douglas DC-3, 18
Dounreay, 165
Dovetail screwdriver, 199
Dragon Quest 3, 206
Drag racing, 220
D-RAM, 122
Dreadnought, 36
Draisienne, 6
Dressings, medical, 97
Drill, dental, 112
Drilling for oil, 157
Drills, hand, 199
Drinking water, 141
Drive belts, motorcycles, 10
Drug detection device, 91
Drum kit, electronic drum kit, electronic drumsticks, 248
Dry cleaning, 190
Dry photographic plates, 254
DSP Block, 14, 36
Dumb-bells that quench the thirst, 225
Dum-Dum bullet, 30
Dungeons and Dragons, 206
DX coding, 257
Dynamics, 55
Dynamite, 34
Dynamo, 155
Dynastart, 10

E

Ears, 110
Earth, diameter, 57
Earthquakes, 150
Earth Resource Technology Satellite (Erts-1), 143
Eau de Cologne, 194
Echogram, 91
Ecological movement, first, 143
Ecological peace-keeping force, 145
Ecology, 140
Edison effect, 236
EH 101, 47
Eiffel Tower, 173
Einstein's rings, 79
Ejector seat, 24
Ektachrome, 256
Electrical core sampling, 158
Electrical machine, first, 60
Electric and electronic musical instruments, 246
Electric car, 2
Electric clock, 175
Electric curling tongs, 197
Electric discharge in gases, 61
Electric guitar, 246
Electric hand drill, 199
Electric heater, 192
Electric iron, 192
Electricity, 155
Electricity and Magnetism, 60
Electricity transmission, 156
Electric lift, 189
Electric locomotive, 13
Electric pinball machine, 205
Electric shaver, 194
Electric telegraph, 241
Electric totaliser, 114
Electric washing machine, 192
Electroacoustic piano, 248
Electrobat, 2
Electrocardiograph (ECG), 90
Electro-encephalogram (EEG), 90
Electromagnetic force, 56
Electromagnetic organ, 247
Electromagnetism, 61
Electrometallurgy, 176
Electron, 65
Electronic clutch, 2
Electronic Discrete Variable Automatic Computer (EDVAC), 115
Electronic drum kit, electronic drumsticks, 248
Electronic flash, 254
Electronic games that interact with television, 206
Electronic gate, 61
Electronic gearbox, 2
Electronic Numerical Integrator and Calculator (ENIAC), 115
Electronic organ, 247
Electronic piano, 248
Electronic pocket calculator, 122
Electronics, 61
Electronic television, 237
Electron microscope, 90
Elementary particles, composite particles, 65
Elements,
110th element, 63
Periodic Table, 63
Elliptic geometry, 52
Embryos, frozen, 103
Embryo transplantation in cattle, 133
Encyclopaedia on videodisc, 232
Endeavour space shuttle, 82
Endoscope, 92
Energy sources, 166
Engine of tomorrow, two-stroke, 158
Engines, and environmental protection, 145
Engines, steam, 153
Envelope, 240
Environment, 140
Enzyme film, 256
Eole, 17
Epidural anaesthetic, 97
Equations, algebraic, 52
Equestrianism, 221
Erasable ballpoint pen, 230
Eraser, 230
Eridani B, 79
Escalator, 191
Esperanto, 228
Espresso coffee, 182
Eurobag cushion, 4
Euclidian assumption, 50
Eurodisneyland, 221
Eurofar, 23
European Communications Satellite (ECS), 244
European Fighter Aircraft (EFA), 46
European River Ocean System (EROS-2000), 141
Ever-Sharp pencil, 230
Evolution of species, 72
Experimental space station, 84
Expert systems (ES), 119
Exploration of the cosmos, 86
Explosives, 33
Eye control of computers, 126
Eye-Dentification, 200
Eye printing, 201
Eyes, 109

F

Facelift, 98
Fairlight I, 248
Falling bodies, 57
False teeth, 110
Fan, 196
Fan sails, 215
Farming, 128
Fastac, 131
Fastest aircraft, 43
Fatless meat, 187
Fax, 242
Feeding bottle, 103
Feeding dairy cattle automatically, 133
Felt-tipped pen, 230
Fenestron, 22
Fermat's last theorem, 52
Fertilisation, test-tube, 103
Fibreglass, 176
Fibre Optic Borehole Earth Strainmeter (FOBES), 152
Fibre optic transatlantic cable, 244
Fighter aircraft, 44–45
jet, 42
Fighter gun, 40
Filariae, 93
Fillings, dental, 111
Film cameras, modern, 259
Film, celluloid roll, 254
Film, laserfilm, 179
Films, 254
classics, colouring, 259
modern, 256
with colour, with sound, 259
Film studio, 258
Fin keel, 213

Fire Ant, 33
Fire extinguisher, 191
Firefighting, oil wells, 158
Firing tube, 29
Fischer-Griess Monster, 52
Fish farming, 134
Five-valve motorcycle cylinder head, 9
Flash, photographic, 254
Flavouring, artificial, 188
Flexible-rim bicycle, 7
Flight simulators, 126
Flights, first,
 aircraft, 17
 airship, balloon, 16
 helicopter, 22
Flint-lock, 29
Float glass, 174
Floppy disks, 124
Flowers, 136
Flushing system, 191
Flute, 245
Flymo, 135
Focus, automatic, 256
Fokker 100, 20
Football, 211
Forceps, obstetric, 103
Forces, fundamental, 56
Ford, Model T, 1
Fork, 184
Formaldehyde, 68
Formula 1, 220
Fosbury flop, 210
Fountain pen, 230
Four-cylinder motorcycle, 9
Four-stroke motorcycle engine, 9
Four-wheel drive, four-wheel steering, 3
Fractures, treatment, 97
Francis turbine, 155
Free climbing, 221
Freedom space station, 84
Freeze-dried food, 184, 188
Freon, 144
Frequency modulation (FM), 237
Front-wheel drive, 4
Froude number, 57
Frozen foods, 202
Frozen meals, pre-cooked, 184
Fruit and vegetables, 137
FS-X, 46
Fuel, and environmental protection, 145
Fuel injection, 220
Fujix TV-Photo system, 257
Fulcrum (MiG-29), 41
Funboard, 217
Functions, mathematical, 54
Fundamental forces, 56
Fungicides, 131
Funicular railway, 15
 sub-glacier, 16

G

Galaxies,
 distant, 79
 most distant, 80
Galileo probe, 88
Gall stones, treatment by laser, 99
Games and toys, 204
 of antiquity, 205
 young millionaire inventors of, 206
Gamete Intro-Fallopian Transfer (GIFT), 104
Gametes, 72
Garden hose, 135
Gas discharge lasers, 179
Gases,
 anaesthetic, 96
 electric discharge in, 61
 expansion, 58
 ideal gas, 59
 liquefaction, 59
Gas generator, 5
Gas-operated hand drill, 199
Gas turbine car, 2
Gay-Lussac's Law, 59
Gearbox, 2
 motorcycle, 10
Gears, dérailleur, 7
Geiger counter, 63
General anaesthesia, 96
General theory of relativity, 57
Generator, MHD, 156
Genes, 72
 artificial, 72
Genetic code, 71
Genetic engineering, 72, 73
Genetic fingerprinting, 72
Genetic manipulation on humans, 105
Genetics, 71
Geological sounding, deepest, 152
Geometry, 50
 definition, 52
Geophysics, 147
Geostationary orbit, 86
Geothermal energy, 166
Germanium, 67
Glasair aircraft, 20
Glass, 174
Glock 17, 32
Glove, computer peripheral, 124
Gluons, 66
GO, 205
Golden number, 50
Golf, 211
Go Motion, 262
Goose liver, 184
Gossamer Albatross; *Gossamer Condor*, 19
Grapefruit, 138
Graphics processors, 121
Gravitational mirages, 79
Gravitational waves, 57
Graviton, 66
Gravity, centre of, 57
Gravity, force of, 56
Great Eastern, 27
Great Western, first transatlantic liner, 27
Greek alphabet, 235
Green Bank (Va), radio telescope, 76
Green Earth experiment, 143
Green energy, use of, 146
Green grapefruit, 139
Griffon (JAS 39), 46
Groups, theory of, 53
Guided missiles, 36
Guitar, 248
Guitar synthesiser, 248
Gulfstream aircraft, 20
Guncotton, 34
Gunpowder, 33
Guns, aircraft, 40, 41
Gutenberg press, 230
Gymnastics, 209
Gynaecology, 102
Gyrocompass, 23
Gyroplane Laboratoire, 22
Gyroscope, 23

H

Hair colour, hair dryer, hair lacquer, 198
Hairstyle video, 197
Hair transplant, 99
Hair-washing machine, 198
Half-court tennis, 221
Halley's Comet, 77
Halley's diving bell, 149
Hand, artificial, 105
Hand drills, 199
Handkerchief, 196
 disposable, 194
Hammer, 199
Hang-gliding, 215
Haricot bean, 138
Harmonica, 246
Harp, 245
Harpsichord, 245
Hard disc, 124
Harrier, 43
Harrow, 128
Harvard Mark 1, 114
Harvester, 130
Hawk-Eye camcorder, 239
Haymaking machines, 128
Hearing, hearing aids, 110
Heart and lungs, 94
Heart, artificial, 95
Heart pump, 95
Heart transplant, 99
Heat pump, 167
Heat-work equivalence, 59
Heavy machine gun, 32
Helicopter-aircraft, 23, 47
Helicopter-balloon, 17
Helicopters, fighter, 46
Helicopter, submarine, 150
Heliogravure, 231
Helium, liquid, 59
Hemopump, 95
Hepatitis-B test, 90
Hepatitis-C, tracing, 90
Hepatitis-B virus, culture, 93
Heredity, 72
Hermès, the European space shuttle, 81
Hertzian waves, 61
Hertz relays, 242
HFA 22, 144
Hidden mass in the universe, 80
Hieroglyphics, 234
High definition television, 238
High pressure steam engine, 154
High Speed Train (HST), 14
High temperature paint, 199
Hipparchos satellite, 78
Hip replacement, 98
Hobie Cat, 213
Hockey, 211
Hollow charge, 33
Holograms, 179
Holographic film, 260
Holographic lenses, 110
Homeopathy, 105
Horn, 245
Horse racing, 221
Horticulture, 135
Hose, garden, 135
Hot-air balloon, 16
'Hot line', 243
Hotol, 21, 81
Hovercraft, 25
HSST, 13
Hubble space telescope, 78
Hula-hoop, 205
Human Lucocyte Antigen (HLA) system, 73
Hurricanes, 146
Hyacinths, to treat pollution, 141
Hydra-Brute Tree Replanter, 135
Hydraulic lift, 189
Hydroactive suspension, 4
Hydrodynamics, 56
Hydro-electric power, 155
Hydro-electric project, biggest, 153
Hydrogen balloon, 16
Hydrogen bomb, 34
Hydrogen, energy source, 166
Hydrogen, liquid, 59
Hydroplane, 25
Hydropneumatic suspension, 3
Hydroponics, 130
Hydrostatic paradox, 57
Hyperbolic geometry, 52
Hypersonic planes, 21
Hypodermic safety device, 106

I

IBM 5100, 122
IBM PC, 122
IBM Personal System/2, 123
IBM/SSI project, 123
Ice-breaker, 25
Ice hockey, 218
Ice skating, 218
Ice yachting, 216
ID checking, 200
Ideograms, 234
IDX 70, 201
Illiteracy, 234
Image animation, computerised, 126
Immune system, 73
Immunotherapy, 105
Incendiary compounds, 34
Incineration, high-temperature, 143
Incubators, paediatric, 103
Induction, 61
Induction coil, 158
Infinitesimal calculus, 54
In-flight refuelling, 24
Information Processing Language (IPL), 118
Information technology, 113
Infra Red Astronomy Satellite (IRAS), 78
Infra-red telescope, 75
Inipol 90, 141
Ink, 229

Insecticides, 131
Instant coffee, 182
Insulin, 106
Intaglio, 230
Integrated circuits, 121
Integrated priming, 29
Intercity experimental (ICE), 14
Inter-Continental Ballistic Missile (ICBM), 36
Interface software, 117
Interferon, 105
Intermediate Range Ballistic Missile (IRBM), 36
Interleukins, 105
Internal memory, computers, 121
International Thermonuclear Experimental Reactor (ITER), 166
Intra-Peritoneal Fertilisation (IPF), 104
Intravenous injection, 106
Inventors,
  record-breaking, 226
  young millionaires, 206
Inverted-wing aircraft, 43
In vitro culture, 72, 73
In vitro fertilisation (IVF), 103, 104
Irish whiskey, 187
Ironing, 191
Iron lung, 96
Iron, materials, 171
Irrational numbers, 48
Isotherms, 147
Isotopic separation, 164

J

Jacket that changes colour, 222
Jacketed bullet, 30
Jade, synthetic, 68
Jarvik, 95
Jazz, 251
Jeans, 196
Jeep, 3
Jet engines, 161
Jet motorbike, 220
Jetnet, 106
Jet powered helicopter, 22
Jigsaw puzzle, 207
Jockey robot, 226
Joined-wings aircraft, 20
Joint European Torus (JET) project, 161
Joule effect, 61
Joystick, 24
Judo, 216
Juke-box, 247
Juke-box laser, 249

K

Kaplan turbine, 155
Karate Do, 216
Kayak, 212
Keck Observatory, 75
Keels, weighted, 27
Kerosene, 157
Ketchup, 186
Kevlar, 69
Kidney stones, disintegration by shock waves, 99
Kidney transplant, 99
Kiev class, 39
Kinetophone, 259
Kit Kat, 182
Kiwi fruit, 139
Kleenex, 194
Knee, mechanical, 105
Knitting machines, 178
Kodachrome, 256
Kodak camera, 254
Kodak disc-film, 257
Kodak I grain, 256
Kong Yun (A 5-K), 44
Kuiper Observatory, 77
KX Z40X pocket photocopier, 203

L

Lacquer tree, 137
Lamarckism, 72
Laminated glass, 174
Laminated windscreen, 5
Languages,
  artificial, 228
  computer, 117
Laser cataract operation, 109
Laserfilm, 179
Lasergraph, 249
Laser printer, 125
Lasers, 179
LASSO atomic clock, 175
Latin alphabet, 235
Lawn mower, 135
Lawn tennis, 220
Lawson's criterion, 165
Lazar houses, 100
Lead-acid storage battery, 155
Lead tetraethyl, 68
Lea Box, 249
Lego, 207
Leica, 257
Lemon, 137
Lenin, 26
Lens coating, 252
Lenses and telescopes, 74
Lenses, photographic, 252
Lens implantation, 109
Leprosy, 100
Leptons, 65
Leyden jar, 60
Library, 231
Librium, 108
Lidar, 147
Life in space, 85
Lift, 190
Lighthouses, 27
Lighting, 189
Light machine gun, 32
Lightning-conductor, 60
Light pen, 123
Light pen for video recorders, 238
Light-sensitive surfaces, first, 253
Liner lounges, stabilised, 223
Liners, 27
LipoScan, 91
Lipstick, 195
Liquefaction of gases, 59
Liquid crystals, 68
LISP, 118
Lithography, 230
Liver transplant, 99
Local anaesthetic, 97
Lock, 192
Locomotives, 11
Logarithm, 54
Logic, mathematical, 53
Logic Theorist (LT) program, 118
Log, mechanical, 27
Long Boiler Locomotive, 11
Long Duration Exposure Facility (LDEF), 84
Long Rifle, .22, 30
Lubrication razor, 194
Lüger pistol, 31
Lumière cinematograph, 258
Luna 9, 87
Lunar Exploration Module (LEM), 85
LY 163-502, 107

M

Macadam, 6
McDonald hamburgers, 186
Machine guns, 32
  aircraft, 41
Machine tools, 172
Mach number, 57
Macintosh microcomputer, 122
  PC - Macintosh interface, 123
Macintosh, raincoat, 196
Mac Paint, 118
MACSYMA, 119
Mac Write, 119
Magellan probe, 88
Magnetic boat, 26
Magnetic confinement, 165
Magnetic monopole, 61
Magnetic photography, 257
Magnetic recording tape, 250
Magnetic trains, 13
Magnetic video tape, 240
Magnetism, 60
Magneto-hydrodynamic generator, 156
Mainstay, 40
Malaria, 93, 101
Manned Manoeuvring Unit (MMU), 85
Man Tended Free-Flyer (MTFF), 84
Map, 148
Marathon, 209
Marconi rigging, 215
Margarine, 189
Matches, 190
Material, laser-resistant, 179
Materials, construction, 171
Materials, shape-memory, 177
Match-lock, 29
Mathematical logic, 53
Matrices, 53
Matter, 57
Mauser rifle, 30
MC ALB, 10
MDR-10 audio headset, 247
Meat, fatless, 187
Meat transport, refrigerated, 184
Meccano, 207
Mechanical lift, 189
Mechanical log, 27
Mechanical television, 237
Mechanised calculation, 115
Medical laser, 179
Medicine, robots in, 127
Mental Flight, 225
Mercedes, 1
Metal detector, 198
Metallurgy, 175
Meteorology, 146
Meteosat, 147
Methane, energy source, 166
Methane tanker, 158
Methanol, energy source, 166
Metric system, 54
Metronome, 246
Mice, genetically modified, 73
Micral, 122
Microbes, 92
Microbiology, 73
Micro cassette, 250
Microcomputers, 122
Micro-metastases, early detection, 91
Micro motors, 158
Microprocessor, 119
Microsatellites, 83
Microscope, 89
Microwave drilling for oil, 157
Microwave oven, 184
MIDI interface, 248
MiG, 41
Military motorcycle, 8
Milk, condensed, 186
Milking machines, 133
Milk refrigeration, 133
Mineral water, carbonated, 184
Mines, 40
Mini, 1
Minié bullet, 29
Miniskirt, 196
Mini submachine gun, 31
Mir, 83, 150
Mirror, 196
Mirror, rear-view, 5
Mirror, anti-glare, 6
Missiles, 35
Mixed Oxide (Max), 164
MLU, 13
Model cars, 208
Model T (Ford), 1
Molecular farming, 131
Molecule perfume, 70
Molotov cocktail, 34
Monoclonal antibodies, 104
Monopoles, 61
Monopoly, 208
Monopropellant rocket engine, 161
Monoski, 218
Montgolfier balloon, 16
Moons, planetary, 87
Moon walk, 85
Morey-Boogie, 217
Morse telegraph, 241
Mortars, construction materials, 174
Mortar, weapon, 32
Moses to the rescue of Venice, 142
MOS integrated circuit video camera, 239
Motor car, 1

Motorcycles, 8
engines, 9
jet, 220
Motor sports, 220
Motorway, 6
Mountain bike, 212
Mountaineering, 221
Mountain range 350 million years old, 152
Mt Palomar telescope, 75
Mouse, computer peripheral, 124
Mouthwash, 111
Movie cameras, 259
Movie projector, 259
Mriya (Antonov 225), 44
MS/DOS, 116
Multiplane camera, 261
Multiple-expansion engine, 154
Multiple-lens telescope, 75
Muon, 65
Musical composition and performance, 251
Musical composition using computers, 125
Musical instruments, 245
Music box, 246
Mussel beds, 134
Mustard, 186
Mutations, 73
Mystery weekends, 206

N

Nail, 199
Narval, 38
Nature conservation, 143
Nautilus (1797), 37
Nautilus, USS (1955), 38
Nautilus (1985), 149
Naval warfare, 36
Navigational aids, 26
Needle, 196
Neon tube, 61
Neuralgia bracelet, 108
Neurocomputer science, 116
Neurosurgery, 98
Neutrino, 65
Neutrinos, solar, 80
Neutron, 65
Neutron and pulsating stars, 79
Neutron anti-explosives device, 34
Neutron bomb, 34
Newcomen's engine, 153
Newspaper, 232
Newtsuit, 149
Next European Torus (NET) project, 166
NH 90, 47
Nickel-iron battery, 156
Nicoret, 104
Nicotine, 68
Nimitz class, 39
Nipkow scanning disk, 237
Nitric acid, 66
Nitrocellulose, 69
Nitrogen, 57
Nitroglycerine, 34
Nivea cream, 194
Noise,
anti-noise noise, 178
Non-Euclidean geometries, 52
Non-stick pan, 186
Nova, 179
Nuclear magnetic resonance (NMR), 64
Nuclear moderator, 164
Nuclear pacemaker, 96
Nuclear physics, 62
Nuclear power stations, 164
robot for, 127
Nuclear reactor, 164
Nuclear submarine, 38
Nuclear weapons, 34
Nucleic acids, 71
Number theory, 48
Numerical control, 114
Nylon, 69
Nylons, 195

O

Objective lenses, 253
Oboe, 245
Obstetric forceps, 103
Oceanographic research submarines, 149, 150
Oceanography, 149
Oceans, energy source, 168
Octant, 27
Ocular implants, 109
Offset, 231
Off-shore drilling for oil, 157
Ohio, 38
Ohm's Law, 60
Oil lamp, 157
Oil prospecting, 157
Oil slick control, 141
Oil tanker, 158
Oil wells, 157
firefighting, 158
OMNIMAX cinema, 260
Oncology, 73
OncoMice, 73
Onyx, 257
Opera, 251
Oral anaesthetic, 111
Orange, 138
Orbital stations, 83
Orchid, 136
Organ, 248
electronic, electromagnetic, digital, 247
Orient Express space plane, 21
OS/2, 117
Osprey (V-22), 23, 43
Othello, 205
Otis lift, 190
Otis mini urban train, 15
Outboard engines, 27
Ovaltine, 182
Oxygen, 57
liquid, 59
Oyster farming, 134
Ozone layer, 144

P

Pacemaker, 96
Pacific locomotive, 13
Page Maker, 119
Painting and drawing, 233
Palm Recognition system, 201
Pan-and-Scan, 260
Pandemics, 100
Panels, prefabricated, 173
Pan, non-stick, 186
Paper, 228
digital, 229
very long, 229
Paperback, 232
Paper clip, 203
Papin's engine, 153
Parachute, 24
Parachuting, 215
Paraffin, 157
Parallelogrammatic fork, 10
Parasites, medicine, 93
Parasitic farming, 131
Parasol for books, 225
Parchment, 228
Paris-Dakar race, 220
Parking meters, 6
Particle accelerators, 65
Particles, 65
Pascaline, 113
Pascal programming language, 118
Paschen's Law, 61
Passenger carriages, railway, 11
Pay phones, 242
PC - Macintosh interface, 123
Pearls, 134
Pear, Williams', 138
Peba, 70
Pedal aircraft, 19
Pelton turbine, 155
Pencil, 229
propelling, 230
Pencil-laser, 99
Penicillin, 104
Percussion, 245
Percussion lock, 29
Percussion technique, medicine, 89
Perfume, 194
Peripherals, computer, 123
Perinorm, 124
Periodic Table, 63
Permanent waving, 198
Personal cassette, 250
Personal computers, 121
Petrol, 156
unleaded, 145
Petrol turbo, 160
Phobos probes, 88
Phoenician alphabet, 235
Phonograph, 247
Photobooth, 255
Photocopy, 203
Photographic gun, 258
Photographic processes, current, 256
Photographic sound recordings, 249
Photography, 252
first positive image, 253
colour, 255
3-D, 256
instant, 257
medical, 91
Photogravure, 231
Photon, 66
Photophone, 242
Phototelegraphy, 242
Photovoltaic cell, 167
Physics, 54
π (Pi), 48
Piano, 245
electroacoustic, 248
electronic, 248
remote control, 248
Pico cassette, 250
Pictionary, 207
Pictures of movement, 258
Pill, contraceptive, 103
Pin, 195
Pinball, 205
Pineapple, 137
Pion, 65
Pistols, 31
Pistol with a firing selector, 31
Piston, metal ring, 154
Piston, anti-pollution, 3
Plague, 100
Planck constant, 54
Plane (tool), 199
Planetary exploration, 87
Plantophon, 135
Plant propagation in vitro, 73
Plastic and rubber, 69
Plastic, biodegradable, 142
Plastics, shape-memory, 177
Plastic surgery, 98
Plate, 181
Plates, photographic, 254
Playmobile, 209
Plexiglass, 69
Ploughs, 128
Pneumatic tyres, 5
Pocket ashtray, 191
Pocket clothes dryer, 192
Pocket photocopier, 203
Pocket submarine, 38
Pocket telephone, 243
Points, railway, 11
Poker, 204
Polacolor, 256
Polaroid, 257
Pollen aspirator, 130
Pollution, 140
Polo, 221
Polonium, 63
Polyamides, 69
Polyesters, 69
Polyethylenes, 70
Polymers, 69
Polypropylene, 70
Polystyrene, 69
Polyurethanes, 69
Pomato, 138
Portable radio, 236
Poseidon project, 158
Positive images, photography,
first, 253
first on paper, 254
Postage stamp video screen, 238
Postal service, 240
Postcard, 124, 240
'Post-it' notes, 202
Postscript page description language, 117
Potato, 139
Pouring lip for paint, 199
Powder and explosives, 33
Power stations using ocean sources, 168
Prefabricated panels, 173
Preselector gearbox, 2
Press-stud, 195
Pressurisation in airliners, 23
Prime numbers, 48
Printer, computer peripheral, 125

Printing, 230
Printing press, 230
Proa, 215
Pro Arm, 10
Probabilities, theory of, 53
Programs, 114, 115
Propeller, jet, 163
Propellers, 27
Propelling pencil, 230
Propfan, 163
Protection of the environment, 143
Proteins, in vitro synthesis of, 73
Proton, 65
PSR 1913+16 binary pulsar, 80
Psychoanalysis, 108
Publishing, 231
Puddling, 176
Pulleys, 168
Pulsars, 80
Pulsating stars, 79
Pulse code modulation, 249
Pulse-jet engine, 163
Punctuation, 228
Punctures,
anti-puncture liquid, 5
PVC, 69
Pyrex, 174
Pythagoras' theorem, 50

Q

Quadruped robots, 127
Quantum effect transistors, 119
Quantum theory, 55
Quarks, 65
Quarter-cam, 239
Quartz clock, 175
Quasars, 79
Quicksnap, 257
Quinine, 108

R

Rabies, 93
Rack-railway, 12
Radar, 35
Radar astronomy, 77
Radar detection of underground water, 130
Radar detection of wind sheers, 146
Radar meteorology, 146
Radiator, 159
Radio, 236
smallest, 237
Radioactive disintegration, 63
Radioastronomy, 76
observatories, 76
Radio control system, 236
Radio-immunology, 90
Radio-interferometer, 77
Radiology, 91
Radiotelephone, 242
Radio telescopes, 76
Radiotherapy, 108
cerebral, 98
Radio tube, 236
Radium and polonium, 63
Radon, 58
Rafale, 46
Rafting, 212
Rails and road: dual-purpose vehicles, 3
Railways, 11
Rain gauge, 146
Rain'x, x, 5
Rally driving, 220
Ramjet engine, 161
Random Access Memory (RAM), 121
Rayon, 69
Razor, 194
disposable, 202
Reactor, nuclear, 164
Real time working, 116
Reaper, 128
Rear-view mirrors, television, 5–6
Reber radio telescope, 76
Recco system, 219
Recoil brake, 32
Recoilless gun, 33
Reconnaissance, air, 40
Record, stereo record, long-playing record, 247
old recordings, new discs, 248
Red Arrows, 41
Refrigeration, 184, 185
milk, 133
Refrigerators, 184
environment-friendly, 143
Refuse Pyramid, 143
Reinforced concrete, 174
Relativity, 56, 57
Relays, 242
Renov-Ink, 125
Repeaters, 30
Reproduction, assisted, 103, 104
Resin for windscreens, 5
Restaurant car, 12
Retroviruses, *see* RNA
Reverse gear motorcycle, 9
Revolver, 30
automatic, 31
Reynolds number, 57
Ribbon speaker, 250
Ribo-nucleic acid (RNA), 71
Richter scale, 150
Rifle, automatic, 32
Rifles, 31
Rigging, 215
Rimfire, 7
RITA, 35
River blindness, treatment of, 107
RNA, 93
Road and rails: dual-purpose vehicles, 3
Robot diver, 150
Robot jockey, 226
Robots, 126
Rocket belt, 23
Rocket engines, 160
Rocket-launchers, 33
Rocket-propelled motorcycle, 9
Rockets, 80
Rocket (Stephenson), 11
Roller skating, 218
Rotary engine, motorcycles, 9
Rotary press, 230
Rotochamber, 159
Rotor, 22
Rotoscope, 261
Rover gas turbine car, 2
Rover safety bicycle, 7
Rowing, 212
Rubber and plastic, 69
Rubber (eraser), 230
Rubik cube, 209
Rucksack, 197
Rugby, 211
Rutherford's atom, 62

S

Saccharine, 187
Safe, 201
Safety bicycle, 7
Safety bindings, skis, 219
Safety fuse, 34
Safety locks, 192
Safety pin, 195
Safety razor, 194
Sail folding system, 215
Sailing, 213
Sailing crossings, regattas and races, 214
Sailing liner, 26
Sail train, 217
Sail travel on rails, 217
Salads, pre-prepared, 202
Salyut, 83
Sampling, 248
Sand yachting, 216
Sänger space shuttle, 21, 82
Sanitary towels, 102
Satellites, 82
Satellites, astronomic, 78
Satellite television, 244
Savannah, first commercial atomic ship, 26
Savannah, first transatlantic steamer, 25
Saw, 199
Saxophone, 246
digital 'sax', 248
Scales, 201
bathroom, 194
SCAMP, 122
Scanner, 91
dental, 112
Schnorkel, 38
Scintiscanning, 92
Scotch tape, 202
Scrabble, 207
Screwdriver, 199
Scubaphone, 150
Sea Dragon, 47
Sea mine, 40
Seaplane, 18
Seat belt, 4
Seaweed farming, 134
Seismograph, 150
Seismographic prospecting for oil, 157
Self-propelled torpedo, 39
Semi-automatic rifle, 31
Semi-conductor memories, 121
Semi-conductors, 61
Semi-crystals, 67
Semyorka rocket, 84
Sera, 94
Set, hair, 198
Set of real numbers, 53
Sewing machine, 192
Sexual regenerator, 225
Shaky Cam, 260
Shampoo, 198
Sheath, contraceptive, 102
Shells, shrapnel, 32
Shinkai 6000, 150
Shipping, 24
Ship's log, waterproof, 215
Shoe, 195
Shock-absorbers, electronically variable, 4
Shooting, 215
Shower for space use, 86
Showscan, 260
Shrapnel shells, 32
Sidecar, 9
Signals, railway, 12
Sign language, 235
Sign-on, 201
Sikorsky S75 ACAP, 47
Silencer, 31
Silent World, The, 260
Silicon, 68
Silicon chip, 120
Simulation, 126
Simulator, driving, 6
Singer sewing machine, 192
Single-cylinder motorcycle, 9
Single frame animation, 261
SI units, 54
Six-cylinder motorcycle, 9
Ski boots, 219
Ski-Doo, 9
Ski flying; skiing, artistic, 219
Ski jump, 219
Ski lift, 219
Ski mittens, 219
Skin cream, 194
Skin culture, 99
Skin graft, 98, 99
Skin, synthetic, 99
Skittles, 211
Skula project, 158
Sky Cam, 260
Skynet-4B, 35
Skyship, 17
Sky-scraper, 170
biggest, 171
Sleeping cars, 12
Sleeping pill, 108
Slow match wick, 34
Slow motion filming, 259
Smallpox, 93
Smog, 143
Snakes and ladders, 205
Snow artificial; snowfall simulator; Snowmax snow inducer, 219
Snowmobiles, 9
Soap, 195
Sodar, 147
Software, 116
Soho space programme, 88
Solar collector; solar energy; solar furnace, 167
Solar thermal power plant, 168
Solid propellant rock engines, 160
Sonar, 35
Sonar golf ball retriever, 211
SOS orbital stations, 85
Sound reproduction, 247
Sound track, optical, 259
Sowing machines, 128

Space, 78
astronomy, 78
probes, 86
shuttles, 81
storage batteries, 156
suits, 86
vehicles, 85
walks, 81
Space Maker car parks, 6
Space objective lenses, 253
Spansule, 107
Spark plugs, glass-capped, 158
Speaker; ribbon speaker; visual speaker, 250
Special effects and trick photography, 261
Spectacles, 109
Spectrography, 70
Speech recognition; speech synthesis, 124
Speed sail, 217
Spinning, 177
Spiral spring, 175
Spirit level, 169
Spring, spiral, 175
Squaring the circle, 50
Squash, 221
SR-71, 43
SSI-IBM project, 123
Stainless steel, 173
Stalin's Organ, 33
Stamps, 240
holographic, 180
Standards, 54
Starflex, 22
Starter motor, motorcycles, 10
'Star Wars', 35
Static RAM, 121
Statistical determinism, 56
Statistics, 53
Stave, 248
Steady Cam, 259
Stealth bomber (B-2), 43, 44
Steamboats, 24
Steam engines, 153
Steel, 173
Steel cutting by water jet, 178
Steel tennis racket, 220
Stepped reckoner, 113
Stereophony, 236
Stereo record, 247
Stereoscopy, 256
Stethoscope, 89
Stock cars, 220
Stock exchange, 203
Stockings, 196
nylons, 195
Storage batteries, 155, 156
Stored program computer, 115
Stores, department, 201
Strategic Defense Initiative (SDI), 35, 179
Strawberry, 138
Strong force, 56
Stun Gun, 31
Submachine gun, 31
Submarine, 36
'classic' submarine, 38
Submarine helicopter, 150
Submersible, 38
Subterranean storage of hydrocarbons, 158
Subtractive synthesis, 256
Sugar, 187
Sulphuric acid, 66
Sumo, 216
Sunglasses, light-sensitive, 109
Sun-tan parasol, 199
Super-cockpit, computer application, 126
Supercomputers, 123
Superconductive ceramics, 67
Superconductivity, 60, 62
Super-copter, computer application, 126
Superheterodyne receiver, 236
Superimposition of cartoon characters, 261
Superkart, 220
Supermarkets, 203
Supernode, 123
Supernova, remotest, 76
Supersonic military aircraft, 42
Super-strings theory, 56
Super-VHS camcorder, 239
Surfing, 217
wind surfing, 218
Surgery, 96
Suspension, 3
motorcycles, 10
Suspension bridge, 169
Sutures, 97
Swatch watch, 202
Swimming, 212
Symbolic Automatic Integrator (SAINT), 119
Symbols, algebraic, 52
Synthesiser, 246
guitar synthesiser, 248
Synthetic diamonds; synthetic jade, 68
Syntony of circuits, 236
Syphilis, 101
Syringe, 106

T

Table tennis, 221
Tactile screen, 124
Talkies, 259
Talking dolls, 209
Talking to chimpanzees, 228
Tampons, 102
Tango, 251
Tank, 33
Tankers, 158
Tape cassette, 250
Tape recorder, 250
intelligent tape recorder, 249
Tarot, 204
Tau, 65
Taxi, 15
Tea, 187
Technicolor, 259
Tedder, 128
Teddy bear, 208
Teeth, artificial, 110
Teeth, extraction of, 111
Teflon, 69
Telecommunications satellite, 244
Telegraphy, 240
Telephone, 241
'notebook' telephone, 243
pay phone, 242
pocket telephone, 243
visiophone, 243
Telephone exchange, 242
Telephonic translator, 242
Telephoto lenses, 253
Telescopes, 75
airborne, 77
Telescopic fork, 10
Tele-shopping, 201
Television, 237
for blind people, 237
rear-view, for HGVs, 6
Television screen, flat, 237
Television set, double scanning, 238
Telex, 243
Telstar, 244
Temperature, 60
Tennis; tennis rackets, 220
Tenpin bowling, 211
Terminal, computer, 123
Testicle, transplant, 99
Test-tube babies, 103
Test-tube fertilisation, 103
Textiles, 176
Thales theorem, 50
Thames clean-up, 140
Theme parks, 221
Therapeutics, 103
Thermal analysis, 91
Thermal printer, 125
Thermochemical cell, 168
Thermodynamics, 59
principles, 60
Thermoelectric effect, 155
Thermography, 91
Thermometer, 89
Thermonuclear reaction, 165
in a test tube, 165
Thermostructural materials, 174
3-D cinema, 260
3-D photography, 256
Throttle valve, 158
Tidal power station, 168
Tides, protection of Venice, 142
Tide tables, 28
Time pill, 108
Time sharing computers, 116
Tinned food, 181
for pets, 135
Tissue Plasminogen Activator (TPA), 94
Tokamaks, 165
Toaster, 184
Toilet paper, 193
Tomato, 138
Tomato tree, 131
Tonic water, 186
Toothbrush, 194
Toothpaste, 111
Topology, 54
Torpedo boat; torpedoes, 39
Torpedo mine, 40
Torus project, 161, 166
Tour de France, 211
Toys, 205
young millionaire inventors, 206
Tracta, 4
Traffic lights, 6
Train à Grande Vitesse (TGV), 14
Trains, high-speed, 14
Tram, 14
electric, 15
Tranquillisers, 108
Transatlantic firsts,
liner, 27
steamship, 25
Transatlantic telephone cable, 244
Transcendental numbers, 48
Transformer, 156
Transgenetic plants, 131
Transistors, 19, 237
Transit I, 82
Translation, automatic, 125
Transmission,
cars, 2, 4
motorcycles, 10
Transmission of electricity, 156
Trans Nicotine Patch System (TNS), 108
Transparent ceramics, 67
Transplants, 98
dental, 111
Transputers and microcomputers, 122
Transrapid, 13
Transuranian elements, 64
Traveller's cheque, 200
Tree replanter, 135
Trees, 136
Treetop raft, 148
Trevithick's steam engine, 11
Trick photography, 261
Trigonometry, 50
Trimaran, 215
Triode, 236
Trivial Pursuit, 207
Trolleys, supermarket, 203
Trumpet, 245
Tsetse flies, 93
Tulip, 136
Tumor Necrosis Factor (TNF), 106
Tungsten, 68
Tuning fork, 246
Tuning, radio, 236
Tunnel effect microscope, 90
Tunnels, 169, 170
Channel Tunnel, 171
Tupolev 144, 204, 19
Tupolev 155, 21
Tupperware, 187
Turbines, steam, 155
Turbocharged diesel lorries, 160
Turbocompressor; turbocompressor for aircraft engines, 160
Turbo engines, 159
Turbofan engine, 162
Turbojet engine, 162
with afterburner, 163
Turbojet helicopter, 22
Turin shroud, 64
Turtle, 37
Two-stroke engine of tomorrow, 158
Two-stroke motorcycle engine, 9
Typesetting machines, 233
Typewriters, 229, 230
Typhoon class submarines, 38
Tyres, 5

U

Uhuru satellite, 78
Ultra-lights, 215
Ultra-sound cataract operation, 109

Ultraviolet Explorer (UVE), 78
Ulysses probe, 88
Umbrella, 196
  car umbrella holder, 223
  light umbrella, 222
Undercarriage, retractable, 24
Underground, 15
Underwater atlas, 148
Underwater cinema, 260
Underwater computer, 150
Underwater museum, 222
Underwater photographic observations, 255
Underwater station Aquarius, 150
Underwater tank, 33
Underwater telegraphic cable, 241
Universal attraction laws, 57
Universal electronic computer, 115
Universal joint, Hooke-type, 4
University microcomputer, 123
Unix, 117
Unleaded petrol, 145
Urea, 68

V2 rocket engine, 161
Vaccination, 92
  anti-pregnancy, 103
Vacuum, 58
Vacuum cleaner, 190
Vacuum flask, 187
Valium, 108
VAN, earthquake forecasting method, 150
Vanilla, 138
Variable transmission, motorcycles, 10
Variable wing aircraft, 44
Varilux Pilote spectacles, 109
Variolation, 93
Vectors; structure of vectorial space, 51
Vegetables, 137
Velcro, 197
Velocipedes, 6, 7
Venetian glass, 174
Venice, protection from high tides, 142
Ventilation, artificial, 189
Vertical pedal bicycle, 7
Vertical Take-Off and Landing (VTOL), 42
Very Large Array (VLA), 77
Very Large Telescope (VLT), 75
VHS, 238
Vibram sole, 221
Video cameras and camcorders, 239
Video cassette recorders, 238
Video Cat, 224
Video Cluedo, 206
Videodisc, 239
  compact, 249
  games, 206
Videodisc encyclopaedia, 232
Video formats, 238
Video games, 205, 206
Video, hairstyle, 197
Video recorders, 240
Video recorder with screen, 238
Video screen, postage-stamp, 238
Video tape, 240
Videotex, 243
Video Walkman, 238
Viewdata, 243
Violin, 245
Viper's bugloss, 131
Virology, 73
Viruses, 92
  synthesis of, 73
Virus, self-destructing, 132
Viscodrive, 3
Viscount V-630, 18
Visicalc, 118
Visi-on, 117
Visual speaker, 250
Vitamins, 108
Volapik, 228
Volcanic eruptions, forecasting, 150
  by computer, 152
Volcanoes and earthquakes, 150
Volkswagen Beetle, 1
Volleyball, 211
Volt battery, 60
Vostok 1, 82
Voyager 2, 87
Vulcanisation, 69

Walking robots, 127
Walkman, wireless, 251
Waltz, 252
Wargames, 206
Warning system, automatic, R. Rhine, 140
Washing machine, 192
Washing powder, biodegradable, 143
Watches, 175
  Swatch watch, 202
Water, constituents, 58
Watercress, 138
Water grabber, 135
Watering kit, automatic, 135
Water level, 169
Watermarks, 200
Water polo, 212
Waterproof ship's log, 215
Waterproof watch, 175
Waterskiing, 218
Water-tube boiler, 154
Water, underground, radar detection, 130
Water-wheels, 155
Watt's steam engine, 154
Wave energy, 168
Wave mechanics, 56
  the atom in, 63
Waves, electromagnetic, Hertzian, 61
Weak force, 56
Weak interaction bosons, 66
Weaponry, laser, 179
Weather cube, 147
Weather radar, 146
Weather satellite, 147
Weight-reduction, 107
Weights and measures, 54
Welding, 176
Well logging, 158
Wet-collodion plate, 254
Wheatless bread, 188
Wheelbarrow, 135
Wheel-lock, 29
Wheels, cast, motorcycles, 10
Wheels without hubs or spokes, 8
Whisky, 187
Whist, 204
White dwarfs, 79
Wide angle lenses, 253
Wide screens and 3-D cinema, 260
Wide-track railways, 11
Williams' pear, 138
Winches, 168
Winchester, 30
Wind engines, 166
Winder, watch, 175
Windmill, 166
Windows, information technology, 117
Window video recorders, 238
Windscreen; windscreen wipers, 5
Wind sheers, radar detection of, 146
Wind surfing, 218
Winter sports, 218
Wired glass, 174
Wireless telegraphy, 236
Wireless Walkman, 251
Wordstar, 118
Work-heat equivalence, 59
Work stations, 116
World literacy year, 234
Wrestling, 215
Write Once Read Many times (WORM), 125
Writing materials, 28

Xerography, 203
X-ray table, transparent, 99
X-rays, 64
  3-D, 91
X-Wing, 47
Xylography, 230

Yellow fever, 93
Yoga for cats, 227
Yogurt, 188

Z

Z1 computer, 114
Z80 microcomputer, 119
Zelda, 206
Zelentchouk telescope, 75
Zeppelin, 16
Zero (mathematics), 48
Zipper, 196
Zoom lenses, 253
Zoptic, 262
Zygote Intra-Fallopian Transfer (ZIFT), 104

PICTURE CREDITS

ANA: 62b(Freeman), 121(Charlier), 208;
Airbus Industries: 19;
Ancient Art & Architecture Collection: 100;
BN: 234
BMW: 10;
British Aerospace: 46;
British Airways: 23;
British Museum/Syndication International: 129t, 231;
British Nuclear Fuels/Calder Hall: 164;
CCGM/UNESCO: 149;
CNRI: 89, 92t, 92b, 101, 105(Castano), 108t;
CNRI-Bories P: 98t;
CNRI-Malzieu: 111;
CNRS-IBMC: 71b;
CNRS-IBMP: 131;
Camera Press: 133b(Ferguson), 222(Rutledge);
Cemagref: 130b;
Cemax: 91;
Christophe L: 30, 32, 251;
Cosmos: 63, 65, 66, 71t, 76b, 77b, 79, 82r, 90, 116, 120t&b, 130t, 202, 208, 239, 247, 249;
DR: 9, 41b, 135, 137, 175, 178, 183, 195, 202, 218, 223, 226, 250, 252, 255, 262;
Dagli-Orti: 234, 235;
Disney: 260;
Dorka: 25t, 39, 51b, 59, 70, 102b, 109, 165t, 166, 193b, 220, 232, 246, 261;
ESA: 78, 127;
Elf Aquitaine: 141, 157;
Environmental Picture Library: 139, 142r;
Ernoult: 210;
Ernoult Alain: 17b, 83tr, 159;
Ernoult Features: 77t, 103(Charmet), 255;
Mary Evans Picture Library: 192, 198, 263(Pratta), 265, 266, 267b, 269, 270, 271, 272, 274, 275t;
Explorer: 11, 17t, 145, 152(Krafft), 193(Howarth), 218(Gerared), 224, 225, 241(Charmet), 258;
Vivien Fifield: 177, 185;
Finlay-Holiday Films: 85;
Gamma: xb(Budge-Liaison), 13t, 14, 26, 27, 43t(James-Liaison), 43b, 50, 58t, 64, 64b(Lochon), 67b, 68, 74, 76(Pierce-Liaison), 821, 83b(Novosti), 84, 87, 95b, 96, 106, 107, 108t, 110, 114, 123, 124, 132t(De Cis-Liaison), 132b, 133t, 136t(Kurita), 136b 167t, 170t, 180, 184, 186b, 199, 201(Novovitch-Liaison), 204(Okoniewski), 207, 212, 215, 217, 219, 225(Streetporter-Liaison), 226, 227, 229, 233, 239, 246;
Giraudon: 51t;
Goivaux: 98b, 99;
Ian Griffiths: xt;
Hachette: 20t, 93, 94, 97, 169t;
Robert Harding Picture Library: vit, vib, viib(Kamal);
The Hulton Picture Company: 129b, 172, 173b, 264, 267t, 268t, 268b, 273t, 273b, 275b;
The Imperial War Museum: 37;
Inapress: 171b, 196;
Jerrican: 113t, 134, 138(Labat), 225, 243, 259;
Keystone: 104;
Lockheed: 163;
London Transport Museum: 15;
The Mansell Collection: 170b, 200, 242;
Météorologie Nationale: 147;
Microsoft: 206;
Musée de L'Armée: 33, 40;
NASA: 86;
National Physical Laboratory: 115;
National Railway Museum: 12;
Odissey: 95t(Gillie);
The Robert Opie Collection: 182t, 186t;
Palais de la Découverte: 49b, 55b, 62t, 67t;
Popperfoto: 173t, 190, 200;
Press Association: 210;
QA Photos: 171t;
REA: 181(Ferrare);
Rapho: 58b;
Rex Features: v, viit, viiit(Bonazza), viiib(Thomas-Fury), ixt, ixb, xib(Jorgensen), xii(Neumeister), 151b, 214;
G. Rivet: 223;
Roger-Viollet: 53;
Rowntree-Mackintosh: 182b;
SFP: 16, 189, 194t&b, 252, 253, 254, 256, 258;
SPK-Berlin: 81;
The Science Museum: 113b, 154b;
Sipa-Press: 25b, 117, 209(Beinat);
The Slide File: 187;
Sodel: 18, 167b, 168, 169b;
Studio X: 13b, 209;
Sydney Freelance: 224b;
Sygma: 125, 126, 144, 165b(Forestier);
Tass: 83tl;
The Tate Gallery: 154t;
Texas Instruments: 119;
Thomson: 244;
Gerrard Vandystadt: 216.